Vacuum Mechatronics

Vacuum Mechatronics

Gerardo Beni
Susan Hackwood
Steve Belinski

Majid Shirazi
Shigang Li
Lakshmanan Karrupiah

Center for Robotic Systems in Microelectronics
University of California, Santa Barbara
Santa Barbara, California

Artech House
Boston • London

Library of Congress Cataloging-in-Publication Data

Vacuum mechatronics / Gerardo Beni ... [et al.].
    p.   cm.
  Includes bibliographical references.
  ISBN 0-89006-456-3
  1. Vacuum technology. I. Beni, Gerardo.
TJ940.V25   1990   90-32359
621.5'5--dc 20   CIP

British Library Cataloguing in Publication Data

Vacuum mechatronics
  1. Vacuum technology 2. Automatic control. Applications
  of computer systems
  I. Beni, Gerardo
  621.55

    ISBN 1-58053-326-4

International Standard Book Number: 0-89006-456-3
Library of Congress Catalog Card Number: 90-32359

10 9 8 7 6 5 4 3 2 1

Cover design by Hilary Brace

DEDICATION

To Welcome Bender, whose aggressive wisdom set
the Viking on the right course to Mars and kept all
of us on the right course in *Vacuum Mechatronics*.

Preface to *Vacuum Mechatronics*

This book is the result of a group effort at the Center for Robotic Systems in Microelectronics (CRSM) of the University of California, Santa Barbara. Since its founding in 1985, the CRSM philosophy has been to contribute innovation through investigating, analyzing and developing new areas of technology – specifically, interdisciplinary areas whose fundamental academic value is currently 'underestimated' and areas whose main development occurs in industry. Vacuum mechatronics is an example of this activity.

While conducting research in vacuum mechatronics, we realized the need for a solid database of facts, principles and procedures for the researcher in this new field. This book is the first step toward this goal. It includes a collection of 30 fundamental reprints and six up-to-date chapters on the main areas.

Steve Belinski is the main contributor to the book. Besides writing two of the chapters, he coordinated the group effort so that the chapters and the reprints form a clear and coherent unit. Majid Shirazi wrote and organized most of chapters 3 and 4, with assistance from Lakshmanan Karuppiah and Degang Chen. Shigang Li was the main contributor to chapter 5.

Many more people at the CRSM have contributed to the development of the book. Degang Chen contributed section 5 of chapter 4 on magnetic levitation. Welcome Bender contributed insightful, constructive criticism. Finally, the book has been made possible by the skillful editorial production of Karen Gundersen.

At the cost of sacrificing appearance, we have opted for publishing in "real time". Thus, the information is current to within two months from the date of this preface. Since the field evolves rapidly, we are planning yearly updates. Information on updates and current vacuum mechatronics developments can be obtained from the CRSM Systems Clinic at 805-961-4912.

Gerardo Beni
Susan Hackwood
Steve Belinski

Santa Barbara, California
June, 1989

Chapter 1

Introduction to Vacuum Mechatronics

Susan Hackwood
Gerardo Beni

Goals of the Book

This book is an introduction to the field that has come to be called vacuum mechatronics. The purpose of the book is to explain why vacuum mechatronics is important, to outline the field, to give examples of the progress that has been made, and to explore future possibilities. This book can serve as a reference for practicing engineers and those planning to enter this research field from other disciplines.

Definition and Scope

The discipline of vacuum mechatronics is the design and development of vacuum-compatible, computer-controlled mechanisms for manipulating, sensing and testing in a vacuum environment. It belongs to the fields of vacuum science, automation in special environments, mechatronics (computer controlled mechanisms) and robotics. The components of vacuum mechatronics systems include intelligent mechanisms, intelligent sensors, and intelligent measuring devices. The system integrating elements are the vacuum chamber and the system computer control. Vacuum mechatronics is relevant to research engineers in integrated circuit manufacturing, surface physics, food processing, biotechnology, materials handling, space sciences and manufacturing.

The Importance of Vacuum Mechatronics

Practical Importance

Vacuum mechatronics is gaining importance due to the increased use of vacuum in applications such as space studies, manufacturing, material processing, medicine, microelectronics, emission studies, lyophylisation, freeze drying and packaging. As the benefits of the vacuum environment (e.g. low pressure, long mean free path length, and cleanliness) become better defined and understood, the desire to implement more processes in vacuum will increase.

The space program has provided much of the forward momentum in vacuum mechatronics due to the numerous vacuum-related challenges which had to be worked out for space missions. These solutions have recently been applied and extended for use in chamber-based production environments, such as those used for coating (e.g. evaporation or sputtering). In this and other vacuum production applications, the sensing, transfer and/or positioning functions provided by the mechatronics equipment is critical to the overall process.

The newly developing field of vacuum mechatronics is also the driving force for the realization of an advanced era of totally enclosed clean manufacturing cells. High-technology manufacturing has increasingly demanding requirements for precision manipulation, *in situ* process monitoring and contamination-free environments. To remove the contamination problems associated with human workers, there is a tendency to move towards total automation for manufacturing. This will become a requirement in the near future, e.g., for microelectronics manufacturing. Automation in ultra clean manufacturing environments is evolving into the concept of self-contained and fully enclosed manufacturing. For example, at the CRSM we are developing a self-contained, automated robotic factory (SCARF™) as a flexible research facility for totally enclosed manufacturing of integrated circuits. The construction and successful operation of a SCARF will provide a novel, flexible, clean, vacuum manufacturing environment. SCARF-type systems also require very high reliability and intelligent control.

Conceptual Importance

Machine design is rapidly becoming a highly complex subject. Not only are the mechanics complicated, but the integration of computer control and sophisticated sensors allows the machine to behave in an intelligent way if it is correctly programmed. The designer must have knowledge of a wide range of technologically advanced subjects, including mechanics, sensor design and computer control. Thus, the machine designer can be thought of as being responsible for the evolution of a species of intelligent machine. Machine evolution cannot be left to the natural selection process that has guided biological evolution. Not only are the time scales inappropriate, but man can utilize tools which are not available in nature, such as CAD and computer simulation. These tools can speed up the design process considerably. What then should be the guidelines for the designer to ensure that the form of the machine fits its function?

One can consider the development of mechatronics and related subjects, such as robotics, as the *environment* driving the evolution of the machines rather than vice versa. Using a biological analogy, this means that the existence of the pond determines, through evolution, the design of the frog so that the animal is optimized for that environment. This guarantees the utility and continuous evolution of the animal for a given environment. In the body of this book we will emphasize how the medium drives the development of machines that have high utility.

The Importance of the Vacuum Environment

A thorough knowledge of the environment opens up new evolutionary paths for machines. New machines can spawn new, cost-efficient commercial ventures. Vacuum science and technology are increasingly more important for the manufacturing of many high-tech products. This is put further in perspective when one considers that vacuum is the natural environment in the universe. Atmosphere is the exception, being confined to Earth and a few other planets. Any space structure is a mechatronic machine operating in a vacuum. Research in vacuum mechatronics is therefore critical for man to realize his goals of not only reaching the stars, but making a home there.

The same considerations can be applied to man's goals of exploring the oceans. The development of specialized machines for operation in the rigorous environment of the deep ocean is also important. However, vacuum is the more significant because the ocean is limited in size and scope, and is intrinsically earth-bound. Vacuum has more cosmic importance.

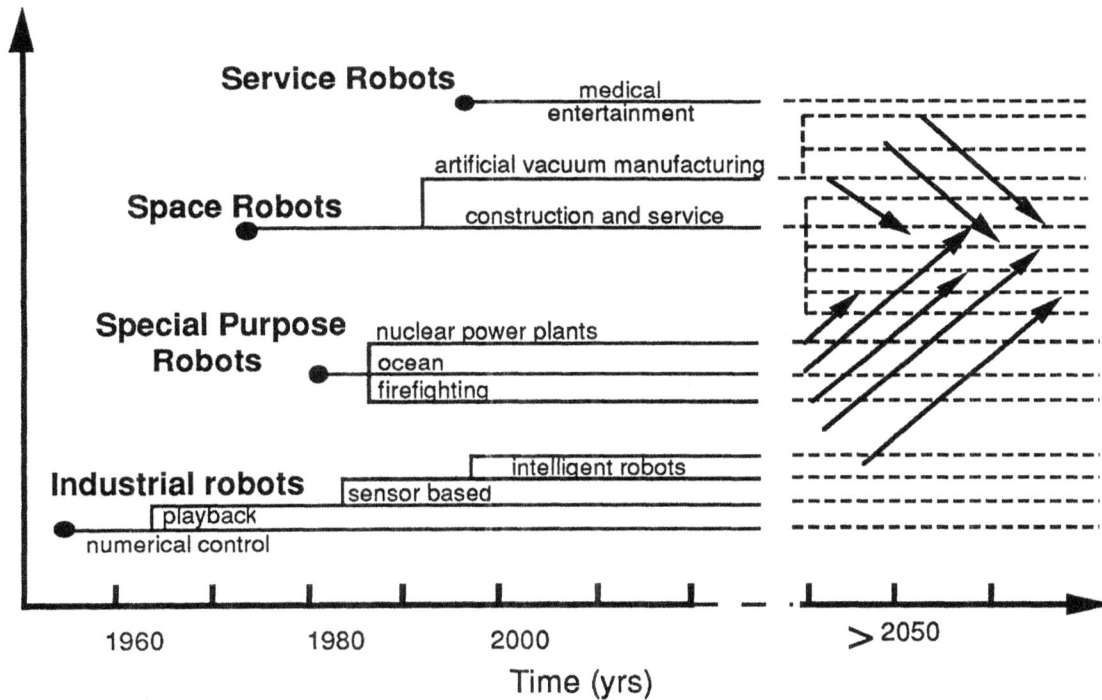

Figure 1.1. Technological evolution to show the ultimate application of industrial robots, special purpose robots and service robots to space.

This is exemplified in the technological evolution of robots, a summary of which is presented in Figure 1.1. During the last two decades, sensory-based robots capable of some intelligent operation have been developed for manufacturing. More recently, robots for use in special environments such as fire-fighting have been a focus of attention. Service robots, e.g., for medical applications and entertainment, have begun to be researched. In the future, when space exploration and exploitation have further evolved, terrestrial environments will become the exception and vacuum environments will become the norm. The robots described above may have to perform similar functions, however, now in the vacuum of space. The diversification of space robots will therefore increase.

Why Vacuum Mechatronics has not Developed Further

Vacuum mechatronics has not developed more rapidly because it is intrinsically a cross-disciplinary field requiring the integration of computers with mechanisms, robots, materials and new actuators (i.e., mechatronics). It is only recently that such integration has become possible. There has been a definite trend towards incorporating computers (microprocessors), as they become smaller, into the machines and even the materials themselves. This is shown in Figure 1.2. The advent of intelligent machines (robotics) was marked by the transition of hard automation to computer controlled mechanisms. The controller was centralized and generally remote from the mechanism. More recently, intelligent mechanisms have evolved which incorporate distributed microprocessor control within the mechanical structure. This trend is continuing with the addition of greater local sensing, control and intelligent operation.

Difficulties in Vacuum Mechatronics Research

The vacuum environment can be classified as natural (space) or artificial (vacuum chamber). However, many of the design criteria and research issues are common to both. The fundamentals

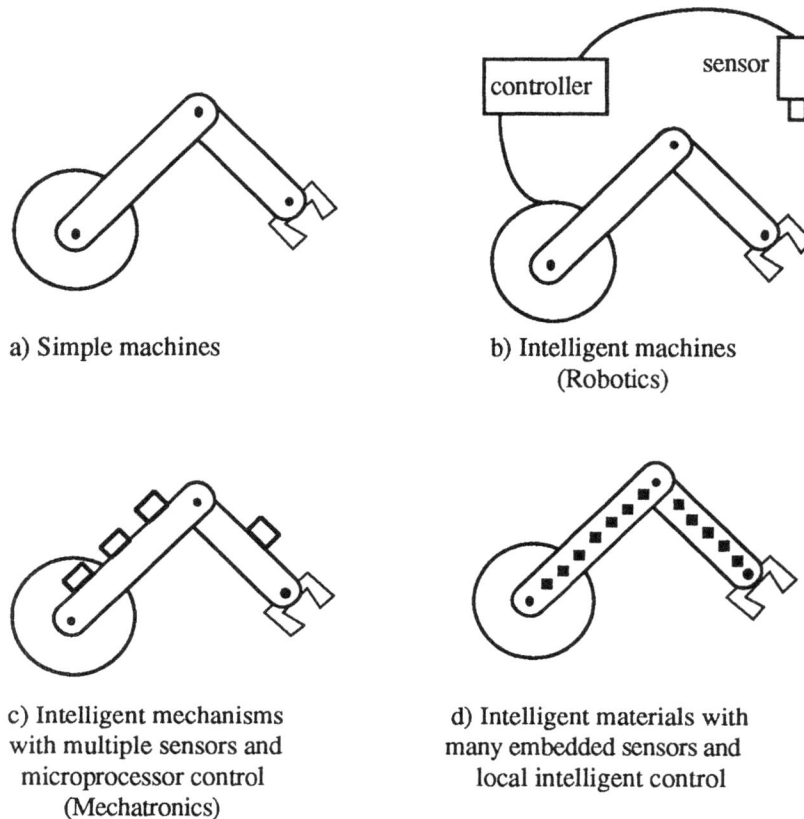

a) Simple machines

b) Intelligent machines
(Robotics)

c) Intelligent mechanisms
with multiple sensors and
microprocessor control
(Mechatronics)

d) Intelligent materials with
many embedded sensors and
local intelligent control

Figure 1.2. The evolution toward intelligent machines has been influenced by the
development of robotics and mechatronics

of mechatronics design for vacuum pose design constraints on the selection of materials, choice of lubricants and on modes of energy transfer.

The earth is, in fact, in a bubble of air surrounded by vacuum. On earth, we build artificial vacuum chambers that appear to us as a bubble of vacuum surrounded by air. In either case, wherever there is an interface, mechatronics is needed. The air/vacuum interface (shown in Figure 1.3) is important, as most of the action takes place there. The interface has to allow transference of materials when it is desirable, and prevent transfer when not desirable. The interface is therefore like a biological barrier. As the skin of a human is the interface between the hostile outside environment and the delicate, internal organs, the vacuum mechatronics devices are the interface for the transduction of actions by humans in atmosphere to actions by machines in vacuum.

<u>What can be transferred across the interface</u>

- Data: Information may be transferred via wires, IR beams, sensors outputs, computer vision, radio waves, etc.

- Matter: Materials must be transferred into and out of the vacuum from atmosphere, and from station to station within the vacuum environment. Load-lock mechanisms are required for this process.

- Energy: This can be via power lines, microwave, lasers, etc.

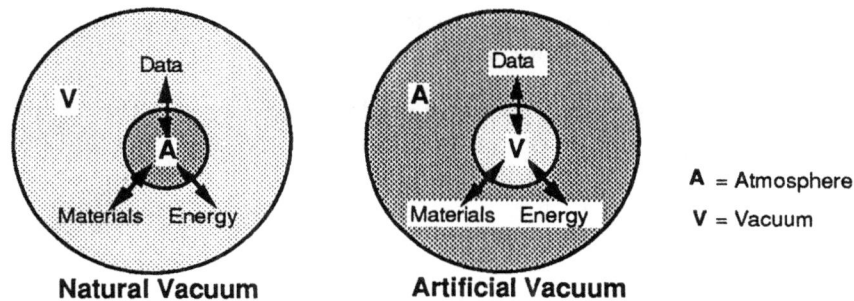

Figure 1.3 The air/vacuum interface for natural vacuum and artificial vacuum must be selectively permeable to data, materials and energy.

The problems encountered in implementing all three of these transferences require considerable thought. For example, many vacuum processes are currently monitored by direct observation through viewing ports, or remotely via cameras and sensors. If vacuum materials are optically transparent, then operators can watch what is taking place inside, even when the system becomes large and complex. However, at the moment, process monitoring can only render partial information. This is a telepresence problem as the operator cannot look directly at what is going on, a situation that is less common in normal factory automation. As vacuum systems become more complex, one must rely more and more on sensory capabilities. Especially important is sensory fusion, which allows intelligent decisions to be made on actions to be taken, without the intervention of an operator.

As process requirements progress from small vacuum chamber (on the order of <1m^2) operations, to larger, more complex structures (>1m^2 and especially >100m^2) they will necessitate the development of new types of interfaces for information transfer. Use of optical viewing ports becomes ineffective; intelligent sensors and sensor fusion would be necessary. Power would have to be fed in through cables, microwave, radiation, etc., and matter transferred through large, efficient load locks.

The development of these large, multi-chamber 'super-vacuum systems' brings about a new class of applications. In addition to the traditional application of vacuum technology to high energy physics, surface physics and instrumentation, new areas will open up in integrated circuit manufacturing, materials processing and biotechnology. In fact, vacuum manufacturing for ICs is the means by which the next generation's intelligent machine can be built. This book will discuss the new class of tasks and describe a new class of machines for the vacuum environment.

The effective use of the vacuum environment will depend on the availability of *vacuum-compatible components*. Mechanisms and machine design research currently includes joints, bearings, energy transmission/control devices, linkages, and fasteners for vacuum as well as actuators, e.g., vacuum rated motors and piezoelectric devices. There is also a need for sensing in vacuum, e.g. encoders for vacuum motors, force sensors, and vision sensors. An *intelligent controller* which can deal with limited sensory information/limited control action possessing fault detection/tolerance capability must be used for vacuum mechatronics control. Real-time multi-sensory data fusion is desired. A computationally very efficient world model is important, because it can be used with active sensing, in workspace understanding and model adaptation, as well as in the expectation and sensory data interpretation during operation.

Since usual robot teaching methods are no longer adequate for vacuum mechatronics, *real-time simulation* capability is highly desirable to assist program control. Some new criteria for optimal trajectory and task scheduling must be introduced. *Reliability* is another important issue in vacuum mechatronics. Besides component design, emphasis must also be placed on the controller, i.e. fault tolerance ability, since frequent repair is undesirable.

In this book, emphasis will be given to the progress made in these areas. The significance of new theories, new applications, the production of new machines and components will be described and discussed.

Self-Contained Manufacturing

A particularly important application of vacuum mechatronics is for the development of self-contained manufacturing systems. The use of these systems will have a significant impact on future manufacturing methods, particularly those requiring isolation for extremes of temperature, pressure, cleanliness, and for isolation of hazardous materials. Vacuum manufacturing is a specific class of self-contained systems which could be used in a wide range of manufacturing processes, from electronics and microelectronics to biotechnology and pharmaceuticals. It can be used on a large scale, as in turbine blade cutting, or on a microscopic scale, as in the manufacture of microactuators. In microelectronics manufacturing, these systems have come to be known as multi-chamber process systems.

Self-contained systems must display the following characteristics:

- Allow isolation from the external environment and from each other;
- Be highly reliable systems that can cope with partial transportation and/or processing machine failure;
- Be reconfigurable systems that offer maximum flexibility for process modification;
- Have distributed control for system reliability, safety and robustness;
- Have a modular mechanical design and standard interfaces for rapid set-up and repair.

Self-contained vacuum manufacturing is of critical importance in microelectronics manufacturing. IC production has increasingly demanding requirements for precision manipulation, *in situ* process monitoring and contamination-free environments. To remove the contamination problems associated with human workers, there is a tendency to move towards total automation for IC manufacturing. This will become a requirement in the near future as dimensions decrease below $1\mu m$ and circuit complexities increase. Automation in ultra clean manufacturing environments is evolving into the concept of self-contained and fully enclosed manufacturing.

It was clear even in the early 1980's that an *integrated manufacturing* capability would be needed by the microelectronics industry [1.1]. Presently, the industry is fairly mature regarding automation and robotics for assembly operations associated with the back-end of the IC manufacturing process. By early 1987, several equipment manufacturers had already displayed self-contained, "stand alone" process tools that are forerunners of larger tool integration yet to come. Drytek (General Signal) and Applied Materials market dry etch and Chemical Vapor Deposition (CVD) equipment, respectively, that are single-wafer-at-a-time tools with multiple process chambers and thus multiprocess capability. Also MTI-Sypher has now marketed a unit with combined deposition (two stations) and etching (one station). The wafers are fed by robots and these tools suggest tool architectures for the further evolution of integrated processes.

Factories of the future will have a facilities architecture under which cells are linked together. If the operations needed to make an entire integrated circuit are combined under the envelope of one unit tool, then we ultimately have a self-contained factory. If the wafers are transported by automation and robotic manipulation, under computer control, we have a self-contained, automated, robotic factory (SCARF). Many companies have embarked on similar paths, including IBM [1.2] and Texas Instruments [1.3].

The SCARF project was initiated at the CRSM in 1987 [1.4]. We are essentially placing the clean room inside a relatively small envelope, evacuating that envelope, maintaining low particle densities and controlling pressure to quickly allow transfer and load locking between wafer storage areas and process chambers. A specific implementation has been designed [1.5], as shown in figures 1.4 and 1.5. The SCARF design integrates small footprint vacuum tools around the central chamber. It is convenient to bring certain process tools together locally, especially those which will be used serially in the process architecture.

A four foot diameter central vacuum chamber is connected directly to deposition modules D1,D2,D3 and D4, and to a complementary four-cell etching chamber E1,E2,E3 and E4. The shaded area represents the radial extent of the central wafer delivering robot. A local robot is needed to serve the etch station, and a linear transport robot is required to connect to the large footprint tools.

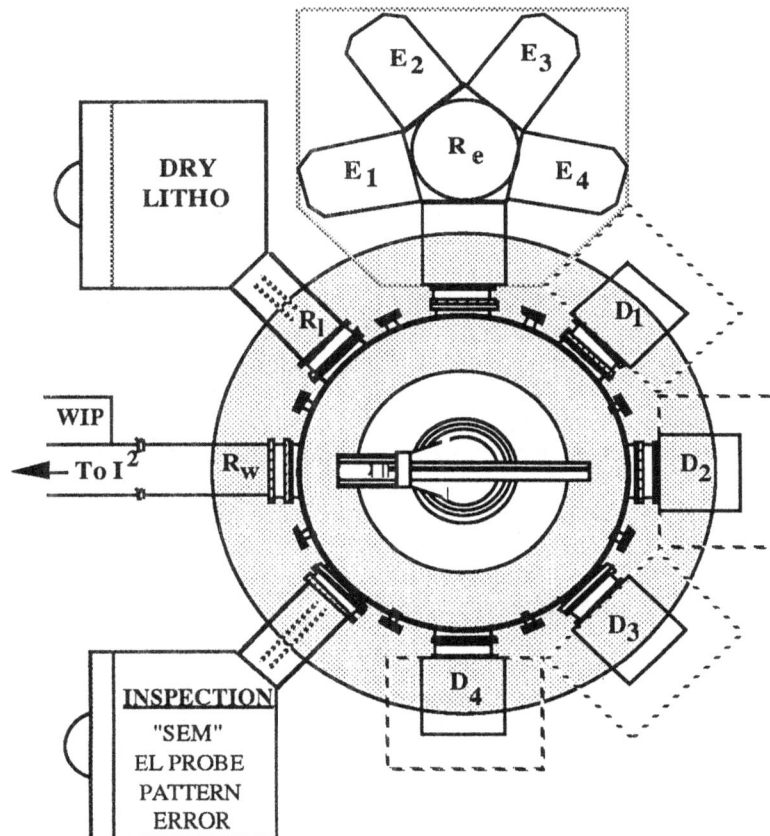

Figure 1.4 Layout of Self-Contained Automated Robotic Factory.

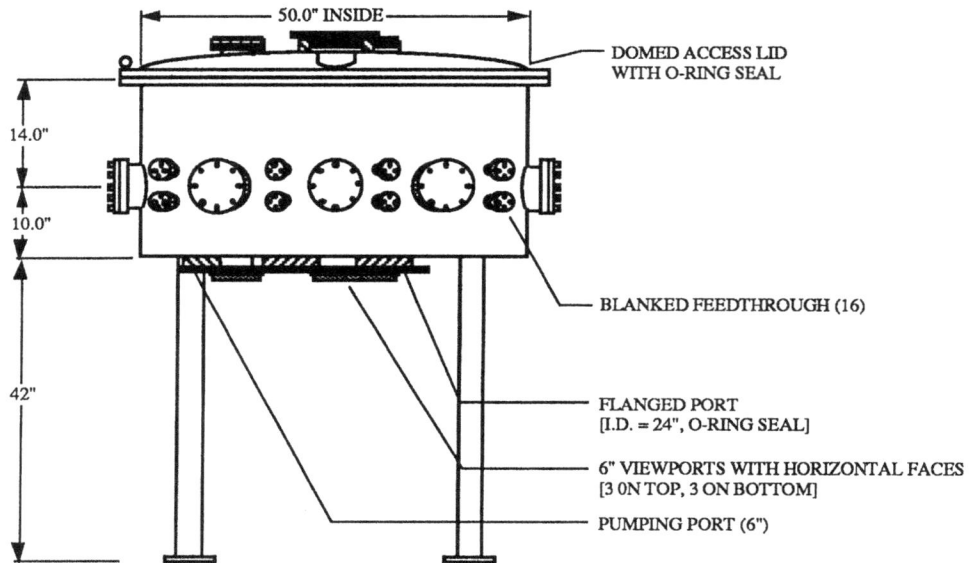

Figure 1.5. Dimensions of SCARF Chamber

Implications of SCARF

1. Fast Cycle Time: Improved cycle time is obtained by placing certain process tools physically close to each other. These tools are often used sequentially, thus providing for fast turn around in a processing "cell". The SCARF goal is to reduce wafer processing cycle time by 3-15x over that which is done in a standard batch facility. Fast cycle times are critical for: prototyping, high and improved yield, competitiveness, and low cost. The component tools and indeed the overall system must have excellent *reliability* to achieve a completely operational SCARF.

2. High Yield: Since small particles (those that impact smaller design rule devices) are more difficult to remove from a standard air-borne clean room, a vacuum design provides the basis for an effective "class 0.1" room for sub-micron sized particles. Small particles settle out quickly in vacuum. Further, the incorporation of in-situ inspection and self-contained sequences reduces time spent for work-in-progress (WIP), and in unnecessary transport.

3. Low Cost Facility Operations: Large batch facilities are becoming more and more expensive. So, as a facility dedicated mainly to prototyping of devices, SCARF allows for the throughput of a *large number of designs per unit time*, albeit with a limited number of wafers.

4. Flexible Prototyping: Since the facility architecture design is flexible with regard to the type of component process tool, we can use future generations of tools without changing the core design of the system.

A simplified processing sequence has been developed to demonstrate "proof of concept". The process allows us to carry out the maximum number of process steps within the SCARF environment, at a relatively early stage of SCARF development. A 4-mask NMOS transistor sequence, featuring self-aligned source and drains using thermal diffusion from CVD deposited sources has been developed.

<u>SCARF Related Research</u>

1. Vacuum Mechatronics research at the CRSM has focused on the design of mechanisms [1.6], particle detection and control [1.7 ,1.8, 1.9], intelligent process control [1.10, 1.11] and distributed control of cellular systems [1.12, 1.13, 1.14, 1.15, 1.16].

2. In-situ sensor systems for inspection using color vision have also been developed [1.17, 1.18, 1.19, 1.20].

Future possibilities

There is no doubt that vacuum mechatronics has an important role to play in the future. At the moment there is no complex/intelligent machine capable of operating in a vacuum. Application areas will grow as technologies require the rigors of vacuum, or vacuum becomes difficult to avoid, as in space. Small systems requiring a vacuum environment, such as for microelectronics manufacturing, are becoming necessary. So are large vacuum systems, which might utilize, as an example, sputter coating sheets of metal or glass. These large systems may require the development of new machines such as vacuum compatible "smart-carts" or new technologies such as mechanically flexible vacuum compatible structures. One can speculate on the limits of vacuum manufacturing; e.g., one might consider what could be manufactured at ultra-high ($>10^{-10}$ Torr) vacuum. The future developments in vacuum mechatronics will certainly unlock some interesting doors.

References

[1.1]   Schicht, H.H., "Clean Room Technology: the Concept of Total Environmental Control for Advanced Industries", *Vacuum*, Vol. 35, No. 10/11, pp. 485-491, 1985.

[1.2]   Bednorz, J.G. et al, U.S. Pat. 4,643,627: "Vacuum Transfer Device", Feb 17, 1987.

[1.3]   Larrabee, G. R., private communication, and J.M.Blasingame, "Microelectronic Manufacturing Science and Technology", Air Force Wright-Patterson, PRDA 87-7 PMRR.

[1.4]   Belinski, S. and S. Hackwood, "Manufacturing in a Vacuum Environment", IEEE International Electronic Manufacturing Technology Symposium, p.269, Anaheim, Oct. 12-14, 1987.

[1.5]   Hackwood, S., "Robotics in Microelectronics Manufacturing", IEEE International Workshop on Intelligent Robots", Tokyo, November, 1988.

[1.6]   Belinski, S., W. Trento, R. Imani and S. Hackwood, "Robot Design for a Vacuum Environment", *Proceedings of the NASA Workshop on Space Telerobotics*, Vol. 1, p.95, Pasadena, Jan. 20-22, 1987.

[1.7]   Chen, D., T. E. Seidel, S. Belinski, and S. Hackwood, "Dynamic Particulate Characterization of a Vacuum Load Lock System", *J. Vac.Sci. and Technol.* , accepted for publication.

[1.8]   Chen, D., S. Belinski and S. Hackwood, "Effect of Moisture Condensation on Particle Count During Pumpdown", *J. Vac.Sci. and Technol.* , accepted for publication.

[1.9]   Chen, D., and S. Hackwood, "Vacuum Particle Dynamics and Nucleation Phenomena During Pumpdown", submitted for publication.

[1.10]   Hu, E., S. Magiarcina, M.Peters, S. Hackwood, G. Beni, "Inference in Intelligent Machines," in *Proc IEEE Conf. on Robotics and Automation*, San Francisco, p. 1966, Apr. 7, 1986.

[1.11]   Mangiarcina, S. and G Beni, "On the Logical and Physical Combinations of Evidence in Intelligent Machines," J. Intelligent Systems, Vol. 1, p143, 1986.

[1.12]   Beni, G., "The Concept of Cellular Robots," IEEE Symposium on Intelligent Control, Arlington VA, Aug. 24, 1988.

[1.13]   Beni, G., and J. Wang, "Patterns in Cellular Robotic Systems", 6th Annual Meeting of the Robot Society of Japan, Nagoya, Oct. 20-22, 1988.

[1.14]   J. Wang and G. Beni, "Pattern Generation in Cellular Robot Systems," IEEE Symposium on Intelligent Control, Arlington, VA, Aug 24, 1988.

[1.15]   Beni, G., and J. Wang, "Cellular Robotic System; Self Organizing Robots and Kinetic Pattern Generation", IEEE Workshop on Intelligent Robots and Systems, Tokyo Oct. 31-Nov. 2, 1988.

[1.16]   Hackwood, S., and J.Wang, "The Engineering of Cellular Robotic Systems", IEEE Symposium on Intelligent Control, Arlington VA, Aug. 24, 1988.

[1.17]   Barth, M., J. Wang, S. Parthasarathay, E.L.Hu, S. Hackwood, and G. Beni, "A Color Vision System for Microelectronics: Application to Oxide Thickness Measurements", *Proc. IEEE Int. Conf. on Rob. and Automation.*, San Francisco, CA., pp.1242-1247, April, 1986.

[1.18]   Parthasarathay, S., D. Wolfe, S. Hackwood, E.L. Hu, and G. Beni, "Color Vision for Microelectronics Inspection", *Proc. SPIE Conf. on Intelligent Robots and Computer Vision,* Vol. 726, pp. 125-130, Oct. 26-31, 1986.

[1.19]   Parthasarathay, S., D. Wolf, E.L. Hu, S. Hackwood, and G. Beni, "A Color Vision System for Film Thickness Determination", *Proc IEEE Int. Conf. on Rob. and Automation*, p.515, Raleigh, NC, Mar. 31-Apr.3, 1987.

[1.20]   Barth, M., S. Hackwood, "A Flexible Small-Scaled Parallel Vision System", *IASTED Int. Conf. on Robotics and Automation*, Santa Barbara, CA, May 25-27, 1988.

Chapter 2

Vacuum Environment and Applications

Steve Belinski

2.1   Vacuum Fundamentals

A vacuum is said to exist when the density of gas in a given volume is reduced below the density of atmospheric gas at the earth's surface. The greater the density reduction, the better (or *higher*) the vacuum. The American Vacuum Society in 1958 defined "vacuum" as a given space filled with gas at pressures below atmospheric, i.e. having a density of molecules less than about 2.5×10^{19} molecules/cm^3. Present vacuum technology provides levels of vacuum spanning about 19 orders of magnitude below atmospheric pressure, with the lower limit constantly decreasing due to technological advances in pumping and measurement techniques.

As pressure decreases, the average distance between molecular collisions, or *mean free path*, increases. Although the mean free path (λ) depends on temperature and the molecular diameter, it can be approximated as:

$$\lambda_{cm} = \frac{5.0 \times 10^{-3}}{P_{Torr}}$$

Eq. (2.1)

Thus, at atmospheric pressure, $\lambda \approx 6.6 \times 10^{-6}$ cm, while in a vacuum of 10^{-8} Torr, $\lambda \approx 5.0 \times 10^5$ cm or 5.0 km. A high vacuum is said to exist when the mean free path is much greater than the linear dimensions of the vacuum vessel. Gas-wall collisions then dominate over gas-gas collisions. Another measure of the level of vacuum is the *time to form a monolayer of gas* on a surface. This is the time required for a freshly cleaved surface to be covered by a layer of gas of one molecule thickness. Figure 2.1 indicates the relationship between pressure and monolayer formation time.

The most common units of pressure are the Torr (mm Hg), Atmosphere (atm), millibar (mbar) and Pascal (N/m^2), which are related as:

$$1 \text{ Torr} = 133 \text{ Pascal (Pa)} = 1.33 \text{ mbar.} = 1/760 \text{ atm.}$$

Although the Pascal is the SI unit for pressure and has technically rendered the use of the unit "Torr" obsolete, the majority of technical articles continue to use the unit Torr as the primary or secondary unit of preference.

The range of vacuum has been roughly divided as follows:

Low Vacuum .... 760 Torr - 25 Torr
Medium Vacuum .... 25 Torr - 10^{-3} Torr
High Vacuum .... 10^{-3} Torr - 10^{-6} Torr
Very High Vacuum .... 10^{-6} Torr - 10^{-9} Torr
Ultrahigh Vacuum .... 10^{-9} Torr and beyond.

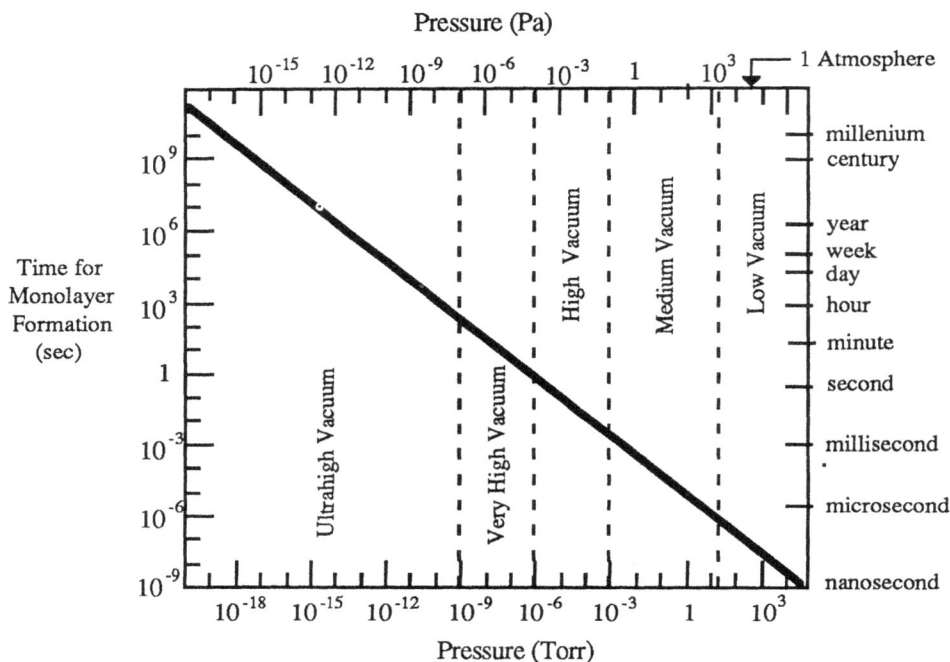

Fig. 2.1 Monolayer formation time vs pressure

In the low and medium vacuum ranges the number of molecules in the gas phase is large compared to the number of molecules covering the surfaces. In the low vacuum range, the gas flow is viscous; mutual interactions of the molecules with one another determine the character of the flow. In the medium vacuum range a region of Knudsen flow occurs, in which there is a transition from viscous to the molecular flow of the high vacuum region. In the high vacuum range, the mean free path is usually greater than the enclosure dimensions and gas molecules are located principally on surfaces. The molecular flow of this region is characterized by molecules which can move freely, with virtually no mutual interference. In the ultrahigh vacuum range the monolayer formation time is sufficiently long that "clean" surfaces can be prepared and studied before a gas layer is formed.

2.2 Artificial Vacuum

Due to the existence of an atmosphere, the pressure level at the earth's surface is maintained at approximately 760 Torr, or *one atmosphere*. Although a few examples of low vacuum are

12

exhibited in nature, a vacuum on earth must in general be produced by artificial means. This involves pumping on a closed vessel while classifying the degree of vacuum by measuring the system's absolute pressure.

Production of Vacuum

The ultimate pressure attained in a vacuum system depends on the rate of gas influx compared to the rate of gas pumped from the system. The influx of gas can vary widely, but is controlled through proper materials selection and system design. The pumping rate is restricted by the pump selection and the design of the orifice between the volume of interest and the pumps. In the molecular flow region, a molecule can be pumped only when it happens to enter the pump orifice. Discussions of gas flow and conductance can be found in many books on vacuum technology. Roth [2.1, pp. 60-133] gives an excellent treatment, including conductance calculations for various shapes and combined shapes.

In practice, the speed of a particular pump is less than the theoretical limit and depends on the pressure of the gas being pumped. All pumps work over a limited pressure range, and outside this range the pumping speed falls to zero. The operational range for various types of pumps is indicated in Figure 2.2. However, pumps of the same type but of different sizes or constructions may have adjacent pressure ranges, so that the pressure range of a specific pumping method may be larger than that of an individual pump.

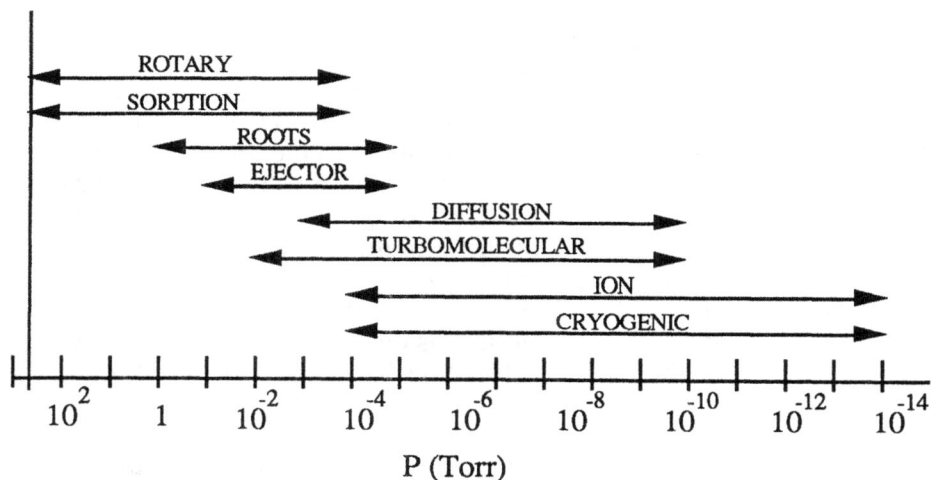

Fig. 2.2 Approximate pressure ranges of selected vacuum pumps.

Vacuum pumps may be grouped into two categories:

> • Compression or gas transfer pumps, which remove gas particles from the pumped volume and convey them to the atmosphere in one or more stages of compression. These include rotary, roots, ejector, turbomolecular and diffusion pumps.

> • Vacuum pumps which condense or in some other manner (e.g. chemically) bind the molecules to be removed on a solid wall, which is often a part of the boundary of the volume being pumped. Examples of this type are: sorption, ion and cryopumps.

The selection of the most effective pump combination depends on the application. Criteria include ease-of-use, maintenance, desired pumping speed and pumping effectiveness for individual gas species. The reprint by Freeman, included in this chapter, discusses some of these issues. Detailed information regarding specific pumps can be acquired from many vacuum technology books, manufacturer's catalogs and representatives. The Recommended Practices Committee of the American Vacuum Society has published articles introducing the new recommended practices for measuring the pumping speed of high-vacuum pumps [2.2] and for pumping hazardous gases [2.3].

<u>Measurement in Vacuum</u>

Pressure

The range of vacuum technology presently extends about 19 orders of magnitude of pressure below atmospheric. This requires the development of devices which can measure low pressures of widely differing magnitudes, from a few Torr to 10^{-16} Torr and beyond. Since no single gauge type has yet been developed that can span the full range from atmosphere to extreme high vacuum, a series of gauges is now available, each having a characteristic measuring range.

In the range of 760 Torr to 1 Torr, pressures can be measured only by using the force resulting from a pressure differential to cause a mechanical motion. These types of gauges (e.g. mercury manometer, diaphragm gauges) were the only types available in the 1950s, when pumping techniques had reached a mature enough state as to require new measuring devices. Although ionization gauges (in which gas is ionized by electron impact from an electron current and the number of positive ions formed is directly proportional to the molecular density) were thought to be capable of measuring lower pressures, those available were known to have a limit of around 10^{-3} Torr. The development of the Bayard-Alpert gauge [2.4] led to ionization gauges effective to the 10^{-9} Torr range, and provided much of the stimulus for vacuum research and development in the 1950-60's. In general, mechanical gauges are useful from atmosphere to 0.1 Torr, thermal conductivity gauges from 1.0 to 10^{-3} Torr, and ionization gauges from 10^{-3} Torr to about 10^{-13} Torr. The reprint by Tilford et al., included in this chapter, presents an orifice-flow-type pressure standard currently in use at the National Institute for Standards and Technology (formerly the National Bureau of Standards).

When selecting a gauge for a definite purpose, the following points should be considered:

> • pressure range of use
> • whether total or partial pressure is to be measured
> • whether the gauge reading is dependent on the type of gas
> • measurement accuracy required
> • type of mounting.

It sometimes becomes necessary to check the calibration of vacuum gauges, although in most cases they are used only to determine the order of magnitude of the system pressure. In the pressure range of 10^{-3} Torr to atmosphere, a gauge can be calibrated by direct comparison to a pressure measurement that results in absolute pressure (e.g. to a mercury U-tube gauge or a McLeod gauge). At pressures below 10^{-3} Torr, direct comparison becomes impractical, and methods must be used to establish a "known pressure" against which a gauge can be calibrated. There are basically two methods for this. The principal *static method* involves the isothermal expansion of a small volume of gas into a larger, evacuated volume. The new, lower pressure can then be calculated and used as a reference. In the *dynamic method*, first proposed by Florenscu [2.5], the pressure is divided across a series of orifices. The pressure difference is then calculated across a known flow conductance at a measured throughput.

Partial Pressure

At pressures above 10^{-3} Torr, the residual gas is likely to have a composition similar to that present before evacuation. In this case, meaningful pressure readings can be deduced by considering the total pressure. At lower pressures, however, gases evolving from components of the vacuum system, in-vacuum transfer mechanisms, or inspection or measuring devices cause the gas composition to become quite different than the original atmosphere. Another contribution to this effect is that pumps used for ultrahigh vacuum systems are selective when pumping gas mixtures. In many applications it is more important to know the relative composition of the residual gas than to know the total pressure.

The partial pressure is the pressure of a designated component of a gaseous mixture exerted on the vessel walls. The sum of the partial pressures of all gases gives the total pressure. In many cases it is important to have knowledge of the main constituent gas and its partial pressure as well as the partial pressures of the less abundant components.

The types of gases are distinguished from one another by their molecular masses. A suitable sensor for partial pressure is a sensitive residual gas analyzer which is small enough to be connected directly to the vacuum system or installed within it. It is basically a mass spectrometer designed for investigating residual gases in a vacuum system. It is more sensitive and more compact than a conventional mass spectrometer, but the mass range and resolution are more limited. There are various types of the instrument, but the general principle of operation is common: the gas molecules are first ionized by electron impact ionization, then accelerated, and finally separated into groups according to their masses.

Several types of residual gas analyzers have been developed (a detailed discussion is presented by Weston[2.6]), although the market is now dominated by the quadrupole mass spectrometer, and to a lesser extent the magnetic sector spectrometer. With the availability of powerful and inexpensive microprocessors, modern instruments include many features which make residual gas analysis much simpler for the vacuum engineer. A typical instrument display of an analysis 'spectrum' is shown in figure 2.3.

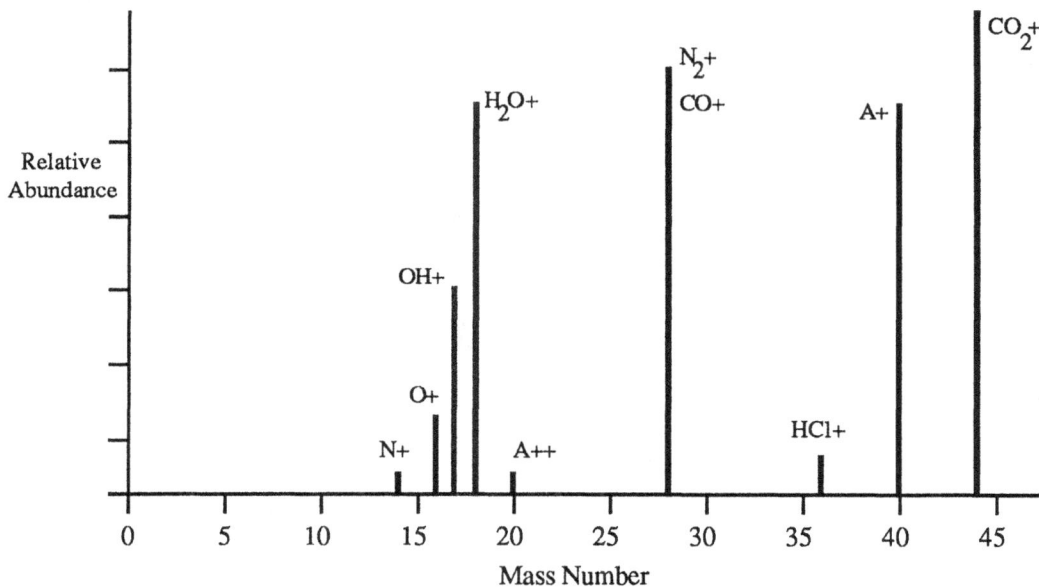

Fig. 2.3 Example of residual gas spectrum

In figure 2.3, peaks are shown which are roughly proportional to the partial pressures of the gases at the various mass numbers. The spectrum will be similar, but not identical, in various instruments. The characteristics of the ionizer, the analyzer, and the detector all affect the nature of the spectrum that is produced. In most residual gas analyzers, more than one type of ion can be produced from a single type of gas molecule. For example, oxygen molecules which are chemically O_2 will produce O_2^+ and O^+ ions. The O_2^+ is usually the most abundant for simple molecules. The array of peaks produced in a gas analyzer by a *single* substance is called a *fragmentation* or *cracking* pattern. Spectra (or *patterns*) of selected materials can be found in vacuum technology reference books, and can be used to help identify which molecules are present in the mass spectrum.

The information produced by a residual gas analyzer is an important aid in the efficient utilization of a high vacuum system, for example, by indicating the presence of a leak or contaminant in the vacuum atmosphere. It can indicate the amounts of gaseous components characteristic of processes occurring within a vacuum and thus can be used to investigate the nature of a process or to monitor process conditions[2.7]. The information can also be used as a tool in vacuum mechanism design by indicating the outgassing load introduced by the component under development.

Leak Detection

No vacuum apparatus can be absolutely vacuum-tight and, in principle, does not need to be. Nevertheless, gas leaks in a system can be very difficult to locate and, when found, difficult to cure. The important point is that the leak rate be comparatively small so as to not influence the pumping rate, gas content and ultimate pressure in the evacuated vessel.

The physical quantity of leak rate is expressed in Torr•liter•second^{-1} (Torr•l/s). The SI unit is N•m•s^{-1}, and mbar•l•s^{-1} is also used. The leak rate is said to be one Torr•l/s if in a closed and evacuated vessel of one liter volume the pressure increases by one Torr in one second. The leak rate of a given system can also be expressed as the pressure rise in a given time for a specific volume, or as the time required for a given quantity of gas to flow into the system. Table 2.1 compares these various quantities. In general, for high vacuum systems, an apparatus with a leak rate of less than 10^{-6} Torr•l/s is considered very leak tight. An apparatus with total leakage greater than 10^{-4} Torr•l/s is not leak tight.

Leak Rate (Torr•l/s)	Pressure Rise in 1 liter vol.	Time for 10^{-3} Torr pressure rise / liter	Time for 1 cm^3 STP gas inflow	Equivalent Opening
10^{-3}	1 µ/s *	1 s	12.7 min	Rectangular slit: 1 cm W x 0.1 mm H x 1 cm D
10^{-4}	6 µ/min	10 s	2.1 hr	Rectangular slit: 1 cm W x 30 µm H x 1 cm D
10^{-5}	36 µ/hr	1.66 min	21 hr	Capillary: 1 cm L x 7 µm dia
10^{-6}	3.6 µ/hr	16.6 min	8.7 days	Capillary: 1 cm L x 4 µm dia
10^{-7}	8.6 µ/day	2.77 hr	87 days	Capillary: 1 cm L x 1.8 µm dia
10^{-8}	0.86 µ/day	27.7 hr	2.4 yr	Capillary: 1 cm L x 0.8 µm dia
10^{-9}	31 µ/yr	11.6 days	24 yr	Capillary: 1 cm L x 0.4 µm dia
10^{-10}	3 µ/yr	116 days	240 yr	Capillary: 1 cm L x 0.2 µm dia

* 1 micron of Hg (µ) = 1 mTorr (10^{-3} Torr)

Table 2.1 Comparison of leak rate specifications (adapted from [2.1])

Despite careful testing of individual components, leaks can be present in the apparatus after assembly due to bad seating of seals or damage to sealing surfaces. To determine whether the poor vacuum is due to a leak or to outgassing, the first step is usually to perform the pressure rise test. For this test the vessel is isolated and the rate of pressure rise monitored. If the pressure rise is constant, as indicated by the first curve in figure 2.4, a leak is present. A rise corresponding to curve 2 is due to gas evolution from inner surfaces or components. Usually, there is a contribution from both leak sources and outgassing (curve 3), making separation of the two sources difficult. Another problem with this method is that gauges are not actually passive elements, but act to pump and convert gas species.

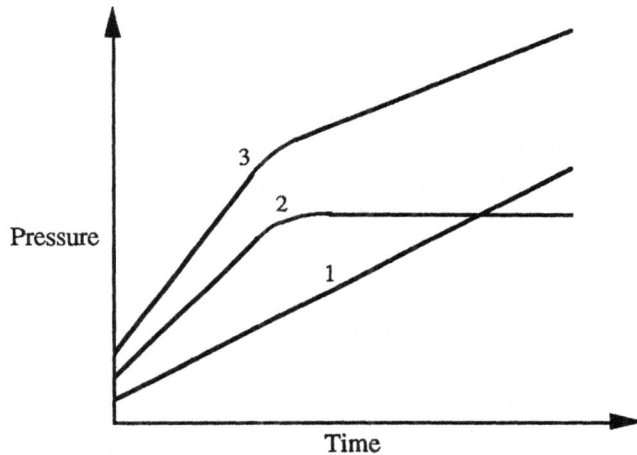

Fig. 2.4 Typical pressure rise curves for an isolated vessel

If one is convinced that the pressure rise is due to a leak, the leak rate (q_L) can be determined quantitatively from the pressure rise (Δp) with time (Δt) in a volume V, according to the relation:

$$q_L = \frac{\Delta p}{\Delta t} \cdot V$$

Eq. (2.2)

Leak detection can be more accurately described as leak *location,* since most instruments do little to establish the presence of a leak without first locating it. Two basic methods are used to localize leaks:

- The "vacuum method", in which the apparatus is evacuated and sprayed from the outside with a search gas. Any leaks are detected by a sensor inside the apparatus.

- The "overpressure method" in which the vessel is filled with a search gas under slight overpressure. Leaks are then detected by a sensor which "sniffs" outside the chamber.

Details concerning the various tests which fall under these two methods can be found in [2.6], [2.7] and [2.8]. For a definitive leak testing reference refer to [2.9], and for a more recent treatment of the fundamentals see [2.10]. In [2.11], Solomon describes the leak testing of a very large system using air as the test gas. Also refer to the *Basic Vacuum Manual* and *Introduction to Mass Spectrometer Leak Detection* volumes published by Varian Associates.

Temperature

As discussed in a later chapter, temperature monitoring can be very important in vacuum, especially when operating active mechanical and electrical devices. Thermocouples and thermistors are commonly employed in vacuum, with special feedthroughs for the wiring which connects to the controller outside the vacuum vessel.

Particulates

In many applications, e.g. microelectronics processing and high precision optical experiments, cleanliness is essential. In fact, an increasing number of manufacturing tasks are being carried out in vacuum because of the ability to maintain high cleanliness levels. Once a vessel is pumped to the molecular flow region (beyond approximately 10^{-2} Torr), the lack of a supporting medium causes particulates to fall out rapidly. However, proper instrumentation is needed in order to characterize and maintain the clean vacuum environment. Contamination issues are discussed in four of the reprint papers of this chapter.

There are many different types of particle sampling instruments. Operation principles include inertial separation, filtration, condensation, electric mobility, optical scattering, etc. The most commonly used type for VLSI facilities is optical scattering, with either white light or a laser. The particle counters used in clean room technology simply become useless in the vacuum environment. For example, the aerosol particle counters, which are designed for sampling in atmospheric pressure, are precluded since the sampling method is based on sucking *air* into the system. Also, the relatively new condensation nuclei counters, which work by pumping the air to be sampled first through a saturation tube and then through the sampling chamber, are clearly not suitable for vacuum applications.

Although surface inspection systems can be used for some vacuum applications, they can only measure particles that have already deposited onto the surface of an unpatterned wafer [2.12][2.13]. In using such a system, a wafer is put into the vacuum chamber, processed as desired, and then taken out for inspection. The actual counting process takes place outside of vacuum. Thus, the surface scanners do not characterize the "vacuum environment" exclusively, and no real-time counting is possible. By means of a surface inspection method, one cannot distinguish those particles deposited during pumping and venting from those deposited during processing or material transport.

The recently developed PM-100 particle monitor (High Yield Technology, Inc.) [2.14] is a new type of particle counter and is found to be the only one that can be used under vacuum. The system includes a sensor head, a preamplifier, and a controller. It provides real-time monitoring by measuring particle flux through a light net [2.15], which gives information on particle motion as well as the number of particles that have flowed through during a certain time interval. The smallest time interval the unit can resolve is 0.01 second and the smallest time interval the user can program is one second. It can be programmed to be reset at any time, allowing the particle count during any particular time interval of interest to be obtained.[2.16]

A principal disadvantage of this device is the wide variation of counting efficiency over the detection range. At the $0.5\mu m$ particle diameter level, the efficiency falls to less than one percent. The unit is thus intended for use as a comparator of relative contamination levels over time rather than an absolute measuring device.

Maintenance and Operation

Bakeout

In virtually all UHV systems, the vessel walls are loaded with condensables such as water vapor in quantities far in excess of their equilibrium values. In an attempt to improve ultimate pressures, vacuum systems can be subjected to a "bakeout". This refers to the practice of temporarily elevating the system temperature during some portion of the pumpdown cycle in order to remove adsorbed gases from heated surfaces faster than is possible at ambient temperature. These gases are removed from the system by pumping during the bakeout. This allows for a lower pressure to be achieved within a given pumping time, or for a shorter required pumping time to reach a given pressure.

Appropriate temperature levels for bakeout will depend on the configuration of the system, the system components (e.g. pumps, transfer mechanisms, sensors), the desired pressure, and the heating method itself. It used to be that temperatures above 400°C were considered necessary, but outgassing studies (see Strausser [2.17]) indicate that the temperature *uniformity* is more important than the temperature itself. A uniform bakeout at 150°C is considered to be effective in most cases, with the caution that leaving even a small fraction of the surface unbaked causes serious degradation. Of course, the operator should determine the temperature limits of the system components and adjust the bakeout accordingly.

The most effective way to uniformly heat a system is by means of a bakeout oven. However, since the system is fabricated of many components having different thermal masses, and the most common material of construction, 304-L stainless steel, has a very poor thermal conductivity of 0.16 W/cm °C, auxiliary heating of larger components is necessary for optimum performance. In many cases a bakeout oven is not available or the system is simply too large to be inserted into such an oven. For these and other (e.g. economic) reasons, heating strips are often used to effect a bakeout. The addition of a shroud, insulation, or both will improve the performance of such an arrangement. Spacing the heaters from the walls of the system can also alleviate the more severe nonuniformity problems. [2.8]

Cleaning

In vacuum technology, cleaning is regarded not only as the removal of undesirable contaminants *lying on* the surface, but must include the removal of all contaminants which are *stuck to* the surface (e.g. oils, greases, particulates) or result from a chemical reaction. Two general types of cleaning are possible: *preventive* and *curative*. Preventive cleaning involves the preclusion of contaminants from entering the vacuum system, while curative cleaning involves removing contaminants which have already entered the system. As much as possible preventive cleaning practices should be followed.

Components which will be part of a final assembly in vacuum can be cleaned with solvents and then baked or 'stoved' prior to installation. There are only a few organic solvents which are appropriate to use in UHV systems, and the quantities should be minimized. Upon initial cleaning with a solvent, a portion of the contaminant of concern will be distributed around the part along with traces of the solvent. As the remaining solvent can be troublesome, it is important to follow the initial cleaning with a dilution. Solvents such as trichloroethylene, acetone and various alcohols such as methanol and propanol can be used for cleaning in different situations. Most vacuum technology reference books give specific recommendations and precautions on the use of such solvents.

After the component has been cleaned and degreased it should be baked under low vacuum, or in a flushing gas, at temperatures as high as 1000°C. The actual temperature will depend on practical considerations, e.g. temperature kept low enough so as not to distort the material. Most of the gas will be removed in the first few hours of the baking process. The result is that outgassing rates can be reduced by two or three orders of magnitude. Of course, care must be used in handling the baked components since touching with the fingers can contribute to a marked increase of outgassing rates. Even after such careful preparation of parts, a bakeout of the total system may be necessary in order to reach UHV levels.

In certain equipment, especially vacuum coaters, it becomes necessary to remove thick layers of deposited materials. In many cases the films can be peeled off to reveal an extremely clean surface, but care should be taken to avoid letting fragments of the material drop into the system. In addition, some of the materials are dangerously pyrophoric under clean conditions, and some must be removed by chemical etch techniques. In all cases appropriate safety precautions should be taken when performing these procedures, e.g. experienced operators, proper ventilation and emergency equipment.

Another method of cleaning the inside of a system is by ion sputtering using an inert gas glow discharge. In this method the vessel is filled with an inert gas and a glow discharge is excited with the metal components serving as a grounded electrode or insulated from the supply. Results on this technique have been reported by Govier and McCracken [2.18] and Jones [2.19], and a review of various methods was published by Holland [2.20].

Maintenance

The majority of routine maintenance procedures for a high vacuum system are related to the pumping and pressure detection systems.

The pumping fluids of mechanical and diffusion pumps are degraded by heat and contamination, and must be regularly changed in accordance with the manufacturer's recommendations. Lamont [2.8] recommends that, while mechanical pump oil may be discarded, the relatively expensive diffusion pump oil used in UHV systems should be recovered and re-processed for economic and environmental reasons. This service is done commercially. Another periodic maintenance item is the regeneration of traps by means of a bakeout cycle. A period of outgassing is also recommended for new ionization gauges, ion sources and electron gun filaments.

2.3 Natural Vacuum

<u>Natural Vacuum on Earth</u>

Although there are no natural high vacuum environments known on earth, some examples of low vacuum are found in nature. Human beings pump to about 740 Torr during respiration, and may achieve pressures as low as 300 Torr by suction. The vacuum action of the octopus, which is able to achieve pressures of 100 Torr, and the mosquito are other examples.

<u>Natural Vacuum in Space</u>

Since the pressure at the earth's surface is due to the presence of an atmosphere, pressure decreases with altitude. The approximate relation, as given by Roth [2.1], is shown in figure 2.5, where a dependency on solar activity is also indicated. The low earth orbit for an artificial satellite is normally above 200 km, where the pressure goes below 10^{-6} Torr. In fact, the pressure at space shuttle altitude is about 10^{-7} Torr. At geostationary altitude, 36,000 km, there is virtually no neutral atmosphere.

In actuality, particle density and temperature are more useful than pressure estimates for predicting environmental effects on spacecraft or satellites. Since these craft are travelling much faster than the air molecules, the actual pressure on the forward surface may be 1000 Torr or greater, while the trailing surface is in high vacuum. In interplanetary space, the velocities are reversed since the solar wind has a velocity of 300 to 2,000 km/s compared to 60 km/s maximum for a spacecraft, but the situation is the same. Again, gas is incident on only one side of the space vehicle.

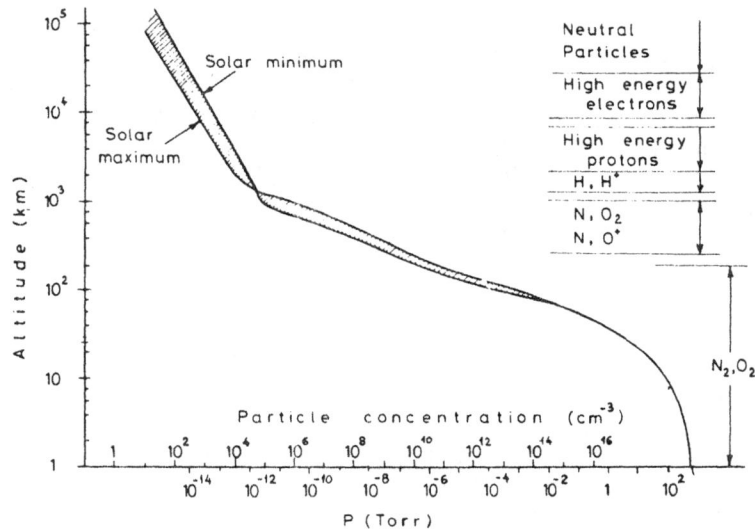

Fig. 2.5 Relationship of altitude and pressure (from Roth [2.1])

Vacuum is one characteristic of the space environment, which also includes zero gravity, radiation and meteoroid activity. Mechanism design for space vacuum is in some ways different than design for artificial earth vacuum. Mechanisms in space will not degrade the *level* of vacuum as is the case when outgassing works directly against the pumping system of a closed vessel on earth. In space, however, the outgassing can create a volume of local contamination which may degrade optics, etc. In the case of space mechanisms, consideration must be given to the method of deployment, length of service, inaccessibility for repair, temperature and radiation effects. For a more thorough discussion of the space environment refer to the reprint by Patrick in chapter 3.

2.4 Vacuum: Advantages and Disadvantages

Advantages of the vacuum environment must be categorized according to the application and the vacuum property used. For example, the low pressure which makes possible a pressure differential force with respect to the atmosphere is the primary advantage in applications for lifting, transportation and forming. Other advantages include low molecular density, long mean free path, long monolayer formation time, and high cleanliness. The low molecular density leads to the removal of active atmospheric constituents and other dissolved gases, and a decrease in energy transfer. The long mean free path means fewer gas-gas collisions, an advantage in many applications, and the long monolayer formation time makes it possible to maintain very clean surfaces.

The disadvantages associated with the vacuum environment are generally not due to the environment itself, but are due to its production and maintenance. Although great advances have been made over the last two decades, and vacuum technology has reached a certain level of maturity, vacuum systems are still quite expensive. As is true with clean rooms (which are generally even more expensive), the cost of producing and maintaining the desired environment is simply added to the cost of the basic process itself. At the same time, the special equipment which is then required for that environment (e.g. clean-room or vacuum-compatible robots) also adds to the cost of the process.

Perhaps the greatest disadvantage in the use of vacuum systems is the limited access due to the container of the artificial vacuum or the space vehicle in the natural vacuum of space. Whereas in the clean room the required special garments are an inconvenience, the vacuum environment precludes any direct human interaction, resulting, in many cases, in very inefficient use of the environment. The development of intelligent vacuum mechatronics systems will increase the effectiveness of present applications and provide the means by which other applications, which would ultimately be more efficient in vacuum, will make the transition. Chapter 5 on vacuum mechatronics control deals with the issue of intelligent control in situations with limited access.

2.5 Applications

Applications of the vacuum environment can be roughly divided into two areas. One of these is concerned with the production of components which as final products are evacuated, or having been evacuated are refilled with another gas or gas mixture (e.g. gas-filled filament lamps or discharge lamps). [2.21] The other main application area involves manufacturing in vacuum components which are not evacuated in their final form. This latter category is continually growing due to the technological advances in vacuum technology, including vacuum mechatronics, which make it easier to take advantage of the environment's special properties.

Since the 1950s, the space program has provided a major stimulus to vacuum technology. Many space simulation chambers, new pumps, gauges and materials were developed rapidly. Although the main impact of the space program on vacuum technology is now over, contributions will still be made in certain areas. And it is still reasonable to expect that manufacturing under vacuum in the microgravity of space will someday become a practical reality.

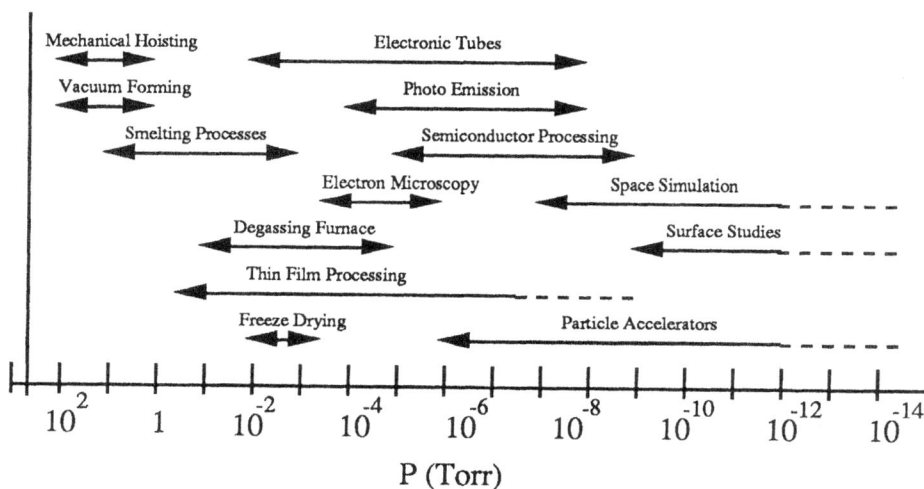

Fig. 2.6 Vacuum applications and corresponding pressure ranges.

The original interest in ultrahigh vacuum on earth came from the requirements of experimenters studying the physics and chemistry of solid surfaces. [2.22] These studies required surfaces cleaned of adsorbed gases over a time period sufficient for study. At 10^{-9} Torr a monolayer of gas will form in about 20 minutes, a reasonable length of time for surface science research.

High cleanliness levels are also required in microelectronic processing steps, which can often total more than 200, with the majority done under vacuum. Coating steps (sputtering, deposition) take advantage of the long mean free path in vacuum. The production of integrated circuits requires many wafer transfers which will eventually be largely handled *in situ* by sophisticated vacuum mechatronics components. Hobson [2.23] predicts that although other vacuum applications require computer controlled transfer mechanisms, the semiconductor industry will carry these developments to their highest level.

Figure 2.6 indicates various applications and their corresponding pressure ranges. In chapter 6 a few current applications are discussed in more detail.

References

[2.1]   Roth, A.,*Vacuum Technology*, Amsterdam.: Elsevier Science Publishers B.V., 1976.

[2.2]   Hablanian, M.H.,"Recommended procedure for measuring pumping speeds", *J. Vac. Sci.Technol. (A)*, Vol. 5, No. 4, pp. 2552-2557, July/Aug, 1987.

[2.3]   O'Hanlon, J.F.,"American Vacuum Society recommended practices for pumping hazardous gases", *J. Vac. Sci. Technol. (A)*, Vol. 6, No. 3, pp. 1226-1254, May/June, 1988.

[2.4]   Bayard, R.T. and Alpert, D., *Rev. Sci. Instrum.*, Vol. 21, pp. 571, 1950.

[2.5]   Florenscu, N.A., *Transactions of the National Vacuum Symposium*, Vol. 8, p. 504, 1961.

[2.6]   Weston, G.F., *Ultrahigh Vacuum Practice*, Cambridge, United Kingdom: University Press, 1985.

[2.7]   Leybold-Heraeus, *Product and Vacuum Technology Reference Book*.

[2.8]   Weissler, G.L., and R.W. Carlson, Editors, "Methods of Experimental Physics", Vol. 14, *Vacuum Physics and Technology*, New York, Academic Press, Inc., 1979.

[2.9]   McMaster, R.C. Ed., "Leak Testing", *NDT Handbook, 2nd Edition, Vol. I:* American Society of Non-Destructive Testing, Columbus, Ohio, 1982.

[2.10]   Reynolds, A.D., *Fundamentals of Leak Testing*, to be published in 1989, contact Mean Free Path Corp., 220 Pegasus Ave., Northvale, NJ 07647.

[2.11]   Solomon, G.M., "Air mass spectrometer leak detection using the special air leak test (SALT) cart", *J. Vac. Sci. Technol. (A)*, Vol. 2, No. 2, pp. 1157-1161, Apr/June, 1984.

[2.12]   Cheung, S. D. and R. P. Roberge, "An Inside Look at Measuring Particles in Process Equipment", *Microcontamination*, p. 45, May, 1987.

[2.13]   Hancock, S., and C. Yarling, "Particle Measurements in Medium-Current Ion Implanters", *Semiconductor International*, p. 144, May, 1984.

[2.14]   Bordon, P., Y. Baron and B. McGinley, "Monitoring Particles in Vacuum Process Equipment", *Microcontamination*, p. 30, Oct, 1987.

[2.15] High Yield Technology Technical Note, "Particle Flux: The New Measurement", September, 1986.

[2.16] Chen, D., T. Seidel, S.E. Belinski and S. Hackwood, "Dynamic Particulate Characterization of a Vacuum Load-lock System", submitted for publication.

[2.17] Strausser,Y. E., *J. Vac. Sci. Technol.* , Vol. 4, pp. 337, 1967 and *Proc. Int. Vac. Congr*, Vol. 4, pp. 469.

[2.18] Govier, R.P. and G.M. McCracken, "Gas Discharge Cleaning of Vacuum Surfaces", *J. Vac. Sci. Technol.,* Vol. 7, No. 5, pp. 552-556, 1970.

[2.19] Jones, A.W., E. Jones and E.M. Williams, "Investigation by techniques of electron stimulated desorption of the merits of glow discharge cleaning of the surfaces of vacuum chambers at the CERN intersecting storage rings", *Vacuum*, Vol. 23, pp. 227-230, 1973.

[2.20] Holland, L.,"Treating and passivating vacuum systems and components in cold cathode discharges", *Vacuum*, Vol. 26, No. 3, pp. 97-103, March, 1976.

[2.21] Carpenter, L.G.,*Vacuum Technology: An Introduction*, Great Britain: Adam Hilger, Ltd., 1983,.

[2.22] Steinherz, H.A., and P.A. Redhead, "Ultrahigh Vacuum", *Scientific American,* Vol. 2, No. 13, March, 1962.

[2.23] Hobson, J.P., "The future of vacuum technology", *J. Vac. Sci. Technol. (A)*, Vol. 2, No. 2, pp. 144-149, Apr/June, 1984.

The National Bureau of Standards primary high-vacuum standard

C. R. Tilford, S. Dittmann, and K. E. McCulloh

National Bureau of Standards, Center for Basic Standards, Gaithersburg, Maryland 20899

(Received 8 February 1988; accepted 3 May 1988)

The theory, design, operating procedures, and estimated errors are discussed for an orifice-flow-type pressure standard currently in use at the National Bureau of Standards. This standard is used to define pressures between 10^{-7} and 10^{-1} Pa. Including the uncertainty of the flowmeter, the estimated total uncertainty varies from 2.6% at the highest pressures, to 1.4% at midrange, and 8.2% at the lowest pressures. Representative calibration results are presented for four different types of hot-cathode ionization gauges.

I. INTRODUCTION

At pressures below $\sim 10^{-1}$ Pa (10^{-3} Torr) (1 Torr = 133.322 Pa) the flow of gas through an orifice of known area can be used as a primary pressure standard. The National Bureau of Standards (NBS) has used this dynamic or orifice-flow technique to develop a standard that is used routinely between 10^{-1} and 10^{-6} Pa and on an exploratory basis to lower pressures. An early version of this system is described in Ref. 1, and a brief description of the present system is given in Ref. 2. This paper will briefly develop the theory of this standard, describe the design of the vacuum chamber, including the orifice, and evaluate the performance of the pressure standard. The flowmeter that is an integral part of this standard is discussed in a separate publication.[3]

II. THEORY OF ORIFICE-FLOW STANDARDS

Orifice-flow standards have been discussed in recent reviews.[4,5] Generally the theory of these devices is described in terms of "throughput" and conductance (generally designated Q and C). In order to avoid ambiguities in the definition of certain flow units[6] the present description will be in terms of more fundamental quantities, but the results are completely equivalent.

From the kinetic theory of gases, the pressure due to molecules of a single gas species incident at a surface is given by

$$P = (2\pi kTm)^{1/2}F, \qquad (1)$$

where k is Boltzmann's constant, T is the molecular temperature, m is the molecular mass, and F is the molecular flux incident on unit area. The pressure due to a multispecies mixture will be a linear sum of such contributions.

If a molecular flow rate $\dot{n}$ (molecules/s) is admitted to a chamber pumped by a "perfect" pump through an orifice of area A, under conditions of constant pressure and molecular flow (molecule–molecule collisions are a negligible fraction of the molecule–wall collisions), the flux incident at the orifice will be given by $\dot{n}/A$ and the pressure at the orifice is

$$P = (2\pi kTm)^{1/2}\dot{n}/A. \qquad (2)$$

If A is a small fraction of the chamber surface area, and the inlet flow is baffled so that molecules must make several collisions with the chamber walls before being pumped out, the pressure will approach uniformity throughout the chamber and T will be very close to the average chamber wall temperature. Thus, knowing A and using a flowmeter to measure $\dot{n}$, the pressure in the chamber could be calculated. However, two important modifications must be made to Eq. (2) to account for deviations from ideal conditions.

Equating the incident flux to the flow rate divided by the area of the orifice is strictly true only in the case of an orifice with zero thickness. Molecular scattering from the edges of a real orifice or nearby structures in a line of sight from the orifice will perturb the flux at the orifice. This can be taken into account by calculating an effective area A_e using the transmission probability approach of Clausing.[7]

A second complication is that perfect pumps are generally not available and usually there is a significant return flux coming back into the chamber through the orifice. Under conditions of constant pressure in the calibration chamber, this return flux is balanced by a corresponding flux back out through the orifice, and the flow rate from the flowmeter into the chamber $\dot{n}$ equals the net flow out through the orifice. The "high" pressure in the chamber P_H is related by Eq. (1) to the total flux out through the orifice, which is the net flux $\dot{n}/A_e$ plus the return flux. The return flux can be calculated from the "low" pressure P_L on the downstream side of the orifice using Eq. (1). Thus, the pressure difference across the orifice can be calculated knowing the net flow through the chamber and the effective area of the orifice:

$$P_H - P_L = (2\pi kTm)^{1/2}\dot{n}/A_e. \qquad (3)$$

A method of correcting for P_L, allowing the calculation of P_H, will be described in a later section.

III. DESCRIPTION OF EQUIPMENT

A. Vacuum chamber

The design of the vacuum chamber, illustrated in Fig. 1, is largely determined by the need to maintain a low background pressure, to provide room to mount the gauges to be calibrated, and to maintain a geometry for which the orifice conductance calculations will be valid and adequate pressure uniformity can be assured. The chamber is in two cylindrical halves, each 27 cm in diameter and 34 cm long, separated by a central wall with the orifice in the center. A chamber of this size permits the incorporation of eight $2\frac{3}{4}$-in.-o.d. (3.5-cm-i.d.) gauge mounting flanges on the upper half at a circumference 13.3 cm above the orifice. A circular baf-

2853   J. Vac. Sci. Technol. A 6 (5), Sep/Oct 1988

25

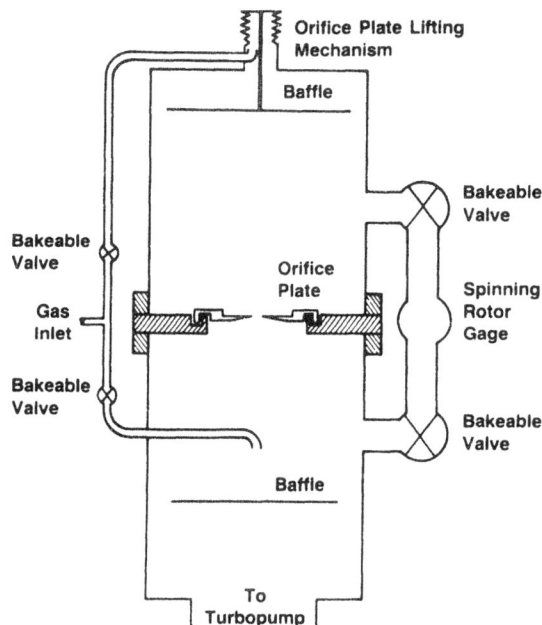

FIG. 1. Schematic of the vacuum chamber. Only one of the eight gauge mounting ports available on each chamber half is shown. A residual gas analyzer mounted in the calibration chamber (not shown) is used to monitor base vacuum conditions and calibration gas purity. The gas inlet is connected to the flowmeters by baked stainless-steel lines.

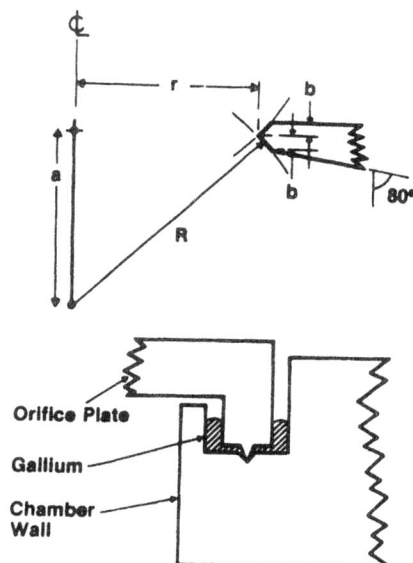

FIG. 2. Detail of the orifice and the gallium-filled groove that seals the orifice plate into the wall between the two chamber halves. In order to obtain a continuous band of gallium around the groove a stainless-steel rod was used to "scrub" or "scratch" the bottom of the groove after filling with the gallium. The lip of the orifice plate was thoroughly passivated.

fle 5 cm below the top of the upper chamber insures that gas molecules entering the top of the upper chamber from the flowmeter will experience several collisions with the chamber walls before entering a gauge or passing through the orifice at the bottom. These collisions are essential to insure the random distribution of molecular velocities, which is necessary for pressure uniformity and is assumed in the conductance calculation. As discussed below, the uniform distribution of molecular flux will be perturbed by the molecules entering the chamber below the baffle and by those escaping through the orifice. This perturbation can be kept within acceptable bounds by using an orifice whose area is a small fraction of the surface area of the chamber. At the same time, the orifice conductance must be large enough to maintain a low background pressure and ensure a high enough flow that gauge pumping and outgassing are minor perturbations. In our system the compromise used is a nominal 1.1-cm-diam orifice, described in detail below, with a nitrogen pumping speed at 23 °C of ~ 11 700 cm³/s, and an area that is 0.03% of the upstream chamber area.

Accounting for the molecules returning back through the orifice into the calibration chamber will be most accurately done if these molecules originate from a uniform flux distribution. This is achieved by making the lower chamber an almost symmetric duplicate of the upper chamber, with a baffle 12 cm above the inlet of the nominal 0.5 m³/s turbomolecular pump attached to the bottom of the chamber. The chamber is not perfectly symmetrical as the baffle is located 24 cm below the orifice. The effective rate of exhaust from the lower chamber is 0.3 m³/s for nitrogen. The turbomolecular pump was chosen for its stable high pumping speed,

low base pressure, and compatibility with most gases.

Ultrahigh vacuum construction practices are employed throughout; the chamber is stainless steel, only metal seals are used, and the chamber and all gauges are baked between 200 and 250 °C after each venting to air. However, the small orifice clearly would restrict the attainment of a low base pressure in the upper chamber. Therefore, the orifice is contained in a plate that seals into a gallium-filled groove, located in the central wall as illustrated in Figs. 1 and 2. During pumpdown or bakeout, the orifice plate can be lifted through a bellows at the top of the chamber, opening a 12.7-cm hole between the upper and lower chambers. For the gallium to form a seal when the orifice plate is lowered back down, it must remain supercooled for prolonged periods below its 29.8 °C melting temperature, a well-known characteristic of gallium. Leaks of 10 cm³/s or larger will have a significant effect on our measurements. In four years of operation we have never observed any evidence of seal failure. Nor have we seen any evidence of gallium vapor in the chamber, which is consistent with the prediction that gallium's room-temperature vapor pressure will be many decades below the detection limit of any vacuum sensor.

When the several meters of ⅜-in.-o.d. (0.95-cm) stainless-steel line connecting the standard to the flowmeters are baked along with the chamber, this system routinely attains base pressures in the low 10^{-8} Pa (10^{-10} Torr) range, the residual gas is almost entirely hydrogen. The base pressure in the calibration chamber typically will increase on the order of a factor of 2 when the orifice plate is lowered and sealed in the gallium groove, although increases as large as a factor of 6 have been observed with 17 hot cathode ion gauges and gas analyzers operating in the calibration chamber. In this latter case a day was required after the orifice plate was lowered for the rate of change of pressure to become $< 0.3\%/h$.

B. Orifice

The orifice proper, as shown in Fig. 2, is a sharp-edged hole at the center of a 15.2-cm-diam stainless-steel plate. The outer edge of the plate includes the lip that fits into the gallium-filled groove in the wall between the chamber halves. A 1.12-cm-diam hole was bored through the center of the plate where it is 0.74 mm thick. The orifice was formed by lapping the hole with a steel ball of 1.587-cm (0.6250-in.) diameter to nominally equal depths from above and below. This technique was chosen because of the ease of calculating transmission probabilities for the spherical surfaces, and the expectation that it would result in a clean edge. Upon final lapping with an unworn ball, the two concave spherical surfaces thus generated met at a sharp circular edge with a measured diameter of 1.1240 ± 0.0008 cm and a computed edge angle of 90.1°. Microscopic examination revealed smooth surfaces and a sharp, burr-free edge. Above the orifice the orifice plate is flat out to a diameter of 8.89 cm, where there is a 0.61-cm step. On the bottom side there is a truncated 80° half-angle cone. Effective area correction factors are calculated for both of these features.

C. Area calculation

Calculation of the orifice effective area, carried out by McCulloh,[8] is based on the molecular transmission probability approach initiated by Clausing.[7] It assumes free molecular flow and cosine law scattering at surfaces. The area is given by

$$A_e = \pi r^2 K_0 K_1 K_2, \tag{4}$$

where r is the measured radius of the lapped hole, K_0 is the transmission probability of the duct formed by the lapped spherical surfaces, K_1 is the transmission probability of the 8.89-cm-diam cylindrical relief in the top of the orifice plate, and K_2 is the transmission probability of the 80° cone in the bottom of the plate.

Estimates based on a generalization of Clausing's theory for cylindrical geometries place bounds on K_1 of

$$0.999\,95 < K_1 < 0.999\,97.$$

The results of Iczkowski et al.[9] were used to place bounds on K_2 of

$$0.9999 \leqslant K_2 < 1.0000.$$

This gives us a mean value for the product $K_1 K_2 = 0.9999$, with a maximum error of 7×10^{-5}.

The calculation of K_0 involves an integral equation that yields the result

$$K_0 = 1 - 2J/(R + a), \tag{5}$$

where R is the radius of curvature of the lapped spherical surfaces, a is the distance from the center of curvature to the center of the orifice, and J is an integral over the surface of the orifice duct of the probability that a molecule striking that surface will not pass into the lower chamber. This integral has not been evaluated exactly, but it is within the limits

$$0.147\,92b < J < 0.148\,01b,$$

where b is the half-thickness of the orifice duct. Using the appropriate dimensions in Eq. (5) we then have

$$0.991\,672 < K_0 < 0.991\,677 .$$

These bounds are uncertain by 0.03% due to the uncertainty of the measured quantities used.

Combining the above we find that at 23 °C

$$A_e = 0.9916\pi r^2 = 0.9838 \text{ cm}^2. \tag{6}$$

The overall uncertainty is $\pm 0.18\%$, with the largest contribution being 0.14% due to the uncertainty in the measured value of r.

IV. CALCULATION OF ABSOLUTE PRESSURES

In the system described here the pressure downstream of the orifice is $\sim 3.7\%$ of the pressure in the calibration chamber for nitrogen. The ratio of these pressures varies, to the first order, with the inverse of the square root of the molecular mass. It is difficult to measure P_L directly since this would require a gauge calibrated below the range of pressure generated by the standard. However, if we can measure the ratio of the pressures upstream P_H, and downstream P_L, from the orifice,

$$R_p = P_H/P_L, \tag{7}$$

then

$$\begin{aligned} P_H &= (P_H - P_L)[R_p/(R_p - 1)] \\ &= (2\pi kTm)^{1/2}(\dot{n}/A_e)[R_p/(R_p - 1)] . \end{aligned} \tag{8}$$

The pressure ratio R_p is measured by a molecular drag gauge [10] (MDG) that, as shown in Fig. 1, can be connected through $1\frac{1}{2}$-in. (3.8-cm) bakable valves to one of the gauge ports in the upper chamber or to a similar port in the lower chamber. Since the MDG is inert and therefore not limited by plumbing impedances, and reproducible at the 1% or better level above 10^{-4} Pa, it can be used to determine R_p quite precisely in the high-vacuum range. This is done during each calibration cycle. Errors in R_p, whether random or due to the systematic pressure gradients discussed below, will be reduced in effect on the generated pressure by the factor $1/(R_p - 1)$, which is ~ 0.04 for nitrogen. A more significant problem is that as the pressure is reduced, the random error of the MDG, typically 10^{-6} Pa or greater, precludes its use for a direct determination of R_p. R_p will be a function only of the orifice conductance and the pump speed, and in the molecular flow range the only variable with pressure can be the turbomolecular pump speed. As detailed below, we have indirectly found the pump speed, and therefore R_p, to be constant to within several percent down to a lower chamber pressure of 3×10^{-8} Pa. R_p will also be sensitive to any leakage through the gallium seal. Apart from changes associated with a failing turbomolecular pump bearing we have not seen significant changes with time of R_p, indicating reliable operation of the gallium seal.

V. EXTENSION TO LOWER PRESSURES

The low-pressure limit of orifice-flow standards is determined by the low flow limit of the flowmeter and by the base pressure of the vacuum chamber. In the case of the NBS system, the base pressure is below the limit established by the flowmeter. Although the system has been operated with gas flow into the calibration chamber down to 10^{-6} Pa, it is typically limited to pressures above 10^{-5} Pa in this mode in order to limit the contributions of flowmeter errors. We have extended the range of the pressure standard by injecting the gas flow into the lower pressure half of the vacuum chamber, downstream from the orifice. The pressure gradients in the lower chamber preclude the direct calculation of the resultant pressure in the calibration chamber. However, we can experimentally determine the ratio of the flows, R_f, into the two chambers required to generate the same pressure. A pressure P is established by a measured flow from the flowmeter $\dot{n}_H$ into the calibration chamber. Using the valves shown in Fig. 1 the flowmeter is then connected to the lower chamber and the flow increased until the same pressure is established in the upper chamber. This measured flow $\dot{n}_L$ is used to derive the flow ratio

$$R_f = \dot{n}_L / \dot{n}_H. \tag{9}$$

Equation (8) is then modified to give

$$P_H = (2\pi kTm)^{1/2}(\dot{n}_L/R_fA_e)[R_p/(R_p - 1)]. \tag{10}$$

Molecular drag and ion gauges in the upper chamber are used to determine R_f. There is a random error in the determination of R_f of the order of 0.1%. More importantly, there is an additional systematic uncertainty due to the two flowmeter measurements required. This effectively doubles the contribution of the flowmeter error in Eq. (10). In the NBS standard, R_f differs from R_p by < 1% although this may be fortuitous since flux and pressure gradients of several percent are known to exist in the lower chamber.

The flow ratio cannot be directly measured for pressures lower than those that can be reliably generated by a measured flow into the upper chamber, so it is typically determined between 10^{-4} and 10^{-5} Pa. However, since the system is clearly in the molecular flow regime at these low pressures, both R_f and R_p will be constant with decreasing pressure if the turbomolecular pump speed does not change with pressure. As noted before, we have indirect evidence that the pump speed changes very little down to 3×10^{-8} Pa.

VI. EVALUATION OF ERRORS

In addition to the errors already discussed in the conductance calculation and the measurement of the pressure and flow ratios, R_p and R_f, several other possible sources of error must be taken into account.

A. Pressure gradients

A high degree of pressure spatial uniformity from port to port on the calibration chamber has been demonstrated by establishing a nitrogen flow to generate a pressure of $\sim 10^{-3}$ Pa in the calibration chamber. The pressure was measured by four ion gauges which were well distributed about the calibration chamber. After recording the gauge readings, the orifice plate was raised and the flow was increased by a factor of ~ 15 to reestablish the same pressure. The relative gauge readings with the orifice plate raised agreed to within 0.1% with the reading obtained with the orifice plate lowered. Since nonuniformity due to axial asymmetry of the flow would be proportional to the flow rate required to generate a specified pressure, the results indicate that any such nonuniformity to be < 0.01% in normal operation with the orifice plate lowered.

The calibration chamber pressure will have a vertical perturbation due to the net flow of molecules between the gas inlet and the orifice. This will have two contributions, one due to the molecules exiting through the orifice and not being scattered back, and the second due to the molecules entering from the inlet. The first contribution is reasonably straightforward to calculate since the orifice is a nearly ideal sink. This contribution will decrease the pressure at the gauge ports from that given in Eq. (8) by the factor $r^2/4r_c^2$, where r is again the orifice radius and r_c is the chamber radius. For the NBS system, this factor is 0.999 57.

The second contribution is more difficult to calculate because the scattering of the inlet flow produces a diffuse source and it will take several collisions before the effects of the initial nonuniformity are negligible. However, estimates of this contribution indicate that it is somewhat smaller than the first, but of the opposite sign.

B. Mounting of the orifice plate

Another small error is the perturbation to the calculated conductance caused by the mounting of the orifice plate. The orifice plate is mounted in the wall between the two chambers with a short duct, 12.7 cm in diameter and 1.27 cm long, between it and the lower chamber. The transmission probability to the lower chamber of molecules exiting the orifice is estimated to be 0.9997 ± 0.0001. This effectively increases the pressure in the upper chamber by a factor of 1.0003, offsetting the effect at the gauge port of the small vertical pressure gradient previously discussed.

We estimate that 0.003% is a generous upper bound on the uncertainties caused by the combined upper chamber pressure gradients and the mounting configuration of the orifice plate.

C. Temperature

The average temperature of the gas molecules is determined by the average temperature of the chamber walls, which is known with an uncertainty of 0.3 K, in the absence of heat sources, notably ion gauges. Depending on the number of gauges, and whether they are tubulated or immersed in the chamber, this uncertainty, in extreme cases, could be as much as an order of magnitude larger. This will contribute an uncertainty to the pressure varying from 0.05% to 0.5%. This does not include thermal transpiration effects for ion gauges being calibrated. These are considered a part of the gauge's calibration and can cause additional errors during later use of the gauge if its thermal environment differs from that at the time of calibration.

D. High-pressure nonlinearities

The calculation of the orifice conductance assumes an absence of molecule–molecule collisions. The number of such collisions increases with pressure, with a corresponding increase in the error of the conductance calculation. We have chosen at this time not to attempt to correct for this effect but rather to restrict the upper pressure limit of the standard to keep this error within bounds. This error amounts to 0.1% at $\sim 8.5 \times 10^{-3}$ Pa for N_2 and at $\sim 2.5 \times 10^{-2}$ Pa for He, and increases linearly with pressure.

E. Flow rate errors

The systematic uncertainties of the flowmeter, detailed in Ref. 3, contribute directly to the errors of the high-vacuum standard.

F. Observed random errors and pump speed changes

Random errors, and systematic errors due to changes in pump speed with pressure, can be evaluated from the results of vacuum gauge calibrations using the primary standard. In all cases these data can at best give an estimated upper bound to the errors of the standard since the data will include not only errors due to the standard, but also those due to short- and long-term gauge instabilities and, in some cases, systematic changes of the gauge sensitivities with pressure. At higher pressures, where molecular drag gauge calibrations repeated over a few day's time are used to estimate random errors, the contribution of gauge instabilities are probably not significant. Higher pressure ion gauge calibrations repeated over a few day's time typically show somewhat larger but still relatively small effects of gauge instability. However, in the case of ion gauge data used to evaluate errors below 10^{-5} Pa the gauge performance probably makes a significant contribution to the observed variation of the calibration data. The amount of such data is limited, and it has been necessary to include data taken under somewhat different conditions and over periods of from one to two months. The magnitude of ion gauge instabilities varies with gauge type,[11] but some can show quite large changes operating over a month's time.[12] We have calibrated 18 gauges below 10^{-5} Pa, using controllers of known reliability and electrometers of known accuracy.

Examples of such data for four different types of hot-cathode ionization gauges are shown in Fig. 3. In each case the differential sensitivity or sensitivity coefficient,

$$S = (I^+ - I_0^+)/I_e(P - P_0) \tag{11}$$

was measured, where I_e is the electron emission current, I^+ is the collector current at pressure P, and I_0^+ is the collector current at base pressure. The primary standard determines the pressure increase, $P - P_0$, above base pressure. The fractional or percentage changes of this sensitivity from its value at low pressures are plotted as a function of pressure to a uniform scale, indicated on the figure. The nude Bayard/Al-

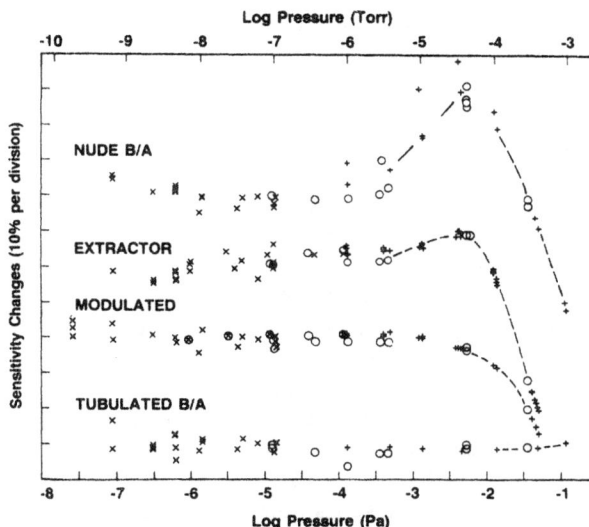

FIG. 3. Representative nitrogen calibration data for four different types of hot cathode ion gauges. Sensitivities are presented as fractional or percentage changes from the average low-pressure sensitivity for each gauge. The ×'s indicate data obtained with flow into the low pressure chamber, the ○'s were obtained with flow into the calibration chamber. Data obtained using molecular drag gauges, previously calibrated *in situ*, are indicated by the +'s, and are included as a reference for evaluating the repeatability of the high-pressure orifice-flow data. The dashed lines are solely to distinguish data for one gauge from another and do not represent measured or extrapolated gauge performance. The data for each gauge were acquired over a 1–2 month period.

pert (B/A) gauge has two thoria coated filaments. Data are shown for one filament operated at an emission current of 1 mA. The average low-pressure nitrogen sensitivity was 0.155 Pa^{-1} (20.7 Torr^{-1}). The peak in sensitivity between 10^{-3} and 10^{-2} Pa is quite typical of this type of gauge, although the magnitude and location of this peak may vary significantly from gauge to gauge. The modulator gauge was operated with the modulator held at grid potential. Its tungsten filament was operated at an emission current of 1 mA. Low-pressure sensitivity was 0.178 Pa^{-1} (23.7 Torr^{-1}). The extractor gauge had an iridium filament operated at an emission current of 1.85 mA. Its low-pressure sensitivity was 0.065 Pa^{-1} (8.7 Torr^{-1}). The glass tubulated B/A gauge has two tungsten filaments, one of which was operated at an emission current of 1 mA. Its low-pressure sensitivity was 0.078 Pa^{-1} (10.5 Torr^{-1}). Constant sensitivity to the highest pressures tested is typical of this type of gauge.

Two of the 18 gauges showed clear evidence of instability. One changed by 8% over a four-month period, the other by 26% over the same period. Neither gauge was used for further analysis. Data for the other 16 gauges show random scatter comparable to that seen in Fig. 3. Some measure of the possible contributions from gauge instabilities can be obtained by comparing the scatter of the high-pressure data for the nude B/A to that of the modulated gauge since they were calibrated at the same time. The larger scatter in the nude B/A gauge is most likely due to instabilities in that gauge. (Note: some data points for the modulated gauge are so close as to be indistinguishable and for a few of the points data were obtained for only one gauge.) While it would be desirable to have data taken over shorter periods of time and under more uniform conditions, the available data still allow the best evaluation of low-pressure errors.

1. Random errors

A measure of the short-term repeatability at higher pressures can be obtained from repeated calibrations of six different MDG's with N_2 at 4.9×10^{-3} Pa. In this case the calibrations were made over a period of 8 h. The pooled standard deviation of the accommodation coefficients of the six gauges was 0.026%, with no significant difference in the random errors associated with any gauge.

A more stringent measure was the repeated calibrations over several weeks of ten MDG's with argon at pressures ranging from 5×10^{-4} to 3×10^{-2} Pa. In this case three to five different calibration series were performed at each pressure, each series taking place on a different day and involving a complete calibration cycle, starting with the vacuum chamber and flowmeter at base vacuum and the orifice plate raised above the gallium-filled groove. The data from two of the SRG's were eliminated because of clearly excessive random errors in the gauges, apparently caused by suspension instabilities. The pooled standard deviations of the remaining eight gauges were 0.10% at 3×10^{-2} Pa, 0.08% at 7×10^{-3} Pa, 0.12% at 3×10^{-3} Pa, and 0.41% at 5×10^{-4} Pa. These are consistent with other sets of MDG data.

At the lowest pressure the above results are probably dominated by the short-term random errors of the MDG's, and MDG data at yet lower pressures will not be useful in assessing errors of the standard. Therefore, we must rely on ion gauge calibration data at lower pressures.

A measure of the low-pressure short-term repeatabilities can be obtained from three calibrations repeated over several hours of a tubulated tungsten filament Bayard–Alpert gauge. In this case separate flowmeter measurements were made for each calibration. At 2×10^{-4} Pa the calibrations differed by a maximum of 0.1%, and at 2×10^{-5} Pa by 0.02%.

A more realistic assessment of the random errors can be obtained from calibrations repeated on different days. For flow directly into the calibration chamber, using Eq. (8), we expect that errors at the lowest pressures will be dominated by random flowmeter errors. Typically, such calibrations repeat to within $\pm 5 \times 10^{-8}$ Pa.

Repeated calibrations with flow into the lower pressure chamber are available for 16 different gauges at pressures from 4×10^{-3} to 3×10^{-8} Pa. Most of the data are below 10^{-5} Pa. Calibrations repeated at the same pressure on different days for most gauges vary by $< \pm 2.5\%$, even down to the lowest pressures. Linear curves fitted as a function of pressure to the sensitivities for individual gauges had standard deviations of the residuals that varied from 0.26% to 3.25% with an average standard deviation of 1.95%. Maximum deviations of individual points from the fitted curves were as large as 7.6%, although for most gauges the maximum deviations did not exceed 3%. Considering that some gauges showed larger variations than other gauges calibrated at the same time, $\pm 2.5\%$ seems a reasonable upper

bound for the possible random errors of the standard with flow into the lower chamber down to 10^{-7} Pa.

2. Changes in pump speed

As discussed earlier, the pressure and flow ratios can be reliably measured only at higher pressures and their use in Eqs. (8) and (10) at lower pressures assumes the pump speed does not change with decreasing pressure. Any change in the pump speed with pressure will cause errors in both R_p and R_f. Errors in R_p will cause relatively small errors in Eqs. (8) and (10). However, errors in R_f will cause corresponding errors in the pressures calculated using Eq. (10), so stability of pump speed is important at pressures below 10^{-5} Pa where the system is generally operated with flow into the low-pressure chamber.

Generally, R_f is determined at a pressure around 10^{-5} Pa and ion gauge sensitivities are determined using Eq. (8) above this pressure and Eq. (10) below. However, for a few gauges sensitivities were determined using both techniques at pressures both above and below 10^{-5} Pa. To within the limits imposed by ion gauge instabilities, the deviations from unity of the ratios of the two sensitivities will be a measure of the constancy with pressure of R_f and the pump speed. Unfortunately, in only a few cases were sensitivities determined within a short period of time (up to five days) using the two techniques. In those cases the ratios of the sensitivities varied from 0.989 to 1.006 over the pressure range 3×10^{-6}–10^{-4} Pa. Additional data exist where the time span between the sensitivity determinations varied from 2 to 18 months. In these cases gauge instabilities are much more of a concern. The ratio of the sensitivities in these cases varied from 0.951 to 1.015 over the pressure range 10^{-6}–10^{-3} Pa.

A less direct measure of possible changes in pump speed can be obtained by examining the constancy with pressure of measured gauge sensitivities obtained from Eq. (10). It is often assumed that ion gauge sensitivities, as defined by Eq. (11), are constant at low pressures. This may or may not be true. However, a constant measured gauge sensitivity implies that either the gauge sensitivity and pump speed are both constant, or that both change with pressure in a manner such that the changes cancel. Neither case can be proven, but it is unlikely that different gauges will have the same nonlinearities, particularly if the gauges are of different designs, so, as more gauges are found to have constant measured sensitivities, to within some bounds, the second possibility becomes increasingly unlikely, and it is more probable that pump speed is constant to within those same bounds.

As can be seen in Fig. 3, there are small trends in the low-pressure data for different gauges. The trends are clearly not the same for different gauges. In order to set bounds on these trends, and possible systematic changes in pump speed, nitrogen sensitivities for each of 16 gauges, obtained with flow into the lower chamber of the standard, have been least-squares fitted to equations of the form $S = A + B \log P$. The number of data points per gauge varied from 4 to 34, the average was 16. The data extended from 3×10^{-8} to 4×10^{-3} Pa; most of it was below 10^{-5} Pa. The B coefficients for the different gauges varied from -2.5% to 3.9% per decade. Only two coefficients were larger than 2.5% per dec-

TABLE I. Percentage uncertainties of the NBS high-vacuum standard with nitrogen. These error estimates are based on gas flow into the calibration chamber above 10^{-5} Pa, and into the low-pressure chamber at lower pressures. Random errors at 10^{-6} and 10^{-7} Pa are included in the allowance for possible errors due to changing pump speed. Since the random and systematic errors cannot be separated in this case they have not been separately summed. The systematic uncertainties have been linearly summed in order to establish an upper bound on the uncertainty.

	Pressure (Pa)				
	10^{-1}	10^{-2}	10^{-3}–10^{-5}	10^{-6}	10^{-7}
Systematic contributions					
Orifice conductance	0.18	0.18	0.18	0.18	0.18
Molecular scattering	1.2	0.12	...	...	...
Flow rate	0.82	0.82	0.82	2.00	2.02
Pressure ratio	0.04	0.04	0.04	0.04	0.04
Flow ratio	...	...	...	0.90	0.90
Assumed pump speed	...	...	...	2.5	5.0
Temperature	0.1	0.1	0.1	0.1	0.1
Total systematic	2.34	1.26	1.14	...	...
Random errors (3 standard deviation)	0.21	0.30	0.30	...	...
Total	2.6	1.6	1.4	5.7	8.2

ade and they were obtained from data sets that extended down to only 10^{-6} Pa. The average coefficient, obtained from the individual coefficients weighted by the inverse of their variances, is < 1% per decade.

Considering the probable significant effect of ion gauge behavior on some of these data, we believe that ± 2.5% per decade is a conservative upper bound on possible pump speed changes below 10^{-5} Pa.

VII. SUMMARY OF ERRORS

The component errors previously discussed have been summarized in Table I for different pressures. The uncertainties at the highest-pressure designated "molecular scattering" are due to the high-pressure nonlinearities in the conductance of the orifice and are assessed for nitrogen. The effects due to pressure gradients in the upper chamber and orifice mounting have been left out as they are considered to be negligible. Errors at 10^{-6} and 10^{-7} Pa have been assessed on the basis of flow into the lower chamber. The flow ratio uncertainty includes random errors and a systematic contribution for the additional flowmeter measurements needed. The uncertainty due to changes in the assumed pump speed is based on the observed bounds of systematic changes in measured ion gauge sensitivities. The uncertainty for temperature errors has been somewhat arbitrarily doubled from the minimum expected value.

The tabulated random errors are upper bounds based on repeated gauge calibrations. These include the random er-

rors of the standard, flowmeter, and gauges. The value for 10^{-3} to 10^{-5} Pa is based on three times the short-term deviations observed in ion gauge calibrations in this range. We expect that longer term random changes, due to orifice plate sealing and flowmeter seal leakage, are probably no different from those included in the higher pressure MDG data.

ACKNOWLEDGMENT

We appreciate the continued support of this work by the Department of Energy, Office of Fusion Energy.

[1] K. E. McCulloh, J. Vac. Sci. Technol. A 1, 168 (1983).
[2] K. E. McCulloh, C. R. Tilford, S. D. Wood, and D. F. Martin, J. Vac. Sci. Technol. A 4, 362 (1986).
[3] K. E. McCulloh, C. R. Tilford, C. D. Ehrlich, and F. G. Long, J. Vac. Sci. Technol. A 5, 376 (1987).
[4] K. F. Poulter, J. Phys. E 10, 112 (1977).
[5] C. R. Tilford, J. Vac. Sci. Technol. A 1, 152 (1983).
[6] C. D. Ehrlich, J. Vac. Sci. Technol. A 4, 2384 (1986).
[7] P. Clausing, Ann. Phys. 12, 961 (1932); English translation appears in J. Vac. Sci. Technol. 8, 636 (1971).
[8] K. E. McCulloh, "Dimensions and Conductance of the Orifice in the NBS Dynamic Expander," Technical Report to the Temperature and Pressure Division of the National Bureau of Standards, 1982.
[9] R. P. Iczkowski, J. L. Margrave, and S. M. Robinson, J. Phys. Chem. 67, 229 (1963).
[10] J. K. Fremerey, J. Vac. Sci. Technol. A 3, 1715 (1985).
[11] S. D. Wood and C. R. Tilford, J. Vac. Sci. Technol. A 3, 542 (1985).
[12] K. F. Poulter and C. M. Sutton, Vacuum 31, 147 (1981).

How to Select High Vacuum Pumps

By John A. Freeman, Product Manager, Vacuum Components,
Balzers, High Vacuum Products, Hudson, New Hampshire 03051

Today's high vacuum semiconductor processes generally use one or more types of high vacuum pump: *diffusion, cryogenic, turbomolecular,* or *ion.* These processes include sputtering, evaporation, plasma etching, reactive ion etching, ion implantation, electron beam lithography, and others. High vacuum pumps are also used in many analytical processes supporting semiconductor production such as transmission and scanning electron microscopy, surface analysis, leak detection, and residual gas analysis.

Technologists who must select, use, or maintain high vacuum pumps rarely come from the field of vacuum technology. Although both literature and courses are available, vacuum technologists typically learn on the job from co-workers, and to a large extent, from equipment suppliers. The purpose of this article is to stimulate more effective communication between pump suppliers and users by reviewing some of the issues involved in the selection process and some of the trade-offs between pump types.

An accurate general statement regarding which pump to use for which application is not possible. However, if we were to establish a simple "composite view" for each pump based on a variety of opinions, the following four statements may emerge:

Diffusion Pumps Use for dirty to semi-clean processes and high pumping speeds where a little oil backstreaming is not important, where a low initial investment outweighs operating costs, and where simple maintenance is a must.

Cryogenic Pumps Use for clean to ultra-clean processes and high pumping speeds where a clean vacuum and low operating costs are important, and where infrequent but skilled maintenance is acceptable.

Ion Pumps Use for semi-clean to ultra-clean processes and low pumping speeds for clean, continuous, infrequently cycled vacuums kept between 1×10^{-6} and 1×10^{-11} torr, with extremely low operating costs and almost no maintenance.

Turbomolecular Pumps Use for semi-clean to ultra-clean processes and low to medium pumping speeds where clean vacuum, freedom from valves and baffles, good helium and hydrogen pumping, a quick start-up, and broad operating range are desired, and where a relatively high initial investment with infrequent but skilled maintenance is acceptable.

As will be seen in the following discussion, there is a considerable amount of truth in these composite viewpoints, but there are many other factors to be considered.

Basic Questions

Prior to the pump selection, there are certain basic process questions that should be at least considered, and answered if possible:

1. What type of gases are to be pumped? And at what throughput?

2. Is water vapor a major factor?

3. How fast must you reach a given pressure?

4. What base pressure and operating pressures are needed?

5. How long is each cycle? And how frequent?

6. Can the process tolerate residual oil vapor or hydrocarbons?

7. What will the temperature be at the pump inlet?

8. Are there particulates? What size, what type(s), and how many?

9. Are atmospheric "dumping" accidents likely? How often?

10. Are stray magnetic fields a problem?

11. Is the equipment vibration sensitive? At what frequencies?

12. How will the pump be mounted?

13. How will valves, baffles, connecting lines, and roughing lines be used?

14. What kind of roughing or backing pump is anticipated?

15. How will the unit be operated and interlocked?

16. When will it be maintained? How? By whom?

17. What are its operating costs?

18. What is the initial investment?

Good answers to these basic questions are important for properly selecting the high vacuum pump, and maximizing the effectiveness of the investment. Generally, the better the answers, the better the selection process. The remainder of this article takes a closer look at the issues.

Type of Gas

Primary issues for discussion with a supplier should include the air, nitrogen, and water vapor pumping speeds, and whether the process will include any toxic, corrosive, or explosive gases such as oxygen, chlorine, fluorine, silane, phosphene, etc. In some cases, the pump's speed and/or capacity for helium and hydrogen, or for argon and other noble gases, could be critical.

Dry air and nitrogen pumping speeds are similar, since air is mostly nitrogen. Pump suppliers generally clearly specify nitrogen pumping speed, and use these values in pump down time calculations.

Water vapor pumping is frequently a key consideration, especially if the process requires frequent pump-downs to less than 1×10^{-5} torr after atmospheric exposure. On small chambers, a turbo or ion pump may be adequate, but on large chambers, thousands of liters per second will be required: this typically requires a cryopump or a diffusion pump with an LN_2 cooled baffle. Water vapor pumping can also be boosted significantly by using a cryogenically cooled surface, such as a Meissner trap, in the process chamber.

Nearly all high vacuum pumps are designed and rated primarily for good air, nitrogen, or water vapor pumping speed. There are times when other gases, usually hydrogen, helium, or argon, can be important. Diffusion pumps pump all gases well, cryopumps have very small helium pumping speeds, and ion

10 and 20 in cryopumps with water vapor pumping speeds of 6800 and 27500 liters/second and argon capacities of 3200 and 10000 standard liters. Such pumps are used for rapid, oil-free pumpdowns of evaporation and sputtering equipment (Balzers Models RCP 501Z and RCP 251).

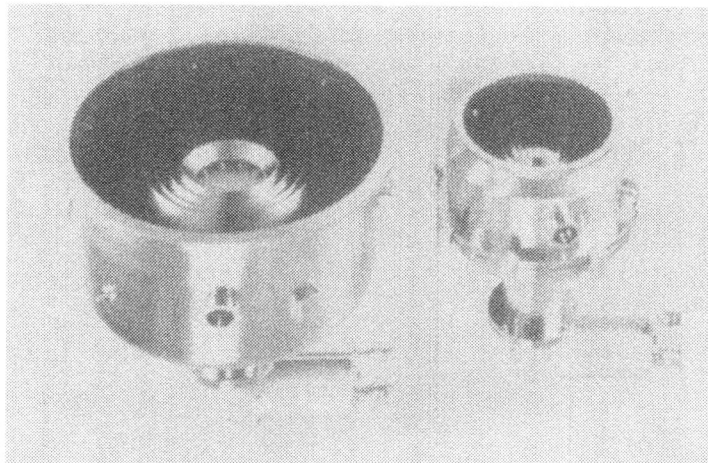

Specially developed turbomolecular pump for plasma etching and other corrosive processes. The pump includes a special protective blade coating, a strong drive for operation from 100 microns to 10^{-8} torr, and an inert gas purge system to protect the bearings (Balzers' Model TPH 330 Plasma Turbo).

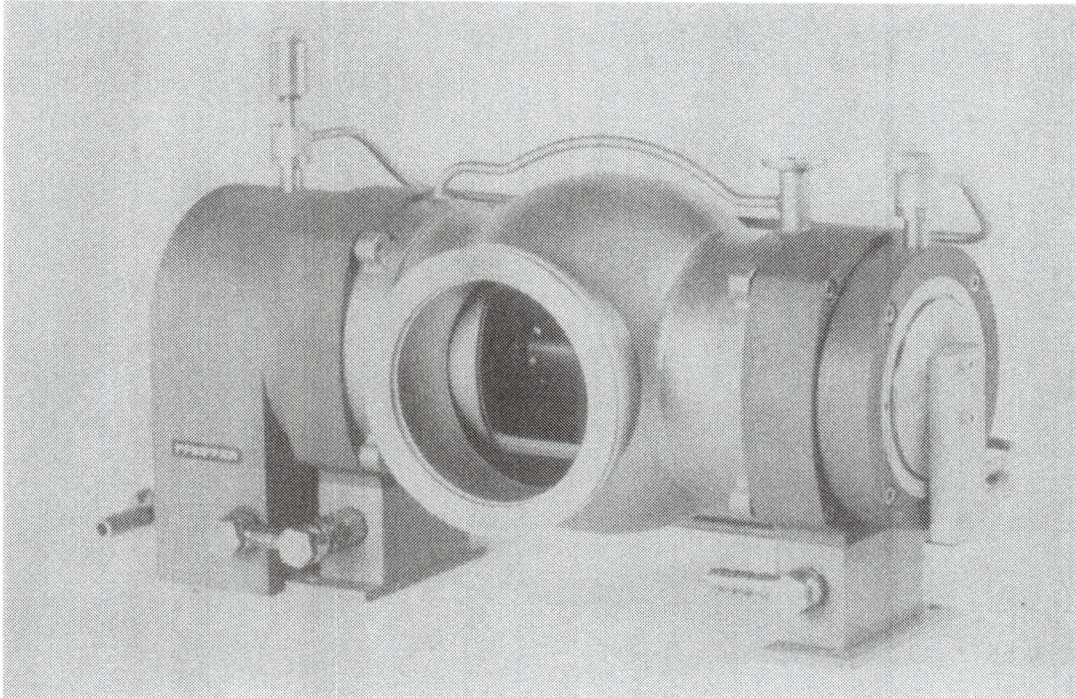

pumps have relatively poor speed for both helium and argon, as well as other noble gases. Turbomolecular pumps each have their own characteristics, depending on the brand and model: some turbos pump light gases very poorly; others pump them extremely well. Capacity for gas handling is another factor, especially with cryogenic pumps. Since they capture and hold pumped gases, they have a finite capacity for different gases, and eventually must be relieved of the stored gas during a regeneration process. A cryo's small hydrogen capacity is usually what determines the length of time between regenerations, so hydrogen-rich processes will require frequent regeneration. Processes with a substantial steady throughput of gas, such as sputtering or reactive evaporation, may also require frequent regeneration, unless the gas load is minimized by "throttling" the pumping speed by a restricting valve between the pump and chamber. This throttling helps avoid frequent regeneration, but also reduces the pumping speed of all other gases, including some reactive ones that you may not want reduced. For processes with high gas throughputs, one should carefully consider diffusion pumps or turbomolecular pumps. If a cryo must be used for other reasons, select one with exceptionally high capacities for the key process gases, so that throttling and regenerations needs are minimized.

Dangerous Gases

Another consideration is whether pumped gases are toxic, or corrosive, or represent an explosion hazard. Toxic gases can form hazardous deposits or generate hazardous vapors in any vacuum pump. In diffusion pumps, they contaminate the oil, in turbos they contaminate the bearing lubrication, and in cryos they accumulate in solid phase over time, then must

all be released as a gas over a relatively short regeneration period. Corrosive gases can directly attack pump parts or form corrosive compounds such as Lewis acids in the pumps, thus damaging or destroying the pump. Diffusion pumps are generally resistant to corrosive attack. However, the aluminum jet assemblies or mild steel construction of some pumps might be more sensitive to attack, so the pump manufacturer should first be consulted. Ion pumps can quickly be destroyed by corrosives, as can cryopumps. With cryos, the corrosives stay nearly inactive until regeneration, when they can react with other gases in relatively high concentrations, attacking pump and system components. If a cryo accidently regenerates too fast for the pressure relief or "foreline" valve, or if the valve is stuck closed, then toxic gases from the pump could exhaust through other valves, ports, or seals, or back into the chamber. The relatively recent development of turbomolecular pumps with special features for reliably pumping corrosive gases has made them a common choice on some process systems, such as plasma etchers. Generally, the problems with pumping dangerous gases are minimized by using diffusion pumps or specially prepared turbos.

Process Cycling

Turbomolecular pumps are generally started at atmosphere, together with the roughing pump, so roughing occurs through the turbo as it accelerates. Cryogenic and diffusion pumps have a fairly long start cycle, so are generally kept in operation under vacuum, then valved in when a "rough" crossover pressure is reached. Even turbos are occasionally valved in at a crossover pressure, especially on equipment requiring extremely fast pumpdowns that would not allow enough time for the turbo to accelerate, such as

Backside of a cryogenically-pumped thin film evaporation system, Balzers' Model BAK 760.

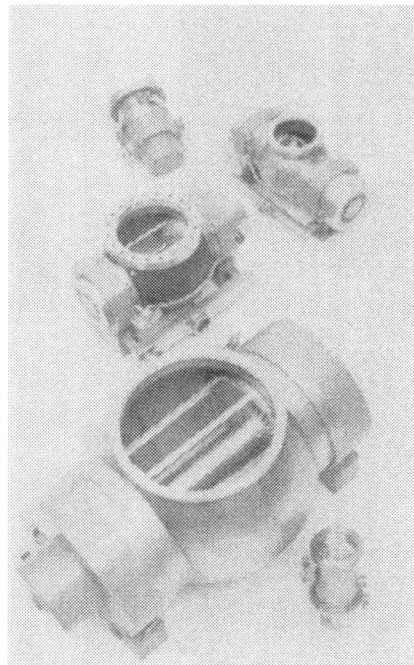

Turbomolecular pumps are available from 27 liters/second to over 6500 liters/second.

loadlocks or fast cycling sample inlets. In rapid cycling systems, a cryo can become quickly saturated and require nearly daily regeneration; although this may not be any problem if the regeneration is done overnight.

Diffusion pumps are usually valved in at around 10 to 50 microns; above this they can lose stability. Cryopumps, since they are often used to keep the process free of oil, should be valved in at relatively high pressure, typically from 200 to 500 microns. The further the roughing cycle continues below 200 microns, the greater is the chance of molecular backstreaming of oil from the roughing pump. (Above 200 microns, the roughing line is in viscous flow, effectively preventing backstreaming.) Ion pumps frequently exposed to high crossover pressures generally suffer from a rapid build-up of titanium deposits, requiring frequent cleaning, or replacement of the pump elements. Therefore, ion pumps should be turned on or valved in at relatively low pressures, and as infrequently as possible. It is extremely important that the high vacuum valve on ion pumps be leak tight. Since roughing to low pressure with an oil filled roughing pump results in backstreaming, ion pump systems are frequently roughed by oil-free sorption pumps or by a turbo.

Pump Down Time

Although the time to reach a given pressure can be a fairly complex issue, there are appropriate general comments. Unless the chamber is small, a fast pumpdown usually requires the high water vapor pumping speed of a cryo, LN_2 cooled baffle, or Meissner trap surface. Since different pumps have different startup characteristics, the time to get them operating or the roughing time to crossover must be considered. Most turbos can be started at atmosphere, but the pumping to about 10^{-1} is still predominantly from the roughing pump. Nevertheless, even with the "roughing" phase, most turbomolecular pumps are fully operational in one to ten minutes and can be cycled quickly in and out of operation. In some applications, such as on leak detectors, this rapid cycling can be advantageous: the system can be turned on minutes before use, then turned off and vented as soon as the process is finished.

Ultimately, the time to reach a given pressure is determined by the total gas load from the chamber, the manner in which this gas load decays over time, and the effective pumping speed for various gases at various pressures at the vacuum chamber, after all conductance losses are considered. These issues become quite involved and can be studied in various texts or suppliers' literature. One issue, conductance, is commonly misunderstood and is so important that it is briefly reviewed here.

Conductance

In the molecular flow region from 10^{-3} to 10^{-11} torr where most high vacuum pumps operate, molecules randomly travel about, bouncing off the chamber walls until they eventually find the pump opening. If the pump is doing its job, most molecules entering it are removed from the system and do not come back, or backstream. The important concept in molecular conductance then is that molecules are not actually "pumped", they are moving freely through the system equally in all directions, and under their own power, until by chance they find the pump.

Diagram of pump system on Balzers' SWS 605 / 6 Wafer-to-Wafer Sputtering System. Wafers pass from an outer load lock pumped by a rotary vane pump, to an inner high-vacuum lock pumped by a turbo, to the cryogenically pumped process chamber. Such staging provides extremely stable and clean vacuum conditions.

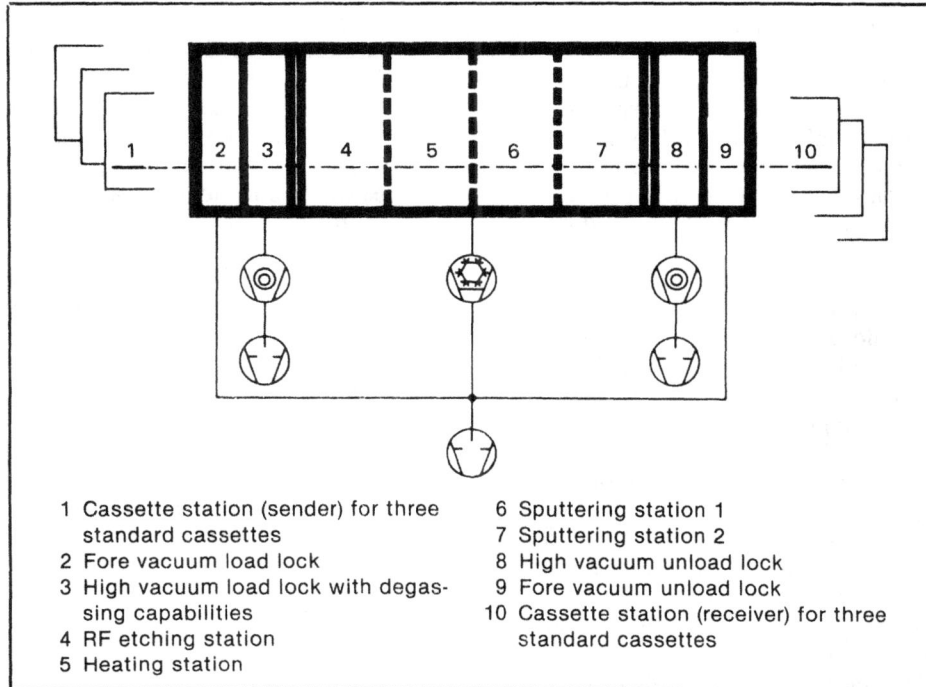

1 Cassette station (sender) for three
  standard cassettes
2 Fore vacuum load lock
3 High vacuum load lock with degassing capabilities
4 RF etching station
5 Heating station

6 Sputtering station 1
7 Sputtering station 2
8 High vacuum unload lock
9 Fore vacuum unload lock
10 Cassette station (receiver) for three
   standard cassettes

As lines become longer and their diameter decreases, conductance decreases. That is, the number of molecules that find their way through the pipe each second decreases. Elbows, baffles, valves and lines all add to the total conductance loss.

Conductance losses can be dramatic, resulting in grossly oversized and overpriced pumping systems. For example, the effective pumping speed for a 2000 liters per second pump attached to the chamber by a 4 diameter × 20 inch long line is only about 210 liters per second. If the pump is ten times bigger, i.e., 20000 liters per second, the pumping speed at the chamber increases only to about 230 liters per second. Another example can be shown with a half-inch diameter line that is two inches long: the maximum rate that molecules can pass through it is only 4.5 liters per second, so the vacuum pump need only be fast enough to take these away. On such a small pipe, the performance of both a 20 liters per second and a 20000 liters per second pump are the same.

The following comments regarding conductance should be remembered:

1. The effective pumping speed is reduced at the chamber by any lines, elbows, baffles, or valves between the pump and chamber.

2. There is an optimum pump size for a given line conductance. Each step up in pump size above this optimum provides progressively smaller gains in pumping speed, until finally there is no gain at all.

3. To maximize the effective pumping speed at the chamber, line lengths should be kept as short as possible, the use of elbows, baffles, and valves kept to a minimum, and the inside diameter of these components should be the same, or bigger, than the ID of the pump intake.

Pumping Speed

Issues to look at in this category are the rated speeds for various gases at various pressures, and both the stability and predictability of these rated speeds.

Pumping speeds are generally rated either in accordance with the American Vacuum Society (AVS) or the International Standards Organization (ISO) standards, and due to differences in the experimental equipment, they are not identical. ISO measurements generally result in lower published pumping speeds, in some cases, nearly 15 percent lower. Also, not all published pumping speeds are by actual measurement; some are arrived at theoretically. If there are questions on how the specifications were determined, ask the supplier.

Pumping speeds are also normally rated over a specific pressure range. Diffusion and turbomolecular pumps provide relatively "flat" pumping speed curves throughout molecular flow to their ultimate pressure. Ion and cryogenic pumps are rated for the peak pumping speed at certain pressures: these speeds generally decrease significantly from about 10^{-7} to UHV.

When the actual pumping speed is unknown, it makes it difficult to eliminate pumping speed as a potential process variable. The pumping speed of diffusion pumps varies in accordance with the oil's vapor pressure, contaminants in the oil, condition of the heaters, and other factors. Cryogenic pumping varies with surface temperatures in the pump, the condition of the refrigeration system and charcoal arrays in the pump, and thermal load. Both cryogenic and ion pump speeds change significantly in accordance with the amount of gas already pumped.

Ultimate Pressure

The ultimate pressure is determined by the total gas load coming from leaks, permeation, and desorption, by the effective pumping speed at the chamber, and by the ultimate pressure of the pump itself. Only the pumps are discussed herein.

Diffusion pumps are generally limited by the vapor pressure of their oil to between 1×10^{-8} torr and 1×10^{-9} torr, although LN_2 baffles can slightly improve these figures. Ion pumps can operate to about 1×10^{-11} torr; however, at these low pressures they have very low pumping speeds relative to their rated peak speed, size, and weight. Frequently ion pumps are used together with a titanium sublimation pump or turbo to boost ultra-high (UHV) capability of the vacuum system. Cryogenic pumps provide good performance throughout molecular flow, but like ion pumps, lose much of their pumping speed by the time they reach their ultimate pressure, in this case about 1×10^{-10} torr. The ultimate pressure of turbomolecular pumps depends on the number of compression stages in the pump, the blade design, the flange seal, and primarily on the type of backing pump. With a single stage rotary vane pump, an ultimate of about 5×10^{-8} is typical. However, in a pump designed for good compression for helium and hydrogen, and if backed by a diffusion pump or another turbo, ultimate vacuums of less than 1×10^{-11} torr are possible, without any loss of pumping speed.

Useful Operating Range

The operating range of the vacuum pump should be chosen to match as closely as possible the operating pressure of the process. Cryogenic pumps can be valved in at relatively high pressures, perhaps several hundred microns, but continuous operation at this pressure to about 1×10^{-3} torr, unthrottled, would quickly saturate the pump and force regeneration. Above 1×10^{-2} torr, diffusion pumps are too unstable for continuous operation, and the risk of oil contamination from the pump is high. Therefore, both cryos and diffusion pumps are generally used between 10^{-3} and 10^{-9} torr. Ion pumps are typically used for continuous operation between about 1×10^{-6} and 1×10^{-10} torr, a range in which they are remarkably trouble-free, sometimes running for years with no maintenance. They are also functioning from 1×10^{-3} to 1×10^{-5} torr, but in this range titanium deposits and flakes rapidly form in the pump elements, which can shorten the time between cleaning or element replacement to perhaps a week or so. Turbomolecular pumps provide their full pumping speed between 1×10^{-3} and 1×10^{-11} torr. Special models for etching extend the high pressure limit to about 100 microns, but with reduced rpm and pumping speed. Near their high pressure limits, turbos should be water cooled, and their low pressure limit is strongly influenced by the backing pump pressure.

Backstreaming and Cleanliness

Certain processes are tolerant of backstreaming oil molecules from the vacuum pump, and other processes are not. Because residual oil vapors can adversely affect thin film electrical or optical properties, adhesion, etching properties, doping properties, or other important characteristics of the final product, many of today's vacuum processes must be kept oil free. Diffusion pumps, since they use oil as a working medium, present a dual problem. First, measurable backstreaming into the process chamber is present even in carefully baffled systems. Second, when a hot diffusion pump is accidently and suddenly exposed to atmosphere, considerable amounts of diffusion pump oil can enter the process chamber. After such an accident, the process fixturing may require complete disassembly and cleaning, and related problems can persist for days afterward. On the other hand, cryogenic pumps, ion pumps, and most turbomolecular pumps are relatively forgiving of an accidental atmospheric "dump". Cryogenic pumps require a complete regeneration of the cryo, but normally the process chamber will remain clean, oil-free, and process-ready. Most turbomolecular and ion pumps can simply be restarted after an atmospheric "dump", also without the process chamber being contaminated. Properly operated cryogenic, turbomolecular, and ion pumps can all maintain the ultra-clean oil-free vacuum environment required in oil vapor sensitive processes. In general then, if cleanliness is critical, the choice is between cryo, turbo, or ion pumps.

Temperatures

Diffusion and ion pumps are both tolerant of high intake temperatures up to hundreds of degrees Celsius. Turbomolecular pumps are limited to maximum intake temperatures between about 100°C and 160°C, depending on the brand and model. Cryogenic pumps may have enough reserve cooling capacity to handle only small radiant heat loads, in which case, they must be shielded by a water cooled baffle or other heat shield. Normally, the cryo cooling capacity is expressed watts at temperatures of 77K and 20K: for example, 65/6 would mean with 65 watts applied to the first stage at 77K, there are 6 watts available at 20K on the second stage. For superior thermal loading resistance, pumps with high cooling capacity are preferable. If cooling capacity information is not in the specifications from the manufacturer, it should be requested.

Particulates

Diffusion pumps and cryogenic pumps are both quite tolerant of dust and particulates falling into the pump. In diffusion pumps, they settle in the oil and may cause some outgassing and a temporary pressure rise, but generally go unnoticed until they form a thick sludge and reduce performance. In cryos, they fall harmlessly to the bottom of the pump. In ion pumps, particulates are harmless unless they are magnetic and attracted to the elements, large enough, or accumulated enough to short out the pump elements, stopping the pump. Tiny dust less than 0.5mm generally can pass through a turbo, but constant bombardment will hurt the pump. Larger particles, screws, glass fragments and the like can instantly destroy or "crash" the turbine blades. To protect turbos, turbo manufacturers sell inlet port screens varying from a relatively large mesh size suitable for protecting the pump only from screws and large objects, to fine multimesh screens designed to filter out anything big enough to damage the pump. The fine mesh screens

300 liter/second turbo on rapid cycling crystal coater. Total of air-to-air cycle time including pumpdown, gold evaporation, and venting is one minute. Unit runs 24 hours a day.

Turbopumped leak detectors are operational within minutes, require no LN_2, and do not contribute hydrocarbons to the process chamber (Balzers Model HLT 100).

provide much better protection, but also reduce pumping speeds about 15 percent.

Magnetic Fields

Actually there are two issues: the effect of the pump on magnetic field sensitive instruments (such as electron microscopes or e-beam lithography equipment) and the affect of strong magnetic fields from instruments on the pump. Generally, only experiments in high energy physics generate magnetic fields strong enough to bother the pumps, so no more need be said here. Diffusion pumps are free of magnetic fields, and in cryopumps they exist only in motor driven cold heads at very low levels. Turbomolecular pumps typically use cobalt-samarium magnets in their drive, generating fields up to a few milligauss at the flange: This field rapidly decreases with distance and is rarely noticed, but people retrofitting magnetically sensitive equipment should be aware of it and prepared to provide magnetic shielding if necessary. Ion pumps present the biggest potential problem with magnetic fields due to their large ferrite magnet assemblies. In magnetic field sensitive instruments, ion pumps can be shielded or mounted away from the sensitive area.

Vibration

Some equipment, such as electron microscopes, are extremely vibration sensitive. Ion and diffusion pumps are vibration-free and do not contribute to vibration problems.

The vibration characteristics of turbomolecular pumps vary considerably by brand and model, with some providing extremely low amplitudes: turbos are widely used on sensitive scanning and transmission electron microscopes, line width measuring equipment, and increasingly on e-beam lithography sys-

tems. Occasionally they must be used with auxiliary dampers. Cryogenic pumps have been developed with special damping systems to minimize their low frequency vibration, but on sensitive applications they require considerable "tuning" to the system. The specially damped models are also relatively expensive. In some applications a cryo can temporarily be shut off, eliminating all vibration for a particularly sensitive momentary process.

Radiation

In semiconductor processing equipment, radiation effects on a pump's lifetime are not a factor. This is generally only an issue where pumps will experience a relatively high chronic radiation, such as on accelerators.

Mounting Considerations

When planning an installation many factors play a part in the total system cost and effectiveness, including the pumping port location, space limits, the pump weight, the location or need of a high vacuum valve, baffle, or roughing line, the space for auxiliary equipment such as a compressor, power supply or system controller, and the complexity of electrical interlocking and control.

Diffusion pumps are vertically mounted, but can be mounted on an elbow for adapting to a side port. Small ion pumps can easily be mounted in any orientation, but the larger 100 L/s to 1000 L/s models are usually top flanged because of their weight, which ranges between 150 and 1000 pounds. Nearly all cryopumps can be mounted in any orientation. With turbos, the mounting flexibility varies by brand and model: some can be top, side or bottom flanged.

Turbos, if used without a roughing line or high vacuum valve, greatly simplify the installation and system control. Ion pumps are sometimes used without a high vacuum valve also, since they can be roughed through their intake port as the chamber is roughed. However, in frequently cycled processes ion pumps should be valved off and kept under vacuum, since frequent cycling shortens their service lifetime. Cryopumps can also be roughed through their intake, and diffusion pumps can be roughed through their foreline, both with the drawback that once roughed, the operator must wait for the pump to reach its operating temperature before high vacuum can be attained. Cryos and diffusion pumps must also be returned to nearly room temperature before venting to avoid problems. To avoid this pump cycling, cryo and diff systems use high vacuum valves and roughing lines, but this, of course, adds to the system complexity and control requirements. Cryos have the further complexity of a regeneration cycle, and diffusion pumps are frequently used with LN_2 cooled baffles, requiring an LN_2 control system.

The size of the pumps themselves can influence the system design. For example, a 20 in ID diffusion pump, baffle and angle valve may require a pump port centered 1½ to 2 feet higher than a similar capacity cryo system, or looking at it differently, the cryo system could be built 1½ to 2 feet lower to the floor. Pump sizes, shapes and weights should all be considered. Finally, both system manufacturers and end users should avoid designing their system so that only one pump from one vendor will fit in it. A second source capability can save a lot of aggravation if the first source becomes a problem.

Diffusion Pump Maintenance

If maintenance, pumping speed and the initial cost were the only issues, perhaps all high vacuum pumps would be diffusion pumps. Diffusion pump maintenance is a relatively simple task, typically done infrequently, in-house, by relatively inexperienced personnel. They are rugged pumps, able to withstand considerable abuse without being damaged. Normally a simple oil change returns the pump to good performance. The frequency of oil changes varies with the process, but since most diffusion pump oils are expensive, they are typically changed only when the pump performance deteriorates or on a long term schedule of once or twice a year, or longer. Special applications may use cheaper oil and/or more frequent changes.

Occasionally the pump must be removed from the system, thoroughly cleaned, and perhaps the jet gaps re-adjusted or heaters replaced. The cleaning procedure for diffusion pumps varies, but nearly always involves solvents, glass bead blasting, and/or hand polishing. The sludge and used oil from the pump may contain hazardous compounds, so care should be taken to avoid contact with the waste or vapors coming from it.

Ion Pump Maintenance

Ion pumps rival diffusion pumps for simplicity, and if properly operated, require very little maintenance. Like diffusion pumps, they have no moving parts. When service is required, it can be done in-house by semi-skilled personnel, providing they pay attention to cleanliness by keeping dirt, oil and other contaminents out of the unit during reassembly. Normal service involves cleaning loose process deposits and titanium deposits from the pump elements and housing. Mechanical cleaning by hand or acid etching is frequently used. In the worst case, all the insulators

Small diffusion pump and butterfly valve combination. Diffusion pumps are available with nitrogen pumping speeds from about 30 liters/second to over 60000 liters/second.

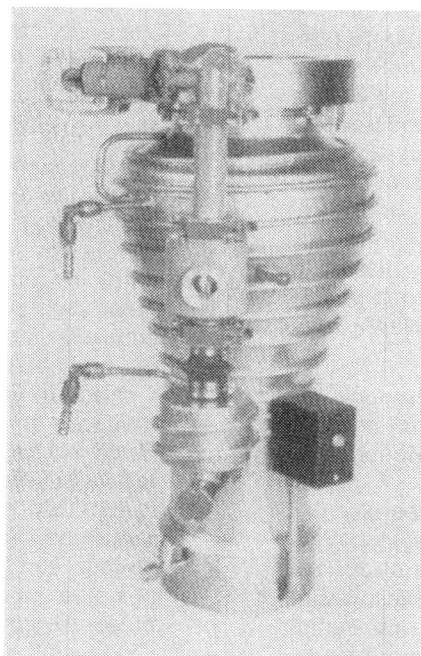

must be cleaned or the pump elements replaced completely, a relatively expensive repair. As with all vacuum pumps, deposits in the pump may be toxic and contact with them, including dust, should be avoided. Large ion pumps are extremely heavy for their size, and normal wrenches brought close to them can be attracted to the powerful magnets suddenly and with tremendous force. Frequently special beryllium wrenches are used: unfortunately, these are expensive, brittle, their dust is toxic.

Major causes of reduced ion pump life are system air leaks, damaged or leaking ceramic electrical feedthroughs, frequent cycling and roughing pump crossovers, crossovers at excessively high pressures, high pressure operation, and the backstreaming oil vapor from roughing pumps forming conductive carbon coatings on the insulators. Sorption pumps or turbos used to rough ion pumps help avoid the carbonization problem. If rotary vane pumps are used, they should be well trapped: newly developed catalytic foreline traps are excellent for this since they eliminate nearly all backstreaming hydrocarbons, are self-regenerating and need almost no maintenance.

Fortunately, most ion pumps require minimal maintenance since they are infrequently cycled and are operated continuously at 1×10^{-6} torr or less

Cryogenic Pump Maintenance

Cryogenic pumps have seals and moving parts in the compressor and cold head, and filtering systems that require occasional service. The maintenance intervals and difficulty vary from brand to brand, but generally some skill, cleanliness, and attention to details are required.

Modern compressor designs feature easily changed charcoal absorber filtration units that seldom require replacement and have solved the oil carryover problems that contaminated cold heads in earlier designs. Cold heads using adjustable poppet valves to cycle the helium through the expander need periodic readjustment to re-time the helium flow and maintain cooling efficiency. A relatively recent innovation, the sliding ceramic valve, never requires re-timing so the cold head efficiency remains high year after year. Occasionally cold head seals need replacement, typically between 6000 and 10000 hours, depending on the cold head design. Providing the re-timing procedure can be avoided, the seal change is straight forward.. Any cold head, once opened to atmospheric moisture and dirt, should be serviced quickly and cleanly to avoid any contamination of the displacer assembly, and to avoid exposing it to high humidity which can cause the displacer to swell so it no longer fits. If this happens, the displacer must be baked to remove the absorbed moisture. The displacer should be handled with gloves so it does not absorb body oils.

The cryopump body generally needs service very rarely, and when it does, it is normally only a replacement of the second stage charcoal array. This replacement is simple, but also should be done in very clean conditions, wearing gloves.

In order to provide the best possible support for their products, some cryo manufacturers have established an exchange program whereby a worn cold head has a substantial trade-in value towards the purchase of a completely factory rebuilt and warranted unit: this approach minimizes downtime and re-establishes the warranty.

Turbomolecular Pump Maintenance

Like cryogenic pump systems, turbomolecular pumps have moving parts, in this case, bearings that support the high speed rotor assembly. They also feature lubrication systems for the bearings. The turbo design and lubrication system varies widely from brand to brand and between models, as do the lubrication intervals and difficulty of a bearing change. Some turbos must be rebalanced after a bearing change, so these pumps must be returned to the manufacturer for the bearing change. Other turbos are designed to easily access the bearings without disturbing the balance, and in these pumps, the bearings can be replaced in the field. However, a word of caution — turbos are high speed precision devices, and the bearing replacement should be performed by a skilled technician, preferably trained in the proper handling techniques by the manufacturer. Also, the work should be performed under clean room conditions, such as in a laminar flow hood: in a turbine rotating at 20000 to 90000 rpm, a little dust in the bearings or improperly installed bearings can result in a reduced pump lifetime. Typical bearing lifetimes range about two to five years, and 15000 to 50000 hours, depending on the process and other factors. Some people change them every six months, others every few years, others wait until the pump gets noisy or begins vibrating or its sound changes in some other noticeable way. In units with visible oil reservoirs, recently changed oil that quickly turns dark again is a potential sign of excess bearing wear.

In the worst case, a turbo "crashes", which means the turbines are destroyed. A crash normally occurs from a large object hitting the rotating blades, or from a bearing wearing to the point where it falls apart. Some turbos can easily withstand accidental atmospheric dumping or implosions, others cannot. The minimum tolerable venting time should be requested from the manufacturer if not in the specifications. Crashed pumps must be returned to the manufacturer for a complete rebuilding with a new turbine or rotor / stator assembly.

Lubrication requirements vary. The reason for replacing or adding lubricant from time to time is to assure the bearings have a good lubricating film and that any particulates that might otherwise remain in the bearings are removed. Typical lubrication intervals range from 1000 to 15000 hours, or annually.

The support provided by a turbo manufacturer can be a key factor in the effectiveness of the pump. A well stocked exchange program permitting the low price purchase of a factory rebuilt and fully warranteed replacement pump, with the trade-in of the worn or crashed equivalent model, makes ownership very easy and minimizes pump related down time.

Initial Cost

In comparing the initial cost of pump types, necessary valves, baffles, roughing lines, regeneration equipment, controllers or other hardware needed for a fully operating pump system should also be consid-

Recent innovations in cryogenics include Balzers' sliding ceramic valve, which eliminates the need to retime the cold head after a seal change.

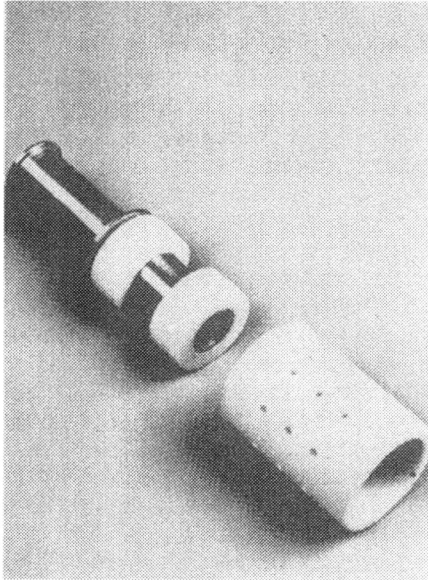

Varian's new "StarCell VacIon" pump for generating ultrahigh vacuums incorporates innovations that make it capable of long-term, stable, and high speed pumping in a wide range of applications. The pumping system includes a matching power unit that provides the exact voltage and current required for different working conditions.

ered. A baffle is normally only needed with diffusion pumps. Ion pumps require no foreline connection or valve. Turbomolecular pumps are frequently used with neither a high vacuum valve nor roughing line, so the cost of these components and a valve controller can be eliminated. Cryopumps frequently require some regeneration equipment. All things considered, diffusion pumps with an LN_2 cooled baffle typically provide the most pumping speed for the money, and turbo and ion pumps the least. However, below 1500 liters per second, there is considerable price competition between all four types, especially when the entire system is considered, and in many cases, a small ion or turbomolecular pump is the least expensive solution. Above 1500 liters per second, diffusion pumps are the least expensive, and turbopumps or specially made ion pumps the most expensive.

Operating Cost

In considering the operating cost, one should look at the total consumption of electricity including the high vacuum pump and backing pump, liquid nitrogen, and cooling water. Diffusion pumps generally use the most electrical energy because of both their heaters and the continuously operating backing pump: very small ones use less than 1 kW with the backing pump, but a typical 20 in pump uses 10 to 15 kW. They also frequently use liquid nitrogen and plenty of cooling water. Cryogenic pumps require only from about 1.5 kW to 5.5 kW for running their compressors, so they can be considerably less expensive to run than diffusion pumps, especially the high capacity models. Cryos normally require no liquid nitrogen and use minimal water cooling or even air cooling for the compressor. Turbomolecular pumps typically have a maximum power draw if only 50 to 500 watts, and much less when running at speed. They do, however, require a continuously running backing pump, which if large enough, can add several kWs. Turbopumps are normally used without liquid nitrogen, and use minimal water cooling or air cooling. Ion pumps are by far the most economical pumps regarding operating costs: they use no water or LN_2, require no backing pump, and typically consume only 1 to 50 watts of power.

Summary

Selecting the type and size of high vacuum pumps involves many issues and should be carefully analyzed and discussed with the potential suppliers. There are advantages and disadvantages to each pump type that must be considered. Long term success using any high vacuum pump depends to a large extent on the operator and pump maintenance, but different pumps resist certain types of problems or mistakes better than others. Sizing the pump is itself a complex issue, requiring both theoretical estimates of pumpdown times and the experience of suppliers or users gained through trial and error on similar systems. Proper attention to issues such as rated pumping speeds, conductance losses, backstreaming, the operating range, maintenance requirements and others discussed above, will most effectively match the pumping system to the process and help maximize the system performance and process reliability at the lowest possible cost. ∎

Acknowledgement

The author wishes to express his thanks to Burdell Keitzman of Varian and Bill Millikin of Balzers for their comments on ion pumps, and to Carol Jung of Balzers for assistance with photography. Also, the ion pump photographs are courtesy of Varian.

Aluminum alloy ultrahigh vacuum system for molecular beam epitaxy

M. Miyamoto, Y. Sumi, and S. Komaki

Seiko Instruments & Electronics Ltd., 31-1, Kameido 6-chome, Koto-ku, Tokyo 136, Japan

K. Narushima and H. Ishimaru

National Laboratory for High Energy Physics, Oho-machi, Tsukuba-gun, Ibaraki 305, Japan

(Received 2 October 1985; accepted 29 June 1986)

As a research project for aluminum alloy molecular beam epitaxy systems, a large aluminum alloy chamber with a special surface finish was constructed. Its outgassing rate and ultimate pressure were measured. The special surface finish was analyzed using Auger electron spectroscopy. The oxide layer on the aluminum alloy was about 40 Å thick. The outgassing rate before baking was measured as 10^{-11} Torr l/s cm^2, and the rate after baking (150 °C, 24 h) was 3×10^{-13} Torr l/s cm^2. This value is approximately one to two orders of magnitude lower than that of a stainless-steel chamber, such as SUS 304. The ultimate pressure was 7×10^{-11} Torr when pumped with a turbomolecular pump and an ion pump. With addition of the newly developed titanium sublimation pump, the ultimate pressure was 6×10^{-12} Torr (6.5×10^{-12} mbar). This ultimate pressure is very low for a large aluminum alloy vacuum chamber. Ceramic coating of SiO$_2$ using a spark discharge in a silicate solution on aluminum alloy was not corroded by gallium. This ceramic treatment was suitable for an ultrahigh vacuum.

I. INTRODUCTION

Aluminum alloy ultrahigh vacuum systems have recently aroused considerable interest. Aluminum alloy vacuum chambers were usually made by extrusion processes. The outgassing rate of A6063 and A1060 chambers made by extrusion has already been reported.[2] The diameters of such chambers are much too small to use in a molecular beam epitaxy (MBE) system, and it is difficult to make larger chambers by this process. As a research project for aluminum alloy MBE systems, a large aluminum alloy chamber with a special surface finish was constructed. Its outgassing rate and ultimate pressure, using a newly developed titanium sublimation pump, were measured.

II. APPARATUS

The aluminum alloy chamber is 1000 mm long and 580 mm in diameter. The main flanges are made of A2219-T852, 700 mm in outer diameter and 40 mm thick. The sealing surfaces are coated with a chromium nitride film by ion plating. The thickness of the Cr$_2$N layer is about 2 μm. The surface hardness of Cr$_2$N is very high and its micro-Vickers (10 g) hardness is about 1200. Cr$_2$N treatment of the aluminum flange gives nearly perfect protection against sticking between the knife edge of the flange and the gasket, and against surface scratching. For the flange seals, a Helicoflex gasket 8 mm thick and 613 mm in diameter is used. Aluminum flat gaskets are used for the smaller flanges. The Helicoflex gasket is fixed in place by four holders which fit into a groove and its outer edge. Anodized aluminum alloy bolts (2014-T4), nonanodized nuts (6061-T6), and a hard-anodized washer (2017-T3) are used without lubrication to tighten the flange. The chamber was made using AC-TIG welding. It was then placed in a larger chamber which was used to produce the special surface finish. After evacuating this larger chamber, it is backfilled with an Ar + O$_2$ mixture without heating and plasma. Argon, which is dilution gas against oxygen to avoid an explosion, is not essential to make a high-quality surface. Oxygen with extremely low water content is essential for good vacuum quality. The main body inner surface is treated in this atmosphere. Large flanges are treated in the same way.

A new kind of titanium sublimation pump has been developed (Fig. 1). This pump has an aluminum alloy (A1100) liquid-N$_2$ trap. The trap is a cylindrical cryopanel, formed using the roll-bond technique. The path for liquid N$_2$ is formed in the aluminum by compression. The trap surface is treated with an alkaline solution. The merits of this pump include light weight and freedom of installation in any orientation. The trap is also used as a shroud in the molecular

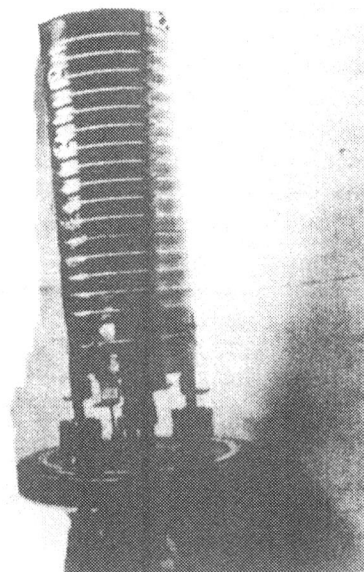

FIG. 1. A new kind of titanium sublimation pump.

43

beam epitaxy system. We call it a "nude-type titanium sublimation pump."

III. EXPERIMENT

The aluminum alloy chamber volume and inner surface area are approximately 240 l and 2.5×10^4 cm², respectively. The surface area ratio of flange (A2219) to body (A1050) is 68% and 32%. The outgassing rate of the aluminum alloy chamber was measured with the following vacuum system (Fig. 2): The aluminum alloy chamber is fitted with a 90° elbow and gauge port, an orifice (diameter 6 mm), a T piece with gauge port, and a turbomolecular pump (270 l/s). These parts are all made of aluminum.

The gas throughput from the experimental chamber was estimated by measuring the pressure on both sides of the orifice, which had a calculated conductance of 3.3 l/s at room temperature. The pressures were measured using two Bayard–Alpert gauges. The outgassing rate of the chamber was estimated from the expression

$$q = (C/A)(P_1 - P_2), \tag{1}$$

where A is the chamber surface area, C is the orifice conductance, and P_1 and P_2 are the pressures measured before and after the orifice, respectively. The ultimate pressure was measured by conducting an ion pump (Fiji Bellows Co. 500 l/s) directly to the chamber, together with a turbomolecular pump (Seiko Seiki 300 l/s) and the new titanium sublimation pump with a liquid-N_2 trap. Measurements were made twice, first without the titanium sublimation pump and then with it. The ultimate pressure P_∞ is given in the following expression:

$$P_\infty - P_0 = qA/S, \tag{2}$$

where P_0 is the ultimate pressure of the pump used and S is the total pumping speed.

IV. SURFACE ANALYSIS

The treated surface was analyzed by Auger electron spectroscopy (AES). Aluminum metal and oxygen depth profiles of the oxide layer were measured (Fig. 3). For comparison with the specially treated surface, a depth profile of an ordinary A2219 surface was also measured. The oxide layer of the specially treated surface was about 40 Å. The layer for the ordinary finish was about 100 Å. The treated surface was also analyzed by ion microanalysis (IMA) and scanning transmission electron microscopy (STEM).[5] Figure 4 shows an $^{16}O^+$ depth profile of this surface and the ordinary-

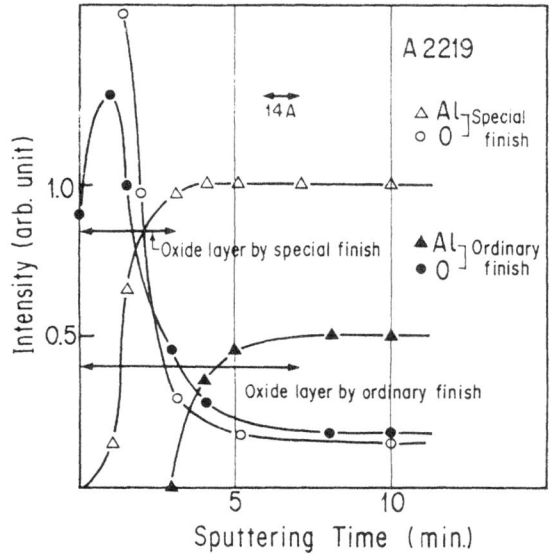

FIG. 3. The depth profile of A2219 surface by AES.

finish surface of A1050 by IMA. Figure 5 is a photograph of the treated surface sectional plan (A1050) by STEM. The oxide layer measured by STEM is different from the thickness measured by AES. In AES, the thickness is measured by sputtering the surface layer, and this thickness has tendency to be smaller than that measured by STEM.[5] But these differences are not well understood. The specially treated surface appears to be covered with a fine, nonporous, dense oxide film, as are the surfaces of the specially extruded aluminum beam ducts.[2]

V. RESULTS AND DISCUSSION

Figure 6 shows the outgassing rate after initial evacuation. The bakeout was done at 156 °C for 25 h. Compared with the outgassing rate results for the specially extruded chambers, the tendency before baking is like that of A1060EX and after baking is like that of A6063EX.[6] Next, N_2 gas was introduced into the chamber. The outgassing rate of the chamber, after being left at one atmosphere N_2 for an hour, was measured after reevacuation (Fig. 7). Air was introduced into the chamber and the outgassing rate of the chamber, after being left at one atmosphere for one month, was measured after reevacuation (Fig. 8). The outgassing rate before baking was about 10^{-11} Torr l/s cm² and the rate after baking

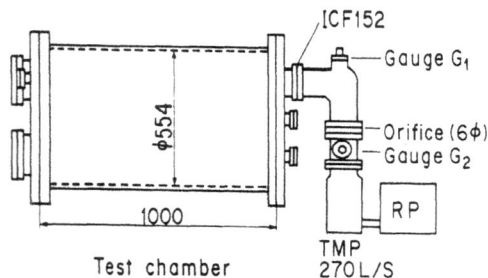

FIG. 2. Schematic of the outgassing rate test system.

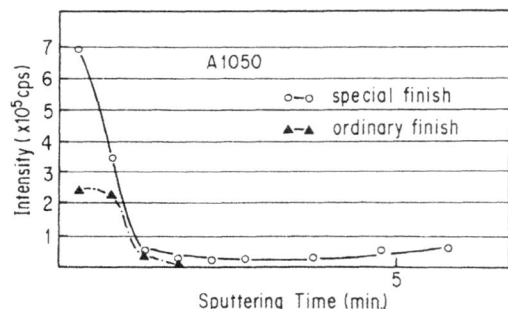

FIG. 4. Oxygen depth profile of oxide layer using IMA.

44

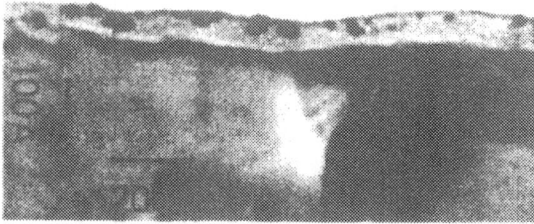

FIG. 5. STEM sectional photograph of the special finish (A1050).

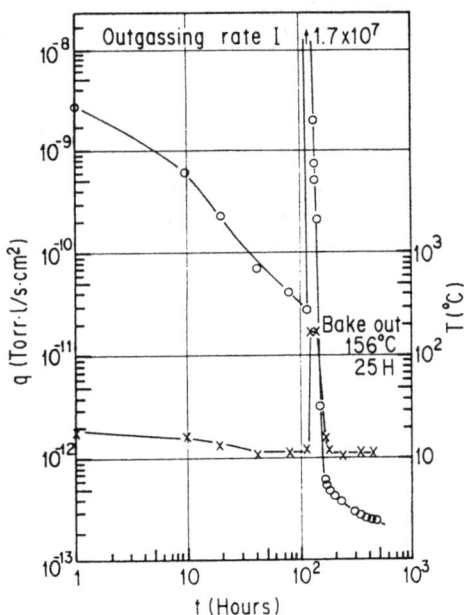

FIG. 7. Outgassing rate after being left at one atmosphere for an hour.

was about 10^{-14} Torr l/s cm². Compared with the initial evacuation, the outgassing rate is lower by one order of magnitude before baking and by one-half an order of magnitude after baking. The outgassing rate of the aluminum alloy chamber is lower by approximately one or two orders of magnitude than that of a stainless-steel chamber, such as SUS 304.[1] The ultimate pressure was measured by connecting an ion pump (500 l/s) directly to the chamber, together with a turbomolecular pump (300 l/s). The evacuation curve is shown in Fig. 9. The pressure was 9×10^{-10} Torr before baking and 7×10^{-11} Torr after baking.

Suppose the pumping speed before baking is 300 l/s and that the outgassing rate is 10^{-11} Torr l/s cm², then

$$P_\infty - P_0 = 1 \times 10^{-9} \text{ Torr.}$$

Suppose the outgassing rate after baking is 2×10^{-13} Torr l/s cm² and the pumping speed is 200 l/s, then

$$P_\infty - P_0 = 5 \times 10^{-11} \text{ Torr.}$$

So the measured values are reasonable. Next, the new titan-

ium sublimation pump was added to vacuum system. The nude-type titanium sublimation pump has efficient pumping conductance. The chamber temperature is kept at room temperature during the titanium sublimation pump operation with liquid N_2. The surface temperatures on the liquid-N_2 paths and on other points were measured using thermocouples with polytetrafluorethylene insulators. All temperatures were nearly -196 °C. The pressure was measured using a modulated Bayard–Alpert gauge (IMR103-IMG40 Balzers). Lower measuring limits with modulator is $\sim 10^{-13}$ mbar and the x-ray limit is $\sim 3 \times 10^{-12}$ mbar. The evacuation curve is shown in Fig. 10. The ultimate pressure was 6×10^{-12} Torr (6.5×10^{-12} mbar) after degassing the

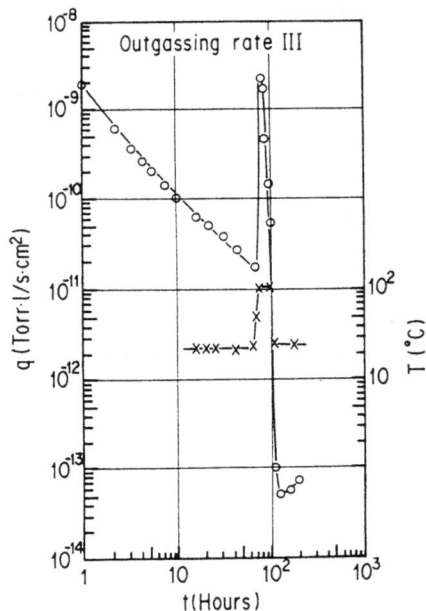

FIG. 6. Outgassing rate after initial evacuation. The outgassing rate and the chamber temperature are shown by ○ and ×, respectively.

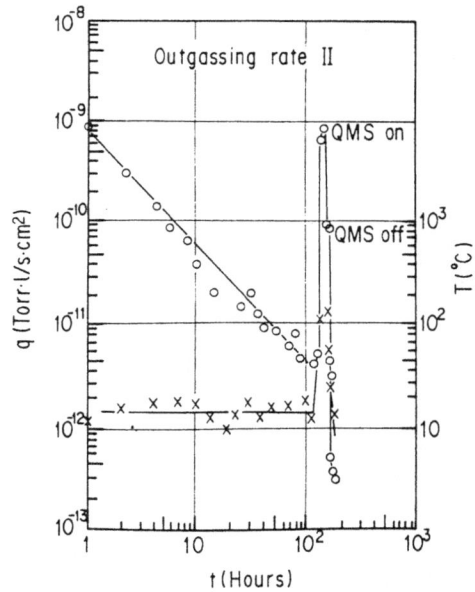

FIG. 8. Outgassing rate after being left at one atmosphere for about a month.

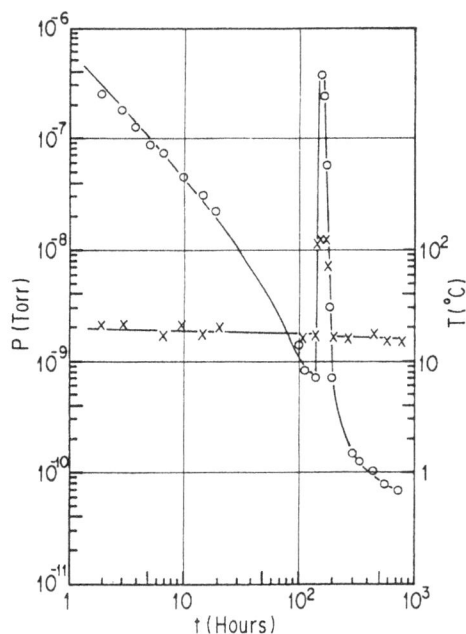

FIG. 9. Evacuation curve.

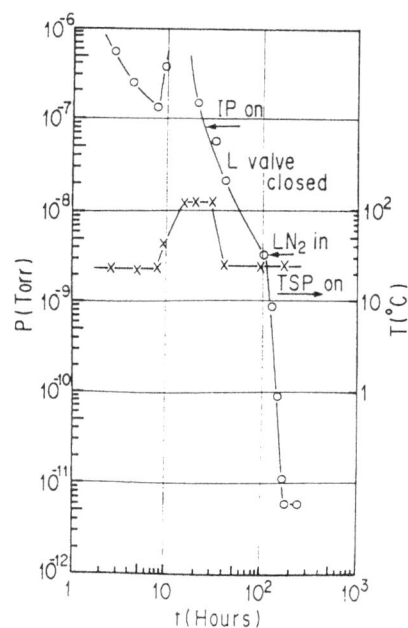

FIG. 10. Evacuation curve adding a new type of titanium sublimation pump.

gauge. If this ultimate pressure is the equilibrium condition between the pumping speed of the titanium sublimation pump and the total system outgassing, the pumping speed S may be estimated by the following expression:

$$S = \frac{3 \times 10^{-13} \ (\text{Torr l/s cm}^2) \times 3 \times 10^{14} \ \text{cm}}{6 \times 10^{-12} \ \text{Torr}} = 1500 \ \text{l/s}.$$

Thus, the pumping speed is estimated at about 1500 l/s. This ultimate pressure is very low for a large aluminum alloy vacuum chamber.

VI. CORROSION BY GALLIUM

An important problem for an aluminum alloy MBE system is corrosion by gallium.[7] Other elements, for example, Cd, Hg, B, In, P, and As, were also investigated. These elements do not corrode aluminum. The problem for gallium is solved by surface treatment. This surface treatment is SiO_2 coating on the aluminum surfaces using a spark discharge in a silicate solution.[8] The result of the test for some specimens against corrosion by gallium is shown in Table I. In Table I

some kinds of surface treatments are shown for comparison. Two sets of specimens were prepared for each surface treatment. Tests of corrosion by gallium for each surface treatment were performed in three steps.

In the first step, two sets of a gallium drop of 1 g weight were put on the SiO_2-coated aluminum alloy and baked at 150 °C for 150 h in nitrogen gas at atmospheric pressure. In the second step, the specimens were baked in vacuum furnace at 160 °C for 48 h from 10^{-5} to 3×10^{-8} Torr. In the third step, gallium was coated about 3 μm in thickness on SiO_2-coated aluminum alloy specimens and a gallium drop of 1 g was put on the gallium-coated specimens and baked in vacuum furnace at 200 °C for 24 h from 10^{-6} to 3×10^{-8} Torr. After baking the specimens were dropped into liquid nitrogen and held at room temperature. Specimens were then taken from liquid nitrogen and held at room temperature for 60 min. Tests of corrosion by gallium were performed in three cycles of these processes of three steps. Evaluation was done after each test cycle. According to Table I, the ceramic coating of SiO_2 on aluminum alloy was not

TABLE I. Several kinds of surface treatments for anticorrosion against Ga. O is no corrosion, × is corrosion, and ⋯ is no test, respectively.

Treatment	Thickness (μm)	Step 1 Nitrogen, 1 atm 150 °C, 150 h		Step 2 3×10^{-8} Torr 160 °C, 48 h		Step 3 3×10^{-8} Torr 1st		2nd		200 °C, 24 h 3rd	
SiO_2 coating	10	O	O	O	O	O	O	O	O	O	O
TiN ion plating	3	×	×	⋯	⋯	⋯	⋯	⋯	⋯	⋯	⋯
TiC ion plating	3	O	O	O	O	O	×	×	×	⋯	⋯
CrN ion plating	3	O	O	O	×	O	×	×	×	⋯	⋯
Ti ion plating	2	×	×	⋯	⋯	⋯	⋯	⋯	⋯	⋯	⋯
Ni + Cr electroplating	7 + 7	O	O	O	O	×	×	⋯	⋯	⋯	⋯
Anodize	10	O	O	O	O	O	O	O	×	×	⋯

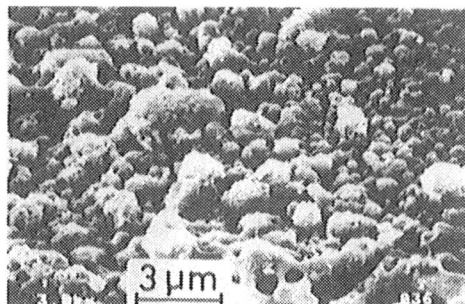

FIG. 11. Surface morphology of ceramic film of 10 μm in thickness.

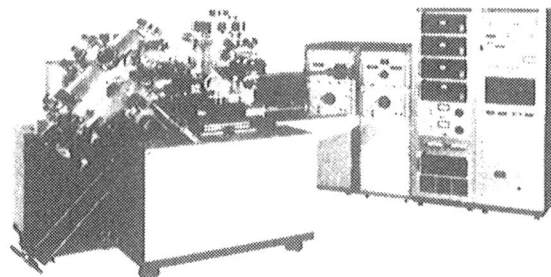

FIG. 12. Newly developed aluminum alloy MBE system.

corroded by gallium in any of the test steps.

This ceramic-coated aluminum alloy plate and wire can be bent without separation of the ceramic film and the aluminum alloy. Further, no separation during thermal cycles from -196 to $400\,°C$ was obtained. Surface morphology of ceramic film was porous as shown in Fig. 11.

The outgassing rate was measured. The typical outgassing rate was several times 10^{-13} Torr l/s cm^2 after a 24-h bakeout.[9] Before bakeout in vacuum, the main component of the residual gas was H_2O and after baking it was H_2. This ceramic treatment was suitable for an ultrahigh vacuum.

VII. CONCLUSION

A large aluminum alloy ultrahigh vacuum chamber with a special surface finish has been developed for an MBE vacuum system. An MBE system with an aluminum alloy vacuum chamber has been developed, and is shown in Fig. 12. This MBE system is composed of a large aluminum alloy chamber with a special surface finish, a new type of titanium sublimation pump and a new type of liquid-N$_2$ shroud. The liquid-N$_2$ shroud was made with the same methods as used for the titanium sublimation pump. This aluminum alloy ultrahigh vacuum chamber makes an efficient MBE vacuum system.

ACKNOWLEDGMENTS

The authors would like to thank E. Isoyama and Y. Katoh (Showa Aluminum Corp.) for presentation of the STEM photograph, and M. Nabae (Furukawa Aluminum Co., Ltd.) for an offer of IMA data. I would like to thank Dr. P. M. Stefan for some suggestions and corrections to this manuscript.

[1]E. D. Erikson, T. G. Beat, D. D. Berger, and B. A. Frazier, J. Vac. Sci. Technol. A **2**, 206 (1984).
[2]H. Ishimaru, J. Vac. Sci. Technol. A **2**, 1170 (1984).
[3]M. Miyamoto, T. Ito, S. Komaki, K. Narushima, and H. Ishimaru, in Proceedings of the 8th Symposium on ISIAT, Tokyo, 1984, p. 187.
[4]J. R. Chen, K. Narushima, M. Miyamoto, and H. Ishimaru, J. Vac. Sci. Technol. A **3**, 2200 (1985).
[5]H. Ishimaru, Hyomen Shori Kenkyu **2**, 1 (1985) (in Japanese).
[6]K. Narushima and H. Ishimaru, J. Vac. Soc. Jpn. **25**, 172 (1982); **26**, 353 (1983); **27**, 457 (1984) (in Japanese).
[7]W. R. Hunter and R. T. Williams, Nucl. Instrum. Methods **222**, 359 (1984).
[8]Japanese patent pending, Serial No. Shou58-17278 (6 April 1983).
[9]K. Narushima, H. Ishimaru, N. Kamata, and S. Komaki, in Proceedings of the 33rd Oyo-butsuri-gakkai, 1986 (in Japanese).

Advances in vacuum contamination control for electronic materials processing

John F. O'Hanlon

IBM Thomas J. Watson Research Center, Eastview Research Laboratory, Yorktown Heights, New York 10598

(Received 12 September 1986; accepted 14 October 1986)

The demand for devices with improved performance and yield and fabricated with reduced processing costs is driving advances in electronic materials processing. A most important consideration is the reduction of defects resulting from particulate contamination during processing. Since many of the steps in device processing use vacuum equipment, their requirements are shaping the design of new tools. In this paper we describe the sources of contamination during processing, the techniques used for their reduction, and note vacuum issues that relate to the use of new processing technologies.

I. INTRODUCTION

Advances in vacuum technique for processing electronic materials are occurring in areas that will improve yield, improve performance, and reduce the overall cost of producing a finished device. A review of traditional silicon production facilities reveals capital equipment costs to be the largest single component, about 53%, of the expenditure (capital equipment, work in progress, facilities, land, and buildings), with the largest elements of the manufacturing cost estimated to be direct materials and manufacturing labor at 45% and 12%, respectively.[1] Yield is the single most important concern in a successful facility and it can be increased by use of larger wafers and smaller die sizes.[2] 150-mm-diam wafers are now widely used, with 200-mm and larger diameter wafers available.[3] A 1-Mbyte dynamic random access memory (DRAM) is made on a die well under 100 mm.[2,4] The minimum linewidth of these devices is 1 μm, but particles that produce defects have been estimated to be as small as $\frac{1}{3}$ to $\frac{1}{10}$ of the minimum linewidth.[5-7] Advanced submicron metal–oxide semiconductor (MOS) devices may require gate oxides 10 nm thick.[8] Burnett[9] predicts that by 1990 manufacturers will have to achieve $\leqslant 0.05$ defects/cm^2 at 0.2 μm and larger to obtain a 10% yield. People and equipment are the two largest sources of contamination in a classical semiconductor fabrication facility and these can be reduced by automation, and improved equipment and facility design. An introduction to our knowledge of contamination and our future needs in microelectronic manufacturing is given in the paper by Cooper.[10]

Clean room facilities are rated in terms of the allowable concentration versus particle size distribution. For example, classes 1000 and 100 refer to a specification of fewer than 1000 or 100 particles per ft^3, respectively, of 0.5-μm diam and larger. New facilities have been constructed in which the allowable size distribution in the most critical areas has been reduced to less than 1 particle per ft^3 of $\geqslant 0.2 \mu$m,[11,12] and in which the maintenance areas are cleaner than the wafer handling areas of 20 years ago. These facilities have been dubbed "class 1," although the phrase currently has no standardized definition.[13] Measurements of particles on wafers show the fraction attributable to the clothes and bodies of workers to range from 20% to over 50%[2,6,14] of the settled particles. The accuracy of these measurements is of some concern. Hoenig and Daniels[15] note that even if the clean room were "class 0," defect-free, narrow-line devices could not be produced with the current exposure of wafers, due to particulates generated by personnel during processing. Advanced wafer handling concepts are needed to reduce the exposure of wafers to personnel-induced defects. Cassette-to-cassette handling, the standard mechanical interface (SMIF) box,[16,17] automated wafer and material handling,[2,3,6] and multiprocess integration—either inline or parallel[18] access are examples of these concepts. Once the exposure to operating personnel is minimized, the largest remaining source of particulates is the processing equipment,[14] much of which is vacuum related. Impure source gases and liquids are the next largest source of particulates.[5]

New vacuum processes such as plasma enhanced chemical vapor deposition (PECVD),[19] molecular beam epitaxy (MBE),[20] metallorganic chemical vapor deposition (MOCVD),[21] and photoassisted chemical vapor deposition[22,23] show promise in fabricating small devices with improved characteristics and throughput. Most deposition processes, however, can generate a large quantity of particulates.

Reduction of particulates is a major concern in processing materials; however, it is not the only concern. Most of these new techniques, as well as some plasma etching techniques, use toxic or flammable gases, some in high concentrations, with the result that a significant investment in safety-related facilities and process equipment is necessary for their proper operation. Herb *et al.*[24] estimate the cost of safety related facilities to be $\frac{1}{3}$ of the capital tool cost.

In this paper we discuss the reduction of particulates and impurities by process equipment, and discuss new process equipment designs that directly depend on improved vacuum technique.

II. PARTICULATES AND IMPURITIES

Device failures resulting from defects and contamination introduced during processing are of mechanical and chemical origin. Particulates cause failures by shorting adjacent lines and producing open circuits, "photolithographic" defects, or by shorting conducting layers separated by thin oxide films, "pinhole" defects.[25] Observed photolithographic defects have a size distribution which varies as $1/x^3$, where x is the defect size.[25] At 1-μm feature size, 0.1–0.3um and larger can cause photolithographic defects. Pinhole defects can be caused by extremely small particles, as small as 0.1 μm for devices of 1-μm feature size, whose probability of causing a short is higher along the edge of an insulator than in the middle.[25] Stapper[26] also notes that yield projections are strongly dependent on the clustering of defects and on wafer-to-wafer yield variations.

Particulates can be stirred up and transported to a wafer by turbulence induced during rough pumping and venting. Any mechanical motion in vacuum, (e.g., a valve closure, or wafer transport mechanism), dirty vent gas, thermal cycling, and chamber cleaning can generate additional particles. Chemically induced contaminants can come from particulates that react and diffuse into a surface and leave no visible trace of a particle, from impurities in a liquid or gas source, or from residual gases in the chamber. Chemically induced degradation is not limited to small devices.[5] Residual impurities can result from atmospheric air, process gases remaining from the previous step, or contamination from improperly cleaned piping, components, and chamber walls. Ukai[27] labels the contamination from flakes, dust, etc., as macrocontamination, and that from residual gases and atomic-scale particulates as microcontamination; however, these names are misleading. It is more appropriate to categorize contamination sources as either particulate or atomic scale. In this section we discuss the effects of and methods for reducing contamination from particulates and atomic-scale impurities.

A. Particulate contamination

The turbulence caused by rapidly removing atmospheric air from a chamber, or venting to atmosphere too rapidly has been known for some time to be an incredibly large source of particulate contamination. Turbulence during the roughing cycle was observed, but not documented, as long as 40 years ago. Hayashi[28] noted the motion of water mist in a glass bell jar as proof of the turbulence induced by fast roughing, while in this country, decorative metallizers realized they could eliminate the dull appearance of aluminum films by throttling the roughing pump. In 1962 Ames et al.[29] applied soft roughing to reduce the particulate contamination in metal and insulating films for superconductive devices. For the last two decades a restricted pumping step has been commercially available from some vacuum manufacturers, and a standard part of approximately one-half the production vacuum deposition equipment manufactured for IBM. However, little quantitative data have been taken to determine precisely the level of flow restriction required to reduce the contamination.

Hoh[30] studied the effect of soft roughing and venting on particulate contamination of Si wafers in a silicon monoxide evaporator. Only one set of flow conditions was used with the maximum throttled velocity in the roughing line corresponding to a Reynolds number 1000.[31] Hoh found that this soft roughing and venting condition greatly reduced the contamination of particulates $\geqslant 2\,\mu$m diameter from an uncountably large number to an average of 25 per 51-mm-diam substrate. Masuda et al.[32] studied the effect of roughing speed, however, their data are too sparse for interpretation. No other statistically significant data have been reported as a function of measure of turbulence, e.g., the Reynolds number, during roughing or venting. This is an area in which detailed measurements are needed to determine separately, the effects of slow roughing and venting on contaminants stirred up from the floor and walls of the chamber, and on the particulate concentration in the chamber atmosphere prior to roughing.

Masuda et al.[32] also measured the fine (0.5–5.0 μm diameter) particle flux to a substrate in static air of class 1 000 000. They expressed the relation between flux and volume concentration as

$$N_s = nv, \tag{1}$$

where N_s is the particle flux (particles/area/time) to the surface of diameter equal to or greater than the given particle size (the cumulative or integrated flux), n is the total number of atmospheric particles per unit volume equal to or greater than the stated value (the cumulative or integrated concentration), and v has the dimension of velocity. Examination of their source air, Fig. 1, shows its diameter dependence to be $n \propto d^{-2.7}$, over a range of 0.5 to 1.0 μm, where d is the particle diameter, while $N_s \propto d^{-3.7}$. Their data show the velocity to vary as d^{-1} and therefore the observed particulate deposition could not have resulted from gravitational settling. Cooper[10] gives the velocity diameter dependence for gravitational settling as

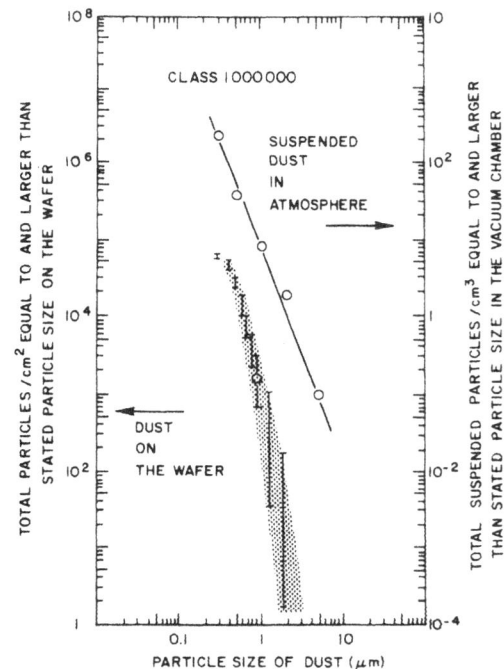

FIG. 1. Comparison of particle size distribution curves between dust laid on a wafer during a 10-min exposure and dust suspended in atmosphere before pumpdown (Ref. 32).

$$\nu = g\left(1 + 2.5\,\frac{\lambda}{d}\right)\frac{\rho d^2}{18\eta}, \qquad (2)$$

where g is the gravitational constant, λ is the mean free path in air, ρ is the particle mass density, and η is the gas viscosity. For gravitational settling at 1 atm $\nu \propto d^2$ for particles of $d \geqslant 1$ μm, and $\nu \propto d$ for particles of $d \leqslant 0.1\,\mu$m. One explanation of the observed data is diffusion of particles from the static air in the chamber. In that case $\nu \propto d^{-1}$ for equal times, and is enhanced at reduced pressures. Electrostatic deposition is another possible explanation, because the particle velocity also increases with decreasing diameter.

Particulate transfer from the process chamber to the substrate can be reduced by the use of a load lock. Transfer is reduced because the substrate is (soft) roughed and vented in a clean lock, where there is little process debris to be stirred. Flaking from the process chamber walls—a major source of contamination—is reduced, because the wall contamination is now made up of successive deposited layers of the same film, rather than alternate layers of atmospheric water and deposited film. The latter will stress and flake from the walls with great ease. Dickson et al.[33] have shown that use of a load lock in combination with heating of the entire deposition assembly and electrodes reduced the particulate contamination in multilayer deposition of amorphous silicon solar cells. However, even the yield in the load-lock system was limited ultimately by the formation of silicon particulates in the films. The yield was reduced to zero after 20–40 runs. They attributed these particulates to the flaking that resulted from stress buildup in the alternate layers of doped and undoped silicon deposited on the electrodes. The differential stresses between the doped and undoped layers caused flaking in the same way as differential stresses caused flaking of alternate layers of water and deposited film, but at a reduced level. By maintaining a constant 240 °C on the electrode and substrate assembly, yield increased significantly as shown in Fig. 2,

Lachenbruch et al.[34] identified two major sources of contamination in a vacuum load-locked system to be backstreaming from the mechanical pump and particulates from inadequately filtered vent gas. They demonstrated mechanical oil backstreaming could be eliminated effectively with a nitrogen purge, while an adsorbent trap acted as a source from which oil could diffuse. A membrane filter was recommended for the vent gas, but high gas flow was observed to abrade the membrane. Soft venting could be used to reduce the differential filter pressure. Lachenbruch also noted that in a reactive-ion etcher, condensable by-products from selective etch chemistries, when unchecked, deposited on the wafer and caused contamination. Spencer[35] has also observed $AlCl_3$ from aluminum etching can deposit as particulates or react with water vapor to form HCl. He noted air leaks, exposure of the etched wafer to humid air, and inadequate chamber heating all contributed to $AlCl_3$ condensation problems.

Wafer handling, system vibration, and any mechanical motion in vacuum all have the potential to generate and transfer particulates to the substrate. Hoh[30] observed the transfer of particulates to substrates in a silicon monoxide evaporator during isolation valve movements. rf discharges can generate particulates. Minute arcs in a discharge originating at a point of contamination can eject small particles, while radicals can polymerize and form small clusters.[36] Wafers are transported in vacuum by mechanisms such as belts, walking beams, turntables, retracting tapes, planetary arms, pantographs, and articulated arms.[18] Each mechanism has advantages and disadvantages and must be evaluated individually for particulate contamination. Urethane O belts were observed to leave a large quantity of particulates on the wafer surface exposed to the belt[34] in a nonload-locked system. Replacement of the O-belt by a spatula-type transport with an edge contact chuck eliminated 98% of the particulates of $d \geqslant 1\,\mu$m. Hardegen and Lane[37] have studied the particulate generation of an articulated arm, and identified the major source to be the pan gears. They reduced the contamination by the use of special lubricants and shielding. Contamination data for other transport mechanisms have not been published. This is a field in which there is a great need for additional data.

Substrate orientation also contributes to particulate deposition. Substrates may be oriented horizontally, process face up or down, or vertically, either singly or in pairs with surfaces to be processed facing outward. Obviously an upward facing substrate will collect more particles via gravitational settling, but not by thermophoresis, diffusion, electrostatic attraction, etching condensates, or polymer deposition. The choice between vertical double sided and horizontal face down is a subject of controversy. Vertical processing has an advantage for productivity.[36]

Ultimately plasma etching and deposition chambers need to be taken apart and cleaned, both physically and chemically. The most troublesome systems are plasma deposition systems, although etching systems certainly produce polymer by-products and condensable particulates. The time and labor of physical and chemical cleaning is a large expense, so an in situ technique is most desirable. The standard technique in a PECVD system is the use of a halogen glow discharge for vaporizing the wall deposits. CF_4 etches silicon, and its oxides and nitrides. However, carbon based halogens also form polymer deposits. NF_3 has been used as a relatively safe source of free fluorine for cleaning aluminum and stainless-steel PECVD chambers. With the exception of aluminum and iron fluorides the by-products formed by its side

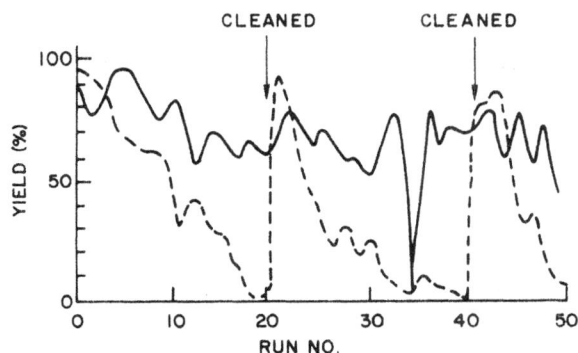

FIG. 2. Effect of debris on yield of solar cells (Ref. 33) — Temperature chamber and wafer holder maintained at 240 °C --- Temperature cycled. In both cases the chamber was cleaned every 20 runs.

reactions are all volatile compounds.[38] NF_3 is an excellent choice for cleaning an aluminum chamber because free fluorine reacts with aluminum to form AlF_3, which is nonvolatile and passivating.[39] From the bulk properties of NF_3 it is known that it decomposes at elevated temperatures to NF_2 and F[38]; the latter reacts readily with steel. NF_3 does clean stainless-steel chambers but the nature of the oxide formed on the chamber walls (passivating or powder) has not been determined.

B. Elemental and atomic-scale impurities

Dividing contaminants into two types, particulates and elemental or atomic-scale impurities seems obvious, but in reality is not such a "clean" division. For example, physical adsorption of H_2O on a wall can cause the flaking of deposited films or the formation of an aqueous acid that will cause corrosion. In this section we discuss the chemical or dopant-level effects of elemental and molecular impurities, and how they can be reduced by clever or specialized vacuum technique.

Residual gases[27,33,40] and sputtering of atoms and small clusters from reactor chamber fittings[27,34] and electrodes[41,42] have been observed to be the cause of impurities in deposited films. Careful vacuum design can overcome most of these problems. The effect of residual gases is diminished by use of a load lock,[34] while electrode design and shielding are necessary to prevent sputter contamination. Metal impurities such as Sn and In have been observed to diffuse from the transparent oxide coating to the adjacent p layer of p–i–n solar cell.[43]

Source gas cleanliness is another concern. The sources and methods for reducing particulates in gas streams have been reviewed elsewhere.[44,45] Assuming particulates can be removed with proper filters, we are left with the gaseous impurities (e.g., O_2, H_2O, and N_2) contained in all source gases or adsorbed on all vacuum walls after exposure to atmosphere. Miura[46] and Mitchell[47] give examples of typical chemical impurity levels and describe analysis techniques for currently available semiconductor process gases. Other papers describe contamination particular to silane[48–50] and disilane.[48,49] Fraas et al.[51] show that specialty hydrides AsH_3 and PH_3 are contaminated with significant quantities of water vapor. They describe a cold-trap technique which allows AsH_3 and PH_3 to be purified to 2.2 and 0.4 ppm of water vapor, respectively.

The solution to the problem of oxygen and water vapor contamination in the supply gases for epitaxial growth is dependent on the material system. Meyerson[52,53] has grown high-quality epitaxial silicon by exploiting the equilibrium conditions of the $Si/H_2O/SiO_2$ and $Si/O_2/SiO_2$ system. He has shown that a partial pressure of less than 10^{-8} Torr (10^{-6} Pa) of both O_2 and H_2O is necessary to keep the surface oxide-free at $T \geqslant 700$ °C. Furthermore, at deposition pressures of 1 to 2 mTorr (0.13 to 0.26 Pa) this corresponds to a supply gas purity of 10 ppm of water vapor or oxygen. A partial pressure of $< 10^{-8}$ Torr (10^{-6} Pa) can be achieved in a turbomolecular pumped ultrahigh vacuum CVD reactor, with strict adherence to UHV practice, and sufficiently clean source gases. A standard O-ring-sealed reactor without a load-lock is incapable of reaching this level of purity in one pump cycle.

This technique for obtaining an oxide-free silicon surface and depositing an oxygen-free epilayer is not appropriate for a very reactive material like aluminum. Aluminum is so reactive that extensive point of use gas purification is necessary to maintain the required gas stream purity. The H_2O/H_2 ratio for equilibrium between $Al_2O_3 + H_2 \rightleftharpoons Al + H_2O$ is 10^{-13} at 650 °C—a level far below that to which a source gas can be purified. Kuech et al.[54] note the primary sources of O_2 and H_2O in $Al_xGa_{1-x}As$ are as impurities in the H_2, AsH_3, and dopant gases used in the compound growth. They purified H_2 via a Pd diffusion cell and passed the AsH_3 through a Ga–In–Al ternary metal bubbler.[55] $Al(CH_3)_3$ was used as the aluminum source, and it readily reacts with H_2O and O_2. This property of the metal alkyl was used to provide in situ gettering of water and oxygen by constructing the reactor with a long ($\simeq 30$ cm) entrance region.[56] By allowing the source gases to react with the alkyl in this region, they were further purified, and devices with increased photoluminescence intensity and decreased impurity incorporation were obtained.[54]

Microcontamination can also arise from impurities on the inside surface of tubing which can become incorporated in the supply gas. AISI 304 and 316 stainless-steel tubing must be cleaned for use in gas distribution lines. The standard tubing available from supply houses has a very rough interior surface which is a source of both particulates and elemental impurities.[57,58] Various electropolishing, acid pickling, and chemical flow polishing treatments are used to etch the surface and remove loose fragments and sites where potential contaminants can lodge. Studies of these cleaning techniques[57–60] all show that electropolished and chemically polished surfaces are very smooth, with as much as a 10:1 reduction in surface area obtained after cleaning.[57] It has been observed that the desire to obtain tubing with smooth inner walls is not always consistent with clean walls.[61] A smooth surface (low effective surface area) is an advantage in tritium service as it reduces the generation of protium (HT).[57] However, the electropolished and chemically flow polished surfaces have been found to contain deposits of heavy metal salts (Cl, S, P),[57–59] which can only be removed with an ion beam. It is likely that chemical or electropolished tubing needs an additional cleaning step (e.g., leaching in DI water or steam) to provide a contaminant-free interior surface. It should be noted that no form of abrasive cleaning or bead blasting should be used on any stainless-steel surface, because the abrasive medium becomes buried in the surface,[59,60] where it is a source of particulates and chemical contamination, such as sodium in silicate beads.

III. NEW EQUIPMENT DESIGNS

High throughput with high yield, low particulates, high equipment reliability, and most of all safe operation are necessary in new processing systems. In any system, reduction of particulates will require great attention to detail, and knowledge of techniques for measuring and analyzing particulate contamination.[61–64] The smaller the particles, the greater will be the difficulty in removing them from the surface, because the binding forces decrease with particle size at a slower rate than the forces of removal.[65] The separate advantages of load locks and automatic wafer handling can be multiplied when used together. Since operator handling constitutes a major source of particulates, it is clear that contamination can be reduced as processes are interconnected. One example of this is a combined steam cleaning–sputtering system described by Rogelstadt and Matarese.[66] In this multichamber, in-line system, alumina substrates were washed, steam cleaned, dried, and sputter coated with Cr–Cu–Cr. Metallized particulate contamination was reduced to 14% of the level obtained when parts were cleaned in a separate operation. The performance was improved because the wafers passed from the cleaning system directly to the sputtering chamber without being handled in atmosphere.

New equipment designs need to incorporate load locks and process automation on all major deposition and etching steps. The simplest load lock is a single lock, separated from the process chamber by a valve whose cross section is a thin slit, and pumped by a cryo or turbopump. A double-entry load lock uses two load locks in series; the first lock is pumped by a roughing pump and the second by a high-vacuum pump.[18] Additional cleanliness can be obtained by use of separate process chamber pumping packages for oxidizing and flammable/toxic gases. The use of dual pumps on the same chamber reduces contamination from the gases used in the previous pump cycle, and reduces pump maintenance.

The reduction in contamination obtainable by automation logically dictates multichamber processing. Multichamber processing can increase throughput and reduce the contamination that would result from crosscontamination of process gases and intersystem transfer of substrates. There are two basic designs of multichamber systems, the in-line system, Fig. 3, and the parallel access system, Fig. 4. In both designs each chamber is dedicated to one process. Multichamber in-line systems have been constructed for multilayer metal depositions,[18,66] and for fabrication of p–i–n solar cells.[43,67] The throughput of an in-line system can be maximized if constructed with load locks at both ends of the system. In addition to using a load lock at each end of the system, the design of von Roedern and Madan[67] has a second lock or isolation chamber, between each of the two end locks and its adjacent process chamber, as well as an isolation chamber between each process chamber. This feature was incorporated to reduce the crosscontamination between adjacent process chambers. Using this production prototype

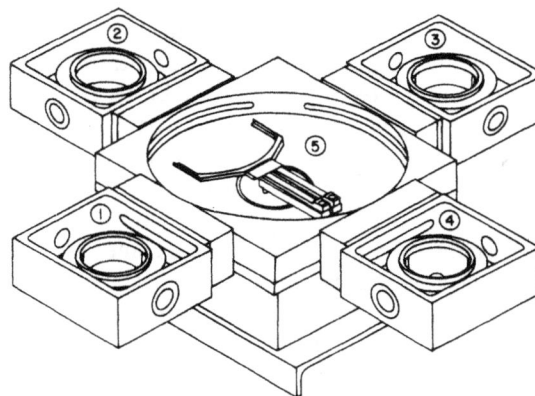

FIG. 4. Outline of a multiple process deposition system (1) Load lock and process chamber, (2)–(4) process chambers, (5) modular vacuum transport and vacuum chamber (Ref. 37).

system they have fabricated solar cells of > 10% efficiency. Parallel deposition systems have been designed for the deposition of amorphous silicon films.[18] The parallel design has the problem that the interior load lock is common to all chambers and thus the potential for crosscontamination is greater than in an in-line system. In both designs, series and parallel access, safety dictates that all processing be terminated, and all power removed from the system to perform maintenance on any one chamber.

Safety is a first-order concern in the design of plasma using toxic and flammable gases. Kumagai[68] has reviewed the vacuum, electrical, and mechanical design considerations which are common to all deposition and etching systems. Supple and Stoneham[69] described the design of a research system in which any one of several chambers could be attached separately to a common manifold and pumping system. Ohlson[70] evaluated personnel exposures to etch chamber residues and pump fluids during maintenance. Health and engineering issues have been discussed by Herb et al.[24] and Okada[71] for plasma etch systems, and by Harada[72] for deposition and etching systems, while Hayashi et al.[73] and Johnson[74] have reviewed the safety and design of MOCVD systems. The use of flow restrictors adjacent to tank valves to reduce the hazard of a catastrophic toxic or pyrophoric gas line failure is a recent technique whose safety advantages have been found to far outweigh any processing difficulties.[75] Control of noxious effluents from plasma reactions is being addressed; however, there is much to be learned. Wet scrubbing techniques have been discussed by Librizzi and Manna,[76] while Mistry et al.[77] describe progress in semiconductor process effluent treatment in Japan, with emphasis on burning methods as well as the iron chloride adsorption technique (Toxoclean) for the removal of AsH_3 and PH_3.

IV. CONCLUSIONS

We have described the sources of particulate and atomic-scale impurities in vacuum deposition and etching systems and reviewed some of the techniques used for their reduction. We note that more data are needed to completely characterize the speed at which chambers should be roughed to

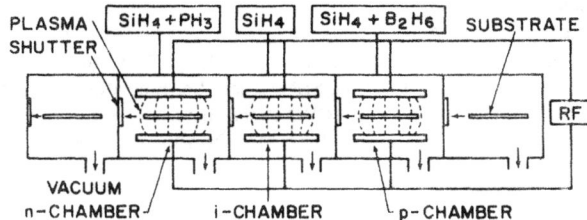

FIG. 3. Consecutive, separated reaction chamber apparatus for the fabrication of α-Si films (Ref. 43).

avoid turbulence, and to characterize the level of particulates generated by the many techniques for intraprocess wafer transport. Automated processing will be necessary to reduce exposure to contaminants generated by personnel; however, the problems of control, maintenance, and repair scale accordingly. The problems of scaling to manufacturing (lack of similitude laws and time) will drive development work on systems of near production type and size. These designs will become chemistry and materials specific and problems will arise if considerable attention is not devoted to the very detailed design issues.

[1] W. R. Bottoms and J. S. Wenstrand, Solid State Technol. **26**(8), 173 (1983).

[2] J. G. Harper and L. G. Bailey, Solid State Technol. **27**(3), 89 (1984).

[3] Microelectron. Manuf. Test., **8**(1), 17 (1985).

[4] W. P. Noble and W. W. Walker, IEEE Circuits Devices Mag. **1** (1), 45 (1985).

[5] J. M. Duffalo and J. R. Monkowski, Solid State Technol. **27**(3), 109 (1984).

[6] T. Hayashi, Jpn. Semicond. Technol. News **3**, 40 (1984).

[7] K. Edmark and G. Quackenbos, Microcontamination **2**,(5), 47 (1984).

[8] D. L. Tolliver, Solid State Technol. **27**(3), 129 (1984).

[9] J. Burnett, Microcontamination **3**(5), 32 (1985).

[10] D. W. Cooper, J. Aerosol Sci. Technol. **5**(3), 287 (1986).

[11] A. Colozzi, M. K. Powers, and L. Roy, Microcontamination **4**(5), 48 (1986).

[12] Semiconductor Int., **9**(6), 15 (1986).

[13] J. Burnett, Microcontamination **3**(3), 21 (1985).

[14] H. L. Brown, Solid State Technol. **26**(4), 239 (1983).

[15] S. A. Hoenig and S. Daniel, Solid State Technol. **27**(3), 119 (1984).

[16] M. Parikh and U. Kaempf, Solid State Technol. **27**(7), 111 (1984).

[17] S. Gunawardena, R. Haven, U. Kaempf, M. Parikh, B. Tullis, and J. Vietor, J. Environ. Sci. **27**(4), 57 (1984).

[18] B. Hardegen and E. L. Jaye, in *Proceedings of Semicon West (1985)* (SEMI, Mountain View, CA, 1985), p. 51.

[19] W. L. Johnson, Solid State Technol. **26**(4), 191 (1983).

[20] A. Y. Cho, Thin Solid Films **100**, 291 (1983).

[21] P. D. Dapkus, Annu. Rev. Mater. Sci. **12**, 243 (1982).

[22] R. Solanki, C. A. Moore, and G. J. Collins, Solid State Technol. **29**(5), 220 (1986).

[23] J. B. Mullin and S. J. C. Irvine, J. Vac. Sci. Technol. A **4**, 700 (1986).

[24] G. K. Herb, R. E. Caffrey, E. T. Eckroth, Q. T. Jarrett, C. L. Fraust, and J. A. Fulton, Solid State Technol. **26**(8), 185 (1983).

[25] C. H. Stapper, IBM J. Res. Dev. **27**, 549 (1983).

[26] C. H. Stapper, IBM J. Res. Dev. **29**, 87 (1985).

[27] K. Ukai, in *Proceedings of SEMI Technological Symposium, Semicon Japan '85* (SEMI, Mountain View, CA, 1985).

[28] C. Hayashi, Ulvac Corp. (private communication).

[29] I. Ames, M. F. Gendron, and H. Seki in *Transactions of the 9th National Vacuum Symposium* (1961) (Macmillan, New York, 1962), p. 133.

[30] P. D. Hoh, J. Vac. Sci. Technol. A **2**, 198 (1984).

[31] J. F. O'Hanlon, J. Vac. Sci. Technol. A **1**, 228 (1983).

[32] Y. Masuda, S. Komiya, and C. Hayashi in *Proceedings of the 8th International Vacuum Congress, Cannes* (1980), edited by J. P. Langeron and L. Maurice (Société Française du Vide, Paris, 1980), Vol. 2, p. 70.

[33] C. R. Dickson, J. Pickes, and A. Wilczynski, Solar Cells **19**(2), 179 (1987).

[34] R. Lachenbruch, O. Gomez, and B. Chapman, in Ref. 27.

[35] J. E. Spencer, Solid State Technol. **27**(4), 203 (1984).

[36] C. Hayashi in 4th International Colloquium on Plasma Sputtering, Nice, 1982.

[37] B. Hardegen and A. P. Lane, Solid State Technol. **28**(3), 189 (1985).

[38] A. J. Woytek, J. T. Lileck, and J. Barkanic, Solid State Technol. **27**(3), 172 (1984).

[39] C. E. Kolb and M. Kaufman, J. Chem. Phys. **76**, 947 (1972); C. D. Stinespring, A. Freedman, and C. E. Kolb, J. Vac. Sci. Technol. A **4**, 1946 (1986).

[40] D. E. Carlson, J. Vac. Sci. Technol. **20**, 290 (1982).

[41] P. E. Vanier, A. E. Delahoy, and R. W. Griffith, J. Appl. Phys. **52**, 5235 (1981).

[42] G. Lukovsky *et al.*, Final Report, SERI/STR-211-2475, 1984.

[43] Y. Kuwano, M. Ohnshi, S. Tsuda, Y. Nakashima, and N. Nakamura, Jpn. J. Appl. Phys. **21**, 413 (1982).

[44] D. L. Tolliver and H. G. Schroeder, Microcontamination **2**(4), 34 (1983).

[45] J. M. Davidson and T. P. Ruane, in Proceedings of Semicon West, 19-23 May, 1986, Burlingame, CA.

[46] K. Miura, in Ref. 27.

[47] J. W. Mitchell, Solid State Technol. **28**(3), 131 (1985).

[48] D. E. Carlson, in *Tetrahedrally-Bonded Amorphous Semiconductors*, edited by D. Adler and H. Fritzsche (Plenum, New York, 1985), p. 165.

[49] R. R. Corderman and P. E. Vanier, J. Appl. Phys. **52**, 5235 (1982).

[50] D. E. Carlson, A. Catalano, R. V. D'Aiello, C. R. Dickson, and R. S. Oswald, AIP Conf. Proc. **210**, 234 (1984).

[51] L. M. Fraas, J. A. Cape, P. S. McLeod, and L. D. Partain, J. Vac. Sci. Technol. B **3**, 921 (1985).

[52] B. S. Meyerson, Appl. Phys. Lett. **48**(12), 24 (1986).

[53] B. S. Meyerson, E. Ganin, D. A. Smith, and T. N. Nguyen, J. Electrochem. Soc. **133**, 1232 (1986).

[54] T. F. Kuech, E. Veuhoff, D. J. Wolford, and J. A. Bradley, Inst. Phys. Conf. Ser. **74**, 181 (1985).

[55] J. R. Shealy and J. M. Woodall, Appl. Phys. Lett. **41**(1), 88 (1982).

[56] T. F. Kuech and E. Veuhoff, J. Cryst. Growth (to be published).

[57] J. T. Gill, W. E. Moddeman, and R. E. Ellefson, J. Vac. Sci. Technol. A **1**, 869 (1983).

[58] M. K. Anewalt, D. M. Drummer, and J. V. Martinez de Pinillos, Microcontamination **3**(4), 52 (1985).

[59] N. Takahashi and K. Okada, Jpn. J. Appl. Phys. **11**, 1580 (1972).

[60] T. V. Rao, R. W. Vook, and W. Meyer, J. Vac. Sci. Technol. A **4**, 1604 (1986).

[61] R. C. Thomas, Solid State Technol. **28**(9), 153 (1985).

[62] B. J. Tullis, Microcontamination **3**(11), 67 (1985).

[63] B. J. Tullis, Microcontamination **3**(12), 15 (1985).

[64] B. J. Tullis, Microcontamination **4**(1), 51 (1986).

[65] A. Khilnani and D. Matsuhiro, Microcontamination **4**(5), 28 (1986).

[66] T. Rogelstadt and G. Matarese, J. Vac. Sci. Technol. A **3**, 516 (1985).

[67] B. von Roedern and A. Madan, Photovoltaic Energy Sources Conference II, Beijing, China, 1986 (unpublished).

[68] H. Y. Kumagi, J. Vac. Sci, Technol. A **4**, 1800 (1986).

[69] R. W. Supple and E. B. Stoneham, J. Vac. Sci. Technol. A **3**, 504 (1985).

[70] J. T. Ohlson, Solid State Technol. **29**(7), 69 (1986).

[71] T. Okada, T. Ukai, and T. Tsukada, J. Jpn. Vac. Soc. **28**(3), 119 (1985).

[72] H. Harada, J. Jpn. Vac. Soc. **28**(6), 494 (1985).

[73] T. Hayashi, K. Kuwahara, S. Ohwa, and S. Komiya, J. Jpn. Vac. Soc. **28**(5), 132 (1985).

[74] E. Johnson, R. Tsui, D. Convey, N. Mellen and J. Curless, J. Cryst. Growth **68**, 497 (1984).

[75] W. E. Quinn and D. Rainer, Solid State Technol. **29**(7), 63 (1986).

[76] J. Librizzi and R. R. Manna, Microelectron. Manuf. Test., 46 (1983).

[77] C. Mistry, Y. Ono, and T. Urata, SEMI Technical Education Program, Safety Aspects of Effluents From CVD Processing Systems, San Francisco, 1986, SEMI, 805 E. Midfield Road, Mountain View, CA 94043.

J. Vac. Sci. Technol. A, Vol. 5, No. 4, Jul/Aug 1987

54

Particle Control for Semiconductor Processing in Vacuum Systems

R. Allen Bowling, Member of the Technical Staff and Graydon B. Larrabee, TI Fellow and Branch Manager, Texas Instruments Incorporated, Dallas,Texas.

The impact of particulate contamination on semiconductor device production is dramatic. The overall problem is expressed in Figure 1, which is a plot of the minimum pattern dimensions for semiconductor devices as a function of killing defect sizes. This is based on a commonly accepted assumption that killing defects range from one fifth to one tenth of the minimum geometry. The continued decrease in minimum device dimensions with increasing device complexity has clearly necessitated and brought about improvements in clean rooms to below class 10. This figure demonstrates that smaller and smaller particles are becoming killing defects and are thus increasing in importance for new devices. The class 10 clean rooms with HEPA filters are ultraclean to the point that the clean room environment itself adds very few particles to the process wafers. The HEPA filter medium is particularly effective at ultrasmall, <0.1 micron diameter, particle filtration. The process wafer also spends a relatively small amount of time outside in the clean room environment. Much more time is spent inside a storage container or inside process equipment. The wafers probably spend less than 20% of their cycle time in the clean room environment. Instead of further improvements in the clean room, control of particles during the storage and processing of wafers is most desirable. This clearly points to improvements in control of particles generated in equipment and processes.

Table 1 shows a listing of some typical semiconductor processes which are carried out in vacuum systems. The large number of these vacuum chamber processes in a typical production area suggests that improvements in particle contamination resulting from vacuum processes could produce the dramatic decreases in killing defects needed for advanced devices.

In the past, very little has been known about the behavior of particles in vacuum. To control particles in a vacuum system, it is critical to have some understanding of particle behavior and generation mechanisms. At atmospheric pressure, very small less than 1 micron diameter particles are known to remain suspended in air almost indefinitely due to gas molecule drag on the particles. We have theoretically modelled the effect of pressure reduction on the gravitational settling of particles. To perform this derivation several assumptions can be made to simplify the problem. We assume spherical unit-density particles at 20°C and zero initial velocity. The following definitions are used in the calculations:

d = particle diameter
η = viscosity of medium
v = velocity of particle
C = Cunningham correction factor (pressure and particle size dependent)
ρ = particle density
τ = particle relaxation time

The drag force and gravitational force on a particle are defined as follows:

Drag force = F_d = 3 $\pi\eta vd/C$

Gravitational force = $F_g = \rho \pi d^3 g / 6$

The Cunningham correction factor is both pressure and particle diameter dependent and has been empirically expressed as

$C = 1 + (2/Pd)[6.32 + 2.01 \exp(-0.1095 Pd)]$ = correction for "slip"

where P = pressure in cm Hg, and d = particle diameter in microns. The Cunningham correction factor corrects the drag force for so-called molecular slip. The correction becomes most significant for small particles and/or low pressures. The small particle can be thought of as "slipping" between the gas molecules under these conditions.

The terminal velocity of a particle is reached when $F_d = F_g$. The expressions from above for F_d and F_g can be substituted and the resulting formula can be reduced to an expression for the terminal settling velocity

$V_T = d^2 g C / 18 \eta = \tau g$

The velocity at any time can de derived from the force at any time:

$F = m dv/dt = F_g - F_d$

and again substitutions and reduction can give the velocity at any time

$v = V_T (1 - e^{-t/\tau})$

The fall distance as a function of velocity and time can be expressed as follows:

$$\text{Fall distance}, S = \int v \, dt = \int V_T (1 - e^{-t/\tau}) \, dt$$

and evaluated to give

$S = V_T [t - \tau (1 - e^{-t/\tau})]$

This expression can be solved for time at the two limits

for $t \gg \tau$, $t = S/V_T$

and for $t \ll \tau$, $t = (2S/g)^{0.5}$

For intermediate values, t can be obtained iteratively. This expression can be evaluated for various values of pressure and particle diameter to obtain a pressure dependence of gravitational settling of particles in vacuum. Shown in Figure 2 is the plot of this expression for the time required to fall 1 meter as a function of pressure for several particle diameters. It is clear from this figure that the gravitational fall approaches a gravitational force dependent limit at about 10^{-5} torr. Below this pressure, all sizes of particles fall at the same average rate, 1 meter in 0.45 second. At this low pressure limit, the particle density and particle shape play no role in the results so that our initial assumptions of spherical unit-density particles do not affect these results. The overall implication of this finding is that when the pressure is reduced in a chamber, particles will rapidly "fall out" of the vacuum chamber.

To study this gravitational fall, a particle study vacuum chamber was constructed with a turbopumped introduction system and high vacuum main chamber. A diagram of this study chamber is shown in Figure 3. The chamber was set up in a class 100 clean room. Silicon wafers, 100 mm diameter, were placed in the introduction system and the chamber was evacuated and then vented to atmosphere. Wafers were removed and the number of added particles were measured using an Aeronca WIS-150 surface particle measuring system. Wafers were placed in the introduction system face-up, face-down, and face-vertical. The results of these studies are shown graphically in Figure 4 as a function of pumpdown pressure. Face-up wafers collected significantly more particles than face-down wafers. Face-vertical wafers, not shown on the graph, were not statistically different from face-down wafers in particle collection. This directly confirms our calculations from above about increased particle fall in vacuum. When the pressure is reduced particles fall onto the face-up surface, and more particles fall as the pressure is reduced further. This implies that for best particle results, vacuum chamber process equipment must utilize face-down or face-vertical processing.

Also using the same particle test vacuum system, we have determined that the pumpdown and venting conditions in the introduction system are also very critical to the particulate contamination levels due to turbulence. This observation has also been reported by other researchers. Our test results are summarized in Table 2. Slow pumpdown and slow backfill result in dramatic reductions in particles on the test wafers. Control of turbulence during pumpdown may be achieved through the use of devices for reducing the effective pumping speed of the pumping system; a throttle butterfly valve or a variable aperture system both work very well. Turbulence during backfill is well controlled by slow venting through a needle valve or similar device. Our observations suggest that control of the venting is slightly more critical than control of the pumpdown. Turbulence may be present over a longer period of the backfill than for the pumpdown.

Control of pumping for reducing turbulence can be theoretically considered by looking at the conditions for turbulence in a tube. The critical pressure P_c, below which turbulence no longer exists, given that turbulence exists for Reynolds numbers above 1000, during pumpdown is given in Figure 5 as a function of tube diameter D and pumping speed S_p. The initial maximum flow of gases out of the chamber is the period of maximum turbulence. Turbulence is present until the pressure falls below about 1 to 10 torr for typical pumping tube diameters. The formula given in Figure 5 for the critical pressure suggests that an increase in tube diameter or a decrease in the effective pumping speed will increase the critical pressure and thus limit turbulence. This figure supports what we have found experimentally, as presented above. Pumpdown must be throttled during the initial pumpdown until the pressure reaches about 1 torr, below which the pumpdown may be rapid.

The following observations have also been noted in our particle studies. However, no specific data will be presented. Particles have been found to be generated in vacuum by mechanical parts such as gate valves, rack and pinion drive systems, and other moving parts. The best advantage of vacuum processes are that the reduced pressures allow these particles that are generated in vacuum to "fall out" rapidly. Even with the intentional addition of particles to the chamber, a short pumpdown period "cleans" the particles from the chamber; wafers subsequently moved through the chamber collect almost no particles.

Pressure differentials between process and introduction chambers when opening connecting gate valves have been found to cause particle movement between the chambers resulting in wafer contamination. The differential is more critical when operating at higher pressures, for example 1-100 millitorr. At lower pressures, a given differential means a smaller number of gas molecules transfer.

In summary, since particles fall in vacuum due to the reduced pressure, a face-up wafer will collect a large number of particles in the introduction chamber where pumpdown takes place. Best results for wafer movement in vacuum with a minimum of particulate contamination have been obtained for face-down wafer handling. These results seem to indicate that face-down wafer processing may be the mode of least particulate contamination. In most cases, vertical wafer processing has been found to have the same advantages. The control of turbulence during pumpdown and venting procedures in semiconductor vacuum process chambers is critical to particle control. Turbulence can be effectively limited by slow pumpdown and venting procedures.

ACKNOWLEDGEMENTS

The authors would like to express special thanks to Benjamin Liu, Cecil Davis, Ed Millis, and Keith Russell for their assistance and consultations on this work.

Portions of this work were sponsored by the Defense Advanced Research Projects Agency under DARPA order No. 5394 and monitored by Mr. Max Yoder, Office of Naval Research, under contract No. N00014-85-C-0286. Approved for public release. Distribution is unlimited.

TABLE 1

PROCESSES USING VACUUM

PROCESSES	OPERATING PRESSURE (torr)
Plasma etching	0.1 to 2.0
LPCVD oxides	0.1 to 2.0
LPCVD nitrides	0.1 to 2.0
LPCVD multilevel glasses (PSG, BPSG)	0.1 to 2.0
LPCVD metals	0.2
LPCVD silicides	0.2
LPCVD polysilicon	0.1
Sputter etching	0.001 to 0.1
Sputtering — metals/doped metals	0.005 to 0.05
Sputtering — silicides	0.005 to 0.01
Sputtering — oxide (quartz)	0.01
Electron beam evaporation — metals	10^{-3} to 10^{-5}
Ion implantation	10^{-6} to 10^{-7}
Molecular beam epitaxy	10^{-10}

TABLE 2

TURBULENCE DURING PUMPING AND VENTING

DESCRIPTION	PUMP	VENT	PRESSURE	NUMBER $0.2-2\mu$ PARTICLES ADDED*		
				FACE UP	FACE DOWN	VERTICAL
MOUNT & LOAD				16	4	5
PUMP & VENT	FAST	FAST	10^{-4} torr	163	14	20
	FAST	SLOW	10^{-4} torr	60	17	12
	SLOW	FAST	10^{-4} torr	65	31	32
	SLOW	SLOW	10^{-4} torr	43	9	3

*Particles added to a 100 mm silicon wafer as measured by an Aeronca WIS-150

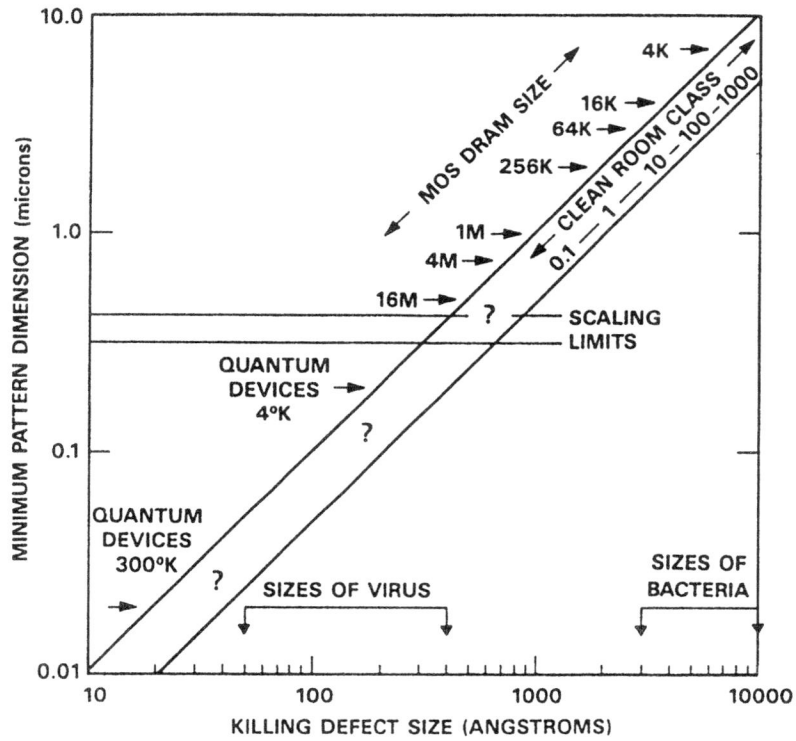

FIGURE 1 **DEVICE PATTERN DIMENSIONS/IMPACT OF DEFECTS**

FIGURE 2. GRAVITATIONAL SETTLING OF PARTICLES

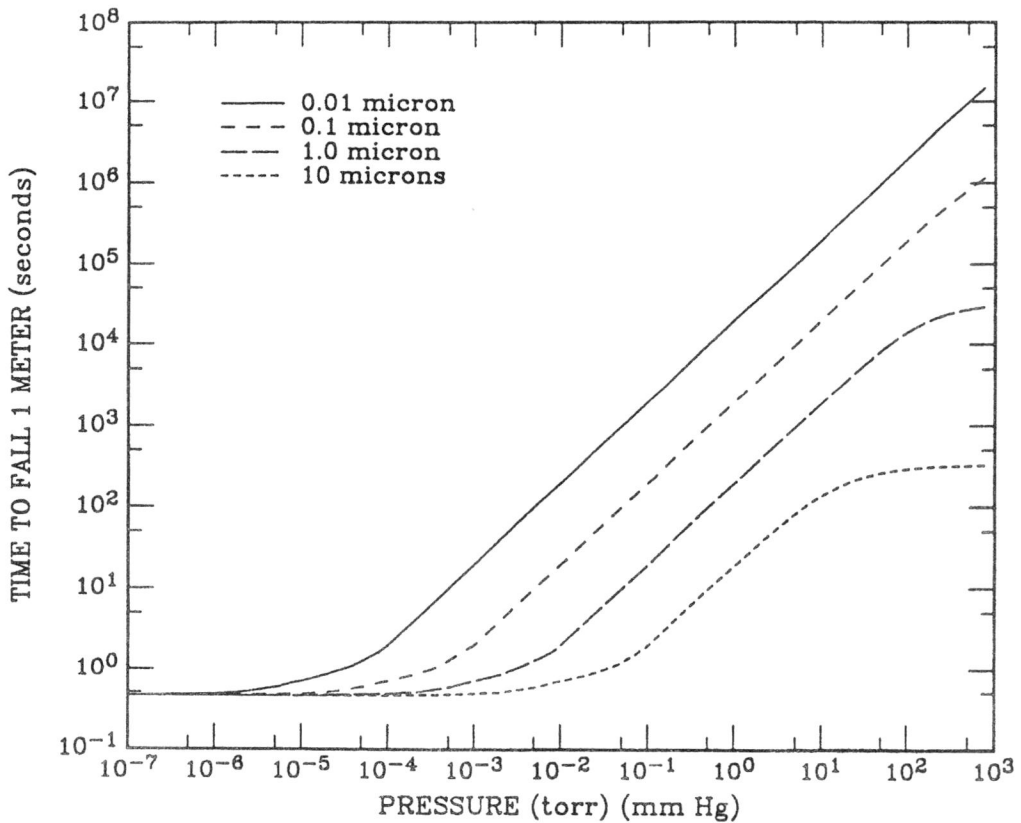

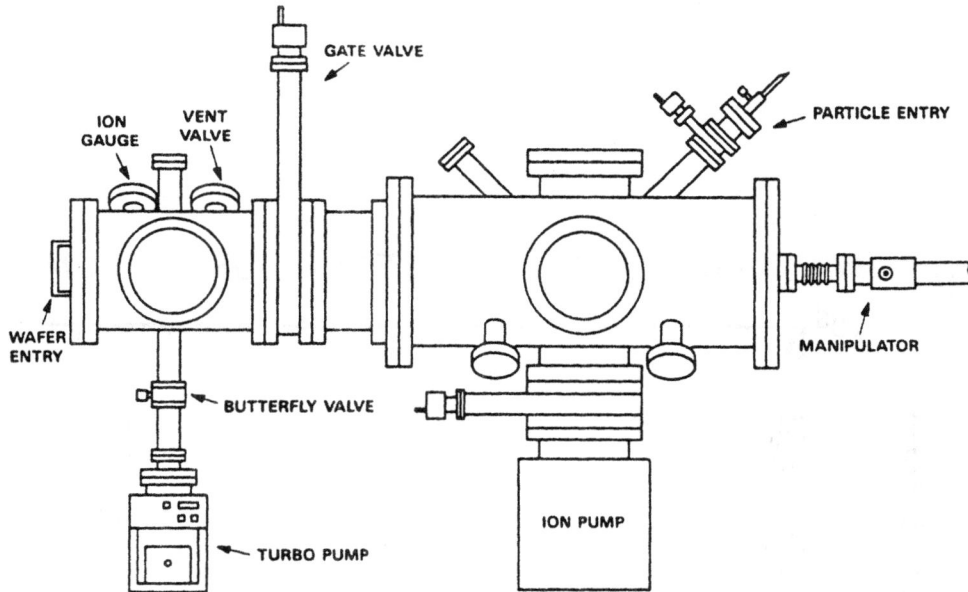

FIGURE 3   PARTICLE STUDY VACUUM CHAMBER

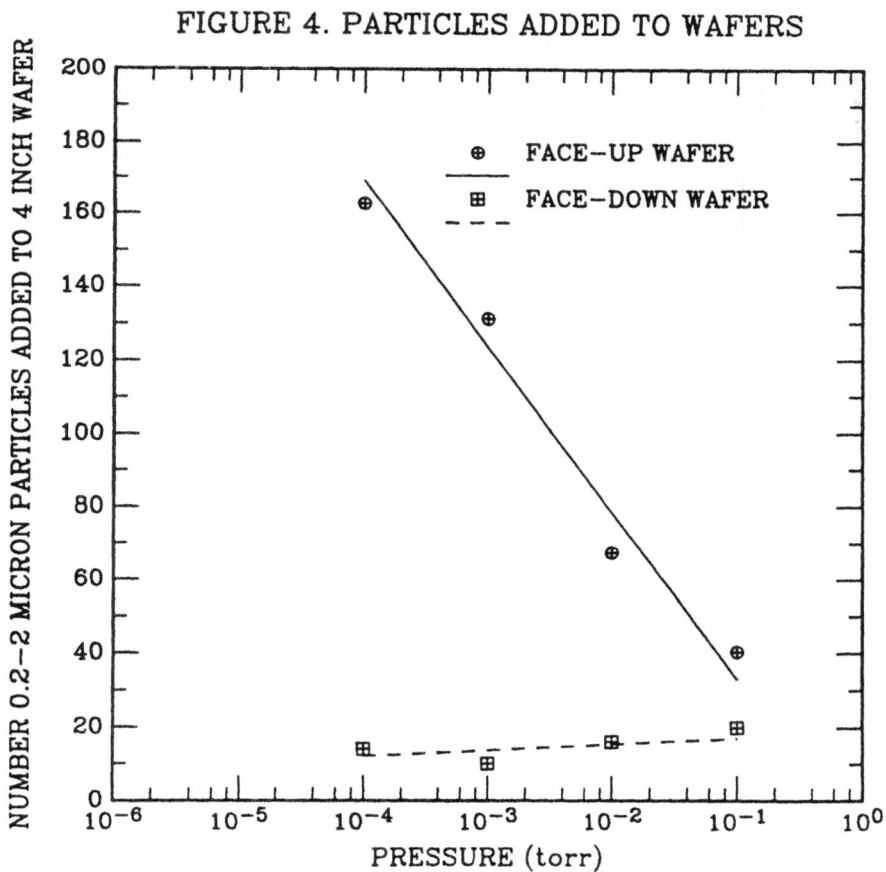

FIGURE 4. PARTICLES ADDED TO WAFERS

FIGURE 5. TURBULENCE DURING PUMPING

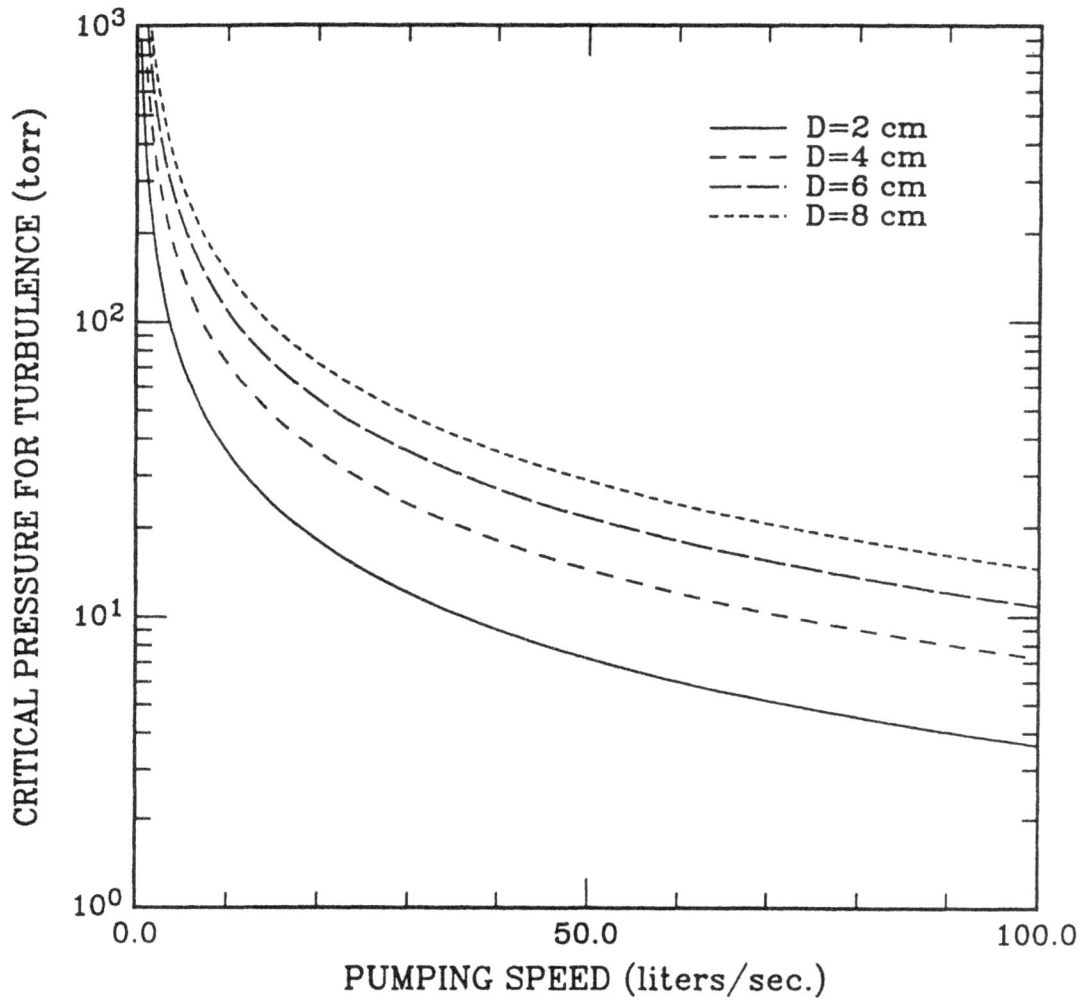

Mechanisms of contaminant particle production, migration, and adhesion

R. J. Miller, D. W. Cooper, H. S. Nagaraj, B. L. Owens, M. H. Peters,[a] H. L. Wolfe, and J. J. Wu

IBM T. J. Watson Research Center, Yorktown Heights, New York 10598

(Received 2 September 1987; accepted 21 December 1987)

Three fundamental processes associated with particulate contamination of manufacturing processes are particle generation, migration, and adhesion. A more complete understanding of each of these processes, and their interactions, can aid in efforts to eliminate product contamination and thereby increase product yields. We present results of particle generation studied in the context of mechanical wear. The production of airborne particles from the wear of stainless-steel components is examined. The results indicate that submicron particles dominate the particle population in terms of number of particles generated. The migration and deposition of submicron particles under clean room conditions is described, with reference to a deposition model that includes particle diffusion effects. Experimental results of particle deposition in a clean room are compared to the model. Measurements of the adhesion under vacuum of dry-deposited particles of known size and composition are also presented. These show qualitative agreement with a model having a linear dependence of adhesion on particle diameter. The results are discussed in the context of increased automation in product handling and more stringent product isolation.

I. INTRODUCTION

We have found it of value to consider product contamination as a sequence of identifiable events. We will consider the contaminant to be a particle, although similar considerations could be applied to ionic or molecular (e.g., organic film) contaminants. The contaminant is generated, usually not directly on the product surface; the contaminant must migrate to the product surface; it must adhere (or perhaps react with the surface); there may be an opportunity for detection and removal; if not removed, it can cause a defect. Effective contamination control measures attack these events singly or in parallel.

The most advanced technologies currently available for contamination control are those of air filtration, airflow control, and advanced equipment for contaminant detection. However, we anticipate rapid product evolution from, for instance, current 1-Mbit dynamic random access memory (DRAM) production to 16- and 64-Mbit DRAM production within several years. Since these advances in product integration will require order-of-magnitude improvements in contamination control for products to be built in facilities currently existing or under construction, the clean room environment and particle detection technologies alone cannot be relied upon for the contamination control improvement necessary. In current semiconductor fabrication technology, contamination from the process equipment and from the processes themselves are prime contributors to product defects, and that is likely to continue as processes for more highly integrated circuits and devices of smaller feature size are brought into manufacturing. Therefore, emphasis must be placed on developing the knowledge base necessary to eliminate contaminant production, improve product isolation from a variably contaminated environment, and achieve more successful product cleaning.

As a means of better understanding contamination of sili-

con integrated circuit products, we consider some of the mechanisms by which particle contaminants are produced, migrate, and adhere to product surfaces. We consider one mode of contaminant particle production specifically, that of particle production by wear. Particle production by wear is relevant to many pieces of semiconductor processing equipment in which it is necessary that moving parts rub against one another. Although the rate of wear is low in most situations occurring in clean rooms or within processing equipment, it becomes evident that a small amount of material lost to wear can become a very great number of micron-sized or smaller particles.

It has also been shown[1] that as submicron particles have increased capability to cause product defects (by virtue of greater product sensitivity with increasing product complexity), greater attention must be paid to diffusion as a mechanism of submicron particle migration. Guided by the previous analysis,[1] we have performed a more detailed treatment of submicron particle motion, which takes into account electrostatic effects as well as diffusion and which makes use of analytical solutions and Monte Carlo simulations of particle motion to predict particle migration toward and deposition onto surfaces.

Third, we examine particle adhesion to surfaces, using electrostatics as a means of particle removal and using the strength of the electrostatic force required for removal as a quantitative measure of the particle's adhesion strength.

Aspects of each of these investigations either have been published or will be published in the near future. References will be made to those papers as they generally will provide a greater level of detail than can be provided here. The additional value of this presentation is that it affords the opportunity to emphasize the interplay between the generation, migration, and deposition in the specific context of contamination control.

2097   J. Vac. Sci. Technol. A 6 (3), May/Jun 1988   0734-2101/88/032097-06$01.00   © 1988 American Vacuum Society   2097

63

II. WEAR AS A CONTAMINANT-PRODUCING MECHANISM

There are numerous instances in which wear can result in contaminant production in a clean room environment, ranging from the sliding of storage boxes or other containers across work surfaces to the wear of mechanical parts of wafer processing equipment. The wear situations are generally light load and low speed, from which wear out in the usual sense is not likely to be a concern. Nevertheless, particle production does occur. In some instances, special constraints to eliminate other forms of contamination increase the occurrence of contamination due to wear. For example, particle generation due to wear increases in some instances where lubricants are eliminated due to concern for organic contamination, vacuum outgassing, or due to high process temperatures.

To investigate the generation of particles by wear, we modified a commercial wear tester (Falex Instruments) to accept a clean wear chamber to perform three-pin-on-disk dry wear testing.[3] A simplified schematic of the wear chamber is shown in Fig. 1. A lower unit contains three wear pins. The lower unit does not rotate, but is capable of moving vertically to apply load. The wear disk is installed in the cap of the assembly, which is rotated during wear testing and is capable of tilting slightly to maintain continuous contact with each of the three pins. We supplied filtered compressed air of controlled humidity to the lower unit, from which it is directed as jets against the three points of pin–disk contact. We used a cartridge-type "absolute" gas filter having a 0.05-μm retention rating to filter the compressed air. Airflow of

~52 1/min provided air velocities of ~60 m/s exiting the slits to entrain the wear particles and transport them down the center of the wear chamber, then laterally to an exhaust sampled by an aerosol particle counter (Particle Measuring Systems, LAS-250X) capable of detecting and sizing particles of 0.2 μm or larger diameter. Since the sampled airflow of the particle counter was only 2.83 1/min, the volume of aerosol sampled was ~5% of the total flow. The mean flow velocity at the probe inlet was ~40% of that at the wear fixture outlet, indicating a diverging flow field at the probe inlet. Under this condition, inertial effects at the inlet would cause some sampling-induced enhancement of the large particle counts in our results.

Inspection of the wear surfaces led us to believe that the air jets were effective in removing wear particles as they were produced, so that they would not be ground finer on subsequent revolutions of the disk. We felt this to be important so that the data would represent the particle size distribution of wear particles as initially generated. We used a new disk and pins, cleaned with acetone, for each test. We also measured the weight loss of the wear specimens. We directed filtered clean air around the exterior of the wear chamber to avoid the influx of aerosol environmental particles that would otherwise add particles to the measured aerosol.

Figure 2 indicates the "optical equivalent" size distribution of particles aerosolized and collected during a wear test of a type-304 stainless-steel disk and type-440C stainless-steel pins. We chose these materials because they are commonly used in tooling and bearing applications. The test was performed at a load of 6.2 N/pin and speed of 24 cm/s. Figure 2 provides a cumulative plot of particles larger than a given size, on a logarithmic scale. This method of plotting facilitates comparison of particle size distribution with the

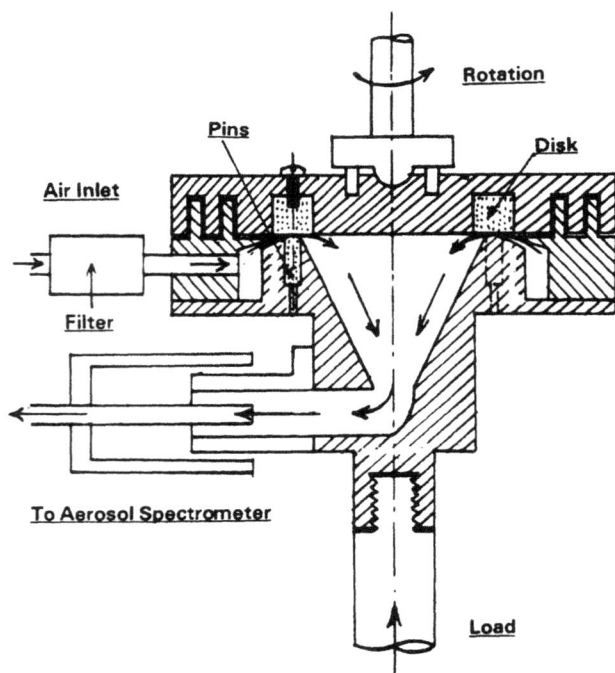

FIG. 1. A simplified schematic of the wear test chamber for sampling wear-generated particles. Filtered airflow directed at the points of pin–disk contact transport the aerosolized wear particles to the optical particle counter inlet.

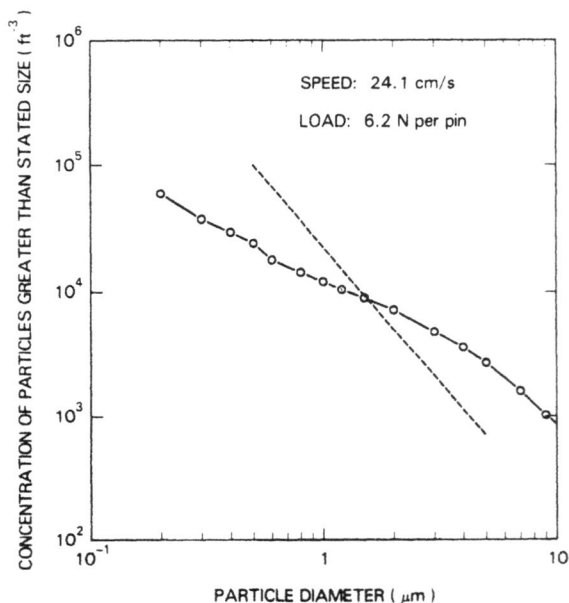

FIG. 2. The "optical equivalent" size distribution of stainless-steel wear particles generated within the wear test chamber. The slope of the dashed line represents the particle size distribution associated with clean room aerosols, according to Federal Standard 209B.

FS-209B federal standard size distribution (dashed line) for clean room aerosols.[2] Particles so large as to not remain airborne long enough to be collected have been neglected. While the optical equivalent size, as provided by the optical particle counter, assumes spherical particles, the wear particles are most likely nonspherical.

The measured weight loss of the specimens indicated that most of the wear occurred to the disk rather than the pins, due to the relative hardness of the pin material. Because of the loss of some larger particles, we were unable to achieve agreement between weight loss of the specimens due to wear, and the number of particles detected by the optical particle counter. Impaction loss in the wear fixture, inertial enhancement at the probe inlet, and gravitational loss in the probe all detract from the accuracy of our data for large particle sizes. Subsequent work is being carried out to assess the size distribution of larger particles by collecting them on membrane filters within the wear chamber and determining their size by scanning electron microscopy.

The particle size distribution observed in these wear tests is somewhat less sloped, implying a greater percentage of large particles, than would be expected from a "normal" clean room aerosol, as typified by the FS-209B curve; however, it is clear from the results that this dry wear process is producing submicron particles, and that a preponderance (by number) of the particle generation is actually in the submicron range. These results are similar to results we reported,[4] in which measurements were made of particles generated by a chain drive which had been removed from a sputtering system and tested in a clean air environment. The results are in contrast to what appears to be a common belief that wear particles are usually $> 1\,\mu m$ in size.

Metal particles have very different refractive index and shape than the polystyrene latex spheres to which optical particle counter is size calibrated, so we considered the possibility that the apparent size of the wear particles differed markedly from their true size. Our calculations and unpublished simulations of light scattering performed by Knollenberg[5] indicate that for particles smaller than $\sim 0.5\,\mu m$, some "oversizing" will occur; i.e., the particle will actually be somewhat smaller than its apparent size, while particles larger than $0.5\,\mu m$ will be sized nearly correctly. This leads to the conclusion that particles whose apparent size was submicron were, in fact, submicron particles.

III. SUBMICRON PARTICLE MIGRATION

With the above understanding of a representative mode of particle generation, we turn our attention to the migration of the particles, focusing specifically on the submicron particles that represent the majority of the particle number concentration. Submicron particles are little influenced by gravity, but are subject to diffusion through the mechanism of Brownian motion, and to electrostatic drift, if charged and within an electrostatic field. Particle diffusion increases as particle size decreases. For particles smaller than $\sim 0.3\,\mu m$, the particle's deposition velocity (the ratio of flux onto a surface versus aerosol number concentration) increases as particle size decreases, in contrast to the particle's gravitational settling velocity. Hinds[6] and Liu and Ahn[7] provide

figures of particle deposition velocity due to diffusion versus particle size for uncharged particles. The increase in deposition velocity for appropriately charged particles in an electrostatic field has also been examined by Liu and Ahn. Cooper[1] has discussed how the combination of increased aerosol concentration with decreasing size and a larger deposition velocity results in the potential for a much larger dependence of particle flux on particle size than is observed for larger particles (larger than $\sim 1\,\mu m$) where gravitational deposition dominates and diffusion is a negligible effect. He has noted that, as the lithographic feature sizes of integrated circuits are decreased for higher circuit density and circuit performance, the circuitry will become more sensitive to defect production by particles of submicron size, and the strong dependence of flux on particle size could become a significant obstacle to maintaining acceptable chip yields while attempting further reduction of feature sizes. There remains some uncertainty over the dependence of cumulative aerosol concentration on particle size for particles smaller than 0.5-μm diameter in clean rooms. Some researchers[8] have obtained results in clean rooms at rest indicating that, for particles smaller than $\sim 0.1\,\mu m$ in diameter, the cumulative particle concentration may increase only weakly as particle size is decreased. Nevertheless, there appears to be general agreement that, at least for particles larger than $0.1\,\mu m$ in diameter, the aerosol concentration is a strong function of particle size. Figure 3 indicates recent measurements[9] of the aerosol particle size distribution in a vertical laminar flow clean room where semiconductor manufacturing is performed. The measurements were made by continuous monitoring over three 24-h periods, with normal manufacturing operations occurring over three shifts in the clean room. The aerosol particle counter used (Particle Measuring Systems, LAS-X) was capable of detecting and

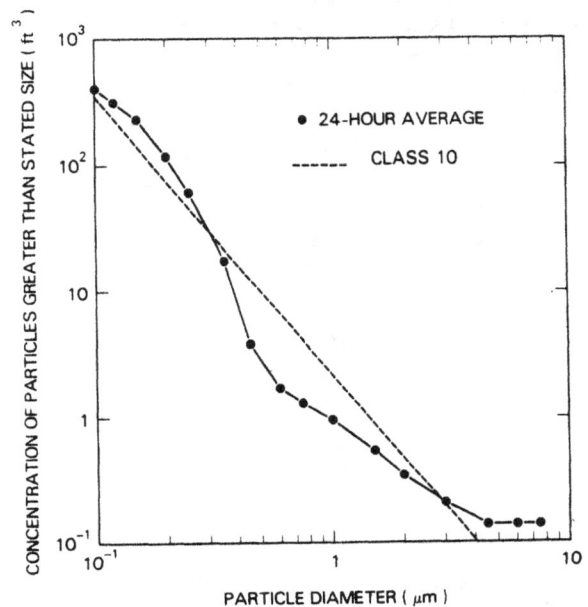

FIG. 3. Aeorsol particle size distribution in a vertical laminar flow semiconductor manufacturing clean room. The particle size distribution essentially conforms to the Federal Standard 209 Class 10 criterion.

sizing particles of 0.1 μm or larger optical equivalent diameter at a 0.005 1/s airflow. The electronic noise level of the instrument is more than an order of magnitude beneath the data. The room aerosol concentration essentially conforms to Class 10 specifications according to Federal Standard 209B. The strong dependence of the particle size distribution is apparent in these data for particles > 0.1 μm in size.

To understand the deposition of submicron particles onto semiconductor wafers, it becomes of interest to include Brownian motion effects in a model of the motion of particles in an airflow directed toward a surface. Guided by the work of Chari and Rajagopalan,[10] we have recently performed such modeling.[11] The flow modelled is diagrammed in Fig. 4. A downward flow at a speed of 50 cm/s, which is typical of airflows in clean rooms, impinges upon a horizontal surface and obtains a viscous axisymmetric stagnation point flow profile. A semiconductor wafer resting on the surface is represented by a circular disk of 20-cm diameter.

For a given aerosol particle size distribution, the model allows the deposition onto the disk to be calculated, including particle diffusion in the airstream. The effect of particle and surface charge are also included in the model. Deposition as a function of radius from the wafer center to its edge can also be determined. Figure 5 demonstrates the deposition velocity as a function of particle diameter for uncharged particles, as derived by this model. The results agree qualitatively quite well with the work of Liu and Ahn.[7] It is evident from the figure that deposition increases rapidly with decreasing particle size, for particles a few tenths of microns in diameter and smaller.

Figure 6 compares the particle deposition that would be expected based on the model and the aerosol concentration of Fig. 3 to actual aerosol particle deposition onto bare silicon wafers exposed as "deposition monitors" in the same clean room where the aerosol data of Fig. 3 were obtained. The number and sizes of the particles deposited on the wafers were determined using a wafer inspection instrument (Inspex 2020). The number and sizes of the particles on these wafers were low enough that coincidence errors, as described by Cooper and Miller[12] would have been unlikely. Only qualitative agreement between the model and the experiment appears to exist, with the actual particle deposition rate exceeding by an order of magnitude that predicted by the model. We believe that electrostatic effects, due to charge

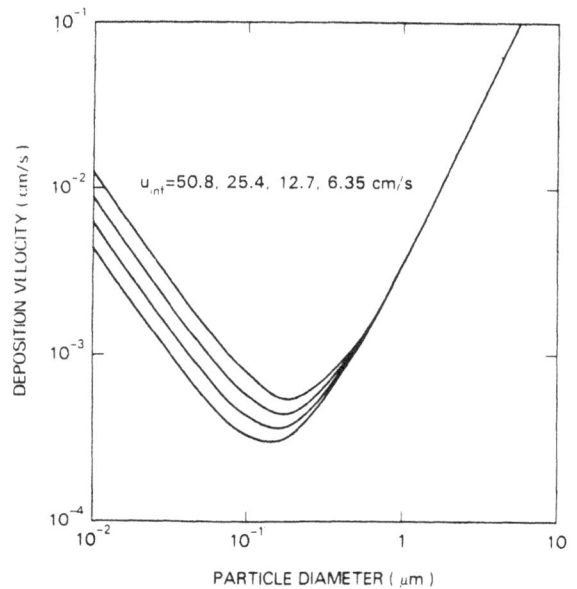

FIG. 5. Deposition velocity as a function of particle size for uncharged particles, as derived from the deposition model incorporating gravitational and diffusional effects.

on the wafers or their holders, may have contributed to the increased deposition. Also, the relatively infrequent detection in the aerosol of a particle of 5-μm or larger diameter (therefore having a large gravitational deposition velocity) had a large influence on the deposition predicted by the model. In the model results, the strong dependence of deposition on particle size for particles smaller than 0.5 μm in diameter reflects the contribution of particle diffusion to the deposition velocity. Some hint of a corresponding increase in deposition is seen in the actual deposition results, but the de-

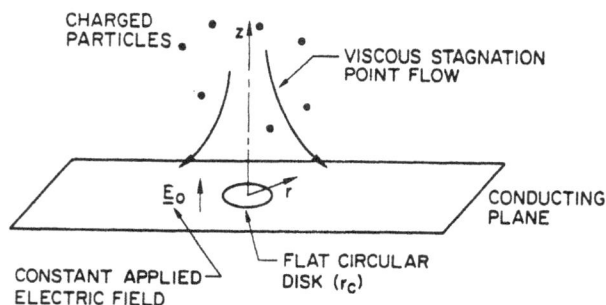

FIG. 4. A schematic of the viscous axisymmetric stagnation point flow profile used to model the effects of Brownian motion and electrostatic forces on particle deposition onto wafer surfaces.

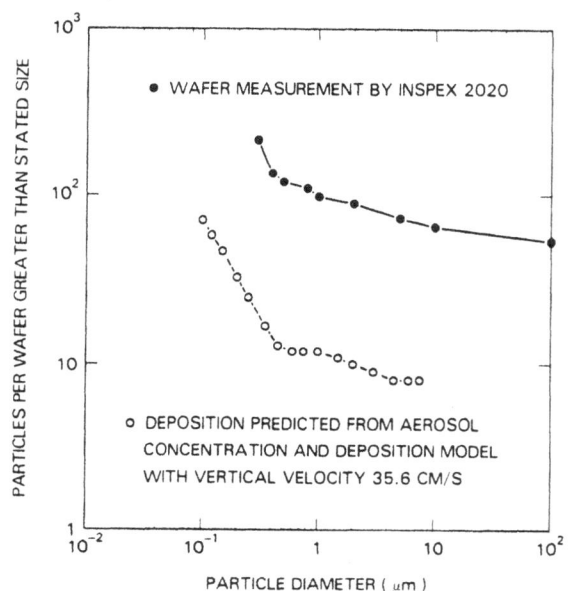

FIG. 6. Particle deposition that would be expected based on the deposition model and the aerosol concentration of Fig. 3 (O), compared to actual aerosol particle deposition onto bare silicon wafers (●).

gree of quantitative disagreement and the inability of the surface inspection system to detect particles smaller than ~0.3 μm in diameter precludes confirming this effect.

IV. PARTICLE ADHESION TO SURFACES

Once a particle has become deposited onto a wafer (or other surface of interest), the likelihood of its removal is governed by the strength of its adhesion to the surface and the amount of dislodgement force that can be applied to it. Adhesion force is generally believed to decrease as particle size decreases, and yet smaller particles are found to be harder to remove than large particles because the amount of dislodgement force that can be applied to them by conventional means decreases in greater proportion with decreasing particle size. Cooper et al.[13] have explored particle adhesion as a function of particle size for dry-deposited particles on metal surfaces, using electrostatic forces to dislodge the particles from the surface. Particles of nickel, silica, and polystyrene latex were used. The surface used was metal to permit the particle to acquire charge through contact to the surface. With a grounded counterelectrode parallel to and minimally spaced from the conductive surface containing the particles, a voltage was applied to the metal surface, to effect particle dislodgment and transfer to the grounded counterelectrode. The surface of the counterelectrode that the particles would contact was made insulating so that the particles would not readily charge oppositely there and return to the specimen surface. After voltage application, the area of the specimen on which the particles had initially been deposited was microscopically examined and photographed to determine removal efficiency. The voltage application was performed within a chamber evacuated to 10^{-6} Torr to permit the use of field strengths which otherwise would have caused the breakdown of air. Checks were performed to ensure that the evacuation and venting of the chamber did not dislodge particles from the specimen surface. Figure 7 shows the particle adhesion strength, determined from the electrostatic force required for particle removal, as a function of particle size. Significant scatter in the data exists, indicating that particles of nominally the same size (and acquired charge) and of the same material have variability in their strength of adhesion; however, the results are not inconsistent with a linear dependence of adhesion on particle diameter for particles larger than 1 μm. Such a linear dependence has been suggested by Hinds.[6]

V. DISCUSSION

By studying particle generation, migration, and adhesion, we are attempting to build an understanding of the processes by which products such as silicon integrated circuits become contaminated by particles. Our results are particularly appropriate to the design of automated product handling equipment and product isolation technology.

Automated product handling has the potential to lessen handling damage to products and the contamination that the product receives from human operators, but it substitutes the potential that contamination from the handling equip-

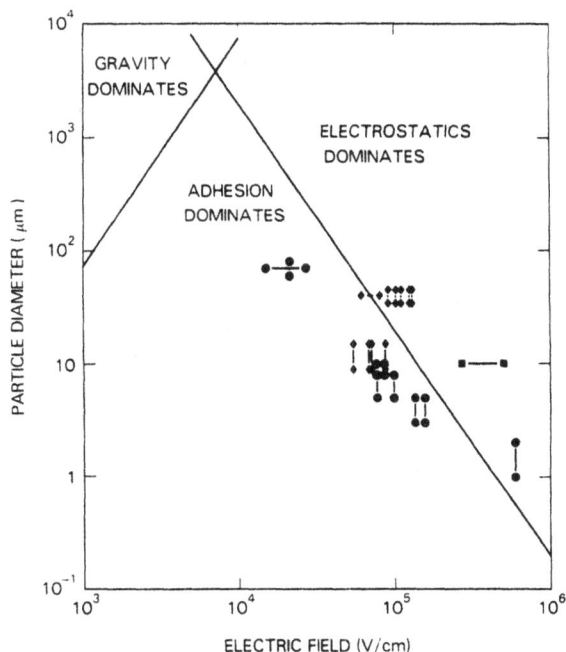

FIG. 7. Particle adhesion strength, determined from the electrostatic force required for particle removal, as a function of particle size, for dry-deposited particles.

ment will be deposited on the wafers. Currently, during the design of wafer transport and cassette transport systems, materials and operating conditions are chosen with an attempt to minimize particle generation, but comprehensive guidelines for materials suitability from a particle generation standpoint are lacking. Engineering studies[14-18] of the particle generation from equipment are then undertaken and modifications made, often to shroud particle generation sources, or to use exhausts or other airflow controls to divert particles away from the product. Our results indicate that a broad distribution of particle sizes is produced by wear processes, including submicron particles. Prevention of the migration and adhesion of these particles to product surfaces must be considered. The development of materials usage guidelines to minimize wear particle generation would be very beneficial in the design of wafer transport equipment and in the design of various components of wafer processing equipment.

Technologies for product isolation offer the opportunity to create a highly controlled, reproducible clean environment for products at potentially lower cost than large comparably controlled clean room facilities. Product isolation also shields the product from random disturbances to the clean room environment. This is particularly important when it is understood that complex integrated circuit process sequences may require that the product spend one to several months "in process." Our results relating to particle migration, deposition, and adhesion indicate that Brownian motion will be a significant factor in the deposition of submicron particles, particularly those 0.3 μm and smaller. Aerosol data indicate that the majority of aerosol particles are in the sub-0.3-μm size range and models of particle diffusion indicate that their deposition velocity onto surfaces is high.

While current products may be only marginally sensitive to such small particles, finer-featured products of higher density will almost certainly be highly sensitive to them. The development of product isolation technologies to provide better control of particles in this size range than is achievable through advances in conventional clean room technology could prove to be a significant advance in contamination and defect control.

[a] Permanent address: Department of Chemical Engineering, Florida State University, Tallahassee, FL 32316.

[1] D. W. Cooper, Aerosol Sci. Technol. **5**, 287 (1986).

[2] Federal Standard 209B, "Clean Room and Work Station Requirements, Controlled Environment," U. S. General Services Administration, Washington, DC, 1973.

[3] H. S. Nagaraj, B. L. Owens, and R. J. Miller (to be published).

[4] H. S. Nagaraj, B. L. Owens, and R. J. Miller (to be published).

[5] R. G. Knollenberg (private communication).

[6] W. C. Hinds, *Aerosol Technology* (Wiley, New York, 1982).

[7] B. Y. H. Liu and K. Ahn, Aerosol Sci. Technol. **6**, 215 (1987).

[8] B. R. Locke, R. P. Donovan, D. S. Ensor, and A. L. Caviness. J. Environ. Sci. **28**(6), 26 (1985).

[9] J. J. Wu, R. J. Miller. D. W. Cooper, and A. P. Szecsy, paper presented at the American Association for Aerosol Research Annual Meeting, Seattle, WA, 1987.

[10] K. Chari and R. Rajagopalan, J. Chem. Soc. Faraday Trans. 2 **81**, 1345 (1985).

[11] M. H. Peters, D. W. Cooper, and R. J. Miller, J. Aerosol Sci. (to be published).

[12] D. W. Cooper and R. J. Miller, J. Electrochem. Soc. **134**, 2871 (1987).

[13] D. W. Cooper, H. L. Wolfe, and R. J. Miller, in Proceedings of the Fine Particle Society, San Francisco, CA, 1986.

[14] L. G. Bailey and G. G. Rogers, Microcontamination **4**(9), 80 (1985).

[15] G. G. Rogers and L. G. Bailey, Microcontamination **5**(2), 43 (1987).

[16] G. G. Rogers and L. G. Bailey, in *Proceedings of the Institute of Environmental Science* (Institute of Environmental Science, Mt. Prospect, IL, 1987), p. 344.

[17] N. A. Brumberger and T. C. Lane, in Ref. 16, p. 348.

[18] T. Ro and K. Unno, in *Proceedings of the 8th International Symposium on Contamination Control* (Institute of Environmental Science, Mt. Prospect, IL, 1986), p. 843.

CONTROL OF PARTICULATE EMISSIONS FROM PLASMA-ETCHING SYSTEMS

Roger B. Lachenbruch and Octavio Gomez

Evaluating the sources of contamination on wafers processed in etching systems is a prerequisite to providing reliably clean processes. This article examines contamination levels caused by: ambient conditions before a wafer is processed; wafer-transport functions; and process by-products. The two automated etching systems evaluated, a model 903 tabletop SiO_2 etcher and a model 1512 floor-standing aluminum etcher, were manufactured by Tegal Corporation, Novato, CA. Both systems are cassette-to-cassette, single-wafer machines with similar pick-and-place wafer transports. However, the 1512 has vacuum load locks, which can generate particulate contamination on wafers.

TEST ENVIRONMENT AND MEASUREMENT EQUIPMENT

Both systems were tested in a 400-sq-ft cleanroom under Class 10 or Class 100 conditions. (Tests were performed in the Class 100 cleanroom for most of the SiO_2-etcher evaluations; additional Class 10 space was used for the aluminum-etcher tests.) During the tests,

full cleanroom attire was worn. The cleanliness of the HEPA-filtered air was determined by a Climet CI-208C airborne-particle counter (Climet Instruments Company, Redlands, CA), and clean monitor wafers were used to evaluate direct particulate-fallout rate. A Silicon Valley Group 8620 wafer scrubber (Silicon Valley Group, Inc., San Jose, CA) was used so that the monitor wafers could be recycled. Operator handling of wafers was virtually eliminated except for the transportation of cassettes between systems.

Wafer contamination levels were measured on a Tencor 200 surface inspection system (Tencor Instruments, Mountain View, CA) and an Inspex 1996 surface inspection system (Inspex, Waltham, MA). Early testing on the 903 etcher was performed with a manually loaded Tencor 200, which was augmented by an Inspex cassette-to-cassette system to eliminate manual handling and to obtain resolution of smaller particles. The Tencor system resolves calibrated latex spheres down to approximately 1.0 μm; the Inspex system resolves 0.3-μm particles. Therefore, the data reported here reflect different resolutions. A Dektak II profilometer

69

(Sloan Technology Corporation, Santa Clara, CA) and a Hitachi S-520 scanning electron microscope (Hitachi Ltd., Tokyo, Japan) were used to validate conformance with specified process results.

The 903 and 1512 systems were set up in the cleanroom, and temperature controllers for the process chamber and vacuum pumps were installed in a service chase outside the cleanroom area. All noncorrosive process gases were provided by house process piping and were filtered at the back of each system. Corrosive process gases were delivered from a house-exhausted cabinet and were similarly filtered.

EXPERIMENTAL PROCEDURES

The air velocity above the 903 and 1512 systems was adjusted to 70 ft/min using a hand-held velocity probe to match the cleanroom airflow. If measurements were found to be out of specification, the flow was balanced or filters were replaced. Airborne-particulate counts were taken and fallout onto monitor wafers was measured to determine the particle count around the systems. The initial count before measuring was less than 30 particles greater than or equal to 0.3 μm in size. The particulate-fallout method was more direct and less variable.

Particle contamination related to process by-products was evaluated on the 903 SiO_2 etcher. Wafers were run through the etch cycle until the particle count reached the predetermined level of contamination at which cleaning was deemed necessary. The power was shut off to prevent any film deposition or possible wafer-surface damage that could interfere with the measurement. A monitor wafer from each processed cassette was measured using the Tencor 200 (1-μm resolution) before and after one etch cycle.

For the 1512 aluminum etcher, groups of 25 aluminum discs (machined 6061 alloy, 125-mm diameter) were processed using timed etch steps. Recyclable dummy wafers were used to reduce costs. Process results for patterned silicon wafers were evaluated after every 100 cycles to ensure that etch rates and nonuniformity were still within specification. The same wafer-measurement technique used for the 903 was repeated for the 1512.

The reactors on both systems were cleaned when the particles that were added to a 125-mm monitor wafer exceeded an arbitrary threshold of 100 particles greater than or equal to 0.3 μm (0.9 particles/cm²).

The baseline contamination level of wafer-transport subsystems was evaluated through the use of 125-mm silicon monitor wafers. These wafers were measured on the surface inspection system before and after exposure through the etcher. All measurements were made with the same wafer orientation to limit the experimental variables. The mean number of particulates added was reported with the standard deviation.

Test Run Number	Duration (hours)	Particles Accumulated	
		Covered Cassette	Uncovered Cassette
1	16.25	9	34.7
2	16.50	12.0	17.0
Average particles per hour	—	0.65 ± 0.01	1.57 ± 0.76

Table I: Comparison of covered and uncovered cassettes in a 903 plasma etcher (particles added ≥1 μm).

Transport Section	Particles Added (>1 μm)	Standard Deviation
Complete O-belt transport (3 wafers) Standard contact with chuck	265.7	43.6
Complete spatula transport (3 wafers)	39.6	44.6
Spatula to stage (3 wafers) Shuttle and chuck	2.6	1.3
Shuttle and chuck (5 wafers) Edge contact with chuck	31.1	28.3
Complete spatula transport (8 wafers)	5.3	5.0

Table II: Comparison of 903 O-belt and spatula transports (wafers face down).

Pump Type/ Condition	Average Added Particles per Cycle (≥0.3 μm)	Standard Deviation
D30AC (27 cu ft/min)		
Run 1	5.20	10.96
Run 4	5.70	13.45
D16AC (14 cu ft/min)		
Run 2	0.42	0.45
Run 3	1.11	1.35
D30AC orifice restriction		
2 mm	0.99	1.13
0.6 mm	0.64	0.52

Table III: The effect of load-lock pump and orifice size on particulate contamination in a 1512 etcher.

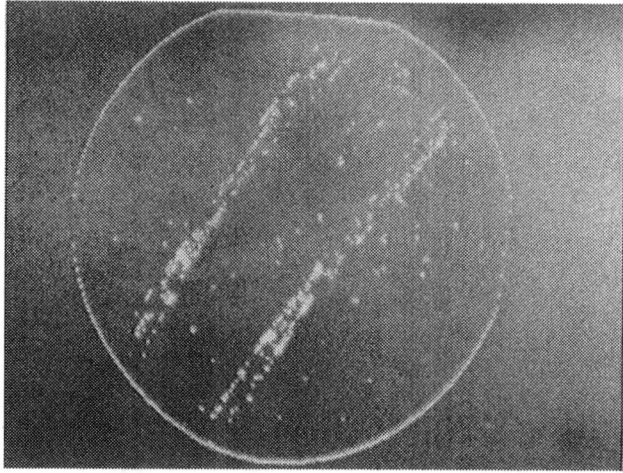

Figure 1: Monitor display of wafer run face down on O-belt transports.

Closed-canopy cassette covers were evaluated to measure the effect of ambient cleanroom air on particulate contamination. Traffic in the cleanroom was eliminated for the testing, which was performed in the original Class 100 section of the cleanroom. It was found that wafers in covered cassettes in static air were less likely to be contaminated than were wafers in uncovered cassettes in dynamic, uninterrupted laminar flow (Table I). This result is important because wafers spend more time awaiting processing than in the actual etching. From these data, it was determined that cassette covers would be used for the 1512 and 903 evaluations.

WAFER-TRANSPORT CONTAMINATION

The baseline contamination level of the transports for both systems was evaluated by turning off process gases and RF power and then cycling test wafers. Individual transport components were also tested in this manner to identify the source, spatial distribution, and level of contamination.

The transport of the 903 oxide etcher consists of a send-and-receive module, which incorporates a staging area, and a central shuttle module. The shuttle interfaces with the lower electrode of the etch reactor. To determine the optimal mechanism for the face-down transport required in some processes, the experimenters evaluated two modules. Because the send-and-receive module was originally manufactured with an O-belt design, the O-belt was compared with the newer pick-and-place, urethane-coated spatula. During the upside-down transportation of wafers, the spatula module was found to be significantly cleaner than the O-belt transport (Table II).[1] This result correlates with the O-belt requirement that the wafer be driven against a physical stop, causing "overdriving" that abrades the O-belts and wafers (Figure 1). Table II also shows the particulate contribution of the other wafer-transport modules.

The 1512 transport is partitioned into two pick-and-place, send-and-receive modules and a shuttle. Vacuum load locks isolate the ambient contaminants that compete with or inhibit the aluminum-etching pro-

cesses. The load lock can stage up to three wafers to allow time for soft vacuum (the gentle evacuation of pressure without turbulence) and atmospheric bleed-back (the movement of pressure up from vacuum to atmospheric without turbulence). From the standpoint of transport particulate contamination, the main benefit of these load locks is not particle reduction but the isolation of corrosive chemistries, such as those used in aluminum etching.

During evacuation, some particulate contaminants on the wafers or on the walls of the load lock may change position. This movement is caused by turbulence during sudden evacuation, which in turn results from impingement by gas vortices. A gentler evacuation maintains laminar airflow and provides an uninterrupted path of escape for airborne particulates. Although this situation would probably be more pronounced for particles larger than 1 μm, these results indicate a uniform distribution of particulate sizes down to the 0.3-μm resolution of the wafer-particle counter.

The highest potential for turbulence occurs during the venting process and when the valve to the vacuum pump is opened for pumpdown. The rate of venting can be controlled with a needle valve; however, the rate of pumpdown is a function of absolute pressure in the load lock. The pumping system and its foreline are most efficient when the vacuum valve is opened. As the pressure drops, the throughput of the foreline and valving decreases and the performance of the mechanical vacuum pump begins to drop off.

This effect can be explained by evaluating the ratio of the inertial forces to the viscous forces in the flowing gas. This ratio, known as the Reynolds number, depends on the size of the load lock and the mass-flow rate of the gas entering or leaving it.[2] For a Reynolds number of 1000 or less, the flow will generally be laminar. The pump size determines this ratio: for a 27-cu-ft/min mechanical vacuum pump, the initial condition in the load lock upon opening the vacuum valve would be predicted as turbulent (Reynolds number = 1270). In contrast, for a 14-cu-ft/min mechanical pump, the initial condition would be predicted as laminar (Reynolds number = 660). The use of two sizes of both load-lock-evacuation pumps and flow-restricting orifices had a direct effect on contamination of 125-mm monitor wafers. This relationship is charted in Table III.

PROCESSING-RELATED CONTAMINATION

Contamination caused by the accumulation of processing by-products for SiO$_2$ and aluminum etching was investigated because such applications represent worst cases. The onset of particulate contamination follows one of two types of profiles, typified by the oxide and aluminum processes. In SiO$_2$ etching, particulate fallout onto the wafer is usually signaled by a sharp rise in particle counts over the course of a few wafers. In aluminum etching, the number of particles generally rises gradually until a threshold is exceeded. The reactor is then cleaned.

Oxide Etching in the 903 System. SiO$_2$ etching requires significant ion bombardment and the formation of a tough carbonaceous polymer to promote selectivity to silicon.[3,4] The ion bombardment removes the polymer coating and etches the unmasked surface. Unfortunately, polymer is deposited on all surfaces in

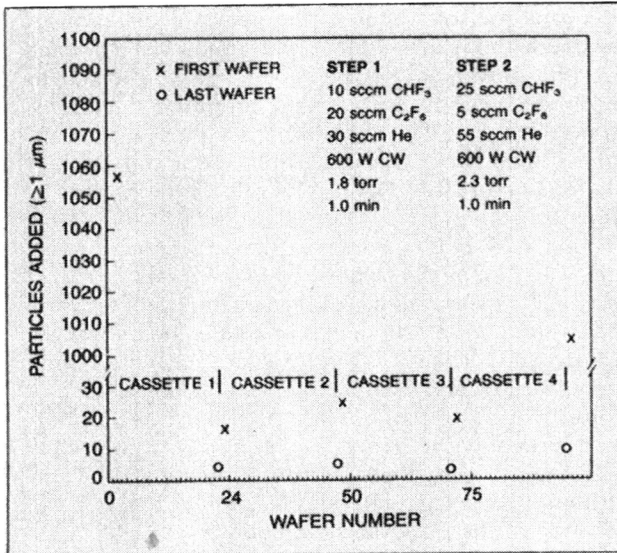

Figure 2: Comparison of first and last wafers in cassettes using a two-step 903 oxide process (125-mm wafers). Clean cycles were not run between cassettes.

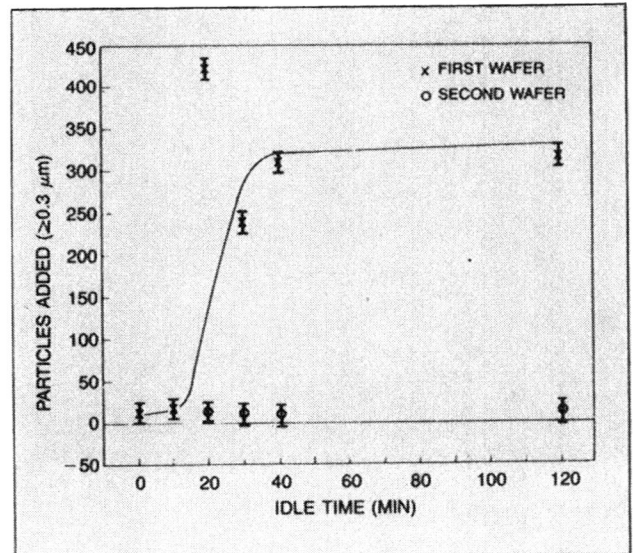

Figure 3: Particles added during pumpdown and purge in a 903 etcher as a function of idle time between measurements with the chamber at atmosphere.

the reactor and eventually cracks and falls off onto the wafer. This polymer buildup can be eliminated through periodic cleaning and reconditioning. The latter step redeposits a thin coating onto the exposed reactor surfaces so that the process remains stable and particle counts are acceptable.

Measurement of wafers and data recording for each cassette in the particle counter caused a delay of approximately 45 minutes between tests. During that interval, the 903 etcher was idle. After this dwell period, the particulate count on the surface of the first wafer exposed to the processing chamber was significantly higher than the count for the last wafer processed from the previous cassette (Figure 2).

To control contamination during dwell periods, the system was put in a standby mode that left the chamber evacuated. Figure 3 shows the contamination of the first processed wafer after varying amounts of dwell time while the chuck was left open. The second wafer was not significantly contaminated by this ambient exposure. After the first wafer was processed, the system returned to its baseline of approximately 5 particles (0.3 μm or larger) added to 125-mm wafers. The use of a vacuum dwell eliminates these contamination sources.

Contamination caused by reaction with room air breaks down the polymer. This process is probably a result of exposure to ambient moisture. For the study shown in Figure 3, the chamber was first cleaned through a plasma process incorporating oxygen. However, residual material, which contained the elements of the polymer, produced particulates at the gas introduc- tion holes. These particulates then collected on the monitor wafer in a pattern reflecting that of the holes. In other cases, where the polymer had deposited on the walls of the reactor, overnight exposure caused the bulk of the polymer to delaminate onto the first wafer processed.

In another experiment, the objective was to predict how many 25-wafer cassettes containing 125-mm sili- con monitor wafers can be processed in the oxide etcher before cleaning is required. Cassettes were run in

succession until a significant particulate-count in- crease was observed on a selected wafer from each cassette. For the oxide process, a sudden catastrophic rise in contaminants determined the number of bare silicon wafers run before cleaning was required. Two similar processes that differ in time of etch and RF power applied were compared for polymer formation.

The first run was performed using 10 std cu ft/min of CHF_3, 20 std cu ft/min of C_2F_6, and 30 std cu ft/min of He with 500 W of 13.56-MHz RF power for 30 seconds. After 18–20 cassettes (450–500 wafers) were pro- cessed, the number of particulates on the wafers in- creased sharply from the initial count of 5 or fewer particles per wafer. The second process used the same gas composition with 600 W of RF power for 2 minutes. In this case, the particle count rose sharply after four cassettes (100 wafers).[5] The total products of processing power and time for these two processes (7.13 million W-sec versus 7.20 million W-sec) agree within 2%. Thus, the maximum number of wafers run in an oxide- etch process before cleaning can be predicted based on a constant gas chemistry.

Aluminum Etching in the 1512 System. Particulate contamination in aluminum etching is a different phe- nomenon. In this situation, chlorinated process gases are used to produce a low-volatility by-product, $AlCl_3$. If an oxide-free surface is exposed, etching can proceed spontaneously in the presence of chlorine without the use of plasma. The evolution rate of $AlCl_3$ is primarily limited not by ion bombardment but by the availability of chemical species at the surface.[6] Other process-gas additives, such as $SiCl_4$, provide sidewall passivation and selective etching.

Because the vapor pressure of this by-product is fairly low (1 torr at 100°C),[7] it is not surprising that condensa- tion can occur at 0.1 torr pressure and 30°C lower electrode temperature.[8] However, the $AlCl_3$ condensate can flake off and contaminate the wafer. This contam- ination buildup over several cassettes of wafers occurs gradually rather than abruptly, as is the case in SiO_2

etching. However, in aluminum etching, the number of wafers that can be run before chamber cleaning is lower and is exponentially affected by reactor temperature (Figure 4). There was an increase of more than one order of magnitude in the number of aluminum discs that could be run when the reactor temperature was increased from 30° to 67°C.

Because of the exponential dependence on temperature, the vapor pressure of the condensible processing by-product was suspected to be the main indicator of the onset of contamination. The slope of the line was significantly lower than that expected from the vapor pressure of aluminum chloride. This finding indicates the possibility that unsaturated or higher-molecular-weight "polymers" may be present.

$AlCl_3$, which is produced as a by-product, is highly reactive to ambient moisture and forms aluminum oxides and corrosive hydrochloric acid as a result.[9,10] Thus, highly reactive process gases and by-products require vacuum load locks to keep ambient moisture from entering the system.[11] Load locks isolate the operators from process gases and prevent the formation of friable oxides, which can crumble and produce particulates.

CONCLUSION

Particulate contamination on wafers during etching can result from: (1) the ambient air in the cleanroom, (2) the etcher's wafer transport, and (3) by-products of the etch process. Covering cassette elevators can help protect wafers from being contaminated before processing. Particulates generated by wafer transports can be decreased by using urethane-coated spatulas instead of O-belts. Process by-product contamination can be controlled by periodic cleaning of the etch chamber; the frequency of cleaning can be predicted depending on the features and process application.

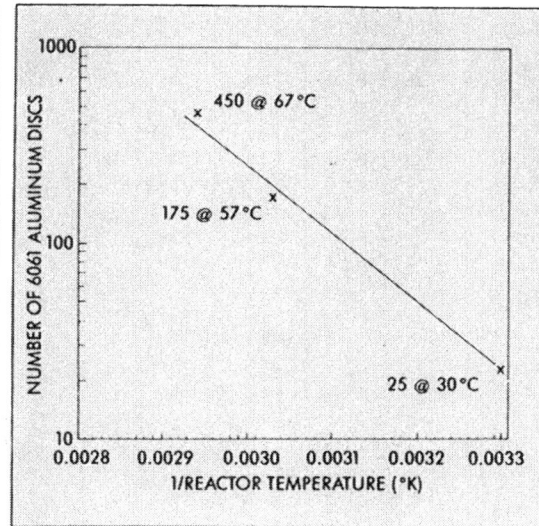

Figure 4: Number of 6061 aluminum discs run on a 1512 etcher before cleaning of the etcher was required (based on accumulation of more than 100 particles ≥ 0.3 μm in size on a 125-mm silicon monitor wafer).

Comparison of the vacuum-load-locked 1512 and the non-load-locked 903 demonstrated no significant benefit of vacuum load locks from the standpoint of wafer-transport contamination. However, the load locks do isolate ambient moisture and contaminants that compete with or inhibit aluminum-etching processes, and moisture control in aluminum etching eliminates the formation of oxides as a by-product that can contaminate wafers. From this perspective, vacuum load locks do have a notable effect on particulate contamination.

REFERENCES

1. Lachenbruch R, Wicker T, and Peavey J, "Contamination Study of Plasma Etching," *Semiconductor International*, 8(5):164–168, 1985.
2. Dushman S, and Lafferty JM (eds), *Scientific Foundations of Vacuum Technique*, 2nd ed, New York, Wiley & Sons, pp 82–84, 1962.
3. Coburn JW, and Winters HF, "Plasma Etching—A Discussion of Mechanisms," *Journal of Vacuum Science Technology*, 16(2):391,1979.
4. Winters HF, "The Role of Chemisorption in Plasma Etching," *Journal of Applied Physics*, 49(10):5165–5170, 1978.
5. Lachenbruch R, Gomez O, and Chapman B, "Particulate Contam- (100) and XeF_2 with W (111) and Nb," *Journal of Vacuum Science Technology, Vol. B: Vacuum, Surfaces, and Films*, 3(1):9–15, 1985.
7. *Handbook of Physics and Chemistry*, 64th ed, Boca Raton, FL, CRC Press, p D-195, 1983–1984.
8. Spencer JE, "Management of $AlCl_3$ in Plasma Etching Aluminum and its Alloys," *Solid State Technology*, 27(4):203–207, 1984.
9. Hess DW, "Plasma Etching of Aluminum," *Solid State Technology*, 24(4):189–194, 1981.
10. Donohoe KG, "In-Line Plasma Etching of Aluminum," in *Proceedings of the Third Symposium on Plasma Processing*, Pennington, NJ, Electrochemical Society, pp 306–325, 1982.
11. Spencer JE, "Process Integrity and the Role of $AlCl_3$ in the Plasma Etching of Aluminum," in *Proceedings of the Fourth Symposium on Plasma Processing*, Pennington, NJ, Electrochemical Society, pp 321–330, 1983.
    ination in Plasma Etching Systems: Sources and Elimination," in *Proceedings of the SEMI Technology Symposium*, SEMICON/Japan, Mountain View, CA, Semiconductor Equipment and Materials Institute, 1985.
6. Winters HF, "Etch Products from the Reaction of Cl_2 with Al (100) and Cu

Vacuum systems for microelectronics

John F. O'Hanlon

IBM T. J. Watson Research Center, Yorktown Heights, New York 10598

(Received 16 August 1982; accepted 15 September 1982)

Sophisticated vacuum systems are required for modern growth, deposition and etching processes, and analytical techniques. These systems have been developed to the point where they are available with optional pumping packages, system designs, and procedures. We discuss the applications of high and ultrahigh vacuum pumping systems and procedures used for device fabrication. Plasma processing in the medium vacuum range is attractive because it is fast, economical, and does not require especially clean vacuum technique. It is now clear that certain device structures must be fabricated under ultraclean plasma conditions. We discuss the current state of ultraclean high flow processing.

PACS numbers: 07.30.Cy, 85.40. — e

I. INTRODUCTION

As Lafferty[1] has observed, from the beginning vacuum environments have always been produced for a purpose. First they were used for scientific purposes, and then for industrial applications resulting from this research. Vacuum technique always has been and always will be a central part of other technologies and not merely an end in itself. The availability of improved vacuum environments has generated new technologies which, in turn, have directed the course of vacuum development. Nowhere is this more apparent than in semiconductor fabrication. Sophisticated vacuum environments are necessary for today's crystal growth, film deposition and etching processes, and analytical techniques which examine surfaces and probe underlying structures.

Vossen[2] has summarized the role of vacuum technique in the semiconductor technologies of today and the near future. About half of the processes used in the present day integrated circuit fabrication are performed in a vacuum environment. Vossen projects over three-fourths of the steps for future VLSI processing will be implemented at reduced pressures. The vacuum environment encompasses systems which operate in the ultrahigh vacuum region below 10^{-7} Pa, and those which operate at pressures above 1000 Pa. The gas flow requirements range from the insignificant gas flow in an analytical instrument to quantities as high as 10^5.Pa-L/s necessary to operate a reduced pressure epitaxy system. Some pumps remove gases or effluent vapors which are highly reactive.

The semiconductor industry is benefiting from the results of continuing process development. Equipment is available for any of the steps required in semiconductor processing. Complete systems for vapor deposition, sputtering, plasma deposition, plasma and reactive ion etching, ion plating, plasma ashing, and electron beam mask fabrication, and several analytical procedures are now available as "off-the-shelf" items. The focus has been on developing the processes and chamber fixturing necessary to produce wafers of consistent uniformity in large quantities with high yields. Paralleling these developments there have been advances in high vacuum technique. In particular turbomolecular and helium-gas-cooled cryogenic refrigerators are now practical alternatives to the diffusion and ion pumps. The rotary vane pump had been perfected to produce low ultimate pressures, but today that requirement has been superseded by the need for compatibility with hostile vapors and gases. Piston and lobe blowers have replaced or augmented mechanical vane pumps for high flow applications in the medium vacuum range. Gauge development was directed toward measurement in the ultrahigh and extreme high vacuum regions. Today there are gauges available which accurately and safely measure pressure and flow of corrosive gases in the medium and low vacuum regions.

The result of these developments and refinements is a choice among pumps, gauges, construction techniques, features, and procedures. Familiarity with their attributes will aid the user in choosing a pumping system and design. This paper describes some properties of high vacuum, ultrahigh vacuum, and high gas flow systems, and discusses procedures and designs which affect system operation.

II. HIGH VACUUM SYSTEMS

Systems with base pressures of 10^{-5} Pa are used for a number of processes, including vacuum deposition from resistively or electron-beam-heated sources and pumping of load-lock chambers. A high vacuum system consists of a mechanical pump used alternately as a roughing and backing pump, a clean process chamber and a high vacuum pump. If the high vacuum is produced with a diffusion or turbomolecular pump, a liquid nitrogen trap will be used as well. The trap serves the purposes of reducing backstreaming from the diffusion pump and increasing the pumping speed for water vapor in both systems. Systems can be designed to reach a base pressure of 10^{-5} Pa in 1 h without the aid of baking. This is the typical, "well understood" high vacuum system. There are however, some effects of the pumping speed, pump type, and operating procedure that are not widely appreciated and which can affect a process.

The effect of high vacuum pumping speed on system performance is often described with the relation $P = Q/S$. That is, an increase in pumping speed results in a lower base pressure, or alternatively, the attainment of an equal pressure in less time. High pumping speed is often equated with cleanliness. Deposition processes are started when the chamber pressure has reached an empirically determined value. It is not valid to retain the same pressure value for process initiation after changing pump or chamber size because all system variables do not scale in the same manner. As an example, consider a system which reaches a pressure of 2×10^{-5} Pa in 60 min. Now replace its high vacuum pump with one whose speed is larger by a factor of 4. The new system will pump to 2×10^{-5} Pa in about 30 min. The volume gas density is the same in each case, but the system walls are not equally clean. Outgassing rates of many materials decrease as t^{-1}, so that wall outgassing in the second case is approximately two times the value obtained with the smaller pump at the same pressure. Thermal radiation during deposition will cause a greater amount of wall outgassing in the system using the large pump than in the system with the small pump. The result is increased film contamination. The solution to this problem is pumping the new system for 60 min before beginning deposition. The essence of this problem is that increasing the size of the pump does not decrease the pumpdown time. Wall outgassing proceeds at its own thermally activated rate which is not affected by either pump size or pump type. The argument that increased pumping speed improves vacuum cleanliness is often used to justify the purchase of a diffusion pump of increased size or its replacement with a cryopump. Increased pump size will improve the base pressure, but it will not increase the wall desorption rate.

In recent years many diffusion pumps have been replaced with cryopumps. One valid reason for this change is the elimination of accidental backstreaming of diffusion pump fluid into the process chamber. Diffusion pumps can be designed and operated in a manner acceptable to many processes, but accidents can never be prevented totally. Let us compare the pumping speeds of a diffusion pump and a helium-gas-refrigerated cryopump of nominally equal (10 in. ASA) throat size. The two pumps chosen for comparison are a diffusion pump with a rated air pumping speed of 5300 L/s with an inlet area of 750 cm^2, and a 3000 L/s (air) cryopump mounted on a flange of 670 cm^2 inlet area.

The air pumping speed of each pump was measured by the manufacturer on a standard AVS test dome with a gauge mounted normal to the direction of gas flow at a distance of 0.5 pipe diameters above the pump inlet. The apparent transmission coefficient a' of the two pumps is 0.62 and 0.39, respectively. The apparent transmission coefficient is the ratio of speed measured on the AVS dome to the maximum speed of the inlet flange. We use the phrase "apparent transmission coefficient" because the orifice of the pressure gauge in the AVS dome is normal to the direction of gas flow. Using an expression derived by Santeler[3] we can obtain the actual transmission coefficient at the plane of the gauge that would be obtained if the gauge faced the chamber,

$$a = \frac{a'}{1 + a'/2} .$$ (1)

This gives actual transmission coefficients of 0.47 (diffusion) and 0.33 (cryo). Using the method of Oatley[4] the pipe conductance between the gauge and inlet flange can be accounted for yielding actual transmission coefficients at the pump

entrance a_p, of 0.53 (diff.) and 0.36 (cryo). The full speed of either pump is not available at the entrance of the process chamber. The diffusion pump will have a valve and cold trap while the cryopump requires only a valve. Again the method of Oatley can be used to calculate the transmission coefficient of the two combinations at the entrance to the chamber. Assume the valve has a transmission coefficient $a_v = 0.7$, while that of the liquid nitrogen trap is $a_t = 0.4$. If all components have equal exit and entrance diameters, the net transmission coefficient a_n of the diffusion pump stack is

$$\frac{1 - a_n}{a_n} = \frac{1 + a_v}{a_v} + \frac{1 - a_t}{a_t} + \frac{1 - a_p}{a_p} .$$ (2)

For this example, $a_n = 0.26$ and the pumping speed is $S_n = a_n Av/4 = 2235$ L/s. The cryopump stack does not use a liquid nitrogen trap. Its net air pumping speed at the chamber is 2530 L/s ($a_n = 0.33$) or 13% higher than the diffusion pump.

Water pumping speeds for diffusion pumps and some liquid nitrogen traps have been measured.[5] The water pumping speeds for cryopumps are only calculated. These calculations assume the projected area pumps at the maximum rate of a surface cooled to liquid nitrogen temperatures. Pumping speeds obtained from the commercial literature yield transmission coefficients of 0.9 to 1.0. One pump advertises a water pumping speed 1.32 times greater than the value obtainable from a perfect pump ($a = 1$) of the same inlet diameter. The approximation used by this author is a transmission coefficient of 0.8 for liquid nitrogen traps and cryogenic pumps. The water pumping speeds of a good liquid nitrogen trap and a cryopump of the same inlet diameter should be equal, provided each is termally shielded from the process heat load.

The argon pumping speeds are a few percent less than the air speeds in each pump. The 7000 L/s helium pumping speed of the diffusion pump described above translates into a net transmission coefficient of 0.28 at the throat of the pump. The helium pumping speed of the cryopump is a function of the temperature and degree of saturation. Where a large amount of helium is to be pumped, the cryopump may be augmented by a turbopump.[6] Hydrogen pumping speeds in cryogenic pumps are also a function of the degree of adsorbent saturation and temperatures of both the cold and warm stages.[7] Hydrogen and helium pumping speeds may not be the same for the two pumps and their level of contamination must be evaluated for each material.

Regeneration is infrequently required in normal use. Regeneration is required when the power fails or dips for an extended time. Regeneration may also be required after helium leak detection. The first air load pumped after leak checking can momentarily warm the cold stage enough to release trapped helium. The desorbed helium thermally shorts the refrigerator, the pressure rises, and the pump ceases to operate. Experience in this laboratory has shown the mean time between regeneration to be determined by these external events and not by normal saturation.

Replacement of a liquid-nitrogen-trapped diffusion pump with a cryopump of equal throat size results in a system with a small improvement in air pumping speed, equal speed for pumping water, and variable helium and hydrogen pumping speeds. The possibility of accidentally discharging diffusion pump oil into the chamber is no longer present, but the system will need regeneration at a time not chosen by the operator. Users should be aware that similitude laws do not show linear scaling of all parameters with pump size, so that some process variables may have to be changed to maintain product quality.

Pumping and venting procedures can affect particulate contamination of wafers during vacuum processing. Dust particles can fall on upward facing wafers or be electrostatically attached to wafers of any attitude and seriously reduce produce yield. One major source of dust in vacuum systems is that raised by the turbulence of rough pumping and venting to atmosphere. This is particularly acute in dusty processes such as the deposition of silicon monoxide or glasses and low pressure chemical vapor deposition. This problem was considered by Ames.[8] He reasoned that it was a nontrivial matter to remove all particulates from the system. Instead of attempting to remove them thoroughly, he found it more reasonable to vacuum clean the system as best as possible prior to evacuation, and then to evacuate the chamber at a controlled slow rate. "Soft roughing" as is it now known, and "soft venting" (the controlled release of the chamber to atmosphere) are used in many production systems in this company, as well as some which are available commercially. The maximum rates for pumping and venting are functions of the geometry and size of the piping and do not lend themselves to calculation. It is known that the turbulent boundary layer is sharper and more able to transfer momentum to particulates than the laminar flow boundary. If turbulence is avoided in the roughing and nitrogen release lines, the agitation of particulates in the chamber disappears. To keep a pipe in laminar viscous flow, Reynolds' number must be less than 1200. A slightly higher flow may be possible, or a reduction may be necessary. However, this must be determined experimentally as gas flow patterns are a function of the roughing and vent line location, baffle location, and chamber geometry as well as the total flow rate. Systems with loose fitting doors which are slammed shut by the impact of initial roughing are notorious for dislodging large amounts of debris and scattering it all over the chamber.

Soft roughing is easily implemented. Consider, e.g., a chamber pumped by a 4-cm-diam roughing line. To keep the flow laminar (Re < 1200) the flow must be less than 1.6×10^6 D (Pa-L/s) where D is the pipe diameter in meters. For a speed of 7 L/s the average pressure in the line should be less than 9500 Pa (70 Torr). The roughing line is constructed with a main valve that is bypassed by a small valve and orifice. Roughing begins by opening the bypass valve and pumping through the orifice or choke. When the small volume of gas on the pump side of the orifice reaches a pressure of 10^4 Pa at the beginning of roughing, it has reached its maximum flow of 2×10^5 $P\pi d^2/4$ Pa-L/s where d is the diameter of the orifice. The flow will remain choked as long as the pressure ratio across the orifice (P_{low}/P_{high}) is less than 0.52. The maximum flow is a function of both pressure and orifice diameter. An orifice of $d = 2$ mm will restrict the flow in a 4-cm-diam roughing line to be viscous and laminar at all

pressures up to and including 1 atm. At a pressure of 70 Torr, the main roughing valve is opened and roughing continues at flows less than a Reynolds' number of 1200. In this example use of a bypass valve and orifice extends the roughing time from 2.5 to 7.5 min.

Soft venting can be implemented in the same manner. A motor-controlled variable leak in combination with a diffuser will speed the rate at which gas is admitted by increasing the flow as the chamber pressure rises. Venting time will be of the same magnitude as the roughing time.

Laminar flow in both the roughing and nitrogen venting lines should be the initial design criteria. The flow can be varied about this design point to achieve the shortest roughing and venting times that do not generate particulate contamination. In one group of systems recently designed by this technique a flow corresponding to a Reynolds' number of 1000 was experimentally found to be the maximum possible flow which would not transport dust to the substrates.

III. ULTRAHIGH VACUUM SYSTEMS

At this time ultrahigh vacuum is not required for any fabrication steps in silicon technology. UHV is routine for many of the systems used for the analysis of device problems. The requirements of bakeout and long pumping times make UHV processing both time consuming and expensive. To a degree there is a trade-off between background pressure and rate of deposition such that high quality insulators and interconnection metallurgy can be deposited with low impurity levels in the 10^{-5} Pa range. This is not true for all processes. Active devices require controlled doping and clean interfaces. Molecular beam epitaxy uses UHV to prevent contamination from unintentional adsorption processes at the semiconductor dopant level. Unwanted gases are a result of desorption from walls, sources, hot filaments, and backstreamed gases from pumps.[9] Niobium films for Josephson circuits must be deposited in near UHV conditions.[10] Niobium is an extremely effective getter for oxygen whose incorporation into the film raises its superconducting transition temperature. The reaction kinetics of tungsten silicide films formed by the annealing of tungsten deposited on silicon depend upon interfacial cleanliness.[11] Tungsten silicide layers are formed by the codeposition of W and Si to avoid the necessity of UHV deposition conditions. These are a few examples of vacuum-deposited device films whose morphology depends upon system cleanliness during deposition.

Ion, turbomolecular, and cryogenic pumps have been used in various combinations along with cleaning and baking to achieve ultrahigh vacuum. There is no one best system or combination. Instead pumps are combined to meet the requirements of a particular application. Ion pumps produce stray ions and magnetic fields, turbomolecular pumps are not used where small amounts of hydrogen cause interference, and helium gas cryogenic pumps are not used where a process will be disturbed by displacer vibration or where helium must be pumped regularly.

Ultrahigh vacuum systems have been avoided for industrial applications because they require bakeout and careful fabrication and operation. The added time and expense cannot be justified when alternative HV processes are available.

Visser and Scheer[12] have demonstrated a cryopumped, load-locked UHV system that does not have to be baked. It consisted of a cryopumped chamber containing an electron-beam source and substrate holder in which the sample was introduced through a load lock. Wall outgassing was reduced not by baking, but by using the 80 K stage to cool all surfaces except those through which the sample was introduced. In this way a pressure of 10^{-7} Pa was reached in 20 min.

UHV conditions are required for depositing films that form active devices. This level of cleanliness is necessary to control impurity and doping levels which affect device performance. When these technologies move into a manufacturing environment, they will do so with a rapid-cycle ultrahigh vacuum technology.

IV. HIGH FLOW SYSTEMS

High flow systems which operate with an externally supplied process gas flow are used for a number of etching and deposition techniques. Gas flows range from 10^2 to 10^5 Pa-L/s over the low and medium vacuum range. The gases used can be inert, toxic, explosive, or corrosive. We arbitrarily divide these systems into two groups which are analogous to high and ultrahigh vacuum systems. We call these clean high flow (CHF) and ultraclean high flow (UCHF) systems. Clean high flow systems include those for depositing and etching insulators and metallurgy by sputtering, low pressure vapor deposition (LPCVD), and plasma/reactive ion etching. Ultraclean high flow systems include those for growth of compound semiconductors and doped a-Si:H by reduced pressure epitaxy (RPE), LPCVD, and plasma-assisted CVD. Clean and ultraclean high flow systems are distinguished by the levels of background contamination in much the same way as high vacuum is distinguished from ultrahigh vacuum. Many of the clean processes are sensitive to the partial pressure of an impurity gas, but not to the level that would affect the doping of an intrinsic or lightly doped semiconductor.

A. Clean high flow systems

Low pressure thermal and plasma processes which use toxic, corrosive, or flammable gases are becoming widely used in VLSI etching and deposition. Several etching processes are also used in packaging technologies. The process equipment differs significantly from that used for atmospheric pressure chemical vapor deposition and wet chemical etching. Exhaust gases from atmospheric pressure systems are either scrubbed or vented safely. However, reactive gases and abrasives emanating from low pressure reactors are exhausted directly into a pump. Conventional pumping systems would be destroyed quickly by polymerization of lubricants, particulates, and acids.

Rotary, Roots, and throttled high vacuum pumps have been used to exhaust plasma systems. Rotary pumps are used where the pumping speed is less than 200 m³/h at pressures > 20 Pa. Roots pumps are added when higher speed or lower pressure (2 Pa) is required. In both cases it is desirable to use a nitrogen bleed to prevent oil backstreaming.

Particulate damage and oil corrosion seriously affect the

TABLE I. Residual gas analysis of mechanical and diffusion pump fluids contaminated by reactive plasma effluents.

Pump fluid	Process gases	Fluid contamination
MP oil[a]	Cl_2, Ar.	C_2Cl_4, toluene.
MP oil[a]	CCl_4, O_2.	CCl_4, O_2, C_2HCl_3.
MP oil[a]	CCl_4, O_2.	C_2Cl_4, CCl_4, toluene, fluoro compound (trace).
MP oil[a]	NH_3, N_2O, SiH_4, N_2, Ar.	Ar, CO_2.
MP oil[a]	Cl_2, Ar, $CBrF_3$	C_2Cl_4, $CBrF_3$.
New MP oil	...	Ar, CO_2, hydrocarbons, no chlorinated compounds.
DP oil[b]	CF_4, Cl_2, CF_3Cl, BCl_3, CCl_4,	C_2Cl_4, C_2HCl_3, phthalate ester, toluene, benzene.
DC-705	BCl_3	Benzene, phenol, and white crystaline HBO_3.
MP oil[c]	CF_3H–O_2, CF_4–O_2.	Acetone, very weak peaks at 72, 86, and 112.
MP oil[a]	CF_4, CCl_4, SF_6, HF, NH_3.	Oil is black, pump siezed. No vapors detected by RGA.

[a] Hydrocarbon oil from mechanical pump backing a diffusion pump.
[b] Solid residue from diffusion pump.
[c] Hydrocarbon oil from mechanically pumped etching system.

operation of reactive gas pumps. Frequent maintenance results from the use of ordinary hydrocarbon mechanical pump oils.[13] Contamination of pump oil by reaction products presents some potentially serious safety issues. Table I shows the results of mass spectrometric analysis of several pumping systems.[14] The data are uncalibrated, so concentrations are not known. Contaminants are listed in order of decreasing peak intensities. The source of the acetone in the mechanically pumped system is not known. It could be a residue from chamber cleaning, or an O_2–hydrocarbon decomposition product. It is clear that there is a need for more analysis in order to establish safe maintenance procedures.

High vacuum pumps can maintain a chamber in the medium vacuum range when the pumping line is suitably throttled. This arrangement is used when the process requires a higher compression ratio (e.g., H_2) than is possible with a rotary or Roots pump. Cryopumps are not used on reactive gas systems because they become a sink for toxic gases. They can be used to pump nontoxic gases such as argon, nitrogen, and oxygen in a sputtering system. Cryopumps should be throttled to keep the cold stage temperature below 20 K where it will pump hydrogen. Throttled diffusion pumps can be used provided the process gases do not react with the diffusion pump fluid at the boiler temperature. Pentaphenylether and silicon appear to be the least reactive fluids. Regardless of which pump is chosen, reactive plasma systems should be thoroughly purged with nitrogen to prevent reactive gas accumulation. Turbomolecular pumps lubricated with a fluorocarbon provide high throughput, moderate pumping speed and, when throttled, have a high compression ratio which will prevent hydrogen backstreaming. When choosing a throttled high vacuum pump, high pumping speed seems necessary to minimize pump-down time. In Sec. II we noted that surface outgassing rates do not scale with pump size. The consequences of high pumping speed and poor wall outgassing are more serious in a plasma system than in a high vacuum system, because high energy neutrals are most efficient at cleaning surfaces.

B. Ultraclean high flow systems

UCHF systems require the cleanliness of ultrahigh vacuum at pressures where the mean free path is much shorter than system dimensions. Ultrahigh vacuum is traditionally achieved by careful component selection and cleaning to reduce outgassing, and pumping to extremely low pressures with entrainment or high compression transport pumps. Ultraclean systems differ from UHV systems in that they contain a plasma, high throughput pumps, and a manifold to mix and deliver process gases. Deposition of clean films by evaporation requires the background flux to be low enough so it will not cause undesirably doping. This is achieved by depositing in UHV. Low background concentrations are also required for plasma deposition of semiconductor films. However, these systems have the added complication of plasma–wall interactions. High energy ions charge exchange to produce energetic neutrals which cannot be screened. When these neutrals collide with walls they stimulate desorption of previously sorbed gases. This phenomena occurs in all plasmas and its severity is related to the desired material purity. Carlson[15] has observed that one reactive discharge containing SiF_4 contaminated solar cells with O, N, and C which were presumably scrubbed from the chamber walls. Kuwano et al.[16] have designed a system for fabricating p-i-n solar cells which contains a separate manifold, deposition chamber, and pump system for each device layer. The system is arranged as separate, connected reaction chambers. The substrate is admitted through load locks which prevent air from ever reaching the walls of the process chambers. In this manner plasma scrubbing cannot contaminate the films with air. Also the walls of each reaction chamber are preconditioned to prevent cross contamination from prior dopants. Contamination has been observed in LPCVD of compound semiconductors from supply gases, wall adsorption in gas manifolds, and carbon from backstreamed oil fragments. The latter has been known to begin as long as one year after system completion indicative of oil creep along foreline walls.

Design techniques for this regime are not unified partly for proprietary reasons and partly because the necessity for such design techniques has neither been recognized nor well documented for this kind of device fabrication. Many of the potential sources of contamination have been considered individually. Process gas can be cross contaminated in the manifold, and chamber walls can adsorb atmospheric gases as well as materials from a prior deposition. Plasma confinement or isolation, and wall conditioning are two methods of reducing the effects of wall contamination. Throttled high vacuum pumps may be necessary in order to achieve compression ratios high enough to prevent hydrogen and light hydrocarbon backstreaming. Counterflow purging has been shown to be effective in reducing volume backstreaming of pump oil, but little data exist on the rate of surface creep in pumping lines. Gaskets and leaks are sources of impurities. Most elastomers are permeable to atmospheric gases, and their permeability can be increased by process gas swelling, e.g., SiH_4 in Viton. Leaks must be reduced to the UHV level even though the system is never pumped to that pressure range. As this technology evolves, these and other methods will be implemented to produce contaminant-free device films.

V. CONCLUSION

We have discussed high and ultrahigh vacuum systems used for fabrication of thin film layers and devices. Effects of pump speed, pump type, and operating procedure which are often overlooked have been described. Plasma deposition or etching of metal and insulating layers has been distinguished from plasma deposition of semiconducting devices. Some problems relevant to each regime have been reviewed.

[1] J. M. Lafferty, Physics Today 34(11), 211 (1981).
[2] J. L. Vossen, J. Vac. Sci. Technol. 18, 135 (1981).
[3] D. J. Santeler, D. W. Jones, D. H. Holkeboer, and F. Pagano, Vacuum Technology and Space Simulation, NASA SP-105 (National Aeronautics and Space Administration, Washington, D.C., 1966).
[4] C. W. Oatley, Brit. J. Appl. Phys. 8, 15 (1957).
[5] A. A. Landfors, M. H. Hablanian, R. F. Herrick, and D. M. Vaccarello, J. Vac. Sci. Technol. (these proceedings).
[6] N. G. Wilson and K. N. Watts, J. Vac. Sci. Technol. 17, 270 (1980).
[7] B. Liu, J. Ren, and X. Cui, J. Vac. Sci. Technol. 20, 1000 (1982).
[8] I. Ames, M. F. Gendron, and H. Seki, Transactions of the 9th National Vacuum Symposium (Macmillian, New York, 1962), p. 133.
[9] R. Z. Bachrach, in Crystal Growth, 2nd ed., edited by B. Pamplin, (Pergamon, Oxford, 1980).
[10] R. F. Broom, R. B. Laibowitz, Th. O. Mohr, and W. Walter, IBM J. Res. Dev. 24, 212 (1980).
[11] C. C. Chang and G. Quintana, J. Electron. Spectrosc. Relat. Phenom. 2, 363 (1973).
[12] J. Visser and J. J. Scheer, J. Vac. Sci. Technol. 19, 122 (1981).
[13] J. F. O'Hanlon, Solid State Technology 24, 86 (1981).
[14] RGA data courtesy of F. Anderson, IBM SCD East Fishkill, New York.
[15] D. E. Carlson, J. Vac. Sci. Technol. 20, 290 (1982).
[16] Y. Kuwano, M. Ohnishi, S. Tsuda, Y. Nakashima, and N. Nakamura, Jpn. J. Appl. Phys. 21, 413 (1982).

MANUFACTURING IN A VACUUM ENVIRONMENT

Steve Belinski and Susan Hackwood

Center for Robotic Systems in Microelectronics
University of California, Santa Barbara

Abstract

The cleanliness requirements for many manufacturing and microelectronic processing tasks are becoming ever stricter. Processes requiring high cleanliness standards are now performed in clean rooms, which are being forced to class 10 and better by the shrinking geometries and increased complexity of semiconductor devices. An alternate manufacturing strategy is to maintain a vacuum environment during the process monitoring, testing, handling, material transfer and assembly operations. This will require an overall manufacturing environment at some level of clean vacuum. A new generation of robots is therefore required for compactness, high precision, and performance specifications that are vacuum compatible. We discuss 1) the advantages and characteristics of the vacuum environment, 2) a specific vacuum manufacturing task for which the Center for Robotic Systems in Microelectronics (CRSM) is developing a robot, and 3) design requirements for vacuum-compatible robots.

Introduction

Many manufacturing operations in electronics and microelectronics require the use of clean rooms. Contamination control is one of the most acute issues facing the IC industry, which has produced incredible product technology, yet still relegates a large percentage of starting materials to scrap. As a result, product yield remains much lower than desired. Today's rigorous cleanliness requirements can be illustrated with the example of the development going on in VLSI manufacturing. Table 1 shows that the critical particle diameter, i.e. the maximum size of tolerable contaminant particles, is projected to be 0.05 μm in the near future.

As of 1985, the most demanding air cleanliness level established in the U.S. Federal Standard 209b is a cleanliness class 100, which refers to a maximum concentration of 100/ft^3 for particles of diameter greater than or equal to 0.5 μm. It is proposed in [1] that the present standards be extrapolated as listed in table 2.

In the near future, cleanliness class 1 and even more stringent environments will be required [2]. The development of ultra-clean environments is not limited to VLSI fabrication; many other material processing, handling and assembly operations require these conditions. Since cleanrooms are normally partial laminar flow, even class 100 and class 10 cleanliness is achieved only under laminar flow hoods. The three main sources of particle contamination are: the outside air (10^7 - 10^8 particles > 0.5 μm/m^3), processing equipment, and humans, who give off about 100,000 dust particles > 0.3 μm and more than 1000 bacteria and spores per minute. An increasing number of specially designed robots are being used in clean rooms not only because of their potential efficiency and productivity, but also to replace one of the principal sources of contamination: human beings.

There has been a slowing of the construction of microelectronics plants in the United States and Europe, while at the same time Japan and other Eastern countries have been adding new, state-of-the-art 150-mm manufacturing plants. James Burnett, Executive Director of the Institute for Contamination Control, predicts that even class 1 facilities will be obsolete by the year 1995 (see table 3). He states that "if the construction of new class 1 (or better) plants is not begun now, we run the risk of losing our high-technology manufacturing industry by default" [3].

Another possible approach to the challenge of high cleanliness manufacturing is to do most, or all of the processing, handling, testing, inspection, assembly and packaging in a vacuum environment, i.e. maintain the vacuum during as much of the manufacturing process as possible. Most lithography and material processing stages are already conducted at ~10^{-6} Torr. Improvement could be made by keeping the components in vacuum during the entire process. This does not necessarily mean high vacuum. In some cases low vacuum may be adequate, as long as it is clean. However, this idea cannot be implemented without the use of robotics.

Presently, epitaxial growth of semiconductor films takes place in the low vacuum range. Sputtering, plasma etching, plasma deposition, and low-pressure chemical vapor deposition are performed in the medium vacuum range. Pressures in the high vacuum range are required for most thin-film preparation, electron microscopy, mass spectroscopy, crystal growth, x-ray and electron beam lithography, molecular beam epitaxy, and the production of cathode ray and other vacuum tubes [4]. These environments are completely unsuitable for the presence of human beings. The special suits that are worn in outer space would be much less appropriate in a specialized vacuum chamber used for a critical processing task. The amount of time needed to produce the desired pressure level once the suited worker had entered the chamber would prove to be quite long. The need for robotics and automated systems inside the vacuum chamber may thus be more urgent than in a normal environment or

Storage density (kbit/chip)	Line spacing on wafer (μm)	Minimum critical particle diameter (μm)	Market dominance period
16	4.0	0.4	1981 - 1984
64	2.5	0.3	1984 - 1988
256	1.5	0.17	1984 - 1988
1×10^3	0.9	0.09	1988 - 1990+
4×10^3	0.5	0.05	1988 - 1990+

Table 1. Line spacing and critical particle diameters for high density integrated circuits (adapted from [1]).

Cleanliness class	Particles per ft^3 equal to and greater than:			
	0.02 μm	0.1 μm	0.5 μm	5 μm
1	10^3	3×10	*	*
10	10^3	3×10^2	10	*
100	†	3×10^3	10^2	*
1000	†	†	10^3	7×10^0
10,000	†	†	10^4	7×10
100,000	†	†	10^5	7×10^2

* Indication not meaningful for statistical reasons.
† Indication not relevant for the definition of cleanliness requirements.

Table 2. Proposal for extrapolating the US Federal cleanliness standards (from [1]).

Facility Type	Process-Equipment Type	Expected Defect Limit*	Feature-Size Limit	Year of Obsolescence
Class 100	100 mm	2.0 at 0.5 μm	2 μm	1986
Class 10	100 mm	1.0 at 0.5 μm	1.5 μm	1989
Class 10	150 mm	1.0 at 0.2 μm	1.0 μm	1992
Class 1	150 mm	0.3 at 0.14 μm	0.7 μm	1995

Table 3. Expected pattern of obsolescence for microelectronic plants in the United States and Europe (*defect limit is given in defects/cm^2, followed by the critical particle size). (adapted from [3])

clean room. A principal advantage would be the ability to manipulate objects within the chamber without opening the vessel and subsequently re-evacuating, thus avoiding a very time consuming process.

As pointed out by Burnett [5], the principal reason to implement automation in micromanufacturing is not to reduce manufacturing costs, but to increase product yield for leading-edge products. Lacking high product yield, a company will not be able to meet market prices and maintain profitability, and will thus be squeezed from the market. Development of automated manufacturing systems based in the vacuum environment will also increase product yield by pushing the level of cleanliness beyond that of the clean room. The availability of vacuum-compatible robots will improve the efficiency of many processes now carried out in vacuum and encourage the use of vacuum processing in new areas.

Robot design for the vacuum environment not only requires skill in conventional robotics disciplines such as servo design, control, sensors and mechanisms, but also knowledge of vacuum engineering, including heat transfer, materials, and tribology. There is much to be learned from the design of mechanisms used for space applications. However, robots designed for operation in space are not entirely suitable for use in vacuum chambers, since they are generally not designed to meet the cleanliness requirements of an enclosed processing chamber. In these chambers, the outgassing from components works directly against the pumping equipment. In space there is no problem maintaining the pressure, but it is known that the outgassing of the space vehicle causes an expanding gas cloud to surround it. Outgassing is also a major concern in space, but it is much more critical to limit outgassing in the artificial vacuum environments produced in chambers on earth. Sources of outgassing in conventional robots include the structural materials, plastics, paint, and most importantly the lubricants. Conventional lubricants have relatively high vapor pressures, and thus exhibit high evaporation rates in the vacuum environment.

Advantages of the Vacuum Environment

It is clear that humans will gradually be removed from the processing area of the manufacturing clean-room. Increased automation will then be implemented, making the move to the vacuum chamber more straightforward. However, if the vacuum environment did not provide added advantages, this move would not be necessary. Traditional reasons for using this special environment are many and include the following [6]:

- To prevent physical or chemical reactions occurring between atmospheric gases and a desired process;
- To disturb an equilibrium condition that exists at room temperature so that absorbed gases or volatile liquids can be removed from the bulk of the material (e.g., degassing of oils and freeze drying), and adsorbed gases from the surface;
- To increase the distance that gas and vapor particles must travel before colliding with one another so that a process particle can reach a solid surface without making a collision (e.g., vacuum coating and the production of high-energy particles);

- To reduce both the number of molecular impacts per second and the contamination times of surfaces prepared in vacuo (e.g., clean surface studies and the preparation of thin films), and
- To reduce the concentration of a component gas below a critical level (e.g., the removal of oxygen, water vapor and hydrocarbons in tungsten filament valves).

In addition to the above advantages, which are on the molecular level, the vacuum environment has distinct properties with regards to particulate contamination. Studies of particle behavior at various levels of vacuum are being carried out at the CRSM and elsewhere.

The motion of particles changes greatly for different pressure levels. In order to control particulates, it is important to understand their behavior at the pressure level at which the process is being carried out. The effect of pressure reduction on the gravitational settling of particles has been theoretically modelled [7]. The goal is to characterize the motion of the particle at various pressures. In general, particles tend to fall when in vacuum because the drag force on the particle is lowered, while the gravitational force remains constant. The results are plotted in figure 1 as the time required for particles of several different sizes to fall one meter as a function of pressure. Note that for pressures below 10^{-5} Torr, all sizes of particles fall one meter in about 0.45 second. Thus, at a suitable level of vacuum, particles will rapidly 'fall out' of the chamber.

The vacuum environment is thus known to have certain advantages at both the molecular and the particulate levels. In order to capitalize on the particulate advantages, they must be fully understood. The maintenance of the vacuum vessel, the pumpdown and venting cycles and the actual particle behavior while at vacuum are all very important. The CRSM is studying each of these areas.

Modification of a Clean Room Robot for use in a High Vacuum Manufacturing Task

Delco Systems Operations has developed a new gyroscope, which is now entering the production stage. The assembly must take place in a high vacuum environment of 10^{-8} Torr. The CRSM is modifying an existing robot so that it can operate at high vacuum and be used by Delco for production of the new gyroscope.

Gyroscope Description

The gyroscope to be assembled is the hemispherical resonator gyro (HRG) [8-10]. It is not a laser based gyro, yet does not have

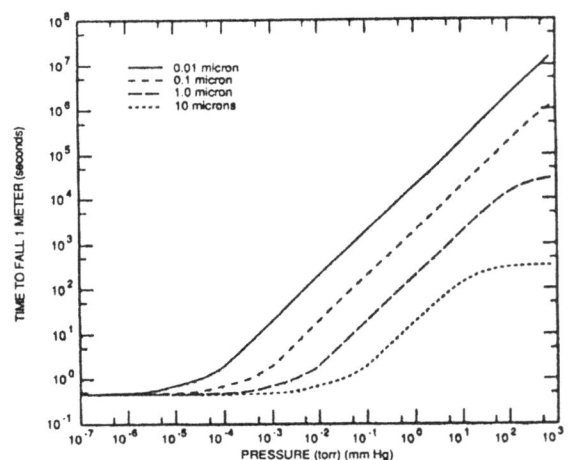

Figure 1. Particle motion vs. pressure level (from [7]).

82

the rotating parts usually found in mechanical gyroscopes. The main components (see figure 2) are constructed of fuzed quartz, which is an inert, stable material having low thermal sensitivity. This makes it an attractive material to use in precision mechanical and optical devices. It also exhibits extremely low internal damping, the damping time constant of the resonator being around 900 seconds. This means that very little energy is needed to sustain the vibration needed for operation, and that a power failure will not cause a loss of positional information until approximately 15 minutes have passed. The positive aspects of using fuzed quartz will deteriorate as the quartz becomes less pure. Even gases which attach themselves to the surface of the HRG components will cause variations from optimal gyroscope performance. The gyroscope thus must be manufactured in a high-vacuum environment and maintained at such throughout its life, a minimum of 20 years.

Gyroscope Production

In order to achieve a relatively high production rate for the gyroscope, Delco has proposed that the entire manufacturing process be performed in a large vacuum chamber as shown in Figure 3. A large central chamber is surrounded by 12 smaller compartments, each of which may be opened to either the outside or the central chamber and contains dedicated automation for a particular process. The central chamber will be constantly maintained at high vacuum (10^{-6}). The alternative to this configuration is a smaller vacuum chamber in which all steps of the assembly would be performed. This would result in an extremely slow assembly task due to the continual and time-consuming actions of venting and re-evacuating the chamber. In order to take advantage of the increased manufacturing efficiency the larger assembly chamber would bring, a means of transferring gyro parts between various stages of assembly is required. A robot positioned at the center of the assembly chamber is the ideal component.

The availability of vacuum-compatible robots is presently limited, although this is likely to change in the near future. As research and development progresses, Delco has decided to have the CRSM modify an existing robot [11,12] for use in their assembly task. The motivating factors in this decision were cost and time. Although it is desirable to obtain a robot which was designed and built specifically for the vacuum environment, the first step is to obtain a

vacuum-compatible robot. This will occur more rapidly by doing a modification to an existing robot. When a 'ground-up' vacuum-compatible robot becomes available it can then be compared to the modified robot.

GMF E-310 Clean-Room Robot

The robot undergoing modification at the CRSM is a GMF (General Motors - Fanuc) model E-310 cylindrical coordinate robot, originally designed for use in clean rooms to class 10. This robot was chosen for its size and configuration as well as its good accuracy and repeatability for a robot of its size. Tests at the CRSM have shown the repeatability to be ±10μm. The E-310 used for the gyroscope assembly task has four degrees of freedom (see Figure 4) consisting of two linear axes and two rotational axes.

The task for the researchers at the CRSM was then to modify the GMF E-310 so that it not only could operate in a vacuum environment to 10^{-8} Torr, but could also operate over long time periods without degrading its surroundings (outgassing, etc). The design requirements called for a modified robot with capabilities as close as possible to the original robot. The robot must have enough horizontal stroke to reach into the compartments and perform specific functions. This is also dependent upon the final configuration of the end-effector, which might also extend into the chamber. An important point, however, is that the stroke must be useful when the robot is configured in the vacuum chamber. In other words, when the robot rotates inside the chamber, the wrist should just clear the inner wall. Then when the R-axis is extended, the wrist will move into the compartment with full stroke capability.

In summary, the principal design requirements for the modification of the GMF E-310 robot for vacuum compatibility were:

- Modification of axes movement range:
  - Z-axis: maintain 300mm stroke if possible
  - R-axis: maintain 500mm stroke if possible; if reduced, resulting stroke must be useful in the vacuum chamber
  - θ-axis: maintain ±150° rotation
  - α-axis: maintain ±180° rotation
- Limit negative effects on the vacuum environment (outgassing, etc)
- Design for ≤ 100°C operating environment

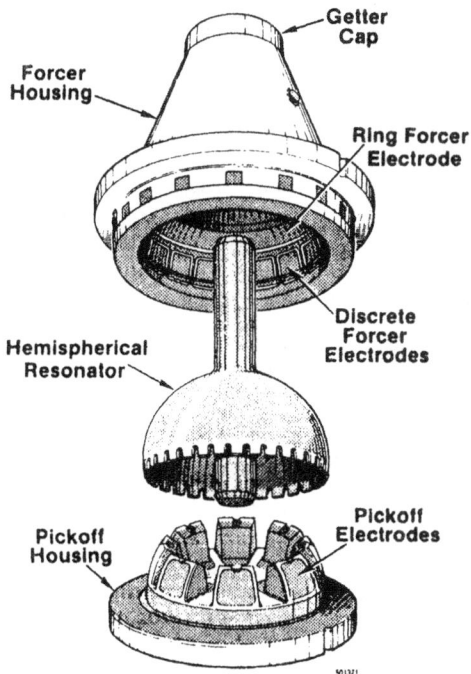

Figure 2. Principal components of the HRG (from [8]).

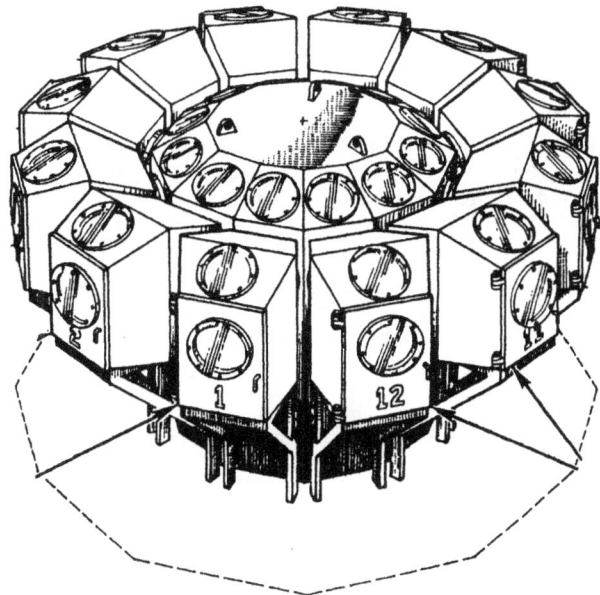

Figure 3. Proposed Assembly Chamber

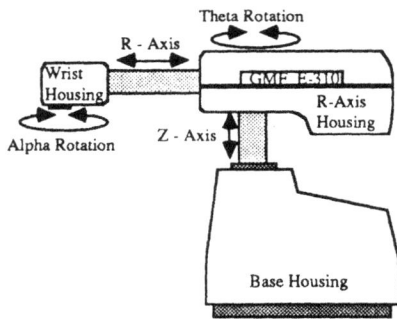

AXIS	MOTION RANGE	MOTION SPEED (max)
Z	0 ~ 300 mm	300 mm/s
θ	-150° ~ +150°	90°/s
R	0 ~ 500 mm	900 mm/s
α	-180° ~ +180°	90°/s

Figure 4. GMF E-310 Robot configuration and specifications.

Design Considerations

The first decision in the modification of the GMF E-310 was between two basic philosophies. The robot could either be totally exposed to the vacuum environment or it could be sealed in a type of "suit" which would allow the inside components to operate at atmospheric pressure, as they were originally designed to do. In order to expose the entire robot to a pressure of 10^{-8} Torr, a number of key changes would have to be made. The major ones would be in the lubrication systems, the surface finish and materials, and the motors. After examining this choice, it was concluded that it would entail a substantial amount of redesign work, and that a total exposure robot would be better designed from scratch. The goal then became one of designing a new housing for the robot which would seal it from the vacuum environment, while accomplishing the design goals. The sealing "suit" would have to be as leaktight as the walls of a high-quality vacuum chamber, yet must also allow the desired motions by sealing two linear (R and Z) and two rotary (Theta and Alpha) motions.

Vacuum Sealing

A differentially pumped 360° rotatable platform was chosen as the rotational sealing mechanism. As shown in the cross section of figure 5, the platform contains three spring loaded seals which are 80% teflon and 20% graphite. Two chambers are formed which are pumped to different levels of vacuum. For this application, the chamber closest to the atmospheric pressure side is pumped to approximately 10^{-3} Torr, while the chamber closest to the vacuum side is maintained below 10^{-8} Torr using an ion pump. Figure 6 shows the appearance of the robot after the modification and indicates

the placement of the rotatable platforms in the design. The two linear axes are sealed with vacuum compatible stainless steel bellows. Note in figure 6 that an additional bellow has been added to the rear of the R-axis housing. Because the robot internal pressure is essentially 760 Torr above its external environment, large forces exist which tend to push the R-axis forward and the Z-axis upward. The rear bellow is used to balance this force, while the force from the vertical bellow is offset by the weight of the robot.

Design Requirements for Vacuum-Compatible Robots

Mechanisms for use in vacuum have existed for years in the form of simple probes, sample manipulators, etc. [13-15]. Many of the early designs for transferring motion into vacuum relied on bellows for both linear and rotary movement. More recently, the use of double-pumped rotary feedthroughs, as described above, has become a popular method of transferring rotations into vacuum. Although they make it easier to introduce rotations to the vacuum environment, they are used as part of overall mechanisms which have non-vacuum-compatible components. For example, in many cases the motors used for actuation are maintained *outside* the vacuum chamber.

Although the state-of-the-art in vacuum manipulators is still mostly limited to feedthrough devices, of which the robot described above is essentially one, the desire to increase manipulation flexibility has resulted in research and development efforts aimed at entirely vacuum-compatible robotics and automation. This is the pursuit at the CRSM. We are developing the technology to design, build and test vacuum compatible robots and end-effectors. In particular, we are interested in

1) building dextrous manipulators suitable for small part handling in a vacuum chamber, and
2) specifying the requirements and measuring the performance of vacuum compatible robots in general.

In order to achieve these goals, a number of elements are critical.

Structural Materials

The structural materials used in the high vacuum environment must have a low vapor pressure and the ability to withstand high temperatures (in some cases 450°C or greater) which may be used during bakeout cycles to promote outgassing and reduce later desorption. Ease of machining and fabrication and the ability to form leak-tight seals (i.e. weldability) is also critical. The principal concern in choosing materials for the robot structure is that of outgassing. When the metal is placed in a vacuum environment, gases which have been adsorbed on the surface and dissolved in the bulk will be evolved at a rate depending on the amount of gas involved, the nature of the sorption process and the temperature. The most commonly used materials in vacuum applications are stainless steel and aluminum. Composite materials, such as those incorporating carbon fiber, show great promise structurally, but exhibit high outgassing rates unless properly coated. Machinable glass ceramics appear to be attractive for many vacuum applications, but have not been used extensively as yet.

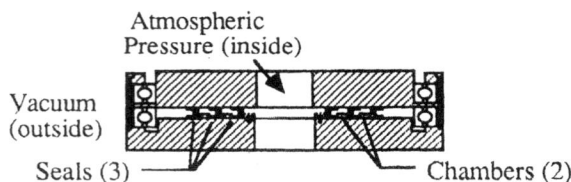

Figure 5. Differentially pumped rotatable platform for sealing Theta and Alpha rotations.

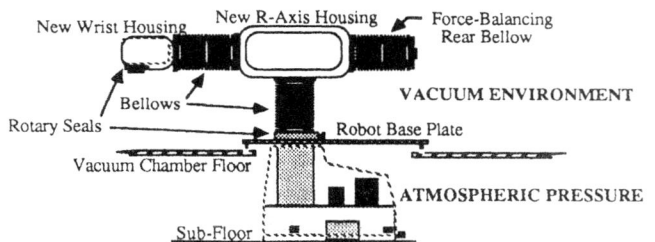

Figure 6. Modified E-310 Robot

Motors

Application of conventional electric motors in vacuum will also lead to problems. Because at high vacuum the gas density is so low that conduction and convection can no longer take place, thermal exchange is carried out mainly by radiation. If power is applied to a motor in vacuum, and no sink is provided, it will heat up until losses due to radiation cause an equilibrium. A temperature of 125°C can be reached in several minutes with the application of the maximum rated voltage to a thermally isolated motor. This problem may be minimized by designing appropriate heat sinking, limiting the voltage necessary to drive the load and reducing or eliminating the holding current when the motor is not running. Even if the temperature effects are controlled, the motor must be constructed of suitable materials and employ appropriate lubrication. The state-of-the-art in vacuum-compatible motors and actuators has not yet reached an appropriate level of sophistication in the United States. The CRSM is cooperating with Yaskawa Electric to develop motors specifically for high-vacuum applications. They have been developing an axial gap pulse motor, which will withstand temperatures to 300°C and vacuum levels of 10^{-11} Torr.

Lubrication

Another major design consideration for vacuum compatible mechanisms is the method of lubrication. Most conventional lubricants have relatively high vapor pressures, and thus exhibit high evaporation rates in the vacuum environment [16-18]. There are two general categories of lubricants: wet and dry. The use of dry lubricants, such as MoS_2 and WS_2, has become more common in recent years for use in high vacuum and beyond. Their advantages include lower vapor pressures and less dust and oil collection than wet lubricants. However, dry lubricants tend to generate more debris and need replenishing more often than wet lubricants. The state-of-the-art in vacuum lubrication is to use a dry lubricant, which has been sputtered or ion plated to the surface, or to use one of the newer low vapor pressure wet lubricants containing the polyfluoralkylether Fomblin Z25. In either case the lubrication system *should* be considered an integral part of the mechanism design and not a process to be added when the design is complete.

Non-Contact Mechanisms

The use of electromagnetics to implement non-contact mechanisms could reduce or eliminate the need for lubrication. There are a number of electromagnetic methods of supporting rotating or moving masses. Those involving permanent magnets have benefited in the last decade from the development of a new class of materials for making improved permanent magnets based on cobalt and some rare-earth elements. Although the control aspect is not trivial, controlled permanent magnets offer the possibility of very small suspension power requirement. It has already been proved that controlled DC electromagnet schemes are capable of operating satisfactorily in the most hostile environmental conditions, including high vacuum and extreme temperature environments, at fairly modest costs. However, the room for further adaption and innovation remains almost unlimited.

Particulate Control

The materials chosen for the design must be not only vacuum-compatible, but are selected and treated so that particle generation is eliminated. In order to characterize how closely the design goals and specifications are met, robot mechanisms will require not only the standard accuracy and repeatability tests, but also tests to determine particle generation (cleanliness) and outgassing (vacuum-compatibility) levels. Methods for monitoring outgassing in vacuum are fairly well developed and can be applied to the testing of robots. However, new techniques are required for measuring the accuracy and repeatability while in vacuum and for characterizing the particle contamination. The position testing requires a measuring device which can be placed in the vacuum environment. Particle contamination measurement, i.e. to characterize clean rooms, is generally accomplished by moving a volume of air through a testing device. This method cannot be used in vacuum. However, contact techniques and non-contact optical techniques *can* be implemented. One such technique involves real-time, in situ, optical monitoring of particle flux. Particle flux is a measure of particles moving towards a surface, and can be measured as the noise present in a multiply reflected laser beam. Researchers at the CRSM are presently using a laser-based particle monitoring device for particulate studies in vacuum.

Conclusions

In order to remain competitive in the IC industry, advanced manufacturing plants must be planned and constructed. We believe this will involve complete-vacuum manufacturing stations. Processing in the clean room has benefited greatly from robotics and automation, serving to increase production and lower costs. The movement into vacuum, however, leaves present robots on the outside looking in. If the industry is to avoid being crippled during this transition, development of vacuum-compatible robotics must keep pace. The CRSM is working to develop some of the necessary technology to allow robots to assist with the advanced processing which will take place in vacuum. These developments will involve the design, testing and verification of vacuum-compatible robots and mechanisms.

Acknowledgements

The work described in this document was performed at the Center for Robotic Systems in Microelectronics at the University of California, Santa Barbara, with funding from Delco Systems Operations and the National Science Foundation.

References

[1] H.H. Schicht, "Clean Room Technology: the Concept of Total Environmental Control for Advanced Industries", Vacuum, Vol. 35, No. 10/11, pp. 485-491, 1985.
[2] T. Lucas and J. Helfrich, "Total Contamination Control and Monitoring in the IC Industry", Microelectronic Manufacturing and Testing, pp. 21-22, Jan. 1986.
[3] J.B. Burnett, "Trends in Microelectronic Plants: Planning for the 1990s", Microcontamination, pp. 22-26, July 1986.
[4] John H. O'Hanlon, "A User's Guide to Vacuum Technology", John Wiley & Sons, Inc., 1980.
[5] J.B. Burnett, "Automation in High-Technology Manufacturing", Microcontamination, pp. 16-20, July 1985.
[6] J. Benyon, "What do we mean by pressure?", Vacuum, Vol. 20, No. 10, pp. 443-444, 1970.
[7] R. A. Bowling and G.B. Larrabee, "Particle Control for Semiconductor Processing in Vacuum Systems", Microcontamination Conference Proceedings, 1986.
[8] E.J. Loper and D.D. Lynch, "The HRG: A new low-noise inertial rotation sensor", 16th Joint Services Data Exchange for Inertial Systems, Los Angeles, Nov. 16, 1982.
[9] E.J. Loper and D.D. Lynch, "Projected System Performance Based on Recent HRG Test Results", Avionics Systems Conference, Oct. 31 - Nov. 3, 1983, pp. 18.1.1 - 18.1.6.
[10] E.J. Loper and D.D. Lynch, "Hemispherical Resonator Gyro: Status Reports and Test Results", National Technical Meeting of the Institute of Navigation, Jan. 17-19, 1984, San Diego, CA.
[11] S.E. Belinski, "Robot System Design for High Vacuum Operation", Master's Thesis, University of California, Santa Barbara, 1987.
[12] S.E. Belinski, W. Trento, R. Imani, S. Hackwood, "Robot Design for a Vacuum Environment", Proceedings of the NASA Workshop on Space Telerobotics, Jan. 20-22, 1987.
[13] P.V. Head, W. Allison and R.F. Willis, "Specimen Manipulators for High Resolution in Ultra-high Vacuum", Vacuum, Vol. 32, No. 10/11, pp. 641-644, 1982.
[14] N.J. Wu, "UHV Universal Manipulator with Direct Cooling", Review of Scientific Instrumentation, 56(5), May 1985, pp. 752-754.
[15] J.J. Zinck and W.H. Weinberg, "Extended Travel Ultra-high Vacuum Sample Manipulator with Two Orthogonal, Independent Rotations", Review fo Scientific Instrumentation, 56(6), June 1985, pp. 1285-1287.

Chapter 3

Vacuum Mechatronics Fundamentals

Majid Shirazi
Lakshmanan Karuppiah
Steve Belinski

3.1 Materials

The selection of materials for any vacuum application depends primarily on the specifications of the vacuum environment. These specifications include: the operating pressure and temperature range, minimum contamination levels, and the tasks to be executed within the vacuum environment. The specifications dictate the necessary properties of the material, e.g., vapor pressure at a given temperature, mechanical strength of the vessel walls and mechanism structures, compatible thermal expansion coefficient, allowable permeation rate, ease of fabrication and sealing, and the need for baking and cleaning to meet the desired specifications. This section presents the properties of materials suitable for vacuum applications, and common properties will be tabulated for ease of material selection for a given application.

Common Properties

Mechanical Strength

Vacuum enclosure material must be able to withstand the pressure differential of one atmosphere. The best structural shape is a sphere but for practical reasons (e.g., cost of fabrication, optimal use of vacuum space, accessibility, and interface to atmospheric environment) the enclosure parts may have cylindrical, planar and hemispherical shapes. Figure 3.1 and Table 3.1 indicate the dimensions of these parts for various vacuum-compatible materials.

Fabrication

The most expensive component of most vacuum systems is fabrication. Current technology reduces development and production periods and maintenance is usually simple, but welding and machining often cause problems. Since the sealing of the various parts of a vacuum enclosure is normally achieved by welding, the weldability of different metals and alloys is a primary concern when the vacuum system has a variety of interfaces. Table 3.2 lists the weldability of different metals and alloys.

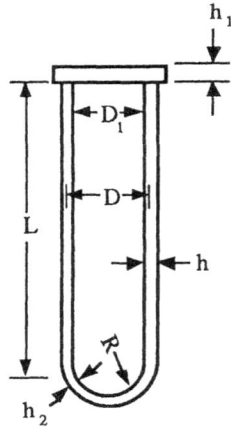

Fig. 3.1 Dimensions of cylindrical, planar and hemispherical parts of vacuum enclosures
(Adapted from [3.1], p. 330)

Material	Cylinders		End Plates		Hemispherical
	D/h	L_c/D	D_1/h_1	h_1/∂	R/h_2
Copper at 20°C	84	10	52	15	600
Copper at 500°C	58	8.5	—	—	—
Nickel at 20°C	100	11	73	8	780
Nickel at 500°C	90	10.5	—	—	—
Aluminum at 20°C	70	9	37	57	470
Aluminum at 500°C	62	8.7	—	—	—
Stainless steel 20°C	105	11.5	89	3	830
Stainless steel 500°C	89	10.5	—	—	—
Glass (hard) 20°C	70	9	16	117	470
Neoprene 20°C	2.5	1.7	10	0.2	30
Teflon 20°C	12	3.8	14	9	—
PVC (Tygon)	3.7	2.1	—	—	—
Perspex	—	—	30	—	—
Mica	—	—	58	15	—

∂ = maximum permissible deflection at center of plate
L_c = critical length for cylinder = $1.11D(D/h)^{1/2}$

Table 3.1 Dimensions corresponding to Fig. 3.1 (From [3.1], p. 330)

Fabrication

Machining should be considered closely when precise assembly is required or when stringent tolerances (for surface finish, dimensions, etc.) are specified for a vacuum-compatible mechanism. Such mechanisms are increasingly necessary for high precision work in vacuum environment manufacturing. An electropolishing process is applied to metals to produce a smooth surface (0.10 to 0.80 μm roughness). The importance of this process is noted in reference [3.4].

<u>Vacuum-Related Properties</u>

Vapor Pressure

Materials used in the vacuum environment should have low vapor pressure even at the maximum operating temperature. Vapor pressure data are summarized in Fig 3.2 a, b and c. Some common metals (Zn, Cd, Pb) have vapor pressures in the range of 10^{-2} Torr at high temperatures (400°-500°C), making these metals, and alloys which involve them, generally unsuitable for use in a high vacuum environment.

Permeation

The overall steady state flow process of the gas phase from one side of a material to the other side is called permeation. The driving force is the pressure differential across the metal envelope forming the vacuum environment. Permeation involves impact and absorption on the high pressure side, followed by solution of the gas within the material, and diffusion of the atoms or molecules onto the outgoing surface. Finally desorption takes place from the wall on the low-pressure side. Permeation of diatomic molecules varies as the square root of the pressure difference. Hydrogen, with the highest diffusion rate, is the main gas involved in permeation. The permeation constant is given by: (diffusion rate)/$(P)^{1/2}$. It is a function of temperature as shown by Figure 3.3. For practical purposes, we need only consider the permeation of hydrogen, since the permeation rate of other gases through most metals is at least an order or two lower than that of hydrogen.

Outgassing

Gases are adsorbed on material surfaces, and also dissolved within materials, mainly during manufacture and processing. This sorption can be both physical and chemical. When the material is exposed to a vacuum environment, these gases will be evolved (outgassed) at a rate that depends on the total gas absorbed, the nature of the sorption process, surface characteristics (such as oxide layers) and the temperature. For untreated metals, the initial outgassing rate at room temperature is on the order of 7.5×10^{-8} Torr•l/s•cm². The activation energy of the adsorption process is important. If a gas is chemi-absorbed with a high activation energy it might not be released; this property can likewise be employed for reducing the pressure in the system.

Surface outgassing for metals depends on surface conditions; i.e., the presence or absence of oxide layers. Sandblasting increases the surface area and breaks up the oxide layer; this in turn increases the outgassing rate for those gases adsorbed or formed on the surface.

During the first 10-100 hours of exposure, surface gases are released into the vacuum environment. The gases dissolved within the bulk material, such as H_2, N_2, O_2, CO and CO_2, diffuse at a rate inversely proportional to the square root of time. This diffusion process increases exponentially with temperature. Metals can be degassed at high temperatures prior to assembly and as part of a complete fabrication. Section 2.2 discusses the bakeout procedure.

Table 3.2 Weldability of metals and alloys ([3.1] p. 346‑7)

Legend:

T‑ torch weld. A‑ arcatom. C‑ carbonarc. H‑ heli (argon) arc. M‑ aircomatic. E‑ electron beam. P‑ cold weld. R‑ resistance weld.

1‑ very easy 2‑ good 3‑ difficult

(1) After cleaning with phosphoric acid
(2) Only OFHC copper
(3) Limitation for large parts
(4) Use electrodes of CrNi‑steel (18 Cr, 8 Ni)
(5) Vacuum‑tight for many cycles up to 480°C
(6) Not recommended for vacuum sealing
(7) Danger of leakage (cracks)

	Zr	W	Ti	Ta	Stainless Steel	Pt	Ni	Mo	Kovar	FeNi	Fe	Cu	CrNi	CrFe	Cr	Co	Brass	Be	Au	Al	Ag
Ag	–	C	–	R3	– / R3	C / R3	C	–	–	C / R3	R3	C	C / R3	–	–	–	–	–	–	A,C,H / M,R3	C,H
Al	–	–	–	R3	–	R3	–	–	–	R2	R2	R2	–	–	–	–	–	–	T,P / R2	–	
Au	–	–	–	–	–	–	–	–	–	–	–	–	–	–	–	–	–	–	–		
Be	–	–	–	–	–	–	–	–	–	–	–	–	–	–	–	–	–	E(3)			
Brass	–	–	–	–	–	–	R2	–	–	R2	R2	R2	–	–	–	–	–R2				
Co	–	–	–	–	–	–	–	–	–	–	–	–	–	–	A	C					
Cr	–	–	–	–	–	–	–	–	–	–	–	–	–	–	A,H (1)						
CrFe	–	–	–	–	–	R1 / C	R2 / C	–	–	–	R2	– / C	C	R2							
CrNi	–	C	R3	C	R2 / C	R1 / C	R2 / C	–	–	– / C	R2	C / R2,R3	–								
Cu	R3	C	R3	C	(7)	R2	(5)	(6)	(6)	C / C	R2	H(2),T / P(5),R3									
Fe	R3	R3	–	R3	–	R1	R1	R3	R3	R2 / C,H	T,R1										
FeNi	–	–	–	C	R1 / R1	C / C	C	R2 / C	C,H / P	R3 / C,H P											
Kovar	–	–	–	–	(7)	(7)	(6)	(6)	A,H(1) / (4,5)												
Mo	R2	C,A	R3	R2 / H	R2	R2 / C	A,C / E(3)	–													
Ni	H	C	R3	R2 / H	R2	R2 / C	C,H(5) / M,R1														
Pt	–	–	–	R2	H,M(5)	R1 / R1															
Stainless Steel	–	R2	–	E(3)	R2 / C																
Ta	–	R2	–	R2																	
Ti	–	C	H,E(3) / R2																		
W	R2	C,A / E																			
Zr	H / R2																				

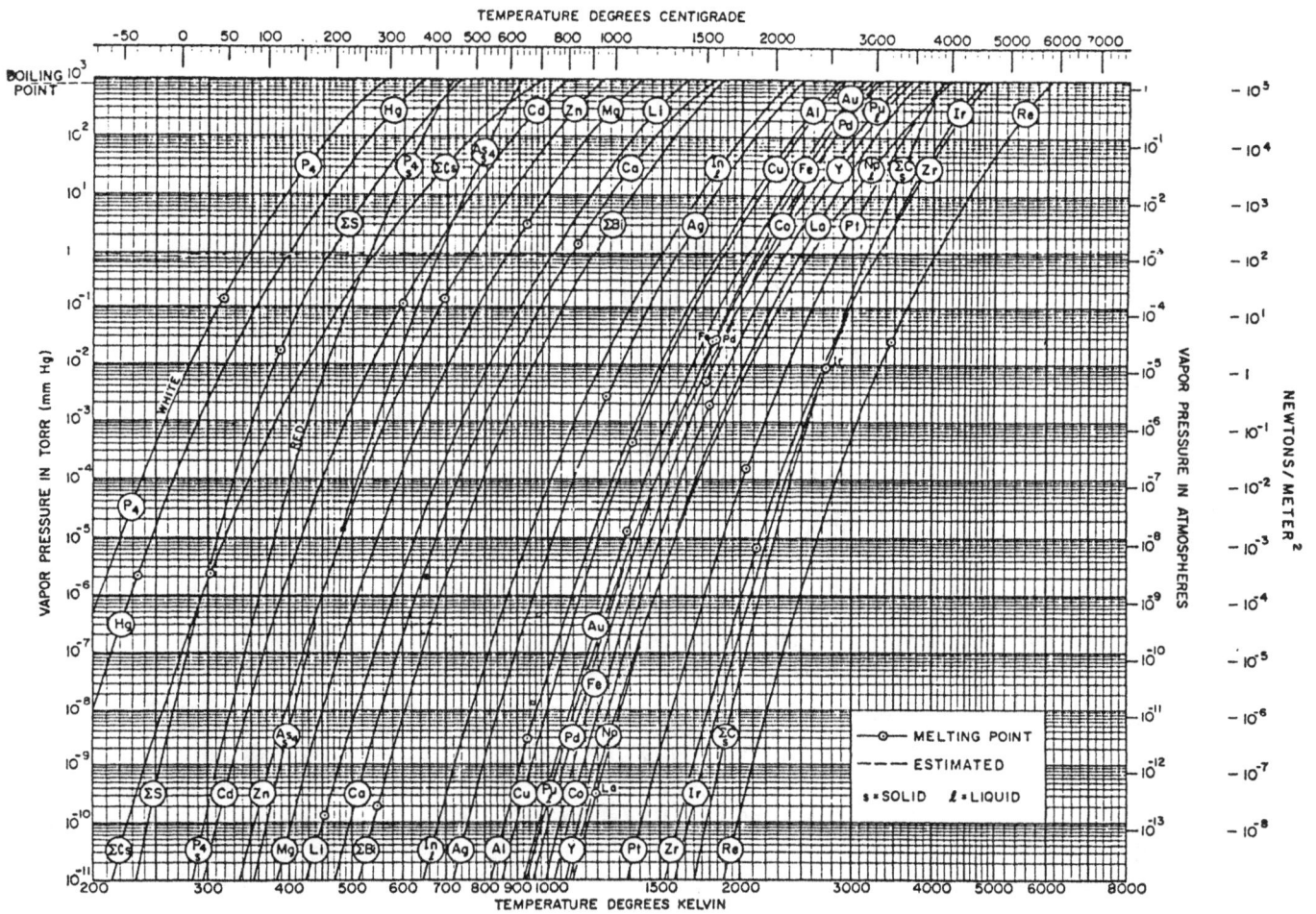

Fig. 3.2a Vapor pressure curves of the elements
([3.27], p. 292)

91

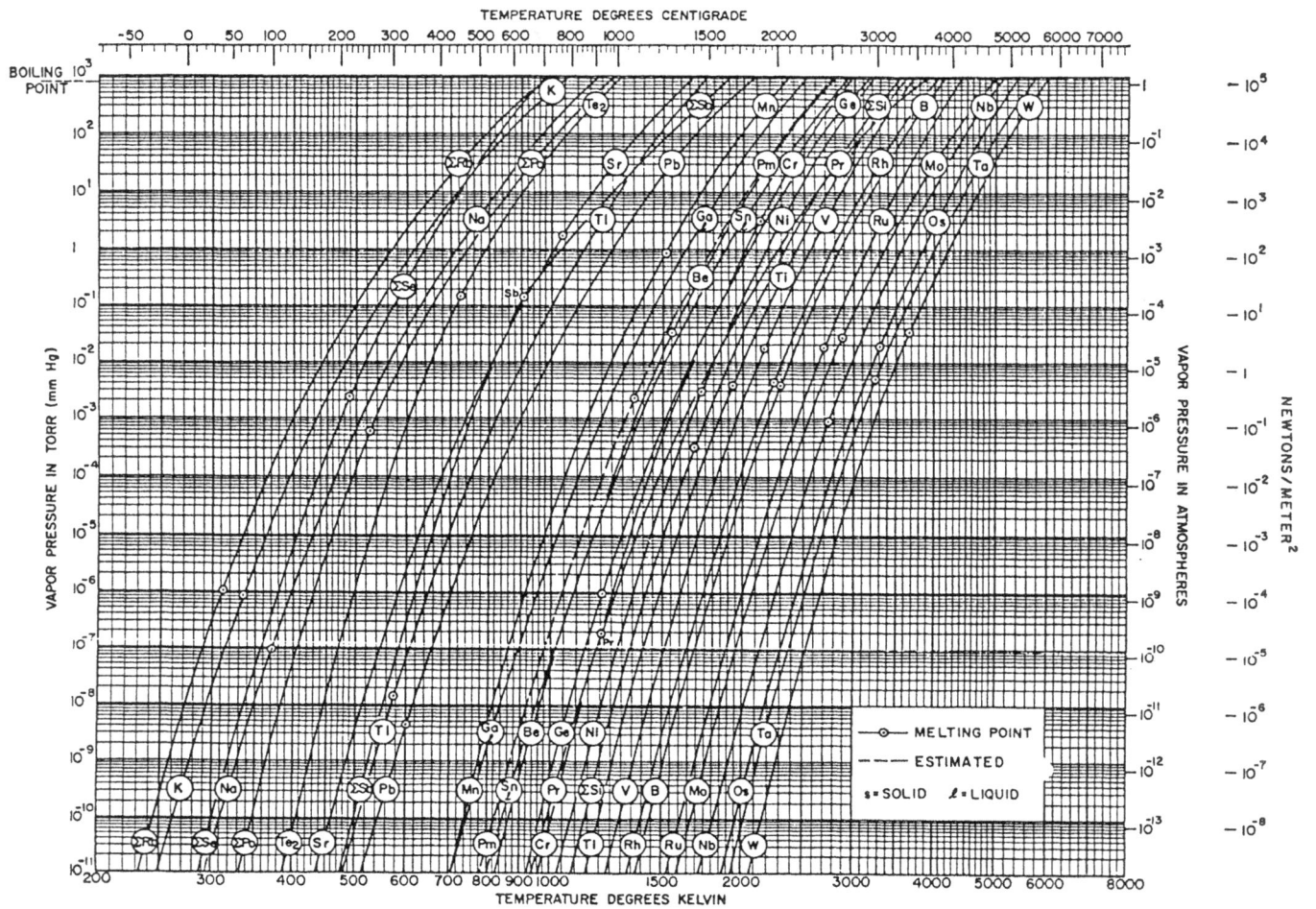

Fig. 3.2b Vapor pressure curves of the elements
([3.27], p. 293)

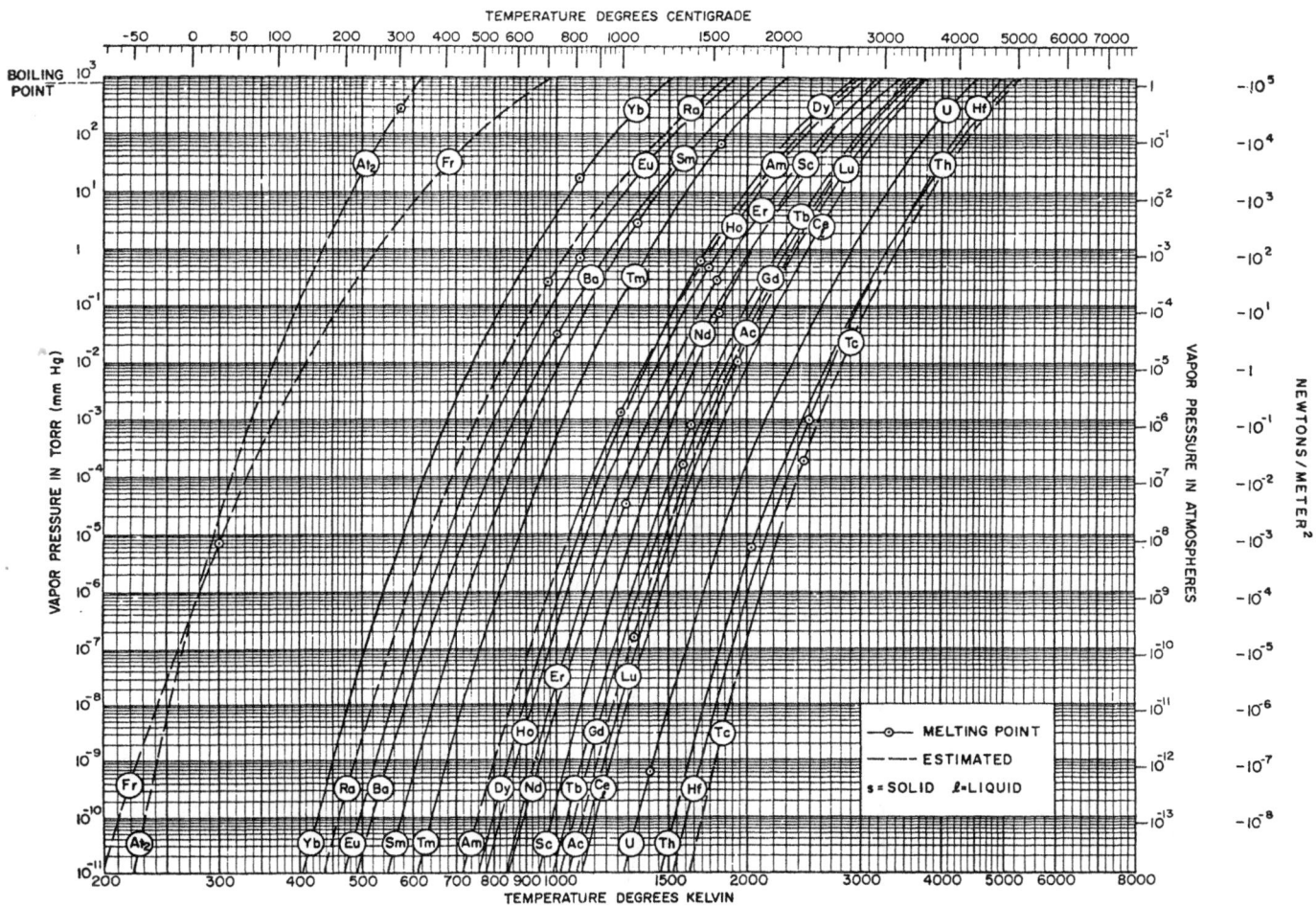

Fig. 3.2c Vapor pressure curves of the elements
([3.27], p. 294)

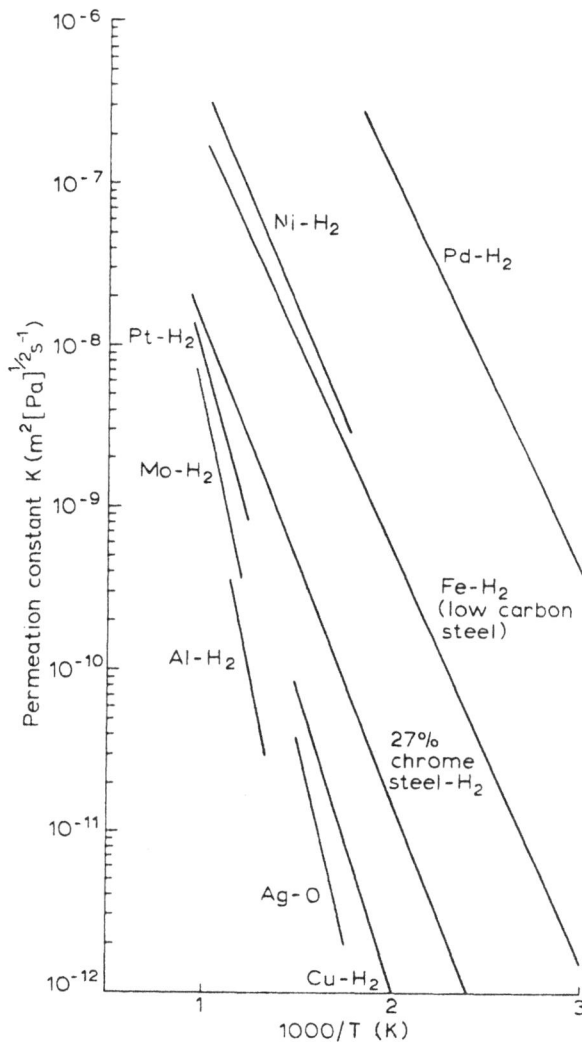

Fig. 3.3 Permeation constant for hydrogen through various metals as a function of temperature ([3.2], p. 43)
Note: 1 Pa = 7.5 x 10^{-3} Torr

Cleaning

Cleaning refers to the removal of contaminants on the material's surface, (e.g. visible dirt, oil, oxides and sulfides) which improves the operational stability of the surface. The degree of surface cleanliness must be greater for high vacuum environments. The aim of the cleaning is to obtain microstructure-free surfaces only, not atomically clean surfaces.

For metals, the most effective cleaning processes are those that remove a controlled amount of the base metal without adding a significant amount of chemicals to the bulk, such as chemical polishing, electropolishing, and chemical etching. Micro-cracks and embedded foreign materials are removed by these techniques. Pickling processes are used to remove metal oxides. This process entails dipping the part in an acid bath for a fixed time, and then rinsing and neutralizing it in an alkaline bath. Oils and greases are removed by chemical cleaning processes. Section 2.2 presents more information on cleaning for vacuum service.

<u>Physical Properties of Materials</u>

This section provides basic background information on various materials for use in vacuum. The reprint by Weston at the end of the chapter provides many more details about the nature of metals, glass and ceramics. In another reprint, Moraw describes his attempt to determine some phenomenological parameters describing the outgassing process for metals. T.J. Patrick, in a third reprint on materials, presents a history of space technology, a description of the spacecraft environment and an analysis of spacecraft materials.

Metals and Metal Alloys

The common metals and alloys used for vacuum work are: stainless steel, structural steel, nickel, tungsten and aluminum. Properties of various metals are as follows:

Stainless steel (304): Prime constructional material for UHV systems. The alloys are strong, non-magnetic, and corrosion-resistant. They are fairly hard to machine and are hardened by cold working. Chromium oxide exists on the surface and can be removed by immersion in nitric acid at $50^{\circ}C$ for 30 minutes.

Structural steel: These alloys are ferromagnetic and are not corrosion-resistant. Thus, the alloy cannot be used in any environment with magnetic analyzers and must be stored in a dry, clean area after cleaning.

Nickel: Has a high melting-point and low vapor pressure, is easily formed, outgassed and spot-welded, and is resistant to a variety of corrosives (water, salt water, alkalies and most organic acids). Nickel is ferromagnetic with a curie temperature of $350^{\circ}C$. It is relatively low in cost.

Aluminum: Aluminum is ductile, with low tensile strength, and is light and corrosion-resistant by virtue of its thin oxide layer. It cannot be baked; its strength decreases rapidly at about $200^{\circ}C$

Tungsten: Has a very high melting-point ($3400^{\circ}C$) and a correspondingly low vapor pressure. It is non-magnetic, and chemically very stable. Applications include: X-ray tube anodes, spring elements, high-temperature thermocouples and welding electrodes.

Ceramics

"Ceramics" refers to a wide range of inorganic nonmetallic compounds which have attained a hard, solid crystalline state as a result of a form of baking called "firing." Ceramics are good insulators, and have better mechanical strength than glass at elevated temperatures. They can be fabricated with close dimensional tolerances. Ceramics are chemically inactive and have low vapor pressure.

There are three types of ceramics:

1) pure oxides
2) silicates and special types of nitrides
3) borides and carbides.

95

Oxide ceramics are produced synthetically; their quality is controlled and their properties are consistent. For these reasons, oxide ceramics are preferable to other ceramics for vacuum applications. Alumina-body ceramics, of which 85% to nearly 100% are comprised of Al_2O_3, are the most frequently employed oxide ceramics. They have a mechanical strength of 18×10^6 kgm^{-2}, a softening temperature of 1650 °C, a coefficient of thermal expansion of about 60×10^{-7}/°C, and a permeation constant of about 10^{-14} m^2s^{-1}. The outgassing rate at room temperature is approximately 7.5×10^{-15} Torr•l/s•cm^2.

Glass

Glass, with silica as the main constituent, is used in the electronic tube industry because it has outstanding sealing properties in comparison to particular metals. The types of glass employed are borosilicate (hard) glass, and soft glass with lead oxide as the additive. Soft glass is used in low vacuum systems. For high vacuum systems, hard glass is used, since it can be baked above 500°C. Outgassing for hard glass is in the range of 7.5×10^{-9} Torr•l/s•cm^2, and this can be reduced to 7.5×10^{-15} Torr•l/s•cm^2 by baking at 500 °C for a period of 24 hours. The intrinsic vapor pressure of glass is extremely low (10^{-15} to 10^{-25} Torr), but gases are trapped and adsorbed at the surface during the manufacturing process and storage.

Other Materials

Over the past few years, synthetic materials such as plastics, elastomers and epoxy resins have been used mainly as gaskets for static and motion seals in demountable systems and for flexible tubing in vacuum systems. Important characteristics include: low vapor pressure, low gas permeability and outgassing, wide temperature range, and elastic properties such as compressibility and lack of permanent deformation after compression. These materials have high outgassing rates ($\approx 7.5 \times 10^{-9}$ Torr•l/s•cm^2) due to gases absorbed during the manufacturing process, and also are permeable to some extent with all gases.

Vacuum-Compatible Material Coatings

Some metals and alloys can be made vacuum-compatible by applying a suitable coating. This is done in cases where the raw material and fabrication cost would make the end product more cost-effective and yet reasonable in quality. For example, structural steel is less expensive than stainless steel, and also easier to fabricate, but it is not corrosion resistant. By coating it with a suitable metal, it can be made vacuum-compatible. Nickel, for instance, has been used to electroplate structural steel for high-vacuum applications. In such applications, the electroplated structure must be handled with care in order to avoid scratches; otherwise the bulk metal might be exposed to the vacuum environment.

3.2 Lubrication

The importance of lubrication to overcome the effects of friction is generally understood for moving mechanisms at atmospheric pressure. In a vacuum environment an additional set of effects must be considered, some of which are very critical to the life of the mechanism as well as the performance of the overall system.

Lubricants are used to reduce friction and wear, and to act as a heat transfer medium. Both aspects will be discussed, especially as their effects may interact in the vacuum mechatronics environment.

Friction produced by the rubbing motion between mechanical parts results in heat generation and the wear of components. The frictional force is proportional to the contact pressure, surface roughness, and the shear properties of the lubricating material, which separates the mating surfaces by means of a viscous fluid (wet lubricant) or low shear strength solid film (dry lubricant). Lowering this frictional force reduces dissipation of mechanical power at the joint and therefore heat generation at the contacting surfaces.

Lubrication is used to reduce wear and friction in machine elements such as thrust and roller bearings, bushings, lead screws, gear trains, etc. In fact, without proper lubrication in a vacuum environment, parts often seize or essentially weld together. Lubricants are also used to increase the heat transfer at joints. In vacuum, heat transfer across a joint may take place by conduction through the contact surfaces, or radiation across the gaps. Convection is absent in vacuum and radiation is mainly effective at high temperatures. Therefore, conduction must be addressed as the useful mode of heat transfer. A wet lubricant with high thermal conductivity and low vapor pressure can be used to fill gaps and increase the heat transfer through the contact area.

The factors involved in the friction between metals are divided into two categories:

> 1) the surface behavior, and
> 2) the external environment.

The first category involves the surface's physical, chemical and metallurgical properties. This includes surface roughness, stability of surface oxides, crystal structure and orientation at the surface. The second category takes into account the influence of the environment on the metal surfaces. When operating a mechanism at or above 10^{-5} Torr, the oxide layer and water vapor on the surface act as a lubricant by separating the sliding surfaces. At lower pressures surfaces are degassed and the oxide layers on the surfaces are removed, resulting in a large increase in friction. For example, copper on copper has a coefficient of friction on the order of 0.5 at atmospheric pressure and about 4.5 at a vacuum level of 10^{-6} Torr [3.11].

In this chapter, the important properties of wet and dry vacuum-compatible lubricants are reviewed. Their limitations and characteristics are specified in tables. The tribological characteristics of composite lubricants and soft metals for the vacuum environment are also tabulated. Two reprints in this chapter provide detailed information on lubrication for vacuum. The reprint by K.G. Roller is an insightful overview for general vacuum service, while the paper by Briscoe and Todd focuses mainly on the lubrication of spacecraft mechanisms.

Vacuum Lubricants

Conventional lubricants usually include oxygen, water vapor and nitrogen gases as components. Oxygen and water vapor provide a protective film on the metal surfaces which decreases friction and wear. When the lubricant is exposed to a vacuum, the water vapor and oxygen evaporate, making the lubricant less effective and leading to increased friction and wear. There are a number of techniques used to reduce friction and wear in vacuum, including the use of low volatility wet lubricants, dry lubricants, and composites which include lubricating pigments.

Wet Lubricants

Wet lubricants for vacuum include oils and greases. They operate by separating the contacting surfaces and creating a low shear film of material between them. They are less volatile than the conventional lubricants used in air and also possess high viscosity. Some critical properties of oils for vacuum applications are:

- low vapor pressure and evaporation rate
  - low surface migration rate
  - chemical stability

The vapor pressure and evaporation rate must be low in order to prevent contamination and the loss of lubrication. The vapors generated from the oil work directly against the vacuum pumps and may contaminate sensitive components such as optical systems. The vapor pressures of some oils and their molecular weights are listed in Table 3.3. The evaporation rate, measured by the weight loss, is sensitive to the chemical composition and structure of the oil. Straight chain structured wet lubricants evaporate less readily than those containing bulky side branching. The comparative weight loss data of some oils after exposure to a 10^{-5} Torr vacuum for 100 hours are shown in Table 3.4. Oils might have other additives such as extreme pressure (EP) additives and other substances to improve anti-wear properties. The evaporation rate is then a function of the vapor pressure of the base oil and the vapor pressure of additives.

Low surface migration rates are important for effective lubrication in the vacuum. They insure that the lubricant does not creep and spread out over the surface of the metal in vacuum. As the surface area exposed to vacuum increases, the amount of lubricant lost due to evaporation also increases. The bearing design should include anti-creep barriers to prevent surface migration. The rate of surface migration is usually highest for silicone oils, followed by fluorocarbons and then hydrocarbons [3.6].

In order to maintain lubricity of the surface, the component design should include an oil reservoir and delivering system to the contacting surfaces. For example, one design of ball bearings for vacuum includes porous ball separators which are impregnated with low vapor pressure oil. The separator acts as an oil reservoir for excess oil and as a metering device. It slows down the evaporation rate by preventing the rapid diffusion of oil to the surface.

Oil	Mol. Wt.	P (Torr) at 20°C	Viscosity
Apiezon C	574	10^{-8}	67.4 cP at 40°C
Apiezon BW	468	10^{-6}	233 cP at 20°C
Convoil 20	(400)	3×10^{-7}	--
Octoil-S	426.7	1.5×10^{-8}	--
Glycerol		$<3 \times 10^{-6}$	--
Silicone, DC 703	570	6×10^{-9}	50 cSt at 25°C
m-bix-(m-phenoxy phenoxy)-phenyl-ether	--	10^{-12}	--
bis-m-(m-phenoxy phenoxy-)-phenyl-ether	--	10^{-15}	--

Table 3.3 Vapor Pressures of Oils (adapted from [3.13], p. 392)

cP = centiPoise = 10^{-12} Poise
cST = centiStoke = 10^{-2} Stoke

Oil	Wt Loss (mg)
Modified diester	3000
Paraffinic petroleum	2500
Petroleum bright stock	2000
Polyalkalene glycol	1800
Phenylmethyl polysiloxane (medium phenyl)	650
Chlorophenyl methyl poly-siloxane	400
Dimethyl polysiloxane (branched)	340
Dimethyl polysiloxane (straight chain)	200

Table 3.4 Comparative Weight Loss Data at 10^{-5} Torr, 140-160°F, 100 Hr
(Adapted from [3.13], p. 393)

Greases are formed by compounding oil with thickeners and other additives. They are primarily used in ball and roller bearings, sliding surfaces, and gears. Greases are also mixed with dry powder lubricants such as MoS_2 for heavy load applications (at gear meshes). The solid component prevents the direct metal-to-metal contact and lowers the wear rate and mechanical noise. For more information on greases and other wet lubricants, refer to [3.5].

Wet Lubricant Application

Prior to the application of fluid lubricants to a surface, the surface must be pre-treated by removing surface components and residues that might outgas, such as glue, oils, greases, paint, dirt and other contaminants. This maximizes the wetting action of the lubricant and also preserves its properties by eliminating the chance of a chemical reaction between the lubricant and chemicals present at the surface. The surface is then exposed to a vacuum and bake-out cycle to free it from oxygen and other chemicals. Before breaking the vacuum, the clean component is then inserted in the wet lubricant for several hours. This will ensure absorption of the lubricant by the part's surface. The excess lubricant can be removed by draining after the part is removed from vacuum.

Dry Lubricants

Solid lubricants are used when the operation of the mechanism in vacuum involves extreme temperatures and the molecular contamination is to be minimized [3.6]. This is especially important when intelligent systems with a large number of sensors are to operate in a vacuum environment, since gases from fluid lubricants may contaminate optical and sensitive components.

In some dry lubricants, inorganic strong layer lattice structures are bonded together with weak Van der Waal-type forces, allowing layers to slide very easily over one another. A large resistance to the force normal to the layers exists, which enables the dry lubricant to take large pressure loads. Graphite and the inorganic lubricants have these types of structures and are classified as "lamellar solids" (layer type structures). Some lubricants of this type are listed in Table 3.5, along with their structures and coefficients of friction.

99

Solid	Crystal System	Coefficient of Friction Air	Coefficient of Friction Vacuum	Pressure (Torr)
MoS_2	MoS_2	0.18	0.07	2×10^{-9}
WS_2	MoS_2	0.17	0.13	3×10^{-9}
WS_2	MoS_2	0.17	0.08	1×10^{-5}
BiI_3	AsI_3	0.34	0.39	5×10^{-7}
LiOH	LiOH	0.37	0.21	2×10^{-9}
$CdCl_2$	$CdCl_2$	0.35	0.16	2×10^{-9}
CdI_2	CdI_2	0.24	0.18	2×10^{-9}
$CdBr_2$	CdI_2	0.22	0.15	2×10^{-9}
Phthalocyanine	—	0.35	0.33	1×10^{-6}

Table 3.5 Coefficients of friction of some lamellar solids (From [3.10], p. 160)

Fluorocarbons such as Polytetrafluoroethylene (PTFE) and similar materials (CTFE, FEP) are also dry lubricants and are widely known as "Teflon" (Teflon is the trade name of DuPont for its line of fluorocarbon products). They have a waxy texture rather than a layer structure. Other materials used as dry lubricants include plastics, nylon, and soft low shear metals such as silver, gold and lead.

Solid lubricants have less evaporative loss and can operate at higher temperatures than fluid lubricants. Thus, they can often provide lubrication for the life of the part under moderate loads. They can also operate in extremely high load conditions with relatively little change in friction as the temperature fluctuates. The coefficient of friction actually decreases as the load increases. Solid lubricants, however, generally have a higher coefficient of friction than fluid lubricants. When used with heavy loads, solid lubricants have a shorter life than the fluid lubricants, due to wear. Dry lubricants cannot be replenished as well as fluid lubricants because they do not flow and spread over the components. They also generate debris and particles which can degrade the operation of mechanical components, such as bearings. The high temperature cure cycles for some dry lubricants may damage the mechanical properties of various materials.

Many natural minerals, e.g. graphite, mica, molybdenum disulfide, and synthetically made materials such as heavy metal dichalcogenides, have the properties listed above. Graphite is not used in vacuum since it requires some moisture (water) in order to have good lubricity. The most commonly used dry lubricants for vacuum applications are PTFE, lead, and MoS_2 because of their performance, availability and lower cost.

Examples of heavy metal dichalcogenides are disulfides, diselenides, and ditellurides of metals tungsten, niobium, and tantalum. They are produced by the direct combination of elements at elevated temperatures and are relatively expensive. Dichalcogenides normally have better friction and wear properties, and are also quite expensive.

The outgassing properties of some dry lubricants are shown in Figure 3.4. The outgassing rates of PTFE and lead are noted in the reprint by Roller. Graphs depicting coefficients of friction vs. load and speed are shown in Figures 3.5 – 3.7.

For more information on the availability of dry lubricants for vacuum, refer to part A of [3.5].

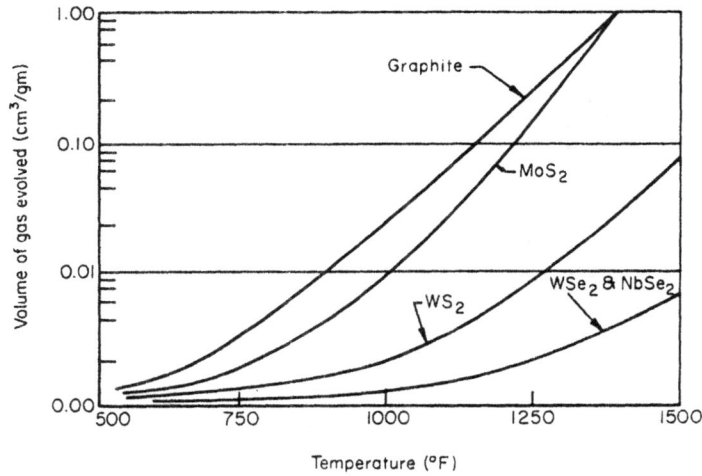

Fig. 3.4 Outgassing characteristics in vacuum of graphite, MoS$_2$, WS$_2$,WSe$_2$, and NbSe$_2$ (From [3.11], p. 119)

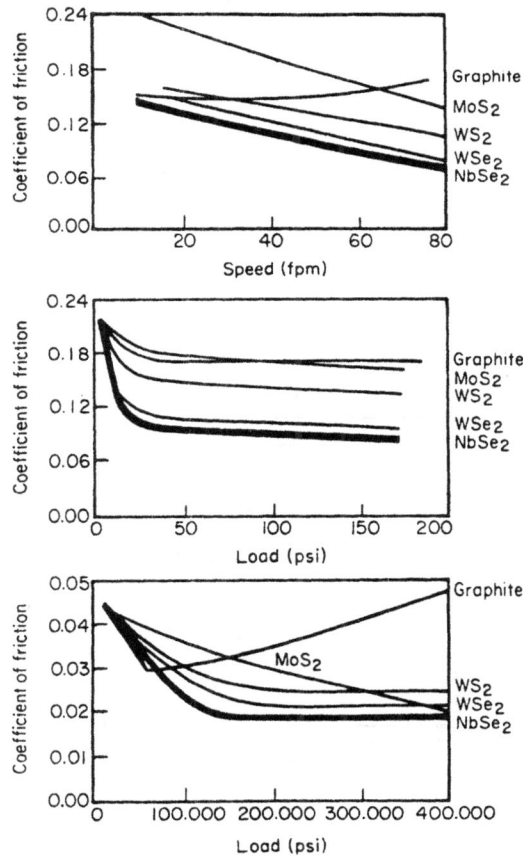

Fig. 3.5 Effects of speed and load on the coefficients of friction of graphite, MoS$_2$, WS$_2$,WSe$_2$, and NbSe$_2$
(From [3.11], p. 117)

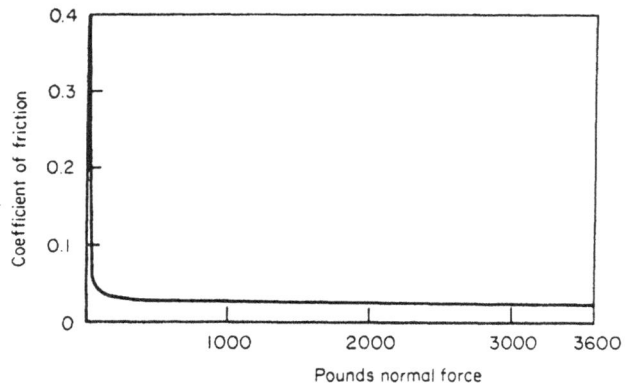

Fig. 3.6 Effect of load on friction of Teflon PTFE (From [3.11], p. 202)

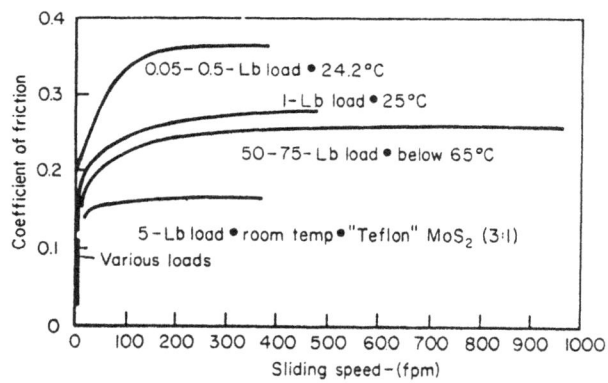

Fig. 3.7 Effect of speed on friction of Teflon PTFE (From [3.11], p. 202)

Dry Lubricant Application

The component part is pre-treated and prepared before the application of the dry lubricant in order to provide good adherence between the surface and lubricant. The pre-treatment is a function of the substrate material, bonding resin employed, and the application procedure. It consists of cleaning the component to remove dirt, grease, oil, surface scale, etc. The surface is then treated by chemical or mechanical means to insure a better mechanical bond with the dry lubricant [3.5].

Dry lubricants are normally divided into three categories, depending on the method of attachment to the substrate:

- unbonded
- resin-bonded
- sputtered

In the unbonded type, the powder lubricant is applied by spraying, brushing or polishing the substrate surface. The resin-bonded film lubricant is a widely used solid lubricant [3.5]. It consists of lubricating pigments in a binder, which is used to prevent the complete loss of the pigment as a result of the relative motion of parts. The film is applied in the form of spraying, dipping or brushing; spray methods have produced the most satisfactory results. This process is followed by a curing process for polymerizing the resin binder. The thickness ranges from 5 μm to 20 μm for spray coating. The life of the coatings is from 7500 to 15,000 sliding cycles under a load of 450 kg/cm^2 [3.6]. The sputtering technique is used for applications where close tolerances are necessary, e.g. ball bearings where thick debris can cause damage. This technique generally exhibits adhesion and wear life which is superior to other types of coating. The sputtering is performed in a low pressure ($<10^{-5}$ Torr) inert gas environment which is ionized by the application of a voltage potential. The positively charged ions of inert gas are accelerated with enough energy to knock off or *sputter* the negatively charged target material. The sputtered material (lubricant) is then deposited on the substrate, which is positioned near the target source. The lubricant thickness is normally in the range of 0.08 to 0.2 μm.

Composite Lubricants

Many plastics and metals can be combined with other lubricants to provide better friction and wear life characteristics. The addition of a lubricant to plastics also increases their dimensional stability and thermal conductivity. Their use is included in the fabrication of ball and roller bearings, bushings, seals, valve seats, gears, motor brushes, slip rings, and electrical contacts. Table 3.6 lists properties of some composite lubricants.

Composition (by wt.)	Density (gm/cm^3)	Coeff.of Friction	Hardness (Rockwell)	Max. Service temp.(°C)	
				Vacuum	Air
90-10 Ag-CbSe$_2$	8.82-9.31	0.10-0.18	M50-70	400	350
85-15 Ag-CbSe$_2$	8.57-9.04	0.08-0.17	M50-70	400	350
80-20 Ag-CbSe$_2$	8.32-8.78	0.07-0.15	M50-70	400	350
70-30 Ag-CbSe$_2$	7.85-8.28	0.06-0.14	M50-70	400	350
90-10 Ag-MoS$_2$	8.45-8.92	0.10-0.19	---	400	260
80-20 Ag-MoS$_2$	7.63-8.06	0.08-0.18	---	400	260
85-15 Ag-WSe$_2$	9.22-9.73	0.10-0.18	M60-80	400	180
75-25 Ag-WSe$_2$	9.07-9.57	0.07-0.15	M60-80	400	180
70-20-10 Ag-PTFE-CbSe$_2$	6.14-6.48	0.06-0.15	H60-80	260	260
70-20-10 Ag-PTFE-WSe$_2$	5.29-5.59	0.06-0.15	H60-80	260	260
90-10 Ni-WS$_2$	7.87-8.31	0.12-0.20	B90-110	540	350
80-20 Ni-WS$_2$	7.72-8.14	0.08-0.15	B85-105	540	350
65-35 Ni-WS$_2$	7.52-7.94	0.05-0.12	B75-95	540	350
95-5 Ni-CaF$_2$	7.35-7.76	0.30-0.40	B60-70	650	650
90-10 Ni-CaF$_2$	7.67-7.04	0.28-0.35	B65-80	650	650
85-15 Ni-CaF$_2$	6.31-6.66	0.28-0.33	B70-85	650	650
90-10 Ni-CbSe$_2$	7.69-8.12	0.12-0.20	B90-110	650	350
80-20 Ni-CbSe$_2$	7.38-7.79	0.08-0.18	B85-105	650	350
85-15 Ni-WSe$_2$	8.02-8.46	0.11-0.20	B110-120	650	350
75-25 Ni-WSe$_2$	8.04-8.48	0.09-0.17	B95-105	650	350
90-10 Ni-MoS$_2$	7.24-7.64	0.08-0.15	B70-80	400	260
85-15 Ni-MoS$_2$	7.10-7.50	0.07-0.15	B70-80	400	260

Table 3.6 Properties of composite lubricants ([3.10], p. 120)

3.3 Energy Transfer

Energy transfer for vacuum environments encompasses heat transfer, heat generation control, and methods of power transmission to the elements inside a vacuum vessel. In an operating mechanism, heat is generated by friction between materials in contact, and by the thermal properties of motors or other devices with resistive elements. This heat must be conducted away from motors, e.g. through the robot joint, when moving a robot arm. The temperature of the motor housing may rise to a critical point and the motor will fail if the cooling method is inadequate.

The system must be designed to allow removal of any excess heat. The rate of heat transfer depends on environmental temperature, material temperature and thermal properties. In typical atmospheric conditions, this task is performed largely by air convection, conduction of heat through the material, and to some extent by radiation. In a vacuum environment heat can be removed *only* through conduction and radiation. At low temperatures, the radiation is negligible and virtually all heat transfer is by conduction.

Another method of reducing a mechanism's temperature is to minimize the heat generation by proper design of its heat-generating components. For example, the generation of heat in a motor is directly proportional to the electrical resistance of the stator and armature coils. It is also proportional to the current applied to the coils. Therefore, by reducing these two parameters, heat generation can be minimized and a more efficient motor obtained.

An additional energy transfer consideration is the supply and transmission of electrical energy to the system inside the vacuum vessel. The supply system of an artificial vacuum is normally outside the vacuum and therefore no constraints exist. However, for natural vacuum (space) the supply system is either a solar collector which converts the solar radiation directly into electrical current, or perhaps an AC/DC supply which is housed inside a small pressurized chamber. There are two main types of power transmitting mechanisms: those that use electrical wires, and those that use microwaves or lasers [3.25]. The first method is used in short-distance connections, and the second method is projected for use in space and military applications.

In this section, general heat transfer modes and methods for cooling vacuum mechanisms are reviewed. Electrical energy transmission methods are also discussed. The reprint in this chapter, by Madhusudana, reviews recent developments in the area of contact heat transfer.

Basics of Heat Transfer

Heat Transfer is the science of predicting energy transfer as a function of time through a temperature gradient. Heat flows from the higher temperature region of a body to the lower temperature region. The rate of flow and temperature variation can be estimated as a function of time using thermal models. There are three modes of heat transfer:

        1) conduction
        2) convection
        3) radiation.

Conduction is a result of the temperature gradient existing in a body, and the heat flux (heat per unit area) is represented as:

$$q = -k\frac{\partial T}{\partial x} \qquad \text{Eq.(3.1)}$$

where q is the heat flux, k is the thermal conductivity of the material and $\partial T/\partial x$ is the temperature gradient in the direction of the heat flow. The minus sign is inserted to satisfy the second law of thermodynamics. The heat transfer rate can be calculated by:

$$Q=qA \qquad \text{Eq.(3.2)}$$

where Q is the heat transfer rate and A is the area perpendicular to the heat flow direction. It is noted that if k or A is increased, the heat transfer Q will increase. Conduction heat transfer is the most significant mode in vacuum when the temperatures of the bodies are low.

Convection is a form of heat transfer which occurs when fluid flow is involved. In this case, Newton's cooling law applies and the heat flow rate is a function of the temperature difference (not the temperature gradient):

$$Q=hA\,(\Delta T) \qquad \text{Eq.(3.3)}$$

where h is called the convective heat transfer coefficient and ΔT is the overall temperature difference between the solid and fluid. The unit of h in SI units is watts per square meter per degrees Celsius. The convective heat transfer coefficient is generally a function of fluid viscosity and the thermal properties (thermal conductivity, specific heat, density). The value of h has been estimated for different situations both theoretically and experimentally. [3.15] In vacuum applications this mode of heat transfer is used only in the form of forced convection.

Another mode of heat transfer is radiation. It is different from the other modes of heat transfer in that no medium is required to carry the heat. Thermal radiation is a part of electromagnetic radiation which is emitted by a body due to its temperature. It lies in the 0.1 to 100 μm wavelength range of the electromagnetic spectrum. Thermal energy is propagated in the form of energy particles (quanta) which travel at the speed of light (300,000 km/sec). The total heat flow between two bodies is:

$$Q=F_\varepsilon\,F_G\,sA\,(T_1^4 - T_2^4) \qquad \text{Eq. (3.4)}$$

where s is the Stefan-Boltzmann constant = 5.669×10^{-8} W/m²•K₄, and where T_1 and T_2 are the absolute temperatures of the first and second bodies in Kelvin, respectively. F_ε is the emissivity function and F_G is the geometric view-factor function. F_ε relates the radiation of gray surfaces to that of an ideal black surface. F_G is dependent on the orientation of the two surfaces relative to each other. Radiation heat transfer is significant in space applications where large temperature differences exist between bodies.

Cooling Methods in the Vacuum Environment

Forced Convection

Forced convection is widely used in cooling electronic devices with long-duty cycles and may be applied to vacuum-compatible motors. The idea is to use a heat-exchanger mechanism which is attached to the motor housing, thereby removing heat from the components. The heat exchanger consists of cold plates, pipes, a radiator and a pumping unit to push the coolant (liquid nitrogen, water, or hydraulic oil) through the system (see Fig.3.8). The system is similar to an automobile engine-radiator cooling system.

The function of the cold plates is to remove the excess heat from the electronic components by raising the temperature of the coolant. The plates must be made of highly conductive and vacuum-

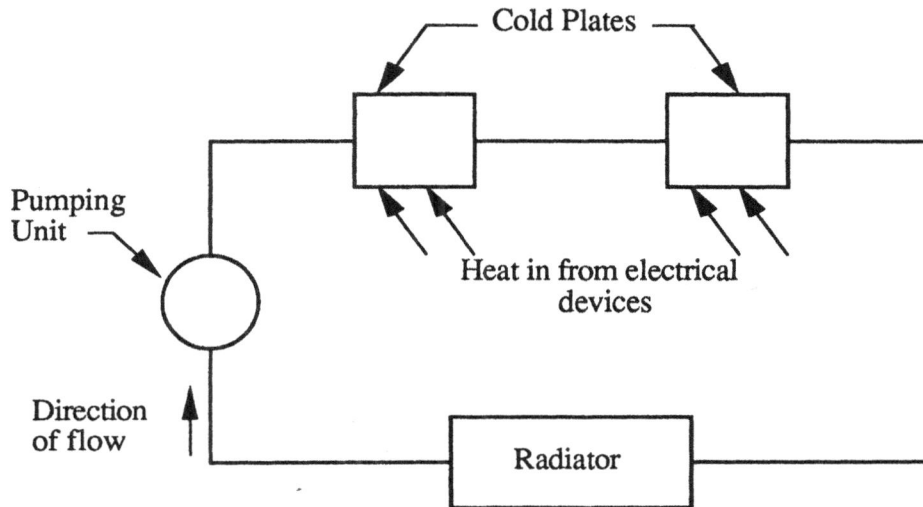

Fig. 3.8 Forced Convection Heat Transfer Loop

compatible materials such as copper, silver, or aluminum. The pipes or thermal bus, as they are sometimes called, deliver the heat (hot fluid) to the radiator where it is rejected. The pump is used to compensate for the frictional losses of the fluid in the thermal bus. Peterson [3.16] documents extensive work that has been performed in designing the two-phase flow systems. Examples of these types of cooling methods can be found in the work by Chaban [3.20] and George [3.21].

Conduction

Another method of removing heat from a component in vacuum is conduction. Components are placed in contact with a colder body such as a cold plate. Thermal resistance, which exists at the joining points of the bodies, is thus minimized to increase the heat transfer rate. Thermal resistance is defined as the ratio of temperature drop at the interface to the heat flux through the joint.

The thermal resistance at the mounting face of the heat-generating elements is a function of the true contact area. It depends on surface properties (roughness, profile height and slope) and the normal pressure on the faces. For nominally flat rough surfaces in vacuum, it has been found that the thermal resistance at the joint is inversely proportional to the pressure load and mean absolute slope of surface profiles [3.17].

Sometimes filler materials are used to enhance the joint's heat transfer capability at contact surfaces. Silicon grease, lithium and molykote lubricants have been used to decrease the thermal resistance of stainless steel joints under vacuum conditions [3.18].

Radiation

A third method of cooling components in vacuum is thermal radiation. This is an effective method when the chamber wall is kept at a very low temperature, using liquid nitrogen or other coolants. The rate of heat transfer is dependent on the emissivity of the component's surfaces, the inner wall of the vacuum chamber, and the temperatures.

107

For a cylindrical vacuum chamber with a body inside the chamber, the following relation exists [3.15]:

$$Q = sA_1 \frac{T_1^4 - T_2^4}{\frac{1}{\varepsilon_1} + \frac{A_1}{A_2}(\frac{1}{\varepsilon_2} - 1)}$$ Eq.(3.5)

where
Q   = total radiation heat transfer
s   = 5.669×10^{-8} W/m$^2 \cdot$K^4
A_1 = surface area of the body,
A_2 = inner surface area of vacuum chamber
T_1 = average temperature of body
T_2 = average temperature of vacuum chamber wall
ε_1 = emissivity of the body
ε_2 = emissivity of the vacuum chamber wall.

When the chamber is much larger than the body or when $\frac{A_1}{A_2} \approx 0$, then

$$Q = sA_1\varepsilon_1 (T_1^4 - T_2^4)$$ Eq.(3.6)

As is seen in the above equation, the total radiation is dependent on the emissivity of the body in the chamber. If the component's surfaces are not emissive, then they must be coated with a high-emissivity material in order to increase the cooling rate. The coating must be vacuum-compatible and have low vapor pressure at the operating temperature. Some common coatings are shown in Table 3.7. Radiation heat transfer in the natural vacuum of space is discussed by Favorskiy and Kadaner in [3.23].

Type	α_s/ε	ε	Maximum temperature
White organic	0.2-0.3	0.9	200-300
Black organic	0.9	0.9	300-400
Anodized aluminum*	0.12-0.30	0.9	350
Waterglass enamel	0.2-0.3	0.8	500 (depends on metal substrate)
Porcelain enamel	0.25-0.35	0.8	600-700
Oxidized stainless steel	0.8	0.85	600-650 in vacuum
Oxidized super alloys	0.8	0.85	600-650 in vacuum
Alumina	0.22-0.4	0.6-0.8	Depends on metal
Refractory enamels	0.5-0.8	0.6	1000
$MoSi_2 + Cr_2)_3$	0.8	0.6	1200

* The 0.12 value is for anodized high-purity aluminum prior to sealing.

α_s= surface absorption, ε = emissivity
Table 3.7 Radiator coatings (from [3.22] p.121)

Minimizing Heat Generation

High thermal efficiency is a mandatory requirement for systems designed to operate in vacuum. It is necessary to minimize the energy losses in devices to reduce the generated heat. Vacuum lubricants can be used to decrease the friction and power dissipation at the sliding faces. Their characteristics and uses are described in Section 3.2. Power losses in electronic components can be minimized by utilizing efficient circuit technologies which reduce both thermal dissipation and device stress. In the future, new superconductor technologies may contribute to this area.

Another method is to decrease the operational cycle time of heat-generating elements. For example, when it is necessary to hold a specific position, a brake mechanism should be used instead of the servo system, since the total energy loss in a motor is a function of the operation cycle time.

Electrical Power Transmission

It is necessary to transmit electrical power to mechanisms, actuators and sensors inside the vacuum. One method is the use of wires with special vacuum-compatible insulation. Plastics and epoxies must be selected by comparing their vacuum outgassing rate, mechanical strength, ease of manufacture, and electrical breakdown strength in vacuum. Inorganic materials with low vapor pressure are sometimes added to epoxies or plastics to reduce their outgassing rate and preserve their dimensional properties during the bake-out cycle. Some experimental data on these materials have been reported by Rosenblum [3.26].

Another method of electrical power transmission is through the use of microwave or laser power beaming. Although this method has not been employed for robotics with a short range of transmission distance, it is suitable for the vacuum environment where the control procedure requires minimal disturbances from wires twisting across the components with relative motion. A number of satellite solar power stations have been conceptually designed to transmit power to earth-orbiting spacecraft by this method [3.24, 3.25]. Extensive studies have shown that microwave transmission efficiencies of up to 74% and laser efficiencies of 71% are possible [3.25]. Both of these methods present possible safety and reliability problems. The leakage of microwaves may cause interference with electrical devices in the area. Laser systems require an accurate tracking system, and slight misalignment could cause damage to the components. Both methods also require a complex power-conditioning and control system.

References

[3.1]   Roth, A., *Vacuum Technology*, Amsterdam: Elsevier Science Publishers B.V., 1976.

[3.2]   Weston, G.F., *Ultrahigh Vacuum Practice*, London: Butterworth & Co., 1985.

[3.3]   Weissler, G.L., R.W. Carlson, "Vacuum Physics and Technology", *Methods of Experimental Physics*, Vol. 14., New York: Academic Press, Inc., 1979.

[3.4]   Weston, G.F., "Materials for Ultrahigh Vacuum", *Vacuum*, Vol. 25, no. 11/12, pp. 469- 484.

[3.5]   McMurtrey, E. L., *Lubrication Handbook for the Space Industry, Part A: Solid Lubricants, Part B: Liquid Lubricants*, TM-86556, Washington DC: NASA Dec., 1985.

[3.6]   Roller, K.G., "Lubrication of Mechanisms for Vacuum Service", *Journal of Vacuum Science and Technology* A, vol. 6, pp. 1161-1165, June, 1988.

[3.7]   Campbell, M.E., *Solid Lubricants A Survey*, Washington, D.C. : NASA, 1972.

[3.8]   Campbell, M.E., J.B. Loser, E. Sneegas, *Solid Lubricants*, Washington D.C.: NASA, 1966.

[3.9]   Bowden, F.P., and D. Tabor, *Friction: An Introduction to Tribology*, Garden City, NY: Anchor, 1973.

[3.10]   Buckley, D.H., *Friction, Wear, and Lubrication in Vacuum*, Washington D.C.: NASA, SP-277, 1971.

[3.11]   Clauss, F. J., *Solid Lubricants and Self-Lubricating Solids*, New York, NY: Academic Press, 1972.
          (Fig. 3.5, 3.6, 3.7 originally appeared in: R.B. Fehr, "Nonlubricated bearings and piston rings of
          polytetrafluoroethylene", J. of Soc. Plastics Eng., 16, 943-948, 1960.)

[3.12]   Kannel, J.W. and K.F. Dufrane, "Rolling Element Bearings in Space", in *20th Aerospace Mechanisms
          Symposium Proc.*, Cleveland, OH, pp. 121-132, May, 1986 .

[3.13]   Booser, E.R. , *Handbook of Lubrication*, Vol. 1, Boca Raton: CRC Press, 1983.
          Table 3.3 and 3.4 reprinted with permission. Copyright CRC Press, Inc., Boca Raton, FL

[3.14]   Paluszek, M., and P. Madden, "A Free-Flying Power Plant for A Manned Space Station", *Proceeding of the
          Tenth IFAC Symposium, Automatic Control in Space*, Toulouse, France, pp. 263-268, June 24-28, 1985

[3.15]   Holman, J.P., *Heat Transfer*, New York, N.Y., McGraw Hill, 1981.

[3.16]   Peterson, G.P. "Thermal Control Systems for Spacecraft Instrumentation", *Journal of Spacecraft, AIAA,*
          Vol. 34, No. 1, pp. 7-13, Jan.-Feb., 1987.

[3.17]   Madhusudana, C.V., and L.S. Fletcher,"Contact Heat Transfer-The Last Decade", *AIAA Journal,* Vol. 24,
          pp. 510-523, March, 1986.

[3.18]   Sauer, H.J., Jr., C.R. Remington, G. A. Heizer, "Thermal Contact Conductance of Lubricant Films",
          Proceedings of the 11th International Conference on Thermal conductivity, Albuquerque, NM, pp.24-25,
          Sept.-Oct, 1971.

[3.19]   Deal, T.E.,"The Surveyor Thermal Switch", *Aerospace Mechanisms, vol. 1, Part A: General Application,*
          pp. 253-259, 1970.

[3.20]   Chaban, E.E., P.H. Citrin, F. Sette, "A Simple Ultrahigh Vacuum Compatible Sample Cooling System",
          Journal of Vacuum Science and Technology A, 6(5), pp. 3018-3019, Sept./Oct, 1988.

[3.21]   George, S.M., "A Simple and Versatile Liquid Nitrogen Cooled Cryostat on a Differentially Pumped
          Rotary Feedthrough", *Journal of Vacuum Science and Technology* A, 4(5), pp. 2394-2395, Sept./Oct,
          1986.

[3.22]   Van Vliet, R.M., *Passive Temperature Control in the Space Environment*, New York: The Macmillan
          Publishing Company, 1965. © Robert van Vliet. Table 3.7 reproduced with permission from Macmillan
          Publishing Company.

[3.23]   Favorskiy, O.N., and Y. S. Kadaner, "Problems of Heat Transfer in Space", *NASA Publication TT F-783,*
          Washington, D.C., August, 1973.

[3.24]   Dietz, R.H., "Solar Power Satellite Microwave Power Transmission and Reception", *NASA Conference
          Publication* 2141, Houston, TX, 1980.

[3.25]   Williams, M.D. and E.J. Conway, "Space Laser Power Transmission Studies", *NASA Conference
          Publication* 2214, Hampton, VA, 1981.

[3.26]   Rosenblum, S.S., "Vacuum Outgassing Rate of Plastics and Composites for Electrical Insulators", *Journal
          of Vacuum Science and Technology* A, 4(1), pp. 107-110, Jan./Feb., 1986.

[3.27]   Honig, R.E. and D.A. Kramer, "Vapor Pressure Data for the Solid and Liquid Elements", *RCA Review,*
          Vol. 30, no. 2, pp. 285-305, June, 1969.

Vacuum/volume 31/number 8/9/pages 351 to 357/1981
Printed in Great Britain

0042-207X/81/080351-07$02.00/0
Pergamon Press Ltd

Space environment and vacuum properties of spacecraft materials

T J Patrick, *University College London, Mullard Space Science Laboratory, Holmbury St Mary, Dorking, Surrey RH5 6NT, UK*

This review of the literature of vacuum aspects of space science and technology during the last decade has focused on (i) the vacuum environment of spacecraft in orbit and under test and (ii) materials for structures, instrumentation and mechanisms of spacecraft. Earlier knowledge of the parameters of the environment for near-Earth orbits has been consolidated. There have been advances in the understanding of the molecular flow around spacecraft. The outgassing of materials remains an important consideration, so criteria for selection still include low mass loss and low propensity to contaminate critical surfaces. There are now good data for the choice of lubricants, both liquid and solid, for mechanisms intended to operate for several years in space. The techniques for spacecraft testing in space simulation chambers are well established and only relatively small special-purpose chambers are now being built.

1. Historical introduction

Space technology has its origins in rocket engineering, but the techniques of vacuum were as essential to progress as skill in electronics and the design of light structures. A proper concern of the designers of the first artificial satellites was the temperature balance of a body exposed to intense solar radiation on one side yet liable to lose its heat to black space on the other. The heat transfer problem is complicated by the conduction of electrical energy from solar panels to heat dissipating circuits which cannot cool by convection unless installed in a relatively heavy gas-filled pressure vessel. Hence was born the need for techniques for testing of space hardware in thermal vacuum chambers whose walls could be heated or cooled. Solar simulators were a further development in which the energy input was an arc lamp whose output was collimated to represent solar radiation, and in such apparatus, which by the mid-1960s had become large and expensive, solar energy systems for satellites were tested and developed. The chambers threw up their own problems, of contamination by diffusion pump oil, but other outgassing and contamination problems, already familiar in vacuum laboratories, were met afresh in the space hardware context, particularly on account of the weight-saving attractions of newly-developed polymer materials in electrical parts. Problems of lubricating the bearings of mechanisms in space were essentially those already encountered in research into friction under vacuum, but the need to adjust the orientation of radio antennae and solar cell arrays gave new impetus to developments, many of them employing new materials, in this area of space engineering and tribology. By the time in the 1960s that successful space missions had departed for the Moon and nearer planets, a pattern had been established in which spacecraft technologists proved their electronic systems in thermal vacuum tests and checked their predictions of thermal radiation balance in a solar simulation chamber. In the ensuing years this pattern for the validation of new designs was applied

repeatedly. Mechanisms began to be tested for lives of several years in vacuum, as geostationary satellites took over telecommunications, and other new spacecraft were used increasingly for meteorological observations, navigation, studies of Earth resources and the monitoring of military activity.

A space science background for this technology developed with it, but pure science objectives were also pursued. For example, radio research led early to the exploration of the ionosphere by sounding rocket and satellite and so, by way of the discovery of the magnetosphere, to today's plasma geophysics with its continuing study of solar–terrestrial relations. The geophysical instrumentation of spacecraft requires calibration in vacuum systems. Satellite access to cosmic uv radiation, X-rays and gamma-rays beyond their absorption in Earth's atmosphere has given birth to new branches of astronomy, for which increasingly large new kinds of orbiting telescope have been developed. These have made even greater demands on vacuum test facilities, and have increased the pressure to exclude materials, often new polymers, whose outgassing products included condensible matter which might contaminate optical surfaces. Now the second generation interplanetary spacecraft are transforming the astronomy of the solar system from an observational to an experimental science.

Amongst goals planned for the next few years, two projects illustrate respectively some space science and technology ambitions awaiting fulfilment. First, the close investigation of a comet. The predicted return of Halley's comet in 1986 has stimulated the European Space Agency, Japan and the Russians to mount missions in which unmanned spacecraft will encounter the comet and return physical data on matter composing it. No previous space flight will have met an environment so hostile as the cloud of interplanetary dust against which these missions will have to be shielded. Secondly, there is the potential of the US Space Shuttle for technological research. This piloted space vehicle which, after orbital flight, will land and be reflown, will be able to carry ESA's Spacelab in which many new experiments may be performed in a microgravity environment. Processing in this situation may even lead to the production of materials not otherwise available.

The paper seeks to review, primarily from the point of view of space instrument engineering, the vacuum technology of materials principally for structures, mechanisms, seals, lubricants and electrical components. As in the early literature in this field, e.g. Goetzel[1], the space vacuum environment of the spacecraft has been included, but it has now been possible to include several new studies stimulated by the development of the Shuttle.

2. Vacuum environment of spacecraft

2.1. The natural space environment. For earth's atmosphere, the variation with height of pressure, density, temperature and composition has long been systematically researched. Atmospheric density at high altitudes is strongly affected by solar activity. Data up to 2000 km are available in the COSPAR International Reference Atmosphere[2] and were given in other more compact texts such as Johnson[3] (now somewhat dated). Witteborn and Simpson[4] list other references, but none more recent. At geostationary altitude, 36,000 km, there is virtually no neutral atmosphere.

Results of planetary exploration by space probe began to appear in the late 1960s after NASA's Mariner 5 had flown past Venus in 1967, and Venera 6 and 7, launched by the USSR in 1969, had landed there. Some discrepancies in the data returned from Venera probes[5] have been cleared up, confirming an atmosphere much denser and hotter than Earth's. Measurements made from Mariners 6 and 7 launched in 1969 to fly close to Mars, improved knowledge of the tenuous Martian atmosphere; Mariner 9 followed in 1971, then Mars 4 and 5 in 1975. In 1976 Vikings 1 and 2 were launched by the US, and, as well as sending back the classic pictures from the planet's surface, yielded data on the structure of the Martian atmosphere[6]. Mariner 10 (1974) made the closest encounter with Mercury and confirmed that small inner planet to be, like our Moon, almost free of any atmosphere.

The outer planets became the targets for fly-by missions in the 1970s. Pioneer 10 was launched from the US towards Jupiter in 1973. Pioneer 11 followed it in 1974 and went on to Saturn. Voyagers 1 and 2 flew past Jupiter and Saturn in 1979 and 1980. The complexities of the atmosphere of these giant planets are a subject of continuing study, and plans for atmospheric entry missions have yet to be implemented.

The neutral atmospheres of Earth, Venus, Mars and Jupiter are surrounded by ionospheres, basic accounts of which have been given by Boyd[7]. Further out are found magnetospheres which contain belts of trapped energetic protons and electrons, also described by Boyd; see also Akasofu[8].

2.2. Ascent. During a rocket launch from earth the ambient pressure around a spacecraft falls quickly (see, for example, Patrick[9]). The rate of fall of ambient pressure is determined by the acceleration–time curve of the ascent trajectory, which does not vary very significantly among vehicles of the same type, and by the pressure–height relation of the atmosphere. Molecular flow conditions are established typically about 3 min from launch, at which time the ambient pressure may be typically 10 μPa at altitude 90 km. One of the lessons from the early history of rocket and satellite launches was that internal compartments need time to pump out to the ambient pressure to avoid the danger of

electrical breakdown (arcing or 'corona' discharge) through the internal atmosphere, if high voltage circuits are switched on too early. Ground rules for the avoidance of electrical breakdowns were compiled by Paul and Burrowbridge[10] and there is a recent article by Nanevicz and Adamo[11]. The rule of thumb is that no high voltage may be switched on to a conductor if the local pressure is likely to be in the range 5 kPa down to 1 Pa. Detailed calculations of spacecraft internal pressures during the ascent have been reported by Scialdone[12], who made allowance for the high initial rates of outgassing, probably water vapour, from parts of the spacecraft.

2.3. External environment in orbit. The environment surrounding an orbiting satellite is determined by (i) the planetary atmosphere (para 2.1. above); (ii) the height in the orbit; (iii) the outgassing and other spacecraft emissions, including intermittent firings of attitude control and orbit changing motors, and leakage from cabins (with dumping of wastes) in the case of manned spacecraft.

Typically a low earth orbit for an artificial satellite has been at altitude above 200 km. The objective of the rocket-driven ascent will have been to take the spacecraft to this altitude and to accelerate it on to the velocity vector required for the planned orbit. At 200 km a satellite will have an orbital speed of 8 km s^{-1}, at geosynchronous altitude 3.2 km s^{-1}.

Scialdone[13] gave an account of his calculation of the environment. There will be differences between a 'condensation' or stagnation region ahead of the satellite and rarefaction region behind. The condensation region is formed by the sweeping up of ambient atoms and molecules by the forward satellite surface. Before collision, these particles have only a thermal velocity much less than satellite velocity, but they are reflected by collision, and form a stagnation region ahead of the satellite, in which region the particle concentration is increased, up to density three times ambient. The region contains the reflected particles, the arriving ambient particles, and outgassed and other particles from the satellite. The rarefaction region trails the satellite, contains molecules from satellite outgassing, and is eventually refilled by thermally motivated ambient particles at about 25 satellite radii behind the satellite. The outgassed or otherwise emitted particles deposited in this wake will be left behind and, unlike those in the stagnation region, cannot return to the satellite to contaminate it. Although the satellite acquires an electric potential (see Garrett[14]), due to impact of charged particles and surface emission due to solar radiation, this will not affect the molecular flow of uncharged outgassed material.

The cabin leakage problem was studied during the Gemini and Apollo programmes by Kovar and Bonner[15] who found evidence of a substantial 'debris' atmosphere.

On the spinning satellite Explorer 32 Silverman and Newton[69] measured pressures using cold-cathode gauges. Pressures varied smoothly from stagnation ahead to rarefaction behind, corresponding to spacecraft ambient modified by outgassing.

The flights planned for the Space Shuttle could include an application of the wake rarefaction effect to produce low pressures, even below 10^{-12} Pa (or 10^{-14} torr). Melfi *et al*[16] analysed a molecular shield model in terrestrial orbit above 200 km, considering outgassing from the shield and instrumentation as well as the free-stream atmosphere and that part of it scattered off the spacecraft (the Shuttle Orbiter). The atmospheric component would be principally atomic hydrogen of density less than 10^3 cm^{-3}, corresponding to a pressure near 4×10^{-12} Pa at 300 K. But the Shuttle Orbiter will be manned, will have an active attitude-control system, and relatively large emissions. Hence such low pressures are more likely to be achieved behind a passive spacecraft; a Long Duration Exposure Facility. Oran and Naumann[17] have discussed the vacuum in the wake of such a vehicle, and gave a graph of the directional flux of H, He and O atoms as a function of angle from the wake axis. It shows 10^6 H atoms $cm^{-2}\,s^{-1}sr^{-1}$ from the wake direction, but 10^{11} on the forward side of the shield. Moore[18] has looked at molecular wake shields of various shapes. Kleber[19] has discussed the problem of pressure measurements in a Shuttle Orbiter environment.

2.4. Internal pressures in orbit. The simplest case is that of a spacecraft compartment directly connected through an orifice with the external space environment. The compartment is evolving gas from its walls of area A_G (cm^2) for which the material is outgassing at a quasi-steady rate Q_G (cm^{-2}). This gas is passing through an orifice of area A (cm^2) for which the pumping speed S is $11.6\,A\,l\,s^{-1}$. Hence, by equating the quantities in pressure–volume units, we calculate the quasi-steady equilibrium pressure as

$$P_L = \frac{A_G \cdot Q_G}{S} = \frac{A_G}{A} \cdot \frac{Q_G}{11.6}.$$

Inserting Q_G in units of torr–$l\,s^{-1}\,cm^{-2}$ yields a pressure in torr.

[Transcribed to coherent (SI) units the equation becomes

$$P_L = \frac{A_G}{A} \cdot \frac{Q_G}{1.16}$$

from which P_L is found in pascals, if Q_G is given in Pa m s^{-1}.]

If outgassing rates as a function of time are available, the compartment pressure for increasing times into orbital flight may be estimated. But often the outgassing data do not run beyond a few hours. Following Schittko[20], it has been suggested[9] that outgassing rates be extrapolated by an inverse square-root law. This approach matches the outgassing model of Guillin[21]; but see also Scialdone[22], and Elsey[23] for discussion of mechanisms.

More complex cases were calculated by Scialdone[12], who developed a computer program which could handle several compartments interconnected in various ways and made due allowance for the nature and temperature of ambient gas, the temperature of internal surfaces, and the reduction of outgassing rate.

2.5. Contamination. The outgassed particles from a satellite are in the stagnation zone ahead of it. Scialdone[13] assembled the theory to derive the density of particles returning to the surface, the mass column-density, and the time to form a monolayer of condensibles. He gathered test data on satellite outgassing after 1 h in vacuum. The calculated monolayer condensation time was a few seconds for manned spacecraft including Gemini, Apollo and Apollo Telescope Mount (Skylab), and also for the scientific satellite IMP-B. But for a Nimbus weather satellite the outgassing rate was some 200 times lower, leading to a result implying that a monolayer would never form. Another point brought out in this and a contemporary paper[24] is that the molecules outgassed produce a concentration which diminishes with distance from the spacecraft surface along the direction of the velocity vector. The decrease depends on altitude. For 100 km and 500 km orbits, the distances for an order of magnitude reduction in outgassing concentration are respectively 0.1 m and 2 m. The ratio of returned to emitted outgassing flux was calculated at one half at 160 km, reducing to one millionth at 1000 km. Another example

of a complex numerical approach to spacecraft self-contamination can be found in Harvey[25]. Scialdone[22] has developed his analysis to predict the gaseous environment for the Shuttle, and has reckoned that the payloads will dictate it except during attitude motor firing and waste water evaporation; his detailed paper quotes the criteria agreed on for limiting gaseous contamination, and lists design and operational criteria for experiment builders. Cooled instruments will be particularly at risk, and first flights of each Shuttle are expected to outgas a dirtier environment than subsequent flights (on account of the 'bake-out' effect of re-entry heating).

The nature and quality of condensibles has been a concern for even longer than outgassing quantities. Zwaal[26] gave IR spectroscopy techniques for the detection of organic contamination. The most feared were human (finger) grease and 'creeping' silicone fluids. Gross[27] lists over 60 IR—characterized contaminants. Contaminants identified by their spectra point back to the need for adequate standards in material selection (4 below), storage, cleaning, fabrication, and testing (2.7 below). A recent review of spacecraft contamination has been given by Jemiola[28].

2.6. Descent. The majority of unmanned spacecraft which have been launched are either still in orbit or have been destroyed by aerodynamic heating after their orbits decayed to altitudes of 30 km or less. Those still in orbit slowly return to lower altitudes, depending on their drag, which, until re-entry is imminent, will be a tiny fraction of their weight. Hence the pressure environment in orbit will always be below the level of a few torr or 1kPa, although high voltage circuits will have been at risk from breakdown below 90 km, 1 Pa.

Manned spacecraft fire retrorockets to achieve controlled re-entry trajectories, along which will be reached (and passed) a hypersonic continuum flow condition with intense aerodynamic heating of an ablative or refractory heat shield. The difficulty of developing a refractory tile system for the Shuttle Orbiter is well known. This same vehicle is expected to encounter water condensation in its payload bay in the phase of aerodynamically controlled flight after re-entry but before landing.

2.7. Test chamber environments. Haefer[29] described '2nd generation' installations for thermal vacuum and solar simulation testing. The facilities required to complete the development of the Shuttle have existed for some time. Sanger and Franz[30] reported that the change to titanium sublimation and ion getter pumps from an oil diffusion system reduced the organic content of residual gas by two orders of magnitude.

Scialdone's analysis[24] calculated that tests conducted in conventional vacuum chambers can result in returning contaminating fluxes comparable to space up to altitude 400 km. For higher equivalent altitudes, chambers could produce contamination exceeding that returned in the space environment. Pressure measurements are not selective and cannot indicate self-contamination. Kleber[31] has also addressed this problem of the meaning of pressure measurements, and Haefer[32] has contributed.

Among new needs to have been met, there have been those of ion propulsion development, and vacuum pipelines permitting X-ray and uv telescopes to be calibrated at a sufficient distance (typically tens of metres) from the radiation source so that rays are almost effectively parallel. Workers planning installations can consult reviews such as Weston[33], Bentley[34], Harris[70] and Henning[35].

3. Criteria for acceptance of materials

The mechanical design of a spacecraft, its systems and instrumentation is largely the application of well established principles in structures using materials which are strong, light, resistant to radiation, and have low outgassing. Subsidiary requirements are low magnetic properties and general resistance to corrosion. Manned spacecraft have additional severe restrictions on flammability and toxicity; see Bennett[36].

A vacuum physicist might define good vacuum properties as low vapour pressure, low outgassing rate, and low gas permeability. The first concern here is with low outgassing. In some circumstances, any outgassing rate is acceptable which does not lead to an equilibrium pressure in an internal compartment high enough to present a risk of electrical breakdown. The actual outgassing rate, itself a function of time in vacuum, can often be found in the vacuum literature, e.g. Elsey[23]. But this approach is rarely adopted. The condensibility of the evolved gases is also important as a source of contamination.

Hence a relatively simple standard test which yields data on both outgassing and condensed products has been adopted by two principal authorities concerned with the flight-worthiness of space hardware, NASA (the National Aeronautics and Space Administration of the USA) and ESA (the European Space Agency). The test procedure is given in several references (Campbell *et al*[37], ESA[38]). The standard criteria for the acceptance of a material are:

Total mass loss, after 24 h at 125°C at 0.13 mPa (10^{-6} torr): denoted TML: <1.0%.

Condensed volatile condensible materials, collected for 24 h on a plate at 25°C: denoted CVCM: <0.1%.

As might be expected it has been found[9] that mass loss correlates with outgassing rate, albeit roughly. Whereas outgassing rate is a physical quantity of a kind, TML is only a percentage figure of comparison. Nevertheless, there is now a large body of TML and CVCM data on materials used in space projects over two decades. Campbell *et al*[37] give a collection of over 3000. Figure 1 and its accompanying table give TML, CVCM values for materials typical of current space practice, and show that similar values are found for comparable materials tested in different laboratories. The rough correlation with outgassing rate is also shown by the table.

Taking a simple view, any material within the criteria is 'space-approved'. In practice, difficult design problems are sometimes solved by choosing materials which are above the limits, but the degree of damage (contamination or loss of mass) which can be tolerated will depend on the relative quantities of good and bad material used, and such cases must be judged on their merits. To quote Dauphin[39], 'a waiver is like a sin in that it must be committed reluctantly, regretted deeply, and not repeated ... if possible'. But the utility of the TML/CVCM figures of comparison continues to appear, as for instance in Guillaumon's report[40] on new paints and silicone varnishes, or Bennett's comment[36] that for the Spacelab ESA tested some 100 materials. Spacelab is the first manned spacecraft built in Europe, albeit captive to the Shuttle.

One way past the arbitrary criteria of ML/VCM has been reported by Thomas[41] as applied in the Faint Object Camera project, which is ESA's contribution to NASA's Space Telescope. The standard test has been replaced by continuous measurement of condensate deposition on a quartz crystal microbalance. This approach has been explored at NASA by Scialdone[42]. Other

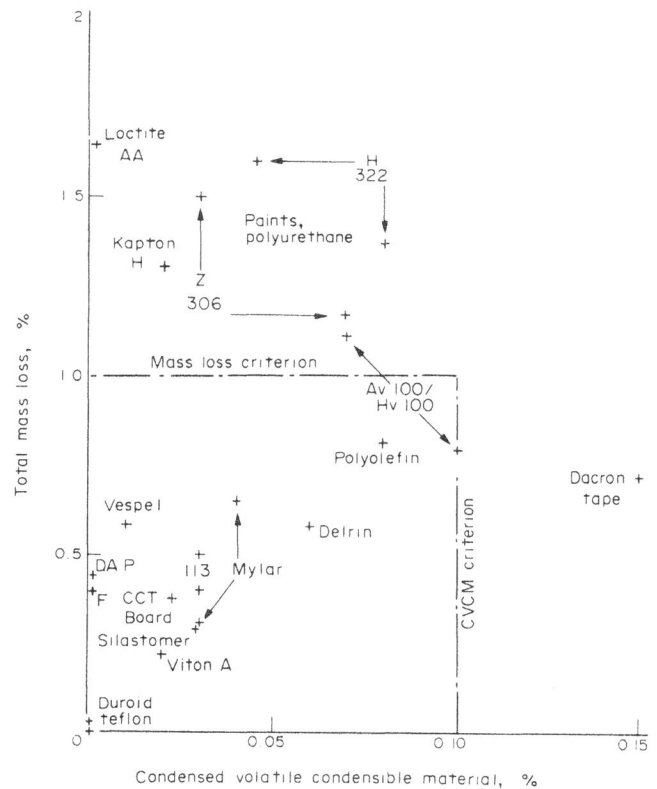

Figure 1. TML (Total Mass Loss) and CVCM (Condensed Volatile Condensible Material) for some typical spacecraft materials which show significant outgassing from Table 1.

refined techniques are exemplified in mass spectrometry, Colony[43], and IR spectrometry, Gross[27].

Evidently there are areas where the requirements for spacecraft materials approach those in laboratory uhv, as given by Weston[44]. Perhaps expensive materials are more often cost-effective in space applications. Other notable differences are (i) space hardware is much less often baked to reduce its outgassing and VCM content; (ii) glass and ceramics are uncommon, the need for insulating materials being satisfied among polymer materials.

4. Structural materials

The light alloys of aluminium, and to a lesser extent titanium, are the metals most likely to be chosen for a spacecraft structure. If space experience has revealed problems they are in other areas, e.g. occasional instances of stress corrosion cracking, rather than in vacuum performance. There is a tendency away from the heat treatable copper-bearing alloys to AlSi and the dimensionally very stable Al 3% Mg alloys. Magnesium alloys, although even lighter than aluminium, are falling from favour on account of their susceptibility to general corrosion and the attendant problems of paint finishes. The higher vapour pressure of magnesium is also against it but has not excluded its use on some projects.

Steel, though three times denser than the aluminium alloys, nevertheless can match them in strength/weight ratio. Usually a stainless alloy (with chromium and nickel) is chosen, the non-magnetic austenitic forms (e.g. 321S20) being preferred. Small components such as screw fastenings are usually of steel or titanium alloy Ti 4Al9V. Carbon steels passivated by cadmium

Table 1. Some spacecraft materials typical of current practice, listed in approximate order of outgassing rate, with Total Mass Loss (TML) and Condensed Volatile Condensible Material (CVCM); after Campbell *et al*[37] or ESA[38] (ESA results denoted E)

Material	Application	Manufacturer	TML (%)		CVCM (%)	Outgassing rate, 10 h Elsey[23] (torr l s^{-1} cm^2)	(Pa m s^{-1})
Steel, stainless, BS 1449 321 S 12	Fittings					1.4×10^{-9}	1.9×10^{-6}
Titanium 6 Al 4 V (IMI 318)	Fittings					1.8×10^{-9}	2.4×10^{-6}
Aluminium 4 Cu	Structure					3.5×10^{-8}	4.7×10^{-5}
PTFE, Teflon	Wire sleeving	Du Pont	0.00		0	2.5×10^{-8}	3.3×10^{-5}
PTFE-glass-MoS$_2$ composite, Duroid	Bearings	Rogers	0.01		0		
Perfluoroether, Fomblin	Oil and grease	Montedison	0.01		0		
FEP, Teflon, film	Thermal insulation	Du Pont	0.02		0		
Fiberglass woven cloth, Betacloth	Thermal insulation	Stevens	0.03		0		
Viton A fluorocarbon rubber	Seals	Du Pont	0.22		0.02	$\sim 10^{-7}$	$\sim 10^{-4}$
Silicone elastomer, 93500	Potting, seals	Dow Corning	0.30	E	0.03		
PETP, Mylar, film	Thermal insulation		{0.30	E	0.03		
			0.65		0.04	4×10^{-7}	5.3×10^{-4}
Epoxy-glass laminate	Circuit board		0.37		0.02		
Epoxy, Araldite F, hot-cured	Potting	CIBA	0.40	E	0		
Diallyl phthalate	Connector bodies		0.44		0		
Polyurethane, Solithane 113/300	Conformal coating	Thiokol	{0.40	E	0.03		
			0.50		0.03		
Polyimide, Vespel	Solid lubricant	Du Pont	0.58		0.01		
Polyacetal, Delrin	Insulating parts	Du Pont	0.58		0.06		
PETP, Dacron, tape 21 D 96	Wire lacing	Gude	0.73		0.15		
Epoxy, Araldite AV 100/HV 100	Adhesive	CIBA	{0.78		0.10	$\sim 10^{-6}$	$\sim 10^{-3}$
			1.10	E	0.07		
Polyolefin, heat shrunk	Sleeving	Raychem	0.80		0.08		
Polyimide, Kapton H, film	Thermal insulation	Du Pont	1.30		0.02		
Polyurethane Z 306	Paint	Hughson	{1.17		0.07		
			1.50	E	0.03		
Polyurethane H 322	Conducting paint	Hughson	{1.39		0.08		
			1.60	E	0.04		
Resin loctite AA/primer N	Thread locking		1.64		0		

plating are not favoured because of some evidence of the growth of cadmium whiskers, whose fracture and (weightless) drift could short-circuit electric circuits.

A widely-used structural component is the honeycomb sandwich panel. This frequently comprises two thin aluminium alloy sheet skins spaced apart by hexagonal cells formed in alloy foil, the honeycomb cells being bonded to each other and to the skins by synthetic resin adhesive. Pinholes in the cell walls allow the cells to vent and outgas, but such a component is a much greater outgassing risk than a plain metal sheet. The problem is one of surface area and effective porosity, and has been studied by Schalla[45] of Lockheed in a study of several different porous materials. A typical 1 m^2 panel will have 3 m^2 internal surface and after 7 h in vacuum will be outgassing 3.10^{-10} torr-l cm^{-2} s^{-1}, i.e. evolving 10^{-5} torr-l s^{-1}, that is about 10^{-8} g s^{-1}, from the panel's interior.

Composite materials, such as those incorporating carbon fibre (carbon fibre reinforced plastic, CFRP, more accurately described as epoxy-resin-toughened carbon fibre) show great promise for structures (see Bowen[46]) but there is no evidence to suggest that outgassing rate will be lower than would be expected from any other material containing perhaps 50% of epoxy resin. Moisture absorption has been a problem and might have to be controlled by storage in a dry gas.

5. Polymers

When, in the first few years of space activity, it was discovered that outgassing was a problem, the polymer 'plastics' materials were recognized as the trouble-makers. Electrical insulating and potting materials, paint finishes and lubricating greases came under suspicion until accepted on ML/VCM criteria. Revisions of lists of approved polymers were frequent and lead to the references (Campbell 1978, ESA 1979) which are current today. Further attention was focused on polymers on account of their degradation under uv and particulate radiations, topics which were discussed by Goetzel *et al*[1] and again recently by Bourrieau and Paillous[47].

In what follows, the polymer materials will be referred to by their customary trade names, with the chemical name given in Table 1.

The most exposed materials on a spacecraft are often the thermal insulation blankets and the paint finishes. The blankets are multiple layers of aluminized plastic foil (Mylar or Kapton) sometimes interleaved with nylon mesh. Glassford and Liu[48] recently published outgassing figures. Recent attention to paints has been given by Lehn[49] and by Giullaumon and Guillin[40]. In another paper on the outgassing of paints and potting materials, Guillin[21] has sought to establish a mathematical model with an activation energy (Arrhenius equation) basis. (See also Henri[50], Heslin[51] and Scialdone[22].) Another concern with thermal control coatings, that of electrostatic charging properties, has been explored for ESA by Bosma and Levadou[52].

For some purposes such as gas storage vessels and valves, elastomer seals are another spacecraft application for polymers. Principles of design have been discussed by Sessink and Verster[53]. Chernatony has made extensive contributions to the literature on the fluorocarbon rubber Viton and its derivatives[54,55].

As an example from the relatively few polymers with outstandingly good outgassing properties PTFE should be mentioned.

Poole and Michaelis[57] have recently drawn attention to improvements found by removing the skin of the extruded material.

6. Ceramics

Not excluding the silica tiles developed for the protection against aerodynamic re-entry heating of the Shuttle Orbiter, ceramic materials have made an undistinguished contribution so far among spacecraft materials. The machinable glass ceramics appear attractive for many applications including space instrumentation. Altemose and Kacyon[58] have considered their vacuum compatibility, and Grossman[59] has given an extended account of their machining properties.

7. Space tribology and lubrication of mechanisms

Spacecraft require mechanisms with such applications as solar panel deployment and orientation, antenna steering, gyroscopes, inertia wheels, instrument booms, shutters, tape recorders, slip rings for power transfer, and cryogenic heat pumps. Bearings, whether plain journals or with rotating elements, are needed. But conventional lubricating oils evaporate and contaminate where they recondense, and so cannot be tolerated unless the mechanism can be sealed in a pressure vessel. The problems are sufficiently formidable to have warranted the setting up in 1972 of the European Space Tribology Laboratory (at Risley, England), and have been discussed by Robbins[60]. Conferences took place in 1975[61] and 1980[62].

Lubricating oils and greases have been developed from vacuum pump oils in order to depress the vapour pressure. The greases have employed molybdenum disulphide and oleophilic carbon. But there has been a long history of successful use of MoS_2 and other lubricants, going back to work reviewed by Clauss[63]. Such developments included proprietary composite bearing materials based on glass-fibre-reinforced PTFE, e.g. Duroid in Bartemp bearings for which many successful applications have been reported (Devine *et al*[64], Patrick[65]). Another significant solid-lubricant is ion-plated lead (Todd and Robbins)[66]. Hadley[67] has reported long life with carbon fibre reinforced polyacetal and has used Vespel polyimide. Ball bearings have been operated in space with both solid and liquid lubricant; see ESA papers[61]. Their thermal conductivity has been studied by Stevens and Todd[68].

Acknowledgements

Thanks for assistance are due to Professor R L F Boyd, CBE, FRS; J A Bowles; L de Chernatony; J Robbins; P H Sheather and W Steckelmacher.

References

[1] C G Goetzel, J B Rittenhouse, J B Singletary, *Space Materials Handbook*, Adison-Wesley, Reading, Mass (1965).
[2] COSPAR, International Reference Atmosphere. Akademie Verlag, Berlin (1972).
[3] F S Johnson, *Satellite Environment Handbook*, 2nd edn. Stanford University Press, California (1965).
[4] F C Witteborn and J P Simpson, *J Vac Sci Technol*, 14 (6), 1251 (1977).
[5] G S Golitsyn and Kerzhanovich, *Cosmic Res*, 9 (6) (1972).
[6] C B Leovy, Martian meteorology, *Ann Rev Astron Astrophys*, 17, 387–413 (1979).
[7] R L F Boyd, *Space Physics*. Clarendon Press, Oxford (1974).
[8] S -I Akasofu, *Space Science Rev*, 21, 489–526 (1978).
[9] T J Patrick, *Vacuum*, 23 (11), 411–413 (1973).
[10] F W Paul and D Burrowbridge, NASA GSFC X 327-69-75 (1975).

[11] J E Nanevicz and R C Adamo, Occurence of arcing and its effects, Space Systems and their Interactions, in *Advances in Aeronautics and Astronautics*, Vol 71, AIAA, New York (1980).
[12] J J Scialdone, NASA GSFC X-327-69-524 (1969).
[13] J J Scialdone, *J Vac Sci & Technol*, 9 (2), 1007–1015 (1972).
[14] H B Garrett, Spacecraft charging: a review, Space Systems and their Interactions with Earth's Space Environment, in *Advances in Aeronautics & Astronautics* Vol 71, AIAA New York (1980).
[15] N S Kovar, R P Kovar and G P Bonner, *Planet Space Sci*, 17, 143–154 (1969).
[16] L T Melfi, R A Outlaw, J E Hueser and F J Brock, *J Vac Sci Technol*, 13 (3), 689 (1976).
[17] W A Oran and R J Naumann, *Vacuum*, 28 (2), 73–74 (1978).
[18] B C Moore, *J Vac Sci Technol*, 16 (3), 946 (1979).
[19] P Kleber, *Vacuum*, 30 (3), 117–120 (1980).
[20] F J Schittko, *Vacuum*, 13, 525–537 (1973).
[21] J Guillin, Proc Symp Spacecraft Materials, ESA SP-145, 139–144 (1979).
[22] J J Scialdone, Proc Symp Spacecraft Materials, ESA SP-145, 101–116 (1979).
[23] R J Elsey, *Vacuum*, 25, Pt 1, 299–306; Pt 2, 347–361 (1975).
[24] J J Scialdone, NASA TN D-6682 (1972).
[25] R L Harvey, *J Spacecraft & Rockets*, 13 (5), 301–305 (1976).
[26] A Zwaal, PSS-15, European Space Reseach Org (1973).
[27] F C Gross, NASA TN D-8451 (1977).
[28] J M Jemiola, Spacecraft contamination: a review, Space Systems and their Interactions, in *Advances in Aeronautics & Astronautics*, Vol 71, AIAA, New York (1980).
[29] R A Haefer. *Vacuum*, 22 (8), 303–314 (1972).
[30] G Sanger and A K Franz, European Space Research Organization, SP-95, pp 383–418 (1973).
[31] P Kleber, *Vacuum*, 25, 191–196 (1975).
[32] R A Haefer, *Vacuum*, 30 (4/5), 193–195 (1980).
[33] G F Weston, *Vacuum*, 28 (5), 209–232 (1978).
[34] P D Bentley, *Vacuum*, 30 (4/5), 145–150 (1980).
[35] J Henning, *Vacuum*, 30 (4/5), 183–185 (1980).
[36] J E Bennett and M D Judd, Proc Symp Spacecraft Materials, ESA SP-145, pp 161–167 (1979).
[37] W A Campbell, R S Marriott and J J Park, An outgassing data compilation of spacecraft materials, NASA Reference Publication 1014 (1978). (See also ref 71.)
[38] European Space Agency, Product Assurance Div, Guidelines for space materials selection, ESA PSS-07 Issue 5 (1979).
[39] J Dauphin and P Guyenne (Editors), Spacecraft materials in space environment, ESA SP-145 (1979).
[40] J C Guillaumon and J Guillin, Proc Symp Spacecraft Materials, ESA SP-145, pp. 63–66 (1979).
[41] R Thomas, Proc Symp Spacecraft Materials, ESA SP-145, pp. 167–174 (1979).
[42] J J Scialdone, *J Vac Sci Technol*, 12 (1), 569–572 (1975).
[43] J A Colony, Mass spectrometry of aerospace materials, NASA TN D-8261 (1976).
[44] G F Weston, *Vacuum*, 25 (11/12), 469–484 (1975).
[45] C A Schalla, *J Vac Sci Technol*, 17 (3), 705–708 (1980).
[46] D H Bowen, Proc Symp Spacecraft Materials, ESA SP-145, pp 49–62 (1979).
[47] J Bourrieau and A Paillous, Proc Symp Spacecraft Materials, ESA SP-145, pp 227–245 (1979).
[48] A P M Glassford and C K Liu, *J Vac Sci Technol*, 17 (3), 696–704 (1980).
[49] W L Lehn, Proc Symp Spacecraft Materials, ESA SP-145, pp 37–47 (1979).
[50] R P Henri, *Le Vide* No 144, 316–330 (1969).
[51] T M Heslin, NASA TN D-8471 (1977).
[52] J Bosma and F Levadou, Proc Symp Spacecraft Materials, ESA SP-145, 189–197 (1979).
[53] Sessink and Verster, *Vacuum*, 23, 319–325 (1973).
[54] L de Chernatony, *Vacuum*, 16 (1), 13–15; (3) 129–134; (5) 247–251; (8) 427–431 (1966). 17 (10), 551–554 (1967).
[55] L de Chernatony, *Vacuum*, 27 605–609 (1977).
[56] L de Chernatony, Proc 7th Int Vac Congress, pp 255–258 (1977).
[57] K F Poole and Michaelis, *Vacuum*, 30 (10), 415–416 (1980).
[58] V O Altemose and A R Kacyon, *J Vac Sci Technol*, 16 (3), 951–954 (1979).
[59] D G Grossman, *Vacuum*, 28 (2), 55–61 (1978).

[60] E J Robbins, Proc 1st Euro Space Trib Symp, ESA SP-111, pp 101–113 (1975).

[61] European Space Agency, Space tribology, Proceedings of the first symposium, Frascati, ESA SP-111 (1975).

[62] European Space Agency, Second Space Tribology Workshop, ESA SP-158 (1980).

[63] F J Clauss, *Solid Lubricants and Self-lubricating Solids*, Academic Press, New York (1972).

[64] E J Devine, H Evans and Leasure, NASA TN D-6035 (1970).

[65] T J Patrick, Proc 2nd Space Tribology Workshop, ESA SP-158, pp 37–38 (1980).

[66] M J Todd and E J Robbins, Paper 1 in Selected ESTL papers on ball bearings for satellites, ESA Trib/1 (1980).

[67] H Hadley, 14th Aerospace Mechanisms Symposium, NASA Conf Pub 2127, 101 (1980).

[68] K T Stevens and M J Todd, Paper 2 in Selected ESTL papers on ball bearings for satellites, ESA Trib/1 (1980).

[69] P J Silverman and G P Newton, *J Vac Sci Technol*, 7 (2), 323–329 (1970).

[70] N S Harris, *Vacuum*, 30 (4/5), 175–181 (1980).

[71] W A Campbell, R S Marriott and J J Park, An outgassing data compilation of spacecraft materials, NASA Reference Publication 1061 (1980).

Materials for ultrahigh vacuum

G F Weston, *Vacuum Physics Division, Mullard Research Laboratories, Redhill, Surrey RG1 5HA, England*

A paper in our Education Series: The Theory and Practice of Vacuum Science and Technology in Schools and Colleges.

The equilibrium pressure attainable in a vacuum system is limited by gas influx from vaporization, desorption, diffusion and permeation. This imposes stringent requirements on the physical properties of the material used for the fabrication of the system, and leads to the selection of a limited number of suitable materials, especially if ultrahigh vacuum pressures are to be achieved. The properties and performance of such materials, namely glass, metals, ceramics and, in some application, synthetics, are reviewed in this article. Data on the physical parameters and preparation procedures, pertinent to their use in ultra high vacuum application, are given.

1. Criteria for ultrahigh vacuum materials

The conditions existing within a vacuum system can truly be described as a dynamic equilibrium. The ultimate pressure which can be reached depends on the one hand on the effective pumping speed of the pump, and on the other hand on the influx of gas from the vacuum envelope and any components contained within the envelope. Since there are always practical limitations to the pumping speed due to the size of the pump, cost etc., the gas influx becomes the prime factor in attaining ultrahigh vacuum conditions, and sets the main criteria to the choice of materials for ultrahigh vacuum use.

The materials must have a low vapour pressure and in order to reduce desorption, be bakeable to temperatures which are usually of the order of 450°C, without losing their mechanical strength or be chemically or physically damaged. If they form part of the vacuum envelope then diffusion of gas through them must be negligible. Also they must withstand atmospheric pressure, and resist corrosion when exposed to air during baking.

Gas influx and mechanical strength are not the only criteria however. The ease of machining or fabrication into suitable arrangements and the ability to weld, braze or otherwise seam with leak-tight joins is also essential. In most vacuum systems there is a need to make electrical contact through the walls and to insulate connections. In some applications viewing windows are required. Thus, in most vacuum apparatus, the envelope will consist of both metal and insulating materials, and suitable methods of sealing the metal to the insulator must be found. Further, since a fairly wide temperature range will be experienced, the thermal expansion coefficients of the various components must be carefully matched, particularly at seals where distortion due to thermal stresses could result in leaks. Lastly, the materials chosen must be readily available at reasonable cost.

Traditionally glass has been employed for small vacuum apparatus, whilst the majority of large systems have been constructed of mild steel or a similar metal chosen for convenience of fabrication and cost. Glass fulfils many of the requirements for ultrahigh vacuum, providing that a suitable type is chosen with a low gas permeability. However, it is mechanically rather weak, and can only be used with softer materials such as grease, wax or synthetic rubber, if demountable seals and conventional valves are required. Because such materials make it impossible to bake the system to a high temperature, an all-glass construction is only suitable if such valves and joins can be avoided. In

particular, it can be satisfactorily utilized in conjunction with metal valves etc., for small systems, especially if a pressure of 10^{-6} Nm^{-2} to 10^{-8} Nm^{-2} is all that is required. For viewing ports there is really no alternative. For lower pressures and for larger systems an all-metal construction is more suitable. Mild steel, although satisfactory for unbaked high vacuum systems, corrodes when heated in air, and also shows permeation to hydrogen at the elevated temperature, *see* section 3.3. It is therefore unsuitable for ultrahigh vacuum systems even when plated. There are other possibilities, but present conditions favour stainless steels for the main vacuum envelope. They fulfil most of the stringent requirements for ultrahigh vacuum, and are fairly easily obtainable at a reasonable cost. For demountable seals and valves, stainless steels can be used in conjunction with softer metal gaskets (i.e. gold or copper) which will allow baking temperatures of 450°C to be employed. Although there are now glasses and ceramics available which can be directly sealed to stainless steel, in general intermediary metals, specially prepared for the purpose, are used to seal the steel to insulating components.

For insulators, ceramics are superior to glass in their properties, particularly mechanical strength and the ability to withstand thermal shock. However cost and difficulty of working, often offset their advantages so that glass is still widely used.

In the following sections the physical properties and methods of preparing these materials will be discussed in some detail with particular reference to their ultrahigh vacuum use.

2. Glass

The term glass can be applied to practically any compound which, after fusion, cools to a solid without crystallizing. However, it is the 'oxide' glasses which are of interest, and then only those containing silica (SiO_2) as the main glass-forming constituent. Other oxides are added as modifiers to give each glass type its particular physical characteristics.

There are a variety of such glasses, but for vacuum systems only a limited number of types are used. These are basically the glasses developed in the electronic tube industry for their sealing properties to selected metals. By suitably matching the expansion coefficients, vacuum-tight seals can be made between the glass and metal, providing the metal surface can be 'wetted'.

The glasses employed, are traditionally classed into two categories, the 'hard' or borosilicate glasses, in which the main

Table 1. Properties of glasses commonly used for vacuum applications

Property	Fused Silica	Pyrex 7440	Tungsten sealing 7720	Tungsten sealing B37	Fernico sealing 7052	Fernico sealing B47	Soda glass 0080	Soda glass S95	Lead glass 0120	Lead glass L92
Chemical composition %										
SiO_2	100	80.8	72.2	75.5	64.3	66.8	73.2	71.5	56.2	56.0
B_2O_3		12.8	15.2	16.5	19.1	21.8				
Na_2O		4.2	3.9	4.0	5.2	3.9	16.8	14.0	3.9	4.5
K_2O			0.3	1.8		4.3	0.3	1.5	8.5	8.0
Al_2O_3		2.2	1.0	2.2	7.1	2.4	1.4	2.2	1.6	1.3
PbO			6.9			0.2			28.7	30.0
LiO					1.2	0.3				
BaO, MgO, CaO					2.7		8.2	10.4		
Viscosity-temperature characteristics										
Strain Pt. °C	990	515	485	455	435	435	470	475	395	390
Annealing Pt. °C	1050	565	525	525	480	490	510	515	435	435
Softening Pt. °C	1580	820	755	775	710	715	710	710	630	630
Working Pt. °C	—	1245	1140	—	1115	—	1005	—	980	—
Expansion coefficient $\times 10^{-7}$ per °C	5.5	33	36	37.5	46	48.5	92	92	89	90
Resistance to thermal shock ¼ in. plate °C	1000	150	130		100		50		50	
Specific gravity	2.20	2.23	2.35	2.25	2.28	2.27	2.47	2.50	3.05	3.07

additive to the silica is boric oxide (B_2O_3) (a glass-forming oxide rather than a modifier) and the 'soft' glasses in which the principle additive is either sodium oxide (Na_2O) giving soda glass, or lead oxide (PbO) giving lead glass. In Table 1 the chemical composition is given of some of the glasses which are in common use in vacuum systems, and which are available from an American and a British manufacturer. Although glasses with similar properties are available from several other manufacturers, the composition may vary slightly from the different companies and from country to country.

Apart from chemical composition the distinction between hard and soft glasses can be found in terms of the viscosity-temperature characteristics. Glass has no specific melting point, but its viscosity decreases monotonically with temperature until it becomes fluid. The soft glasses 'soften' and can be worked at lower temperatures than hard glasses. As a result, systems constructed with soft glass envelopes cannot be baked above 350°C without risk of deformation under the atmospheric pressure. The hard glasses on the other hand are quite safe at 400°C and some, such as Pyrex, can be baked above 500°C. Mainly for this reason the hard glasses are used for ultrahigh vacuum systems. For certain applications, such as transparency to ultra-violet radiation, or high temperature working, other glasses may be required. If they form part of the envelope then care must be taken to see that the properties of these glasses are also compatible with ultrahigh vacuum technology.

2.1 Physical properties. The physical properties of glass of importance to ultrahigh vacuum application are those affected by temperature, since temperature plays a vital role in the outgassing of the vacuum system. Two properties are concerned, one is the viscosity, which is a measure of the mechanical rigidity of the glass, and the other is the expansion coefficients, which determine the stresses and strains which are set up when uneven temperature distributions or contact with other materials occurs.

As already mentioned glass has no definite melting or freezing point, but loses its solid-like character as it is heated, by virtue of the continuous decrease in the value of the viscosity, η. The viscosity–temperature curve depends on the composition of the glass, and in Figure 1 some typical curves of log η against temperature are given for glasses of similar composition to those described in Table 1. Four viscosities on the viscosity–temperature curve have been defined by the American Society for Testing Materials (ASTM) to represent the texture of the glass as it changes from solid to liquid and these are generally accepted internationally. The temperature corresponding to these viscosities are designated the strain point, the annealing point, the softening point, and the working point. The strain point represents the temperature at which internal stresses are relieved in a few hours, defined by viscosity value of $10^{14.5}$ poise. The annealing point is the temperature at which internal stresses are relieved in a matter of minutes, defined by a viscosity value of 10^{13} poise. The softening point is defined in terms of the elongation of a standard fibre under its own weight, and corresponds to the temperature at which the viscosity is $10^{7.6}$ poise for glasses with a density around 2.5 g cm^{-3}. The working point is the temperature at which the glass is soft enough to be worked by normal fabricating techniques, and is defined by a viscosity value of 10^4 poise. These temperature points have been inserted in Table 1, for the various glasses. The temperature at which the vacuum envelope would deform under atmospheric pressure, depends on the shape and thickness of the glass and the length of time it is held at the temperature, but as a guide the strain point can be taken as the upper temperature limit for safe bakeout.

The change in viscosity with temperature, however, is of less significance than the effect of thermal expansion. Glass when heated tends to expand, and although the expansion is relatively small compared with other materials, the effect can cause stresses and strains within the glass, which because of its brittle nature, can result in fracture. It is of particular concern in

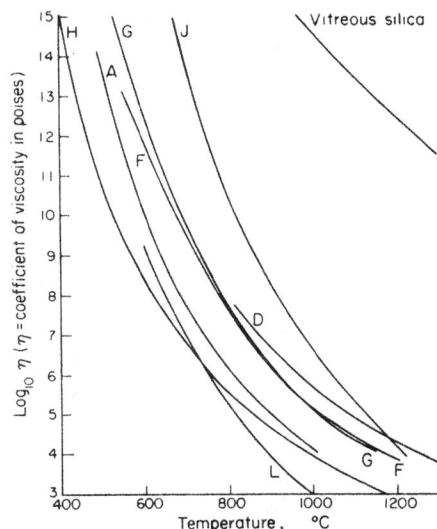

Figure 1. Viscosity–temperature curves for glasses in the table, according to Douglas.[1]

Glass type		Composition %								
		SiO_2	Al_2O_3	B_2O_3	PbO	CaO	MgO	Na_2O	K_2O	BaO
A	Soda glass	70.5	1.8	—	—	6.7	3.4	16.7	0.8	—
D	Borosilicate glass	80.1	3.0	12	—	0.2	—	3.9	0.3	—
F	Borosilicate glass	71.0	7.4	13.7	—	0.3		5.3	2.4	—
G	Borosilicate glass	71.6	5.7	11.0	—	3.6	0.6	3.5	3.9	—
H	Lead glass	56.5	1.5	—	29.0	0.2	0.6	5.6	6.6	—
J	Hard glass	54.5	21.1	7.4	—	13.5	—	—	—	3.5
L	Soda vapour resistant glass	22.6	23.7	37	—	10	—	6.5	0.2	—

connection with rigid seals between the glass and other materials, such as metal and ceramics, and also where dissimilar glasses are joined. Some typical expansion curves expressed as $\Delta L/L$ against temperature are shown in Figure 2. In general the expansion is greater for the softer glasses, and in the borosilicate glasses decreases as the amount of B_2O_3 decreases. Below 300°C the curves are essentially linear, and a constant expansion coefficient can be ascribed to the glass; values of this expansion coefficient for the various glasses are given in Table 1. At higher temperatures the rate of expansion increases, and it is appreciably higher as the annealing temperature is approached.

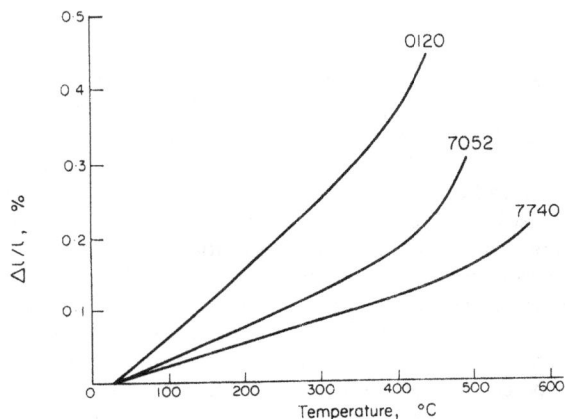

Figure 2. Expansion–temperature curves for three Corning glasses.

The values of the expansion coefficient are only reproducible and reversible for well-annealed glasses. Poorly annealed and strained glass will give somewhat higher values of coefficient and exhibit irreversible irregularities in the expansion/temperature curve.

Stresses are not only set up at seals, as a result of expansion, thermal gradients across the glass can also be a danger. If one side of a glass plate is hotter than the other, then the heated side will experience a compressional stress whilst the cooler side will be under tension. The tensile stress which is the cause of failure in glass, set up in this case, will depend on the temperature difference and on the glass properties, particularly the coefficient of expansion. In general, the lower the expansion coefficient the higher will be the temperature gradient which the glass can withstand. Thus for a constrained plate, the temperature difference causing a tensile stress of $7 \times 10^6\ Nm^{-2}$ is about 50°C for Pyrex, but only about 15°C for soda glass. Although large steady temperature gradients are not likely to occur in practice, large transient gradients may well occur, for example, when a glass condensation trap is first immersed in liquid nitrogen. The strength of glass is greater under momentary stress than under prolonged stress, so that the resistance to thermal shock cannot be assessed from the static characteristics. It depends not only on the expansion coefficient but on the shape of the sample, its thickness, and whether the stress is incurred by sudden heating or cooling (the latter is the more damaging). An empirical testing schedule used by Corning, where a plate of given dimensions is heated and then plunged into cold water, gives an indication of the resistance to thermal

121

shock. The highest temperature to which the plate can be heated, without damage on cooling, is taken as the criterion, and the values for the Corning glasses have been inserted in Table 1. In general soft glasses are unsatisfactory as cooling traps or other parts of the vacuum envelope subject to thermal shock. Its low price and ease of working, however, has seen its widespread use for electronic tube envelopes.

2.2 Permeation of gases through glass. From the early experiments in a vacuum, it was known that gas could permeate through a thin glass wall, and there are several references[2] to the measurement of the permeation of gas through silica and glass in the 1920's and 1930's. However, it was considered that, for practical purposes, the rate at which gas 'leaked' into a vacuum system of an electronic tube from the atmosphere was so small at room temperature that the effect could be ignored.

With the attainment and measurement of ultrahigh vacuum in glass systems, permeation of gas through the walls from the surrounding atmosphere was recognized as a contributory source of gas influx, limiting the ultimate pressure. For example, in 1954 Alpert and Buritz[3] reported that, in their Pyrex glass system, permeation of atmospheric helium (the equilibrium pressure of helium in air is $\sim 5.3 \times 10^{-1}$ Nm^{-2}) through the walls was the predominant source of residual gas. They observed that, in a sealed off volume of 400 l, the pressure rose from 2×10^{-7} Nm^{-2} to 2×10^{-6} Nm^{-2} in about 10 h.

The microstructure of glass has the general form of SiO$_4$ tetrahedra sharing oxygen atoms, as in crystalline quartz, but in an irregular arrangement making a more open structure in which gas atoms can be sited, see Figure 3. The addition of modifiers Na$^+$ and K$^+$, etc. occupy some of the sites within the

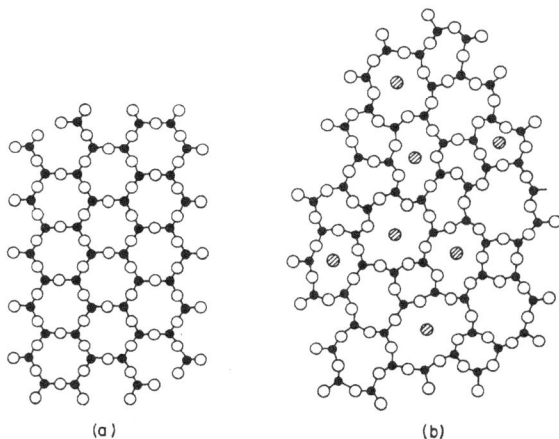

Figure 3. Two dimensional schematic representation of the structure of (a) quartz crystal (b) glassy form of silica, with modifier atoms shown as hatched circles.

pockets surrounded by the silica or silica–borate structure. Thus it might be expected that the permeation of gas through glass will depend on the 'porosity' of the microstructure, and be reduced by the presence of modifiers. It can also be expected that the size of the gas molecules will be important.

The quantity Q of gas permeating through a solid wall or membrane of thickness d and area A is given by the equation:

$$Q = \frac{AK(p_2{}^n - p_1{}^n)}{d} \tag{1}$$

Where p_1 and p_2 are the gas pressure each side of the wall and K is the permeation constant.

For glass where n is found to be 1

$$Q = \frac{AK\,\Delta p}{d} \tag{2}$$

Using SI units, with Q expressed as Nm s^{-1} and p as Nm^{-2} both normalized to 25°C, then K will be given in m^2 s^{-1}.

It can be shown that the value of K increases exponentially with temperature according to the following equation

$$K = K_0 \exp\left(\frac{-E}{RT}\right) \tag{3}$$

Where E is the activation energy, R the gas constant and K_0 a constant of proportionality. It is thus convenient to present data on permeability as log K against $1/T$ plots.

Measurements of permeation confirm the general premises mentioned above. The effect of gas molecular diameter is demonstrated in Table 2, where the permeation constant K for gases passing through fused silica at 700°C, according to Norton[4] is listed against the relevant atomic or molecular diameter. It is seen that helium with the smallest atomic diameter has the highest permeation rate, whilst argon, nitrogen and oxygen are too large to permeate appreciably. For practical purposes, silica can be considered as impervious to the latter gases. The results on hydrogen compared with neon, however, indicate that atomic diameter is not the only factor. Norton[4] suggests that the greater permeation rate for hydrogen is related to surface and solubility effects.

Table 2. The atomic or molecular diameter in atomic units and the permeation constant for gases through fused silica at 700°C (Norton[4])

Gas	Atomic/molecular diam. (au)	Permeation const. K m^2 s^{-1}
Helium	1.95	1.7×10^{-11}
Neon	2.4	3.5×10^{-13}
Hydrogen	2.5	1.7×10^{-12}
Deuterium	2.55	1.4×10^{-12}
Oxygen	3.15	less than 10^{-18}
Argon	3.2	less than 10^{-18}
Nitrogen	3.4	less than 10^{-18}

The permeation rates of helium through glasses of different compositions have been investigated by several workers, and give general agreement. The values of K as a function of $1/T$ plotted in Figure 4 are taken from the data of Altemose[5] converted to SI units. The figure shows the values for Corning glasses listed in Table 1, and also for Corning 1720, a special aluminosilicate glass particularly suitable for high vacuum as far as permeation is concerned.

The figure shows that the permeation rate for all the glasses are lower than for silica, and thus permeation of gas other than helium can be ignored. In general, the permeation rate decreases as the percentage of glass network formers, SiO$_2$ and B$_2$O$_3$, decreases, and correlation between K and the weight per cent of glass former and also the glass density was demonstrated by

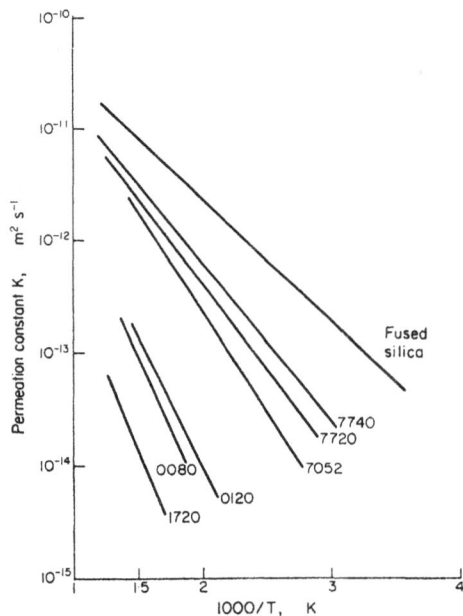

Figure 4. Permeation constant for helium through some of the Corning glasses listed in Table 1 as a function of temperature (from Altemose.)[5]

Norton.[4] However, as Altemose[5] pointed out, such a correlation does not satisfy the results for lead and soda glass. He suggested that mole per cent rather than weight per cent should be used on the grounds that it was the packing density of the atoms rather than their mass which was the controlling factor. The plot of $\log K$ against mole per cent of $SiO_2 + B_2O_3 + P_2O_5$ for a large selection of Corning glasses according to Altemose is shown in Figure 5.

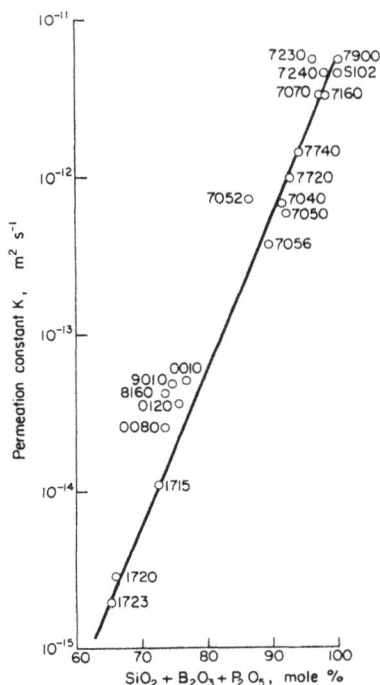

Figure 5. Permeation constant for helium through some Corning glasses at 300°C as a function of the mole percent network formers (from Altemose).[5]

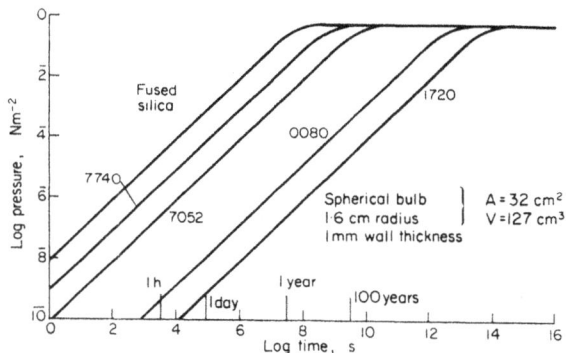

Figure 6. Accumulation of helium in a glass vacuum bulb as a result of permeation from the atmosphere at 25°C for various glasses.

The influx of helium from the atmosphere and the effect on the ultimate pressure can be calculated from the value of K. To illustrate the effect, the rise in pressure in a sealed-off system, assuming equilibrium conditions, is plotted in Figure 6 as $\log p$ against $\log t$ for a 1.6 cm radius bulb (with a 1 mm wall thickness), made of various glasses. Thus, at room temperature the pressure in such a bulb made of silica will rise to 10^{-5} Nm^{-2} in a matter of minutes, for Pyrex it will take a few hours, for molybdenum-sealing borosilicate glass a few days, and soda glass several years. Pyrex is not, therefore, a suitable glass for ultrahigh vacuum systems, but it should be pointed out if the temperature is raised to, say, 400°C, then even for soda glass in the example given, a similar pressure rise will occur in less than an hour.

2.3 Outgassing of glass. The permeation of gas through glass implies a certain solubility of the gas in the glass, which will diffuse out into the vacuum system at a rate depending on the concentration and temperature. Gas trapped in the open network structure of the glass between the constituent atoms, is said to be physically dissolved. However, in glass there is also the possibility of chemical solubility, whereby the presence of gases in the glass is due to chemical reaction. In this case larger molecules can be dissolved, and also the quantity can be greater. Infrared spectroscopy and other techniques have shown that water vapour, carbon dioxide, oxygen and sulphur dioxide are dissolved in this way, the gases being dissolved during the manufacturing process whilst the glass is in a molten state. Water vapour constitutes the major component of dissolved gas with a solubility some two orders higher than helium. Unlike helium or other physically dissolved gases, the amounts of water vapour dissolved in glass are greater the higher the percentage of alkali modifiers, i.e. the solubility in soda or lead glass is considerably higher than in borosilicate glass.

In addition to the gas dissolved within the glass there will also be the gas adsorbed on the surface. Again water vapour is the main constituent, which appears to be bound fairly tightly, probably as surface hydrates.

The adsorbed and dissolved gases constitute a source of gas influx for glass envelope systems which would be unaccepted for ultrahigh vacuum in an unbaked system; Alpert[6] quotes a figure of gas influx from an unbaked borosilicate glass system of 10^{-5} Wm^{-2}. The gas influx can be enormously reduced by an outgassing bake. Early investigations established that as glass is heated in a vacuum a rapid evolution of gas occurs at temperatures around 200–300°C. Further heating produces a slower but more persistent evolution of gas. Both gas evolutions are primary water vapour and it is generally accepted that the major part of the initial evolution is due to surface adsorbed layers whilst the more persistent evolution is due to gas diffusing from the interior.

A general review of the measurements made on gas desorption from glass is to be found in Dushman.[2] Of interest are the studies of Todd[7] who substantiated the general hypothesis by showing that after an initial period, the gas evolution was inversely proportional to the square root of the baking time in accordance with a diffusion process. He further established that the constant of proportionality was exponentially dependent on temperature and also depended on the glass composition. An indication of the gas evolution from different glasses can be seen in Table 3, where the amount of gas evolved from within the glass during various bakeout conditions have been calculated from data presented.[7] The amount is expressed in $J\,m^2$, but if one considers a litre volume then the figures would be equivalent to the pressure rise in Nm^{-2} assuming that the gas is coming from $10\,cm^2$ of glass surface.

By baking a glass system at high temperature for a period of 24 h, the surface adsorbed layers of gas, and sufficient absorbed gas is driven out to reduce the subsequent outgassing rate at room temperature to the order of $10^{-12}\,Wm^{-2}$.

3. Metals

Metals can be used in ultrahigh vacuum systems as internal components for their electrical, mechanical, or heat conducting properties, or as an integral part of the vacuum envelope. As internal components, the specific use will dictate the most suitable choice of metal. Thus for grids, where fine wires of high mechanical strength are required, molybdenum or tungsten is employed, whereas for plate electrodes requiring complex shapes, softer metals which are more easily fabricated such as nickel and iron are more appropriate. As far as the vacuum requirements are concerned, the metal for internal components only has to satisfy two conditions; the vapour pressure of the metal at its working temperature must be below the desired ultimate pressure, and also the metal must be capable of adequate degassing so that it does not constitute a major source of gas influx. When using the metal as an integral part of the vacuum envelope there is a third criterion; gas from the surrounding atmosphere must not permeate through it.

3.1. The vapour pressure of metals.
Recently Honig[8] has compiled vapour pressure data from the literature, which covers most of the common elements down to the ultrahigh vacuum

pressure range. For convenience for this review values have been extracted for those metals either used *in vacuo* or likely to occur as impurities and are given in Table 4. The data are presented as temperature values of the respective metal corresponding to the given vapour pressure.

At room temperature few metals have vapour pressures above $10^{-9}\,Nm^{-2}$ and although care should be taken to avoid materials which may contain impurities such as sodium or potassium, the selection of metals for components is not seriously restricted on this score. One metal, however, which is often overlooked is cadmium, and for ultrahigh vacuum systems, cadmium plated screws etc., should not be employed.

As the temperature of the system is raised other metals will have vapour pressures above $10^{-9}\,Nm^{-2}$, and since it will normally be necessary to outgas the system at elevated temperatures, up to 450°C, zinc, lead, tin and similar metals should be excluded. If the temperature of the component in its application is raised to higher temperatures for example for thermionic emission, then the selection of the metal is more limited. Even a tungsten filament emitter running at 2000°C will have a vapour pressure of $10^{-7}\,Nm^{-2}$ in its vicinity, which as Alpert and Buritz[3] have pointed out, could represent a lower pressure limit value in an ion gauge of $10^{-10}\,Nm^{-2}$, due to the presence of the tungsten atoms. Nevertheless, for most vacuum applications there is a wide range of metals whose vapour pressures are low enough, even at elevated temperatures, not to constitute a problem down to $10^{-11}\,Nm^{-2}$ pressure, many of which are readily obtainable and are economic to use.

3.2 Outgassing of metals.
The outgassing properties of metals warrants more careful consideration, not because metals differ greatly in their outgassing rates, but because of the processing necessary to reduce the outgassing rate to an acceptable level. As with glass, gases are adsorbed on the metal surfaces, and also dissolved within the metal, mainly during the manufacture and processing of the raw material. The gases may be physically or chemically sorbed, both in the case of surface adsorption, and absorption throughout the bulk. When the metal is placed in a vacuum environment these gases will be evolved, at a rate depending on the total gas sorbed, the nature of the sorption process, and the temperature. In particular the activation energy of the sorption process will be important.* For untreated metals, particularly if the surface is contaminated by a thin

---

* If gas is chemisorbed with a high activation energy it may not be released, and indeed metals exhibiting strong bonds with the active gases, if freshly formed or cleaned in vacuum, can be employed for reducing the pressure in the system.

Table 3. Gas evolution from glass calculated from data by Todd[7]

Glass type (Corning)	Gas evolved during 1 h bake in Jm^{-2}		Gas evolved during 10 h bake in Jm^{-2}	
	at 600 K	at 800 K	at 600 K	at 800 K
7740	8.3×10^{-4}	6.2×10^{-2}	2.6×10^{-3}	2.0×10^{-1}
7720	1.5×10^{-3}	8.4×10^{-2}	4.7×10^{-3}	2.7×10^{-1}
0080	1.8×10^{-3}	3.3×10^{-1}	5.7×10^{-3}	1.05
0120	6.5×10^{-3}	2.8×10^{-1}	2.1×10^{-2}	9.2×10^{-1}
1720	2.0×10^{-6}	2.1×10^{-3}	6.3×10^{-6}	6.7×10^{-3}

Table 4. Vapour pressures of metals expressed as temperature K for a given pressure in Nm^{-2}

Metal		Melting pt	Temperature K Giving				
			$P = 1.33 \times 10^{-9}$	$P = 1.33 \times 10^{-8}$	$P = 1.33 \times 10^{-7}$	$P = 1.33 \times 10^{-6}$	$P = 1.33 \times 10^{-5}$
Ag	Silver	1234	721	759	800	847	899
Al	Aluminium	932	815	860	906	958	1015
Au	Gold	1336	915	964	1020	1080	1150
Ba	Barium	983	450	480	510	545	583
Be	Beryllium	1556	832	878	925	980	1035
C	Carbon	—	1695	1765	1845	1930	2030
Ca	Calcium	1123	470	495	524	555	590
Cd	Cadmium	594	293	310	328	347	368
Ce	Cerium	1077	1050	1110	1175	1245	1325
Co	Cobalt	1768	1020	1070	1130	1195	1265
Cr	Chromium	2176	960	1010	1055	1110	1175
Ca	Caesium	302	213	226	241	257	274
Cu	Copper	1357	855	895	945	995	1060
Fe	Iron	1809	1000	1050	1105	1165	1230
Ge	Germanium	1210	940	980	1030	1085	1150
Hg	Mercury	234	170	180	190	201	214
In	Indium	429	641	677	716	761	812
Ir	Iridium	2727	1585	1665	1755	1850	1960
K	Potassium	336	247	260	276	294	315
La	Lanthanum	1193	1100	1155	1220	1295	1375
Mg	Magnesium	923	388	410	432	458	487
Mn	Manganese	1517	660	695	734	778	827
Mo	Molybdenum	2890	1610	1690	1770	1865	1975
Na	Sodium	371	294	310	328	347	370
Ni	Nickel	1725	1040	1090	1145	1200	1270
Pb	Lead	601	516	546	580	615	656
Pd	Palladium	1823	945	995	1050	1115	1185
Pt	Platinum	2043	1335	1405	1480	1565	1655
Re	Rhenium	3453	1900	1995	2100	2220	2350
Rh	Rhodium	2239	1330	1395	1470	1550	1640
Sb	Antimony	903	447	498	526	552	582
Se	Selenium	490	286	301	317	336	356
Sn	Tin	505	805	852	900	955	1020
Sr	Strontium	1043	433	458	483	514	546
Ta	Tantalum	3270	1930	2020	2120	2230	2370
Th	Thorium	1968	1450	1525	1610	1705	1815
Ti	Titanium	1940	1140	1200	1265	1335	1410
W	Tungsten	3650	2050	2150	2270	2390	2520
Zn	Zinc	693	336	354	374	396	421
Zr	Zirconium	2128	1500	1580	1665	1755	1855

layer of oxide, the initial outgassing rate at room temperature is of the order of 10^{-4} Wm^{-2} i.e. of an order greater than found for untreated glass. Figure 7 gives the outgassing rates for some of the (untreated) metals used in the fabrication of vacuum systems, as measured by Blears *et al.*[9] Values for neoprene and Araldite are given for comparison. The outgassing rate decreases with time, and according to Dayton,[10] can be expressed over the first 10 h or so by the formula

$$q_t = q_0/t_h{}^n \qquad (4)$$

where q_t is the outgassing rate at time t_h (in hours) q_0 is about 10^{-4} Wm^{-2}, and n may vary from 0.7 to 2 but is frequently in the neighbourhood of 1.

As with glass this large outgassing rate, which is mainly water vapour, is ascribed to gas physically adsorbed on the metal surface. However, from the total quantity of gas evolved several monolayers are involved; some experiments suggest more than 100 monolayers. In the case of metals having a more or less porous layer of oxide on the surface, the gas is likely to be sorbed throughout the oxide film. It can then be expected that the degassing rate will be controlled by diffusion of the gas through the pores or grain boundaries of the coating. This can explain why the time of removal is prolonged compared with glass, and with the theory. After the first 10 –100h when the bulk of the surface adsorbed gas has been evolved, the degassing rate decays exponentially to a much lower value, when the gas diffusing from the bulk material becomes the controlling factor, i.e. the rate is then inversely proportional to the square root of the time.

The gases within the bulk metal are commonly H_2, N_2, O_2, CO and CO_2, and not water vapour. They are mostly taken up during the molten stage, and arise from the furnace gases during

125

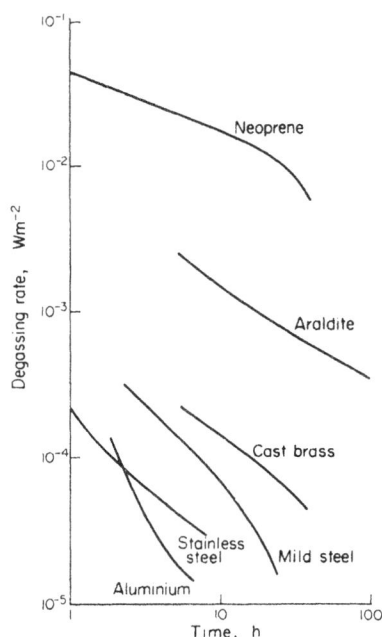

Figure 7. Degassing rate of some (untreated) metals compared with araldite and Neoprene according to Blears *et al.*[9]

the melting process, and/or from ambient gases during casting. Some of the gases listed above react with the metal to form compounds, whilst others are physically held in solution. The mechanism will depend on the metal–gas combination. When the melt solidifies the gases are only partially released leaving considerable gas still entrapped in the metal. Indeed typically the gas content of metals is some 10–100 % of their volume at s.t.p., which is of the same order as the water vapour content of glass.

The gas content of metals is not always attributable to the melting process, since gases can diffuse into most metals in the solid state. The process is relatively slow at room temperature, and metals which have been degassed can generally be stored for periods of days without an appreciable increase in gas content. The diffusion, however, increases exponentially with temperature. The rare gases are an exception, and do not dissolve in any metal under purely thermal conditions, even when the metal is molten. They can only be absorbed if the gas atoms bombard the metal with a high energy, either as ions or energetic neutrals.

A considerable amount of information exists on the solubility of gases in metals, mainly in the solid phase; a summary can be found in Dushman.[2] Some general rules which can be applied when considering the solubility of gases in metals are given below under the headings of the gases involved.

Hydrogen. For hydrogen solubility, the metals can be divided into two main groups; those which form solid solutions and those forming hydrides or pseudo-hydrides. In the first group the solubility is proportional to the square root of the partial pressure of hydrogen and increases with temperature. Metals in this group listed roughly in order of increasing solubility are Al, Cu, Pt, Ag, Mo, W, Cr, Co, Fe, Ni. In the second group, the metals forming true hydrides are the alkali and alkali–earth metals i.e. NaH, and CaH_2, etc. and also elements of groups

IVb, Vb and VIIb such as B, C, S, Si, As. The so-called pseudo-hydrides have the form MH_m where m is not an integer, and include the metals Mn, Ta, V, Nb, Ce, La, Zr, Ti. Their hydrogen solubility can be orders of magnitude greater than for metals in the first group and they are of interest for their gettering properties. The solubility of the second group decreases with increasing temperature.

Oxygen. Oxygen is soluble to some extent in most metals, but, except in case of the noble metals, an oxide phase also appears when the limit of solid solubility is exceeded. It is difficult therefore to distinguish between the solution of oxygen and that of the oxides. During the molten state, many metals will take up large quantities of oxygen, which are precipitated as oxide when the metal solidifies. As a result, the quantity of oxygen in the metal, capable of being released into a vacuum system, can be large compared with the amount which can be dissolved in the solid metal.

Nitrogen. Nitrogen dissolves only in those metals which form nitrides at higher temperatures, for example Zr, Ta, Mn, Mo and Fe. It has been shown to be insoluble, within experimental limits, in Co, Cu, Ag and Au. For the metals like Mo and Fe, the solubility is small of the order of 1 % or less by weight, and there is no tendency to form nitrides in the solid phase. On the other hand Zr when heated dissolves large quantities of nitrogen to form the nitride.

Carbon monoxide and dioxide. A similar behaviour to nitrogen is encountered with CO. It is dissolved only by metals such as Ni and Fe, which are capable of forming a carbonyl. CO is not soluble in copper. The observed evolution of carbon monoxide and dioxide on heating some metals in a vacuum may be accounted for by diffusion of carbon and its reaction with the oxides in the metal, and not direct solution of CO or CO_2.

Metals can be degassed in a similar way to glass by baking in the final vacuum system. However, since the rate of diffusion is exponentially dependent on temperature, for the dissolved gas, it is advantageous to go to the highest temperatures that metals are capable of withstanding, rather than the most convenient for the vacuum system as a whole. This outgassing can be carried out before assembly in the vacuum system, since, as has already been stated, re-absorption of gas at room temperature is a relatively slow process.

Thus the outgassing of metal parts can be carried out in stages, used alone or more usefully in combination, as follows:

1. Melting the raw material in a vacuum to provide gas free ingots.

2. Heating the preformed parts in a vacuum before assembly.

3. Heating the assembled parts in the final vacuum system.

The vacuum melting of metals can now be carried out on an industrial scale and most metals are available in a vacuum-melted form. Details of the techniques and the apparatus used for the degassing of the crude metal during melting are to be found in the comprehensive book by Espe.[11] However, gas-free vacuum-melted metals are relatively expensive, and are therefore normally limited to applications where freedom from oxygen or deoxidant impurities are essential, e.g. cathode nickels and glass-sealing alloys. For the majority of vacuum applications there is little to be gained from using gas-free ingots, since, in fabricating the components, it is difficult to avoid contamination by oil etc. or to avoid oxide formation, especially if hot-forging, hot pressing or welding are involved.

The degassing of the fabricated components prior to their assembly, on the other hand, is almost essential. The fabricated components are first chemically cleaned to remove surface oxide or other contaminant layers, and degreased to remove oil, which may be deposited on the surface during handling. The component is then heated in a vacuum of 10^{-2} to 10^{-3} Nm^{-2} to a temperature of the order of 1000°C. Alternatively the component can be heated in a flushing gas which is either not readily absorbed or easily removed at a later date. For example, hydrogen stoving is used extensively for electron tube parts. The hydrogen reduces any oxide present, and because of its high diffusion rate is driven off fairly readily during the subsequent tube processing. It is also a more economic stoving process as the absence of vacuum seals makes it possible to feed the components continuously through the heated furnace on a moving belt system. However, the removal of the residual hydrogen, which is not detrimental in an electron tube, can be a problem in ultrahigh vacuum systems, so that hydrogen stoving is not recommended, even when followed by vacuum stoving.

The actual temperature of the stoving furnace and the stoving time are dictated by several factors. Clearly the temperature must not be so high that appreciable evaporation of the metal takes place, or that the melting point is approached. In practice, a lower limit is set by distortion of the component (creep limit) or in some cases by the design of the stove. In general, the temperature will be about the annealing temperature, and as a result of the stress equalization the component may warp and have to be trued up later. In the case of stressed parts such as supports and springs, the temperature must be kept below the annealing temperature. For most metals, temperatures from 900 to 1000°C are adequate, and these can be attained with resistance wound furnaces employing nichrome or Kanthal windings. Higher temperatures are desirable for tungsten and molybdenum, requiring more elaborate furnaces; Norton and Marshall[12] found that nitrogen, which formed more than 50% of the sorbed gas in molybdenum, was not released until temperatures above 1200°C were reached. Titanium and copper must be stoved at lower temperatures, 500 to 700°C, and only oxygen free high conductivity (OFHC) grade copper is suitable for high vacuum applications. Aluminium, because of its low melting point cannot be stoved much above the normal bake-out temperature of the completed system and, therefore, there is little to be gained by a pre-outgassing process. Hydrogen is not necessary to reduce the metal oxides since most of the metal oxides will decompose at the stoving temperature, giving off oxygen. The exceptions are the few oxides such as Al_2O_3, MgO and ThO_2 having low dissociation pressures.

The stoving time should be as long as practicable, but since most of the gas is removed in the first few hours, a stoving time of 8 h at the equilibrium temperature is usually sufficient. Taking into account the heating and cooling time, this allows stoving to be carried out overnight.

As a result of vacuum stoving, the outgassing rate in the final vacuum system may be reduced by several orders of magnitude. Flecken and Nöller[13] observed an 88–97% reduction in the amount of gas desorbed after stoving stainless steel at 800°C

for 1 h followed by exposure to the atmosphere for 1 h. Care must be taken in storing and handling the components between stoving and assembly, since as Varadi[14] has shown, touching the component with fingers can markedly increase the degassing rate, and surprisingly some of the increase could be attributed to gas re-absorbed into the bulk metal.

Even with careful handling, surface contamination can occur, and it can be expected that gas adsorption will take place on the surface of the stoved parts. Thus a further degassing process *in situ* is necessary to achieve ultrahigh vacuum pressures. Also for small components such as filaments and grids, higher temperatures can be reached in the vacuum system than can be conveniently achieved in a vacuum furnace. Heating the metal in the vacuum system for degassing purposes can be carried out in one of four ways: (1) baking in an external furnace; (2) direct passage of current through the component; (3) eddy-current heating by H.F. induction; (4) electron or ion bombardment. Method (1) is generally limited to temperatures below 500°C, whilst methods (2), (3) and (4) allow much higher temperature outgassing to be utilized. Eddy-current heating is restricted to components mounted in a glass or other insulating material envelope, but nevertheless is a useful method of outgassing sheet metal electrodes to high temperatures in experimental or measuring electronic devices; for details see Espe.[11] Heating by direct passage of current can be applied to filaments, and components of fairly high resistance, for example helical wound grids. The component should be heated to a temperature in excess of that reached during subsequent operation. Electron bombardment of a component is probably the most versatile method of heating metal parts to temperatures above 500°C. Providing a suitable electron source is available, it imposes no restriction on the target material or its shape and can be conveniently controlled by circuitry. Ion bombardment not only heats the target, but also removes the surface by sputtering. In applications where surface cleaning is required it can be used with advantage, although usually a gas pressure above 10^{-2} Nm^{-2} is required to obtain the necessary ions.

Methods (2), (3) and (4), however, are normally restricted to relatively small components. For large components, especially the metal vacuum envelope, baking with an external furnace provides the only feasible method of thermal outgassing the assembled system. Since much of the surface adsorbed gas is water vapour, baking at 200–250°C over an extended period will appreciably reduce the subsequent degassing rate, and indeed some vacuum engineers suggest that baking at a higher temperature is unnecessary. However, most workers recommend baking to 400–450°C for at least 16 h to obtain ultimate degassing rates which are compatible with ultrahigh vacuum pressure attainment. A number of values have been reported for outgassing rates from metal after such bakeout processes, most of them being concerned with stainless steel. Values vary from 10^{-10} to 10^{-12} Wm^{-2} at room temperature, and one can reasonably expect to attain the upper limit for most metal systems where the precautions and processes described in this section have been adhered to. It must be remembered, however, that if the system is let down to air, especially if air is not dried, a re-bake may be required on subsequent pump-down to remove surface adsorbed gas.

3.3 Permeation of gases through metal. If the metal forms part of the vacuum envelope the further criterion arises, that of permeation of gas through the metal. Unlike glass the diffusion takes place through the crystal lattice, and only those gases which are soluble in the metal will permeate through it. Thus, helium and the other inert gases will not permeate through any metal even at elevated temperatures, whereas hydrogen and oxygen will permeate to some extent through most metals. Permeation of the diatomic molecular gases varies as the square root of the pressure difference, i.e. n in equation (1) is $\frac{1}{2}$. This indicates that the molecules dissociate on adsorption at the high pressure surface and diffuse through the metal as atoms, recombining on the vacuum side on desorption. Hydrogen, having the highest diffusion rate is the main gas involved in permeation through metals, and in Figure 7 the log of the permeation constant for hydrogen through several metals has been plotted against the reciprocal of the temperature. The values have been derived from curves given in a survey by Norton.[15] Since K for a metal system is dependent on $(p)^{1/2}$, it is expressed as $m^2 \, (Nm^{-2})^{1/2} \, s^{-1}$ when p, Q, etc. are in SI units.

It can be seen from the figures that the permeation rate of hydrogen through palladium is about two orders higher than for any other metal. The permeation of other gases through palladium on the other hand is negligible, so that palladium makes a useful filter for obtaining pure hydrogen; usually employed in the form of a heated tube. The permeation rate of hydrogen through nickel and iron is also relatively high. For example at pressures below $10 \, Nm^{-2}$ it is higher than the permeation rate of helium through silica under similar conditions, particularly at elevated temperatures (c.f. Figure 3). The 'glass-sealing' metals of the fernico type have similar high permeation rates. A recent review on permeation and outgassing of vacuum materials by Perkins[16] lists values of K for various stainless steels and iron–cobalt–nickel alloys. The permeation

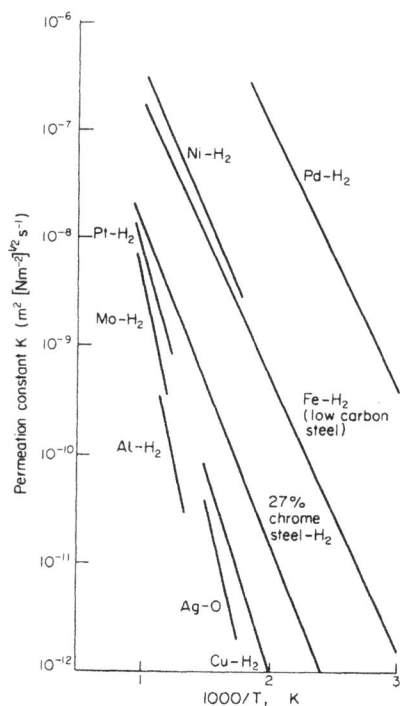

Figure 9. Accumulation of hydrogen in a metal vacuum bulb as a result of permeation from the atmosphere at 25°C for various metals.

constant for stainless steel is some two orders lower, nevertheless hydrogen permeation through chamber walls constructed of stainless steel due to the partial pressure of hydrogen in the atmosphere cannot be ignored if ultrahigh vacuum pressures are to be achieved. This can be illustrated by considering the influx of hydrogen from the atmosphere for the same example as used in Section 2.2, i.e. the size of a sealed off spherical vacuum vessel of 1.6 cm radius with a wall thickness of 1 mm. The rise in pressure with time in the vessel made of various metals is plotted in Figure 9. In spite of the low partial pressure of hydrogen in the atmosphere ($5 \times 10^{-2} \, Nm^{-2}$) the pressure rises, at room temperature, to $10^{-5} \, Nm^{-2}$ in less than 1 h for iron, 2 or 3 days for stainless steel, and about 100 years for copper. At higher temperatures the permeation increases markedly, and if the stainless steel sphere were to be baked at 450°C then the pressure would rise above $10^{-4} \, Nm^{-2}$ in less than a second (*see* dotted curve, Figure 9).

In the case of iron, the influx of hydrogen can also occur as a result of the chemical reaction of water with the exterior walls. The reaction produces hydrogen, some of which can permeate to the interior. Since the abundance of bound hydrogen, as water vapour, in the atmosphere is about 10^5 times that of free hydrogen, rusting is probably a much greater source of hydrogen influx than permeation of the free hydrogen from the air.

The permeation rate of other gases through most metals are at least an order lower than for hydrogen and for practical purposes can be neglected. However, the permeation of oxygen through silver is of interest. Because of the high solubility of oxygen in silver, its permeation rate is much higher than any other gas, including hydrogen. As a result a heated silver tube can be used to admit oxygen to a vacuum system. The oxygen so produced is spectrographically pure.

3.4 Physical requirements. If the metal is to be employed as the vacuum envelope, then there are further criteria imposed on it which concern its physical and chemical properties. Clearly the envelope must be mechanically strong enough to withstand the pressure due to the atmospheric environment, and one does not wish to employ metals which would entail excessively thick walls to satisfy this criterion. Also it must maintain its strength when heated to baking temperature, and indeed should not be distorted by temperature cycling. The need to fabricate the shape of the vacuum vessel and to provide leak-tight joins by welding and brazing, further restricts the choice of metals.

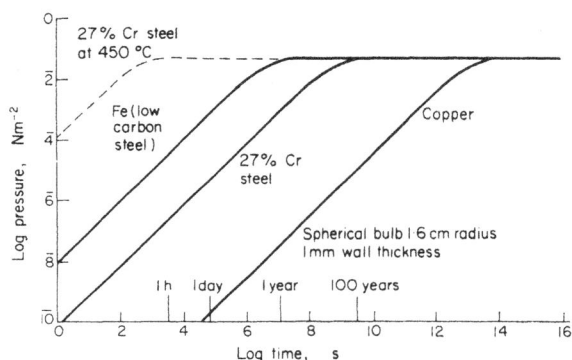

Figure 8. Permeation constant for hydrogen through various metals as a function of temperature (from Norton).[15]

Another important feature is the chemical stability of the metal. If the metal reacts with the atmospheric gases when heated then erosion of the surface and mechanical weakening may occur. Copper is unsuitable for this reason; oxidation at elevated temperature causes scaling which flakes off to expose fresh surfaces. As has already been pointed out, rusting of mild steel gives rise to appreciable hydrogen permeation, which makes it unsuitable for ultra-high vacuum systems. Finally in many applications a vacuum envelope is required which is non-magnetic.

Bearing these requirements in mind, present-day technology favours stainless steel as the most satisfactory metal for ultra-high vacuum. It is also a fairly common material and reasonably inexpensive. Stainless steel is a term commonly used to indicate any or all iron alloys which resist atmospheric corrosion. It generally refers to low carbon steels containing 10–15% chromium, and two main classes are recognized; (1) those containing chromium as the only major alloying constituent and (2) those containing both chromium and nickel as the major alloying elements. The latter group known as the Austenitic stainless steels, American AISI 300 series, are of most interest particularly those containing 18% chromium and 8% nickel (18/8 stainless steels). They can be considered as non-magnetic, having magnetic permeabilities of less than 1.02. They resist corrosion up to 800°C and do not harden when hot worked. On the other hand they can develop high mechanical strength when suitably processed and maintain their strength at elevated temperatures. For argon-arc welding the steels are either 'stabilized' by the introduction of a small percentage of titanium or niobium, or a specially low carbon content Austenitic steel is used. Table 5 gives the maximum percentage composition of the various stainless steels in common use in the UK together with the American equivalent or near equivalent type numbers. The low carbon content steel is preferred and used extensively in the States, type 304. However, it is difficult to obtain in the UK so that EN58B or EN58F is normally used. These appear to be perfectly satisfactory for most ultrahigh vacuum systems. The EN58B is better from the non-magnetic point of view having a lower magnetic permeability, but EN58F is more readily available and cheaper.

4. Ceramics

The term 'ceramic' describes a wide range of inorganic non-metallic compounds which have attained a hard solid crystalline state as a result of firing. They have advantage over glass as a constructional material for insulation application, in their better mechanical strength, particularly at elevated temperature, their better electrical properties, and in their capability of being fabricated with close dimensional tolerances. Like glass they are practically chemically inactive, and have a low vapour pressure. Advances in the technology and manufacture of ceramics have produced improved quality components at reasonable cost, and as a result they are finding increasing use in vacuum systems and devices.

There are three basic types of ceramics: pure oxides, silicates, and special types of nitrides, borides, and carbides. The last group have been developed specifically for high temperature applications in rockets etc. and have not found application so far in vacuum technology.

Although the oxides can be single-phase crystalline compounds, most ceramics also contain a certain proportion of glassy phase which bonds the crystal aggregates together. Since the ceramic is formed by a sintering process it is usually porous in structure, a 10% pore volume being quite common, and the glassy-phase also serves to seal the pores and render the ceramic gas tight. In general the silicate ceramics have a higher quantity of glassy-phase as, for example, in porcelains where up to 70% can be of this phase. The amount of glass-phase present has a marked influence on the ceramic properties, in particular the mechanical strength and electrical properties. The method of fabrication can also affect the properties giving differing microstructure to the resulting ceramic.

In general, the oxide ceramics are obtained synthetically from chemically prepared materials, whereas the silicates are obtained from naturally occurring minerals, although to some extent they may be purified by preparatory processes. Because of this the silicates tend to show more variability, and even when they are substantially of the same chemical composition can have differing properties according to the source of the materials.

Since the vacuum technologist requires high quality materials with controlled characteristics which are reproducible, the oxide ceramics are usually preferred. However there are certain properties of the silicates, for example their dielectric constant, which makes them desirable for certain applications. Also they are easier to make and therefore less expensive.

The manufacture of ceramics depends to some extent on the type, but the basic stages in the schedule apply to most ceramics. It involves grinding the ingredients to fine particles, and then adding sufficient moisture, or in some cases organic binder, to

Table 5. Composition and type nos of austenitic stainless steels

British type	American equivalent AISI no	C	Si	Mn	Ni	Cr	Others	Description
EN58A	320	0.16	0.20 (min)	2.0	7–10	17–20		18/8 steel
EN58B	321	0.15	0.20 (min)	2.0	7–10	17–20	Ti ≮4 × C	18/8 steel Ti stabilized for welding
EN58E	304	0.08	0.20 (min)	2.0	8–11	17.5–20		Low carbon 18/8
EN58F	347	0.15	0.20	2.0	7–10	17–20	Nb ≮8 × C	18/8 steel Nb stabilized for welding
—	430	0.12	1.0	1.0	—	14–18		Magnetic

give the mix a clay-like plasticity. The clay is then worked to de-aerate it and produce a homogeneous mass. The ceramic part is moulded, pressed, or extruded from the homogeneous clay and left to dry. When dry it is sintered in a high temperature furnace to give the final hard state. During this firing process shrinkage of the component takes place, which has to be allowed for. In some cases it is possible to give an intermediate lower temperature firing, which permits the ceramic to be machined before the final hardening. For more details of the various processes the reader is referred to Espe.[17]

4.1. Types of ceramics used for vacuum components. Only a limited number of ceramic types from the large range available are suitable for vacuum components. If the ceramic is to be used for the vacuum envelope, necessitating ceramic–metal or ceramic–glass seals, then the range is further restricted.

The main impetus to the use of ceramics as insulators instead of glass in vacuum systems arose from the work carried out in Germany and later in the USA during the 1940's, on suitable ceramics with low dielectric loss factors for microwave tube components. The ceramics developed were based on steatite, a magnesium metasilicate ($MgSiO_3$) body ceramic, and the main advance was in the successful ceramic to metal seals which were evolved. Typically, steatite is made by combining 70–80 % talc ($3MgO.4SiO_2.H_2O$) with 20 to 30 % china clay ($Al_2O_3.2SiO_2.2H_2O$) to which alkali or alkaline–earth oxides are added as flux. The final ceramic, which is fired around 1400°C, consists of the $MgSiO_3$ crystals bonded by a glass, high in alkali oxides. Steatites have a limited temperature range for firing, $\pm 10°C$, and require accurate temperature control in manufacture. An improved material is obtained by enriching steatite with magnesium compounds to give another type of ceramic known as Forsterite (Mg_2SiO_4). Forsterite has a wider firing temperature range and also lower dielectric loss. The thermal expansion coefficient is high at $11 \times 10^{-7}/°C$, i.e. similar to that of soft glasses, and it can, therefore, be sealed to chrome–iron and also titanium to which it is more closely matched. Another ceramic used in the 1940's and early 1950's was zircon porcelain ($ZrO_2.SiO_2$). Zircon has a very low coefficient of thermal expansion, closely matched to molybdenum, and as a result has a good heat shock resistance. It has, however, a high dielectric constant.

At that time pure oxide ceramics were too expensive for general use, although their superior properties were appreciated. With the increased demand, and improved ceramic technology, such ceramics have become more readily available, especially

alumina ceramics which are almost exclusively used for vacuum applications today. A variety of alumina (Al_2O_3) body ceramics are available, having an alumina content of from 85 % to nearly 100 %. They are mechanically stronger than most other ceramics, and in spite of their higher thermal expansion coefficient than for zircon porcelain, they are able to withstand high temperatures. The mechanical strength, and dielectric properties improve with increase in purity, but the purer alumina-body ceramics are more difficult to make, and therefore more expensive. The purity also affects the ease with which the ceramic can be metallized. The sintering temperature for high alumina–body ceramics is higher than for silicate ceramics, from 1700°C to 1850°C.

Other oxides that have found use in vacuum applications, are zircon (ZrO_2), and beryllia (BeO). Beryllia has advantage over alumina in having a high thermal conductivity, but in powder form it represents a serious health hazard, and it must, therefore, be used with great caution.

An interesting and recent development has been the production of glass ceramics, known as Pyroceram* or Cervit.† By heat treating glass, it has been found possible to convert it to a crystalline ceramic state, particularly if nucleating agents are added to the glass body. The component can be fabricated by normal glass-making techniques and converted by further heat treatment. The resultant material is an opaque ceramic of virtually the same dimensions as the original glass article, but with greater strength and resistance to thermal shock.

4.2. Physical properties. As with glass, the important physical property of a ceramic in its vacuum application is its mechanical strength, and its behaviour with temperature, especially where ceramic–metal or ceramic–glass seals are involved. Like glass, ceramics are brittle in that they fracture under strain without any elongation or flow. Also, it is the tensile stress which is of most concern, since the compressive strength is some 10 to 20 times higher than the tensile strength. For most silicate ceramics the tensile strength is similar to that of glass, 4 to 10×10^6 Kg/m^2, but for the alumina body ceramics tensile strengths up to 26×10^6 Kg/m^2, can be attained; typical values are given in Table 6, where the characteristics of the common ceramics used *in vacuo* are listed. The strength depends on the porosity of the ceramic and also the shape and total cross-sectional area; small diameter fibres are stronger than larger diameter rods.

* Trade name of Corning Glass Works.
† Trade name of Owen-Illinois glass.

Table 6. Physical properties of ceramics used in vacuum technology

Ceramic	Main body composition	Coefficient expansion $\times 10^{-7}$	Softening temperature °C	Mechanical strength/ tensile strength kg/m²	Specific gravity
Steatite	$MgOSiO_2$	70 → 90	1400	6×10^6	2.6
Forsterite	$2MgOSiO_2$	90 → 120	1400	7×10^6	2.9
Zircon porcelain	ZnO_2SiO_2	30 → 50	1500	8×10^6	3.7
85% Alumina	Al_2O_3	50 → 70	1400	14×10^6	3.4
95% Alumina	Al_2O_3	50 → 70	1650	18×10^6	3.6
98% Alumina	Al_2O_3	50 → 70	1700	20×10^6	3.8
Pyroceram 9606		57	1250	14×10^6	

Ceramics do not exhibit the 'slow' monotonic changes in viscosity with temperature experienced with glass. Nevertheless, because of the glass phase the melting point of a ceramic is often rather indeterminate. A softening temperature is usually defined by the collapse of a cone of given dimensions made from the ceramic. Typical values of softening temperature are given in Table 6. As a practical guide, ceramics used as part of the vacuum envelope should not be heated in excess of a temperature which is 400–500°C below the softening temperature. Since the softening temperature of the ceramics in common use in vacuum systems are above 1200°C, this limitation does not normally present a problem to the vacuum engineer.

4.3. Permeation of gases through ceramics. Gas permeates through ceramics in the same way as it permeates through glass, via the pores in the micro-structure. Consequently the permeation rate of gas through ceramic will depend on the packing density of the ceramic 'crystals' (porosity) and whether or not a glassy phase is present. It will also depend on the size of the gas molecule involved, helium having the highest permeation.

The manufacturing process, size of particles etc., as well as chemical composition will affect the permeation; for example some ceramics are especially designed to be porous, and can be used for leaking gas into a vacuum system at a controlled rate.

In spite of the increasing use of ceramic as the vacuum envelope in tubes and systems, very little information has been published on helium permeation. Most experiments have shown helium leak rates which are lower than for glass without giving details of the permeation constant. Manufacturers claim their material to be vacuum tight, using a helium leak detector with sample discs. There is, however, one article by Miller and Shepard[18] where the permeation of helium and air through Pyroceram 9606 is compared to that through a 97% alumina ceramic. The curves are reproduced in Figure 11. Permeation of oxygen, nitrogen and argon through extruded alumina ceramics tubes at elevated temperatures have been measured by Budworth and his co-workers.[19] Below 1500°C the ceramics were impermeable to within the limits of measurement, i.e. 5×10^{-13} m² s⁻¹. At temperatures above 1500°C, where permeation

Changes in temperature sets up stresses and strains at the interface with other materials due to the expansion coefficient. However, because of the higher strength, especially under compression, it is possible to make successful metal–ceramic seals without the expansion coefficients of the two materials being as closely matched as is required for glass. If they are closely matched, a more versatile and reliable seal can be obtained than is possible with glass. Typical thermal expansion curves are given in Figure 10. Unlike glass the curves are almost linear up to the softening temperature. Expansion coefficients

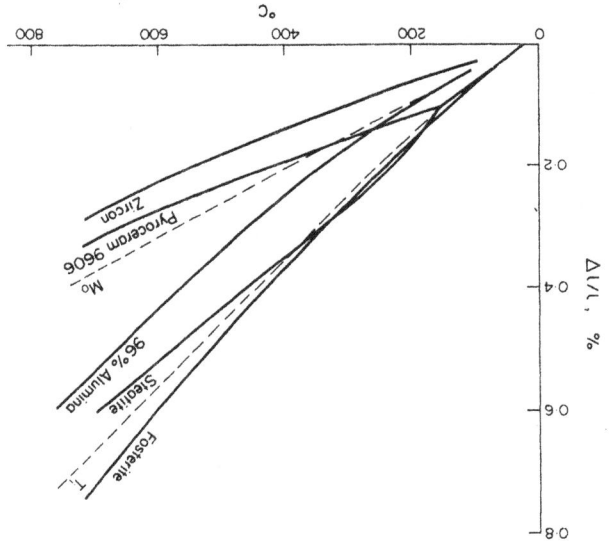

Figure 10. Thermal expansion characteristics of some of the ceramics commonly used in vacuum technology.

of the ceramic types are also listed in Table 6. Generally the resistance of a ceramic to thermal shock is greater the smaller the expansion coefficient, although the tensile strength is also of importance. From this point of view they are more resistive to thermal shock than glass.

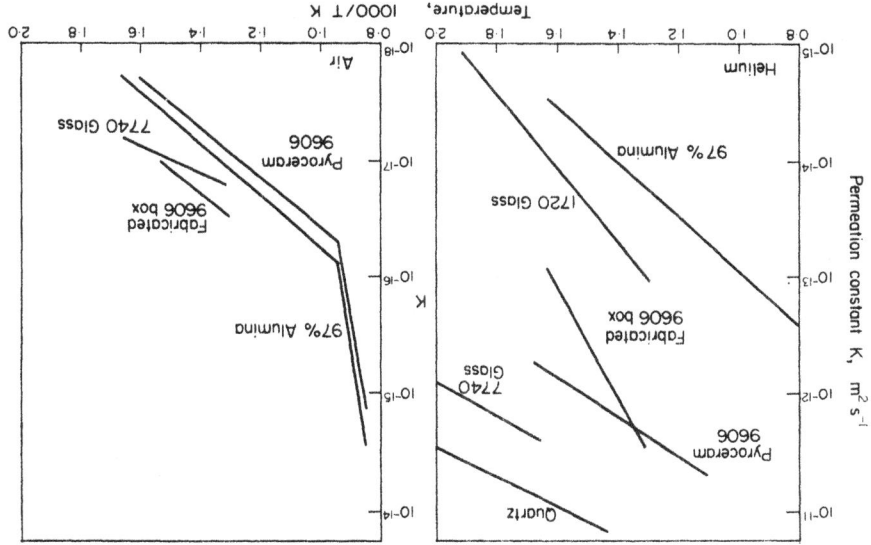

Figure 11. Permeation constant for helium and air through alumina ceramic and glass (from Miller and Shepard).[18]

constants of 10^{-11} m^2 s^{-1} were measured, changes in the microstructure of the tubes took place in some of the materials, giving rise to selective permeation of oxygen. Much higher permeation values were obtained with hot pressed discs of similar composition with measurable values at room temperature (Budworth[20]). This suggested that the method of manufacture was the important factor.

In general, providing the ceramic is designed for vacuum application and has no faults such as cracks or holes, helium permeation does not appear to be a problem in the attainment of ultrahigh vacuum. The situation is further improved since much greater wall thicknesses are possible in ceramic systems without danger of thermal stresses causing fracture.

4.4. Outgassing of ceramics.

Because of their glassy phase, silicate ceramics behave similarly to glass with regards to gas desorption. Water vapour, carbon dioxide and carbon monoxide are the main gases evolved, the gases coming from adsorbed surface layers and from the bulk of the material. Hydrogen is also usually detected, and Gibbons[21] stated that it was the principal gas given off from alumina ceramic chambers designed for a synchrotron storage ring. Clearly the porosity of the ceramic will affect the outgassing properties. For internal components, the degassing is facilitated by using ceramics with high porosity, the converse to the permeation situation. In general the silicate ceramics with a high percentage of glassy phase, such as porcelain, contain the most absorbed gas, comparable to glass. Steatite, Forsterite and alumina body ceramic have a much lower gas content, and can be further improved by forming the component in a vacuum environment. The advantage of ceramics over glass is the possibility of degassing the component prior to assembly, in the same way that metal parts are pre-stoved. This is probably only necessary where the ceramic part forms an appreciable part of the final vacuum system. In the final vacuum system bake-out temperatures above 500°C can be employed. However, Norton[22] has pointed out that high baking temperatures can cause evolution of gas if metal oxide impurities are present due to dissociation. For example Fe_2O_3, which is a common impurity in silicate ceramics, has a dissociation pressure of about 10^{-4} Nm^{-2} at 800°C, and can cause prolonged evolution of oxygen when baked at elevated temperatures.

In general, therefore, a ceramic component should be vacuum stoved at about 1000°C before assembly, and baked at 450°C or above for several hours in the final system. The resulting outgassing at room temperature, should then be of the order of 10^{-11} Wm^{-2} or better.

5. Other materials

In recent years a number of synthetic materials—plastics, elastomers, epoxy resins etc.—have found use in high vacuum applications. These materials all exhibit high outgassing rates from gases absorbed during manufacture, and also are permeable to some extent to all gases. The values of degassing rate and permeability are considerably higher than for metal, glass and ceramic. Although the outgassing rate can be reduced by prolonged pumping, or in some cases a low temperature bake, the materials rapidly re-absorb gas when exposed to the atmosphere, especially water vapour, and values comparable to a baked metal or glass system cannot be achieved. Such materials are, therefore, generally unsuitable for ultrahigh vacuum applications. However, because they are cheap and have chemical and physical properties which are particularly desirable, for example their high elasticity, methods of employing them to give the minimum gas influx have been advocated, particularly in continuously pumped systems where pressures much below 10^{-6} Nm^{-2} are not required.

Several workers have investigated degassing and permeation characteristics of such materials[23,24,25,26] and data on a wide range of commercially available plastics and elastomers have been published.

In Table 7, some typical values of outgassing rates at room

Table 7. Outgassing rate for synthetic materials

Material	Hours pumping Before bake	Inc 24 h bake	After exposure to air for 24 h	Baking temperature °C	Outgassing rate W/m²
Araldite ATI	51				3.4×10^{-4}
		80		85	Not detected
			51		6.0×10^{-5}
Mycalex	52				2.7×10^{-6}
		83		300	Not detected
			51		1.0×10^{-6}
Nylon 31	51				1.1×10^{-4}
		82		120	8.0×10^{-7}
			51		8.9×10^{-6}
Perspex	51				1.3×10^{-4}
		102		85	7.8×10^{-6}
			51		6.6×10^{-5}
Polythene	262				4.0×10^{-5}
		496		80	6.6×10^{-6}
			94		2.3×10^{-6}
Viton A.	51				1.3×10^{-4}
		101		200	2.7×10^{-6}
			48		4.0×10^{-5}
Polyimide		12 h bake		200	6.6×10^{-8}
		12 h bake		300	4.0×10^{-8}
			5 h exposure to air		5.3×10^{-5}

temperature are given for materials of most interest to the ultra-high vacuum engineer. The data are mainly taken from the work of Barton and Govier[27] but some other results are also included. They show the effect of baking and exposure to air. For all the materials given the outgassing rate is initially high 10^{-3} to 10^{-4} Wm^{-2}, and even after pumping for 50 h it is not reduced much below 10^{-5} Wm^{-2}. The exception is PTFE where values below 10^{-7} Nm^{-2} can be obtained without baking. A large percentage of the gas given off is water vapour, and baking at the maximum permissible temperature, restricted by deformation or decomposition, reduces the outgassing rate by at least an order. The materials giving the lowest values are in general those which can be baked to the highest temperatures. For this reason PTFE, Viton A, Mycalex, and Polyimide have all found uses in vacuum systems going down to pressures below 10^{-6} Nm^{-2}. The most interesting of these is Polyimide, since it can be baked to temperatures around 300°C to give outgassing rates lower than obtained with Viton 'A', $\sim 4 \times 10^{-8}$ Wm^{-2}. However, polyimide has a similar chemical composition to nylon, and is to some extent hygroscopic, so that on exposure to air, water vapour is readily adsorbed. The unbaked degassing rate can therefore be fairly high. Polyimide, like nylon, is a fairly hard resistant material and is especially suitable for gaskets. It has a high expansion coefficient, and since it will suffer permanent distortion if compressed more than 20%, rather more care has to be taken in design of joints where it is to be used. It can also be used for sealing small leaks in the vacuum envelope. For this purpose the Polyimide is dissolved in a suitable solvent to form a viscous liquid which is applied to the leak by brushing.

The values of the permeation constant for the gases; hydrogen, helium, oxygen and nitrogen, passing through some of the synthetic materials useful for vacuum work, are listed in Table 8. The data were taken mainly from Bailey[30] and are from measurements by Barton at AERE. They are in general agreement with results quoted by other workers, although it can be expected that variation will exist on materials from different manufacturers.

Since the diffusion of gas is via the pores in the structure, the permeation is proportional to pressure, and value can be compared directly with those for glass or silica. The values for

helium permeation through synthetic materials are at least an order greater than for fused silica, and it will be noticed that the hydrogen permeation rate does not differ greatly from that of helium. Indeed there is not the market effect of molecular diameter found with glasses, so that nitrogen permeation from the atmosphere is as serious as helium permeation. The high value for PTFE is associated with the difficulty of manufacturing the material with a higher (less porous), density. As far as is known the only published values for the permeation of gas through Polyimide are those of George[32] quoted by Perkins.[16]

The main use of plastic materials in high or ultrahigh vacuum applications is as gaskets at demountable joints in the vacuum envelope. In this application a minimum surface area is exposed to the vacuum, and the diffusion path through the material is fairly large. Commonly the gasket is made in the form of a toroid, known as an O-ring, and is squashed between flat flanges one of which would contain a locating groove, probably of trapezium cross section. Details of such arrangements are given in most books on vacuum technology. Using gaskets of Viton 'A' and also Polyimide, Hait[26] has given pump down characteristics with a baked stainless steel system which reach an ultimate pressure below 10^{-7} Nm^{-2}. In general, however, metal gaskets are preferred for ultrahigh vacuum systems. Another use of plastic in vacuum, for which PTFE is particularly suited is for bearings or sliding bushes, where a low coefficient of friction in vacuum is required.

Before leaving this section on vacuum material, mention should be made of mica, which is used extensively in vacuum devices, such as electron tubes.

Mica is a naturally occurring mineral, which is in the form of transparent laminations. It is either K-Al-silicate or K-Mg-Al-double silicate, and it can also be made synthetically in commercial quantity. It is a good insulating material, resistivity 10^{16}–10^{17} Ω.cm, with a high dielectric constant. One of its special properties is the ease with which it may be split into thin sheets, which can be stamped out into relatively complex shapes. It is, therefore, particularly suitable as spacers between electrodes, which need to be accurately located.

Unfortunately the laminated structure of mica results in gas being trapped between the layers, which is difficult to drive off. Baking at too high a temperature drives off the water of

Table 8. Permeability constants at ambient temperature of synthetic materials

| Material | Permeability constant K is m²/s at 23°C | | | | |
	Nitrogen	Oxygen	Hydrogen	Helium	Argon
Polythene	9.9×10^{-13}	3.0×10^{-12}	8.2×10^{-12}	5.7×10^{-12}	2.7×10^{-12}
PTFE	2.5×10^{-12}	8.2×10^{-12}	2.0×10^{-10}	5.7×10^{-10}	4.8×10^{-12}
Perspex	—	—	2.7×10^{-12}	5.7×10^{-12}	—
Nylon 31	—	—	1.3×10^{-13}	3.0×10^{-13}	—
Polystyrene	—	5.1×10^{-13}	1.3×10^{-11}	1.3×10^{-11}	—
Polystyrene*	6.4×10^{-12}	2.0×10^{-11}	7.4×10^{-11}	—	—
Polyethelene*	6–11×10^{-13}	2.5–3.4×10^{-12}	6–12×10^{-12}	4–5.7×10^{-12}	—
Mylar 25-V-200*	—	—	4.8×10^{-13}	8.0×10^{-13}	—
CS2368B (Neoprene)	2.1×10^{-13}	1.5×10^{-12}	8.2×10^{-12}	7.9×10^{-12}	1.3×10^{-12}
Viton A	—	—	2.2×10^{-12}	8.2×10^{-12}	—
Kapton (polyimide)**	3.2×10^{-14}	1.1×10^{-13}	1.2×10^{-12}	2.1×10^{-12}	—

* Data taken from work of Brubaker and Kammermeyer.[31]
** Data taken from the work of George.[32]

crystallization and causes the mica to crumble. The only satisfactory method of outgassing it is prolonged baking at say 200°C–300°C. Synthetic mica is better from this point of view, but on the whole mica is not a very satisfactory material for ultrahigh vacuum applications.

The permeability of gas, normal to the cleavage plane is low, at 400°C, K for helium is less than $10^{-17}\,m^2\,s^{-1}$, i.e. it is lower than for the best glass. This factor, coupled with its strength in the same direction, has promoted its use as vacuum windows, which can be thin and therefore transmit a high percentage of incident radiation.

The thermal expansion coefficient of mica is high, 80–130 $\times 10^{-7}/°C$ parallel to the cleavage plane and 160 to 250 $\times 10^{-7}/°C$ perpendicular to it. It is most nearly matched to the soft glasses and metals with similar expansion coefficients such as nickel–iron or chrome–iron, and can be sealed to the materials with a 'solder' glass (a glass of low melting point).

References

[1] R W Douglas, *J scient Instrum*, **22**, 1945, 81.
[2] S Dushman, *Scientific Foundations of Vacuum Technique*, 2nd Edn, p 494. Wiley, New York (1962).
[3] D Alpert and R S Buritz, *J appl Phys*, **25**, 1954, 202.
[4] F J Norton, *J Am Ceram Soc*, **36**, 1953, 90.
[5] V O Altemose, *J appl Phys*, **32**, 1961, 1309.
[6] D Alpert, *Advances in Vac Sci Technol*, Vol 1, p 31, *Proc. First Int Congr on Vac Techniques*, 1958, Pergamon Press, London (1960).
[7] B J Todd, *J appl Phys*, **26**, 1955, 1238.
[8] R E Honig, *RCA review*, **23**, 1962, 567.
[9] J Blears, E J Greer and J Nightingale, *Advances in Vac Sci Technol*, Vol 2, p 473. *Proc First Int Congr on Vac Technol*, 1958, Pergamon Press, London (1960).
[10] B B Dayton, *Trans 8th Nat Vac Symp and 2nd Int Congr on Vac Technol*, 1961, Vol 1, p 42, Pergamon Press, London (1962).
[11] W Epse, *Materials of High Vacuum Technology*, Vol I (English trans), Pergamon Press, London (1966).
[12] F J Norton and A L Marshall, *Trans Am Inst Min metall. Engrs*, **156**, 1944, 351.
[13] F A Flecken and H G Nöller, *Trans 8th Nat Vac Symp* and *Proc 2nd Int Congr on Vac Sci Technol*, 1961, Vol. 1, p 58, Pergamon Press, London (1962).
[14] P F Varadi, *Trans 8th Nat Vac Symp* and *Proc 2nd Int Congr on Vac Sci Technol*, 1961, Vol 1, p 73. Pergamon Press, London (1962).
[15] F J Norton, *Trans 8th Nat Vac Symp* and *Proc 2nd Int Congr on Vac Sci Technol*, 1961, Vol 1, p 8. Pergamon Press, London (1962).
[16] W G Perkins, *J Vac Sci Technol*, **10**, 1973, 543.
[17] W Espe, *Materials of High Vacuum Technology*, Vol II, Pergamon Press, London (1968).
[18] C F Miller and R W Shepard, *Vacuum*, **11**, 1961, 58.
[19] D Hayes, D W Budworth and J P Roberts, *Trans. Br Ceram Soc*, **62**, 1963, 507.
[20] D W Budworth, *Trans Br Ceram Soc*, **62**, 1963, 975.
[21] W F Gibbons, *Proc 4th Int Vac Congr*, 1968, Inst. Phys. Conference Series 5, part 1, p 255.
[22] F J Norton, *J appl Phys*, **28**, 1957, 34.
[23] R Geller, *le Vide*, **13**, 1958, 71.
[24] K Diels and R Jaeckel, *Leybold Vakuum Taschenbuch*, Ch 13, Springer, Berlin (1958).
[25] D J Santeler, *Trans 5th Nat Symp Vac Technol*, 1958, 1.
[26] D F Othmer and G J Frohlich, *Ind Eng Chem*, **47**, 1955, 1034.
[27] R S Barton and R P Govier, *J Vac Sci Technol*, **2**, 1965, 113.
[28] B B Dayton, *Trans 6th Nat Symp Vac Technol*, 1959, 101.
[29] P W Hait, *Vacuum*, **17**, 1967, 547.
[30] J R Bailey, *Handbook of Vacuum Physics*, Vol 3, part 4, Pergamon Press, Oxford (1964).
[31] D W Brubaker and K Kammermeyer, *Ind Eng Chem*, **44**, 1952, 1465. **45**, 1953, 1148 and **46**, 1954, 733.
[32] D E George private communication quoted by Perkins.[16]

7. Appendix

Conversion tables for vacuum quantities

Gas pressure

	dyn/cm²	N/m²	torr (mm Hg)	Millibar
1 dyn/cm²	1	10^{-1}	7.5×10^{-4}	10^{-3}
1 N/m²	10	1	7.5×10^{-3}	10^{-2}
1 torr (or mm of mercury)	1333	133.3	1	1.333
1 millibar	10^3	10^2	0.75	1
Atmospheric pressure	1.013×10^6	1.013×10^5	760	1.013×10^3

Gas quantity

	Torr litre	Nm (joule)	μl	mole
1 torr litre	1	0.133	10^3	5.38×10^{-3}
1 Newton meter or joule	7.5	1	7.5×10^3	4.04×10^{-2}
μl	10^{-3}	1.33×10^{-4}	1	5.38×10^{-6}
Mole	1.86×10^4	2.48×10^3	1.86×10^7	1

Outgassing rates

	torr l/cm² s⁻¹	Wm⁻²	μl/cm² s⁻¹
1 torr l/cm² s⁻¹	1	1.333×10^3	10^3
1 Newton meter/m² s⁻¹ = Watt per square meter	7.5×10^{-4}	1	0.75
Lusec per square centimeter μl/cm² s⁻¹	10^{-3}	1.333	1

Vacuum/volume 36/numbers 7–9/pages 523 to 525/1986
Printed in Great Britain

0042–207X/86$3.00 + .00
Pergamon Journals Ltd

Analysis of outgassing characteristics of metals

Michał Moraw, *Polytechnic of Wrocław, Institute of Electron Technology, Wybrzeże Wyspiańskiego 27, Wrocław, Poland*

An attempt to determine some phenomenological parameters describing the outgassing process has been made. A correlation between the material and desorption activation energy has been found. The possibility of extrapolation in time of outgassing characteristics has been shown. The prediced characteristics based on the phenomenological parameters have been compared with the experimental data and the accuracy was within 10%.

The emission of gases from solid surfaces into vacuum is a process consisting of simultaneous phenomena of gas transport from inside the surface and of desorption of gas molecules from this surface. The latter is the predominant phenomena in metals[1–3]. This is due to the complicated structure of a metal surface. Under normal conditions a large number of gas molecules are adsorbed at the surface.

Desorption from a surface may be described by the equation:

$$\frac{dn}{dt} = -\frac{n}{\tau} \tag{1}$$

where: dn/dt = instantaneous desorption rate; n = number of gas molecules adsorbed on unit surface; τ = average time of sojourn of the molecule.

This time of sojourn is related to the activation energy by

$$\tau = \tau_0 \exp\frac{E}{RT} \tag{2}$$

where: E = activation energy; R = gas constant; T = temperature; τ_0 = vibration period of adsorbed molecule.

The activation energy on the surface depends on the kind of interaction between the solid state atoms and gas molecules but it is influenced by the arrangement of interacting elements too. According to the suggestion of Dayton confirmed by the work of Evans[4], Mathewson[5], Everett and others[6], the actual metal surface is rough and covered with porous chemical compounds. Such a surface may offer a large number of various arrangements of interaction between surface atoms and gas molecules. Thus it may be assumed that a large number of possible activation energy levels exists. Macroscopically, the activation energy of the gas–surface system is represented by an average value of the energies of all adsorbed gas molecules.

Such a model predicts changes of activation energy in the course of material outgassing. When the sample is placed in the vacuum vessel, first the weakly bound molecules escape from the surface, thus the average activation energy must increase. Because the desorption rate decreases with time, a decrease of slope of the function describing the time dependence of activation energy may be expected. The value of this energy increases due to the decrease

in the number of weakly bound molecules. Thus the activation energy may be considered to vary with time and its value may be considered to contain a time-varying term and also a component determined by the actual conditions (labelled 'characteristic activation energy'). Taking into account an exponential relation between molecule time of sojourn and energy one may assume the simplest function of activation energy change to be of the form:

$$E = E_c + \gamma RT \ln\frac{t}{\tau_c} \tag{3}$$

where: E = average instantaneous activation energy; E_c = characteristic activation energy; γ = coefficient of activation energy change; and

$$\tau_c = \tau_0 \exp\frac{E_c}{RT}. \tag{4}$$

This assumption will be justified by the comparison of curves derived from the relation (3) using experimental data.

Under assumption (3) the time of sojourn τ may be written as follows:

$$\tau = \tau_c \left(\frac{t}{\tau_c}\right)^\gamma \tag{5}$$

and equation (1)

$$\frac{dn}{dt} = -\frac{n}{\tau_c}\left(\frac{t}{\tau_c}\right)^{-\gamma}. \tag{6}$$

This equation may be solved as follows:

$$\frac{dn}{dt} = -\frac{n_0}{\tau_c}\left(\frac{t}{\tau_c}\right)^{-\gamma} \exp\left[-\frac{1}{1-\gamma}\left(\frac{t}{\tau_c}\right)^{(1-\gamma)}\right] \tag{7}$$

where n_0 = initial number of adsorbed gas molecules.

The minus sign in front of the right-hand side indicates the direction of the gas stream only.

The values of γ should be inside the range

$0 \leqslant \gamma < 1$.

The function (7) describes only curves with monotonically

increasing slope. It is obvious that not all the experimental characteristics follow this condition. However, each experimental outgassing characteristic may be approximated by a sum of expressions (7)

$$\frac{dn_k}{dt} = \sum_{k=1}^{m} \frac{n_{0_k}}{\tau_{c_k}} \left(\frac{t}{\tau_{c_k}}\right)^{-\gamma_k} \exp\left[-\frac{1}{1-\gamma_k}\left[\frac{t}{\tau_{c_k}}\right]^{(1-\gamma_k)}\right]. \tag{8}$$

In practice, a limitation in the number of summed components is important.

An analysis of 30 experimental outgassing characteristics (copper, aluminium, steel, etc.) shows that the number of summed components may be limited to two:

$$\frac{dn}{dt} = \frac{dn_1}{dt} + \frac{dn_2}{dt} = \frac{n_{0_1}}{\tau_{c_1}} \left[\frac{t}{\tau_{c_1}}\right]^{-\gamma_1} \exp\left[-\frac{1}{1-\gamma_1}\left(\frac{t}{\tau_{c_1}}\right)^{(1-\gamma_1)}\right]$$
$$+ \frac{n_{0_2}}{\tau_{c_2}} \left(\frac{t}{\tau_{c_2}}\right)^{-\gamma_2} \exp\left[-\frac{1}{1-\gamma_2}\left(\frac{t}{\tau_{c_2}}\right)^{(1-\gamma_2)}\right]. \tag{9}$$

Such an approximation shows an accuracy better than 10%. The approximation formula contains six parameters of outgassing characteristics:

$$n_{0_1} \cdot \gamma_1 \cdot \tau_{c_1} \left(E_{c_1} = RT \ln\frac{\tau_{c_1}}{\tau_0}\right)$$

$$n_{0_2} \cdot \gamma_2 \cdot \tau_{c_2} \left(E_{c_2} = RT \ln\frac{\tau_{c_2}}{\tau_0}\right).$$

Table 1. Parameters of outgassing characteristics

Material	Ref	Experimental data of outgassing characteristics Instantaneous outgassing rate $\frac{K_1}{K_{10}}$† (mbar ls^{-1} cm^{-2})	Slope of characteristic $\frac{\alpha_1}{\alpha_{10}}$†	Phenomenological parameters of outgassing characteristics $\frac{\gamma_2}{\gamma_1}$	$\frac{\tau_{c^2}}{\tau_{c_1}}$ (s)	$\frac{E_{c_2}}{E_{c_1}}$ (kJ mol^{-1})	$\frac{n_{0_2}}{n_{0_1}}$ (cm^{-2})
Mild steel	1	$\frac{7.2 \times 10^{-7}}{6.7 \times 10^{-8}}$	$\frac{1}{1}$	$\frac{0.65}{0.35}$	$\frac{4.19 \times 10^5}{1920}$	$\frac{104.4}{90.5}$	$\frac{4.87 \times 10^{17}}{1.13 \times 10^{17}}$
Mild steel (slightly rusty)	1	$\frac{8 \times 10^{-7}}{1.8 \times 10^{-8}}$	$\frac{3.1}{1}$	$\frac{0.57}{0.55}$	$\frac{1.8 \times 10^5}{290}$	$\frac{102.3}{86.6}$	$\frac{9.84 \times 10^{16}}{1.9 \times 10^{19}}$
Stainless steel 18/9/1	8	$\frac{1.1 \times 10^{-8}}{1 \times 10^{-9}}$	$\frac{1}{1}$	$\frac{0.65}{0.3}$	$\frac{5.06 \times 10^5}{1680}$	$\frac{104.8}{91}$	$\frac{6.98 \times 10^{15}}{1.58 \times 10^{13}}$
Stainless steel NS22S (mech polished)	3	$\frac{2.2 \times 10^{-9}}{6.1 \times 10^{-10}}$	$\frac{0.5}{0.7}$	$\frac{0.4}{—}$	2.63×10^5	103.2	2.98×10^{15}
Stainless steel NS22S (electropolished)	3	$\frac{5.7 \times 10^{-9}}{5.7 \times 10^{-10}}$	$\frac{1}{1}$	$\frac{0.65}{0.4}$	$\frac{4.63 \times 10^5}{2350}$	$\frac{104.6}{91.8}$	$\frac{4 \times 10^{15}}{5.06 \times 10^{14}}$
Stainless steel ICN472	3	$\frac{1.8 \times 10^{-8}}{2 \times 10^{-9}}$	$\frac{0.9}{0.9}$	$\frac{0.6}{0.31}$	$\frac{4.92 \times 10^5}{2380}$	$\frac{104.8}{91.8}$	$\frac{1.19 \times 10^{16}}{2.26 \times 10^{13}}$
Stainless steel ICN472 (sanded)	3	$\frac{1.1 \times 10^{-8}}{1.4 \times 10^{-9}}$	$\frac{1.2}{0.8}$	$\frac{0.57}{0.52}$	$\frac{6.5 \times 10^5}{2100}$	$\frac{105.5}{91.5}$	$\frac{8.34 \times 10^{15}}{3.28 \times 10^{13}}$
Molybdenum	3	$\frac{6.9 \times 10^{-9}}{4.9 \times 10^{-10}}$	$\frac{1}{1}$	$\frac{0.54}{0.25}$	$\frac{1.76 \times 10^5}{2800}$	$\frac{102.3}{92.2}$	$\frac{2.59 \times 10^{15}}{1.3 \times 10^{15}}$
Titanium	3	$\frac{5.3 \times 10^{-9}}{4.9 \times 10^{-10}}$	$\frac{1}{1}$	$\frac{0.54}{0.45}$	$\frac{1.76 \times 10^5}{3200}$	$\frac{102.3}{92.5}$	$\frac{2.59 \times 10^{15}}{1.12 \times 10^{13}}$
Copper (fresh)	3	$\frac{5.3 \times 10^{-8}}{5.5 \times 10^{-9}}$	$\frac{1}{1}$	$\frac{0.67}{0.2}$	$\frac{1.08 \times 10^6}{2950}$	$\frac{106.7}{92.3}$	$\frac{4.04 \times 10^{16}}{3.13 \times 10^{13}}$
Copper (mech polished)	3	$\frac{4.7 \times 10^{-9}}{4.8 \times 10^{-10}}$	$\frac{1}{1}$	$\frac{0.68}{0.32}$	$\frac{8.85 \times 10^5}{2830}$	$\frac{106.2}{92.2}$	$\frac{3.59 \times 10^{15}}{2.72 \times 10^{14}}$
OFHC (fresh) copper	3	$\frac{2.5 \times 10^{-8}}{1.7 \times 10^{-9}}$	$\frac{1.3}{1.3}$	$\frac{0.75}{0.3}$	$\frac{6.55 \times 10^5}{1120}$	$\frac{105.5}{90}$	$\frac{2.18 \times 10^{16}}{2.59 \times 10^{13}}$
OFHC copper (mech polished)	3	$\frac{2.6 \times 10^{-9}}{2.2 \times 10^{-10}}$	$\frac{1.1}{1.1}$	$\frac{0.72}{0.3}$	$\frac{8.4 \times 10^5}{1920}$	$\frac{106}{91.3}$	$\frac{1.56 \times 10^{15}}{1.85 \times 10^{14}}$
Aluminium (fresh)	3	$\frac{8.4 \times 10^{-9}}{8 \times 10^{-10}}$	$\frac{1}{1}$	$\frac{0.76}{0.12}$	$\frac{4.9 \times 10^6}{3200}$	$\frac{110.4}{92.5}$	$\frac{8.38 \times 10^{15}}{1.28 \times 10^{14}}$
Aluminium (degassed 24 h after 24 h in 1 mbar)	3	$\frac{5.5 \times 10^{-9}}{4.1 \times 10^{-10}}$	$\frac{3.2}{0.9}$	$\frac{0.7}{0.53}$	$\frac{7.4 \times 10^6}{480}$	$\frac{111.4}{87.9}$	$\frac{3.52 \times 10^{15}}{2.13 \times 10^{16}}$
Aluminium (degassed 24 h after 3 h in air)	3	$\frac{8.8 \times 10^{-9}}{6.3 \times 10^{-10}}$	$\frac{1.9}{0.9}$	$\frac{0.69}{0.58}$	$\frac{3 \times 10^6}{440}$	$\frac{109.2}{87.7}$	$\frac{5.06 \times 10^{15}}{4.7 \times 10^{16}}$
Aluminium compound	1	$\frac{2.3 \times 10^{-7}}{4.7 \times 10^{-8}}$	$\frac{0.75}{0.75}$	$\frac{0.56}{—}$	$\frac{1.33 \times 10^6}{—}$	$\frac{107.2}{—}$	$\frac{3.2 \times 10^{17}}{—}$

† (Value after 1 h)/(value after 10 h).

The upper (subscript 1) parameters describe the outgassing characteristic during its initial part, whereas the lower ones (subscript 2) describe the long time outgassing.

Results of analysis for some materials are given in Table 1.

It appears that the values of two sets of parameters (labelled 1 and 2) differ considerably, especially τ_{c_1} and τ_{c_2}. It is also seen that the values of the activation energy E_{c_2} are different for various materials:

steel—average	104 kJ mol^{-1};
copper—average	106 kJ mol^{-1};
aluminium—average	110 kJ mol^{-1}.

It should be noted that no significant differences were observed for E_{c_1} whose values fall within the range 86–93 kJ mol^{-1} for all metals tested. The amount of gas adsorbed at the surface $(n_{0_1} + n_{0_2})$ is equivalent from several to several hundred monolayers (for rusty steel even more). Proper surface treatment (polishing, sandblasting, etc.) decreases these values.

The above analysis is phenomenological and the calculated bond energies and gas amounts are averaged for various gases desorbed from the surface. An error is also introduced by the normalized measuring method recommended by AVS[7,8]. Apart from the above limitations some hypotheses may be made:

A complicated structure is present on the surface which causes strong binding of the adsorbed gas molecules (activation energy above 100 kJ mol^{-1}). Properties of this structure depend on the kind of bulk material.

Simultaneously, a weakly bound quantity of gas molecules (activation energy below 100 kJ mol^{-1}) is present. This weakly bound gas may be adsorbed on the less-developed surface layer, e.g. on surface contamination.

Routine treatment of the surface greatly decreases the amount of adsorbed gas. However this treatment does not change other average parameters of the desorption process.

The proposed analysis of the desorption characteristics may be useful when long term outgassing data are required. A preliminary extrapolation experiment in which 10 h data were used for prediction of 20 h data is shown in Figure 1. The obtained results seem to confirm this approach.

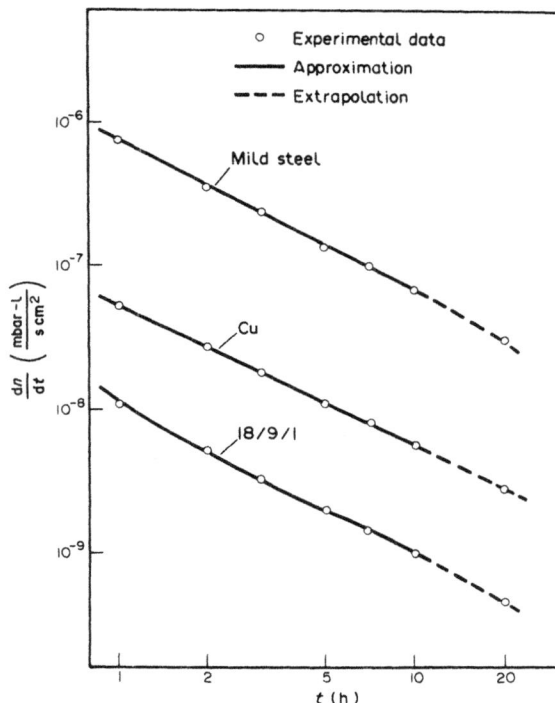

Figure 1.

References

[1] B Dayton, *Trans 8th AVS Vac Symp*, p 42 (1961).
[2] A Schram, *Trans 9th AVS Vac Symp*, p. 301 (1962).
[3] A Schram, *Le Vide*, **103**, 55 (1963).
[4] J R Evans, *Metallic Corrosion Passivity and Protection*, 2nd Edition, Edward Arnold, London (1946).
[5] A G Mathewson, *Vacuum*, **24**, 505 (1974).
[6] D H Everett and F S Stone, *The Structure and Properties of Porous Materials*, p 91, Academic Press, New York (1900).
[7] R J Elsey, *Vacuum*, **25**, 299, 347 (1975).
[8] A Hałas, M Moraw and H Prasoł, *Elektronika*, **24**, 14 (1983).

Lubricating of mechanisms for vacuum service

K. G. Roller

Ball Aerospace Systems Division, Boulder, Colorado 80303

(Received 20 August 1987; accepted 15 October 1987)

Seizure of metal parts which occurs during sliding or rolling in a vacuum environment can be prevented by use of appropriate lubricants. Fluid lubricants, having low evaporation rates, dry solid film lubricants, soft, low shear strength metals, and composite materials are customarily used, but ion implanted metals and ceramics are available for extremely harsh environments. Bearings, bushings, gears, cams, guides and slides, solenoids, commutators, motor brushes, and sliprings can be lubricated to yield vacuum service lives comparable to those achieved in normal atmospheric pressure applications, or even longer due to the absence of O_2 and H_2O. The use of fluid and dry lubricants may require the incorporation of seals to prevent molecular or particulate contamination of test or production hardware within the vacuum system. The most important parameter to be considered in selecting the appropriate lubricant is the operating temperature, which drastically affects the evaporation and contamination rates. Fluid based lubricants are available having evaporation rates of 10^{-9} to 10^{-13} g cm^{-2} s^{-1} and metals, composites, and dry lubricants having 10^{-10}–10^{-14} g cm^{-2} s^{-1} loss rates at room temperature.

I. INTRODUCTION

Operation of mechanisms involving the relative motion of components in contact with each other in a vacuum often requires that lubricants be used in conjunction with or on the moving components in order to facilitate operation and to prevent failure. The lubricants can be incorporated into one or more of the contacting surfaces or be applied to non-self-lubricating surfaces. The lubrication serves chiefly to reduce friction and wear, and to prevent catastrophic failure of the moving parts. Some lubricants also serve as a heat transfer medium, distributing heat so that temperatures do not become excessive. The materials most likely to need lubrication are metals, although many plastics also need lubrication in order to roll or slide freely when in contact with themselves or another material.

Mechanisms fail when the lubricant fails to perform properly or when it becomes sufficiently depleted or deteriorated. Sudden, catastrophic failures usually occur when adhesion between the contacting surfaces becomes so great that por-tions of metal (or other materials) are torn from one component and adhere to the mating component. Abrasive wear mechanisms and lubricant breakdown or deterioration may cause a more slowly occurring failure characterized by increasing friction and heating. In the case of motors, sliprings, encoders, potentiometers, and switches, the failure may be characterized by unacceptable electrical noise or the buildup of insulating films between contacting surfaces.

Mechanisms operating in an industrial or laboratory air atmosphere may operate satisfactorily for relatively long periods of time with no apparent lubricant present. This is because oxygen, sulfur, water, and multitudes of organic substances are present which adhere to or react with the surfaces to provide some degree of lubrication. They separate two metal surfaces so that the nascent or clean, bare metals cannot touch and adhere or abrade appreciably. In low vacuum, sufficient numbers of adsorbed water molecular remain on the surfaces so that failure does not occur rapidly. However, when the ambient atmospheric pressure

TABLE I. Characteristic differences between wet and dry lubricants.

Characteristic	Wet	Dry
Volatility, P_v (Torr) at 303 °C	10^{-4}–10^{-11}	10^{-12}–10^{-14}
Temp. range (K)	225–375	4–700 (substrate limited)
Hertz stress limit	Fatigue life imposes limit	Wear life imposes limit
Torque	Viscosity dependent	Dry generally lower
General problem areas	Containment	Wear
	Design complexity	Environment sensitivity
	Torque	
	Assembly/handling	Assembly cleanliness
	Contamination	Assembly/handling
		Production complexity
Environment problems	Particulates	Humidity
	Temperature	Dirt

139

TABLE II. When to use wet lubricants in vacuum.

When longest life is required
When highest reliability is required
When heat transfer is required
High-speed applications
When temperatures are moderate (near room temperature)
When contamination of nearby surfaces can be tolerated (or controlled)
When radiation environments are relatively benign

TABLE III. Low evaporation rate oil vapor pressures.

Type	Vapor pressure at 296 K (Torr)
Mineral	10^{-8}
Silicone	10^{-8}–10^{-9}
Fluorocarbon	10^{-10}–10^{-12}
Polyfunctional ester	10^{-10}
Polyalphaolefin	10^{-10}
Polyphenylethers	10^{-12}

drops lower than $\sim 10^{-5}$ Torr at room temperature, operation of the mechanism will desorb the various gases to the extent that they can perform no further lubrication function and failure is imminent. It is at this stage that special lubricants are required to prevent the failure of the moving parts.

It is the function of the lubricant to separate the normally contacting materials by means of a low shear strength material. The material may be either fluid (wet) or solid. Table I lists some of the characteristic differences between wet and dry lubricants.

II. WET LUBRICANTS

Wet lubricants used in vacuum environments include both oils and greases. They are, or can form, low shear strength materials which separate rubbing or rolling surfaces by maintaining layers of molecules between the surfaces in static situations and by supporting moving surfaces by means of an electrohydrodynamic film of oil (much as a water ski is supported by the water it is sliding across). Table II lists conditions when wet lubricants are suitable for use. Those fluids having solid particles in them can support the loads by means of the solid particles as well. Nearly any liquid can be used as a lubricant, except for some of the cryogenic fluids such as liquid helium. For vacuum use, those liquids (oils) having low vapor pressure are most useful because the evaporative weight loss is a strong function of the vapor pressure. Bisson[1] cites the Langmuir equation for evaporation in vacuum as

$$G = (P/17.14)(M/T)^{1/2},$$

where G is the evaporation rate (g cm^{-2} s^{-1}), M is the molecular weight of the substance (g mol^{-1}), T is the temperature (K), P is the vapor pressure (mm Hg), which can also be written as

$$P = C \exp (L/RT),$$

where C is a constant, R is the gas constant (cal mol^{-1} K^{-1}), T is temperature (K), and L is the enthalpy of phase change (cal mol^{-1}).

The value of G, the evaporative or sublimative weight loss rate, for oils useful as lubricants in vacuum ranges from 10^{-9} to 10^{-13} g cm^{-2} s^{-1} at room temperature. Typical ranges for M and L are 500–1200 g mol^{-1} and 17 000–36 000 cal mol^{-1}, respectively. The numerical values of C and L for a given substance are readily obtained by performing weight loss rate measurements in vacuum at two temperatures, inserting the values into the Langmuir equation, and solving the resulting simultaneous equations for C and L.

Other properties, such as good chemical stability and low surface migration rate are important to oils for the lubricating of mechanisms in vacuum. High chemical stability is necessary so that the lubricant molecules do not break down under heat and stress. Low surface migration rates ensure that the lubricant stays where it is put and does not spread out to increase the surface area and hence the mass rate of evaporation. Barrier films[2] are commercially available which prevent surface migration and have been shown to increase bearing life by a factor of 13 in gyroscope bearings[3]. The rate of surface migration is usually highest (this depends somewhat on viscosity) for silicone oils, followed by fluorocarbons and then hydrocarbons. Labyrinth seals outboard of lubricated parts impede the vapor phase loss rate of lubricants from the vicinity of the contact and therefore reduce the total lubricant mass loss and contamination of the space.

Oils are the most widely used lubricants in vacuum at moderate temperatures (225–325 K) in bearings, dc motor brushes, sliprings, and other electrical contacts, because they are relatively inexpensive to purchase and apply, and because they are traditional and fairly well understood. Because evaporative weight loss (and the concomitant contamination by the oil molecules of nearby surfaces) is one of the most important lubricant depletion mechanisms, it is important to select oils having low evaporation rates. Table III shows the rates for several useful oils at room temperature.

Oils usually contain additional substances such as extreme pressure additives, and antiwear additives to impart desired characteristics. Other types of additives may be present. Because the base oil and additives will have different vapor pressures, it has been found that the use of more than two volatile components makes the prediction of evaporative weight losses unnecessarily complicated.

Low vapor pressure, chemically stable oils are used to impregnate ball-bearing separators for long-term vacuum service. The porous separator stores excess oil, and acts as a reservoir and metering device. It also retards evaporation because the oil must diffuse to the surface before evaporation or surface migration can remove it from the ball-race contact region. Oils are also used in and on electrical contacts such as motor brushes, slipring wipers, and other nonstationary contacts. If the contacts are metallic the lubricant prevents galling, cold welding, and reduces friction. In carbon and graphite brushes the lubricant modifies friction, reduces electrical noise, and reduces the wear rate.

Oils compounded with thickeners and other additives form greases. Table IV lists a few types of material used to thicken oils into greases. The greases are typically used on sliding surfaces, gears, and bearings of all types where there

TABLE IV. Types of thickeners used to make greases of the oils.[a]

Quartz powder (fumed silica)
Modified Teflon
Clays
Complex

[a] At times, molybdenum disulfide is also added to give additional desirable properties (separation of surfaces, load carrying capacity).

are no porous members to retain oil. The grease acts as a reservoir for the oil and allows the oil to flow from the thickener to the loaded contact region. In highly loaded contact regions such as gear tooth meshes, it is often beneficial to incorporate dry solid lubricant powders into the grease. The solid particles in the lubricant separate the teeth so that metal-to-metal contact occurs less frequently, and so that lower wear rates occur and mechanical noise is reduced. NASA CR-161109[4] Part B is an excellent source book for fluid lubricants.

III. APPLICATION PROCEDURES FOR FLUID LUBRICANTS

For best results, special procedures should be used in cleaning devices for use in vacuum, especially if the devices require lubrication. Table V lists representative types of devices which require lubrication. In addition to removing particulate contaminants which might interfere with mechanical motions, it is desirable to remove molecular contaminants, such as manufacturing oils, plasticizers, finger salts and oils, and unreacted components of resins, paints, and varnishes. Materials such as rubber grommets or seals, cork gaskets, adhesive, and other high weight loss materials should be removed or replaced with suitable low vapor pressure materials. These materials not only contaminate the volume in vacuum space, they may prevent the lubricant from wetting the surfaces where oil is needed, or they may chemically react with the lubricant.

Excellent cleaning of the parts can be obtained by vigorous solvent washing of the components followed by a Soxhlett extraction process, and vacuum baking for several hours at elevated temperature. Parts to be lubricated should be immersed in the lubricant after the vacuum bake and prior to breaking the vacuum, so that the lubricant comes into intimate contact with the clean surface before oxygen and or-

TABLE V. Types of components using wet lubricants in vacuum.

Bearings
Gears
Potentiometers
Solenoids
Encoders
Motors—commutators and brushes
Sliprings—commutators and wipers
Bushings
Cams
Valves
Hinges
Recording media (tapes, drums, disks, etc.)
Screws/inserts

ganic molecules present in the bleed-up gas can come in contact with these surfaces. After the parts are coated with lubricant, they may be removed from vacuum and drained to remove the excess. The final desired quantity of lubricant on the component can be obtained by then adding more lubricant or by rinsing the component with a diluted solution of the lubricant. Solvents used in the cleaning processes must be compatable with the materials. Alcohols, xylenes, hexanes, trifluorotrichloroethane, and azeotropes of these materials have been found generally acceptable solvents which do not leave undesirable films after evaporation. Inhibited methyl chloroform is acceptable for materials other than aluminum.

IV. SOLID LUBRICANTS

Operation of mechanisms in vacuum at high and low temperatures and where the molecular contamination of optics or other sensitive surfaces is to be minimized requires the use of solid lubricants, because these have evaporative weight loss rates much lower than the best fluid lubricants (Fig. 1). Dry solid film lubricants, like fluid lubricants, are a controlled contaminant interspersed between contacting surfaces. They provide fairly low friction and reduce mechanism seizure. In some (Fig. 2) the particles are arranged like a deck of cards, so that they can support high normal forces, yet have a low shear strength in the direction of motion. Many naturally occurring minerals (graphite, mica, molybenum disulfide) have this property, as well as many more synthetically made materials (various dichalcogenides and intercalation compounds[5]). Polytetrafluoroethylene (PTFE) and similar materials (CTFE, FEP, carbon monofluoride) are dry solid lubricants having a waxy texture rather than a layered structure. Low shear strength metals such as gold, silver, and lead are at times used in vacuum as lubricants as well.

Mica and graphite are not used as lubricants in vacuum. Graphite requires the presence of water (or a water substitute) in order to be a lubricant—otherwise it is abrasive. The

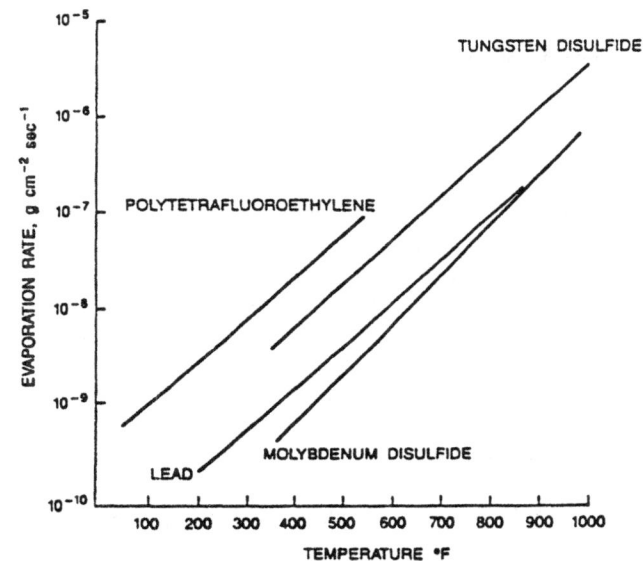

FIG. 1. Evaporation rate for four dry lubricants.

J. Vac. Sci. Technol. A, Vol. 6, No. 3, May/Jun 1988

141

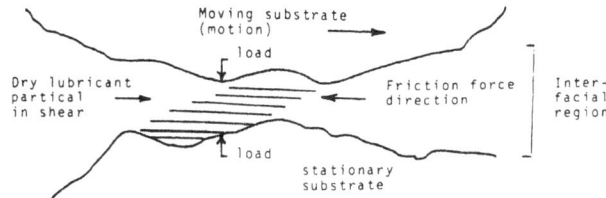

FIG. 2. Showing how dry lubricants lubricate. Deposition of lubricant on substrate. Shearing of lubricant in the direction of sliding. Low coefficient of friction (μ):

$$\mu = \frac{\text{friction force}}{\text{load}} = \frac{\text{shear strength} \times \text{area}}{\text{pressure} \times \text{area}} = \frac{\text{shear strength}}{\text{pressure}}.$$

The shearing occurs in the thin dry lubricant film while the load or pressure is borne by the substrate metal.

most commonly used dry lubricants are PTFE, lead, and molybdenum disulfide. Although the synthetic dichalcogenides (WS_2) and graphite intercalation compounds (graphite/$FeCl_3$) are better lubricants, higher costs and the lack of reproducibility in manufacturing of these products precludes extensive use.

Dry solid lubricants generally have a shorter service life than oils or greases, because once the lubricant coating wears through, metal contacts metal and failure occurs. There is no replenishment mechanism similar to oil flow or creep to recoat the surfaces. The low contamination potential and relatively little change in friction with temperature often more than compensate for the shorter service life expectancy. Part A of NASA CR-161109[4] lists most of the dry solid lubricants and suppliers.

V. APPLICATION PROCESSES FOR SOLID LUBRICANTS

Dry lubricants can and are applied to surfaces in many ways and in a range of thicknesses. Clean surfaces are required to obtain good adherence and smooth coatings. Cleaning with solvents is usually sufficient for adhesively bonded and electroplated lubricant coatings. Coatings applied by evaporative processes, ion plating, and sputtering techniques usually have their surfaces ion sputtered to achieve additional cleaning.

The least complicated application technique for solid lubricants is burnishing the part to be lubricated by rubbing lubricant onto the part until a smooth uniform film is obtained. A similar coating can be obtained by tumbling the parts in powdered lubricant or by high-velocity impingement of the part by lubricant entrained in a gas in a process similar to sandblasting. These processes leave coatings having a thickness of $< \frac{1}{4}\mu$ up to $\sim 2\mu$ in thickness and the wear life is limited to a few tens of sliding cycles and a few thousands of rolling friction cycles. The life is dependent upon coating thickness, bearing load, speed, and contact configuration.

A bonded film lubricant has longer life. The bonded lubricants are comprised of lubricating pigments in a binder. Application is by brushing, dipping, roll coating, or preferably, spraying. The coating is applied 5 to $10\,\mu$ thick and cured. Life of these coatings are from 7500 to 15 000 sliding cycles

TABLE VI. Life as a function of load for epoxy bonded MoS_2.

Load (g)	Life (m of sliding) Film thickness of lubricant (μ)	
	1.25×10^{-6}	12.5×10^{-6}
100	25	1000
1000	2.25	25
10 000	1	6.25

under a 450-kg load. Table VI shows how life is affected by load for a bonded film lubricant, while Fig. 3 shows the variation in friction coefficient with thickness.

The most tenacious coatings are those which are applied by vacuum deposition processes. Evaporated metal and electroplated coatings are usually the thicker, less expensive, and least tenaciously adhering and may fail due to blister formation in the coating. Metals can also be applied by ion plating, a process which embodies the processes of evaporative coating and sputtering. The ion plated coatings may have a wide range of thickness. Sputtering processes can be used to apply nonmetallic as well as metals. Sputtering is an energetic process in which the lubricant atoms or molecules are electrically accelerated to the part to be coated and penetrate it up to a few atom layers deep. Continued sputtering deposits additional material until the lubricant coating is 0.08–0.2 μm thick. Metals and nonmetals can be cosputtered to achieve even better properties than either material by itself. Adhesion and wear life of the sputtered coatings are vastly superior to other types of coating and the coefficient of friction is usually lower than for the other types of coatings. Molybdenum disulfide (often cosputtered with nickel[6]), lead, gold, and PTFE have all been sputter coated onto parts for vacuum use.

VI. COMPOSITE LUBRICATING MATERIALS

Matrices of two or more materials, one or more of which are lubricating pigments, are made up in forms which can be machined to the desired configuration, or molded to the desired configuration. These are self-lubricating materials and are often formed into bushings, ball bearing separators, mo-

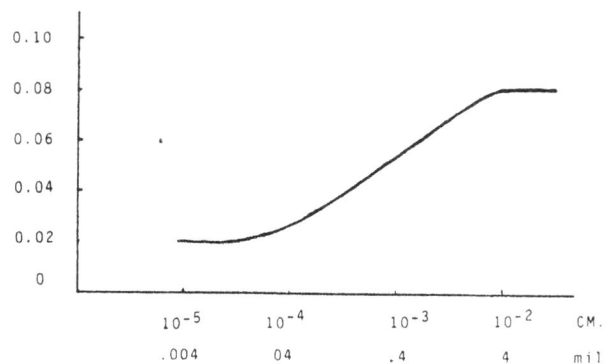

FIG. 3. Friction coefficient as a function of lubricant thickness for epoxy bonded MoS_2.

tor and slipring brushes, cams, and gears. The bulk of the matrix is usually a metal or resin and the remainder is lubricant or electrically conductive metal. Electrical contact wipers made of silver, molybdenum disulfide, and small percentages of graphite or copper provide self-lubricating electrical contacts. Ball-bearing separators made from glass fiber, PTFE, and molybdenum disulfide allow bearings to operate in vacuum with no other lubricants for $200–400\times10^6$ revolutions before the sacrificial separators wear out.

PTFE impregnated fabrics are another composite-type dry lubricant material which can be used in vacuum. An adhesive backing on the fabric allows it to be fastened to a substrate. These materials often are used in spherical bearings and as slide strips.

Typically, the composite materials will operate from cryogenic temperatures up to near the melting point or decomposition temperature of its constituents. Polyimide resin based dry lubricants and composites have operated in vacuum at 800 °F (700 K) for ~ 1 h, and at 700 °F (640 K) for 15 h in a ball-bearing assembly (intermittent operation, high-radiation environment). Higher temperature operation in vacuum may be possible with materials currently being evaluated.

VII. CURRENT AND FUTURE EXPECTATIONS

Work is progressing in several areas to provide improved lubricants and materials for use in vacuum.

TABLE VII. Material/ion combinations for property improvement.

Base material	Property	Implantation ions
Aluminum alloys	Corrosion	Mo
	Hardness	N
Beryllium alloys	Hardness	B
	Wear	B
Ceramics	Hardness	Y, N, Zr, Cr
	Wear	O, N
	Toughness	O, N
Cocemented	Wear	N, Co
tungsten carbide	Hardness	N, Co
Copper alloys	Corrosion	Cr, Al
Steel, high alloys	Corrosion	Cr, Ta, Y
	Wear	Ti + C
	Hardness	Ti + C
	Fatigue	N
	Friction	Sn, Ag, Au
Steel, low alloys	Corrosion	Cr, Ta
	Wear	N
	Hardness	N
	Fatigue	N, Ti
	Friction	Sn
Super alloys	Wear	Y, C, N
	Corrosion	Pt, Au, Ta
Titanium alloys	Corrosion	N, C
	Wear	N, C, B
	Hardness	N, C, B
	Friction	Sn, Ag
	Fatigue	N, C
Zirconium alloys	Wear	C, N, Cr + C
	Hardness	C, N
	Corrosion	Cr, Sn

In fluid lubricants studies, polyalphaolephins have been shown to be excellent lubricants for electrical contacts and bearings. Polyphenylethers, long used as diffusion pump oils because of their low vapor pressures and stability, are being modified so that they will wet metals and therefore possibly be good lubricants for bearings and other mechanisms.

New dry lubricant pigments known at graphite intercalation compounds are being developed and tested. These materials are created by sputtering foreign molecules such as $FeCl_3$ into the lattice layer configuration of graphite. By increasing the distance between the layers, lower shear strength materials are created which will still support normal bearing loads. Also, new binders for the dry solid pigments are being tested which can reduce costs as well as increase the wear life of available bonded lubrication systems.

The most exciting possibility for the future is that of not having to have special lubricants for vacuum use. New ways of processing steels and ceramics are under development. Several companies now have processes which coat steels so that the hardness values are in the Rc70 to Rc80 range, typical of topaz and some tungsten carbide alloys. Materials of that hardness tend to cold weld and have adhesive wear failure less than the softer materials. Electrical contact materials in the form of sliprings have operated continuously in hard vacuum with no lubrication and almost no discernible wear for more than six months. Work performed at various laboratories recently has shown that ion implantation techniques can modify materials in significant ways so that lubrication may not be required for vacuum or other environments. Table VII shows some of the properties improved by ion implantation. Ion implantation of bearing steels, for instance, can increase corrosion resistance, reduce wear and friction, and increase the fatigue life.[7] Ceramics, suitably implanted, show increased toughness, hardness, and less wear than untreated ones. Combinations of steel races and ion implanted ceramic balls are currently commercially available, as are bearings having silicon nitride balls and races. These are capable of operating at temperatures up to about 1300 K and hold promise for unlubricated vacuum operation as well.

[1]E. E. Bisson, "Advanced Bearing Technology," NASA SP38, 1964.
[2]W. F. Nye, Nyebar Types C and H, Summary Catalog, 1969–1970.
[3]V. G. Fitzsimmons, C. M. Murphy, J. B. Romans, and C. R. Singleterry, Lubr. Eng. 24, 35 (1968).
[4]Midwest Research Institute, Lubrication Handbook for the Space Industry, Part A: Solid Lubricants, Part B: Liquid Lubricants, 1978.
[5]W. E. Jamison, in ASLE Proceedings: 3rd International Conference on Solid Lubrication, Denver, CO, 1984, ASLE SP-14 (American Society of Lubrication Engineers, 1984), p. 73.
[6]B. C. Stupp, "Performance of Conventionally Sputtered MoS₂ Versus Co-sputtered MoS₂ and Nickel," in Ref. 5.
[7]P. Sioshansi, Mach. Des. 58, 61 (1986).

J. Vac. Sci. Technol. A, Vol. 6, No. 3, May/Jun 1988

143

CONSIDERATIONS ON THE LUBRICATION OF SPACECRAFT MECHANISMS

by: H. Mervyn Briscoe* and Mike J. Todd**

1. INTRODUCTION

The wide variety of lubrication techniques now available to spacecraft engineers very often makes the choice of an optimum process for a specific application very difficult to select. The ever increasing demand for the reduction of costs in spacecraft engineering has led to the unthinking application of commercial processes with the minimum of either intellectual or practical justification, and sometimes disastrous results.

The purpose of the paper is, therefore, to try to focus space tribology and to propose a number of precepts to guide designers in its application. Of the many techniques available all, without exception, have limitations in performance. Two European processes will be discussed in more detail and their limitations identified. Some performance results on a recently introduced liquid space lubricant will be given.

2. APPLICATIONS AND REQUIREMENTS

Table 1 is a non-exhaustive list of important space mechanisms with brief details of their lubrication needs in which four broad types of application can be identified.

- low speed sliding contact
- high speed sliding contact
- low speed rolling contact
- high speed rolling contact

For each of these there is a number of solutions but the selection of the optimum will depend wholly on the physical details of the mechanism. Which brings us to the first precept of space tribology.

Precept 1   The optimum lubrication system for a space mechanism is an integral part of the mechanism design and not a process to be added when the design is complete.

Even today when space engineering is more than twenty years old the precept is ignored. The authors have personally experienced two instances during the past year where the lubrication system was expected to make a poor design concept work.

*European Space Technology Centre (ESTEC), European Space Agency, Noordwijk, The Netherlands
**European Space Tribology Laboratory (ESTL), UKEA, Risley, United Kingdom

3. <u>LUBRICANT CLASSIFICATION</u>

They may be categorized as:

DRY

(a)   Dichalcogenides (lamellar solids)
(b)   Solid lubricant composites and transfer film lubricants
(c)   Soft metals

LIQUID

(a)   Hydrocarbon oils
(b)   Synthetic oils
(c)   Greases

It is quite impossible in a short paper to review all available techniques and processes so the authors will concentrate on those available in Europe and with which they are most familiar.

4. <u>DRY LUBRICANTS</u>

4.1 <u>Dichalcogenides</u>

These are a group of metal compounds which exhibit a lamellar structure to which current theory attributes their good lubrication properties. In fact only a limited number may be classed as lubricants, amongst which are WS_2, WSe_2, MoS_2, $MoTe_2$, NbS_2, $NbSe_2$, and $TaSe_2$.

Of these MoS_2 has shown itself to have the best performance in a space environment as a result of years of testing. It has good wear characteristics, low friction and in general performs better in vacuum than in moist air. Its coefficient of friction is load dependent, although some recent work has introduced dispute on this point, and ranges from 0.4 to lower than 0.1. There is a large literature on MoS_2 and ESA is funding a continuous bibliography on its applications to space prepared by A.R. Lansdown. (Ref. 1-5).

MoS_2 may be applied in a number of ways and those of interest to space engineers are reviewed briefly below.

4.1.1   Bonded Films

A bewildering variety of bonded films has been invented, now more than 100, but only a few of them have found general commercial application. The binders may be organic or inorganic, heat cured or simple air drying agents mostly based on cellulose. Some inorganic binders based on ceramics require curing temperatures of $1000°C$ which limits the materials to which they can be applied.

The thickness of a bonded film may commonly be of the order $10\mu m$ which is enough to degrade the precision of many high grade mechanisms. The debris from the film may also lead to very erratic torque, particularly in rolling bearings. The control of such processes is very operator dependent which makes it difficult to achieve consistently uniform results and tends to preclude their use in critical mechanisms. The authors would not now use a bonded film in an application where high precision was a requirement.

4.1.2   Burnished Films

A moderately adherent thin film of MoS_2 may be achieved by burnishing onto a thoroughly degreased and clean metal surface. Rubbing with chamois leather or a wire brush, or tumbling in a ball mill are established methods of burnishing. It is difficult to achieve consistent results even when the process is most carefully controlled, but the process is readily acceptable for very simple single operation devices such as latches, hinge pins and pyrotechnics. It can be applied to rolling bearings where the life is a few thousand revolutions and loads are moderate. Running-in should be an integral part of the process.

4.1.3   Sputtering

The most recent technique for applying MoS_2 films is by sputtering, first developed by Spalvins and Przybyszewski at NASA Lewis Research Centre. Since then a good deal of work on sputtered MoS_2 in ball bearings has been reported in the USA, chiefly by Spalvins (6) and by Christy and Barnett (7) and also in Europe by Bergman (8). In Europe the process has been developed and refined both at the Laboratoire Suisse de Recherches Horlogeres (LSRH) and at the European Space Tribology Laboratory (ESTL) in the U.K.

The very significant advantage to be gained by the process is the achievement of a very thin, about 1μm, film strongly adherent and of controlled composition. However, it has been shown by ESTL that the performance of sputtered films varies widely depending upon the sputtering technique and the stoichiometry of the film. During the past two years films of MoS_2 sputtered by LSRH and ESTL have been compared for performances in 42 mm diameter ball bearings, preloaded to 40N and run at 100 rpm. Results are discussed in para.4.4.

4.2   Solid Lubricant Composites and Transfer Film Lubricants

The number of polymer type materials falling under this heading is now considerable and quite beyond the scope of this paper. In practice only a very few are suitable for use in a space environment and in Europe only PTFE, often filled with MoS_2 and reinforced by glass fibre polyimide and polyacetal (e.g. Delrin) are commonly used for their tribological properties. The disadvantages of polymers in a space environment are:

- high coefficient of expansion
- low load capability
- high wear rate
- limited dimensional stability
- poor thermal conductivity

But these are balanced by the advantages of:

- cheapness and simplicity
- low coefficient of friction
- low outgassing

The lubricating action of these polymers and composites may be by the transfer of a film to the mating material or they may simply provide wear resistant components, but the two actions should not be confused. In consequence they may be used in space either as the base material of one component of a rubbing pair, e.g. gears, or to provide a transfer film to lubricate a rolling or sliding metal pair, e.g. ball bearings.

PTFE + MoS$_2$ + glass fibre has for many years been used as a cage material in small, lightly loaded, ball bearings. This composite has been studied in detail by Stevens and Todd (Ref.9). Loads must be limited to ensure that the Hertzian contact stress does not exceed 1.2 x 10^3 MN/m^2 at 20°C. The MoS$_2$ appears to have the function of preventing the transfer film of PTFE becoming too thick. Failure is usually by wear-out of the cage, creating large quantities of debris. In oscillatory motion a build up of transfer film at the end of the arc of travel leads to excessive torque peaks. In addition the thickness of the transferred film can be dependent upon speed of operation.

Both polyimide and polyacetal have been used for lightly loaded gears in a space environment running against stainless steel, titanium or aluminium. Work in progress at ESTL, to be reported in the near future, has shown that Vespel, a polyimide, gives the lowest wear rate of all polymers tested. Polyacetal filled with carbon fibre has also been tested and shows lower wear rate than the unfilled polymer but still greater than the polyimide. Maximum loads for all these materials should be limited to 10 N/mm tooth width for gears of module 1.

4.3 Soft Metals

It has long been known that the soft metals such as Ba, Au, Ag, Pb, In and others are capable of providing lubricating films in certain circumstances. In this paper we shall confine ourselves to the use of lead in rolling bearings and gears in space. The work was initiated by the Royal Aircraft Establishment in the UK in the 1960's and a very extensive test programme was carried out by Marconi Space and Defence Systems on 19 mm diameter bearings at 3000 rpm, some of which ran for eleven years. The programme was extended by ESA to bearings of 90 mm diameter running at 100 rpm and six pairs completed more than 60,000 hours without failure. In all these tests the only limitation was the wear debris from the lead bronze cage, and the search for a better cage material is continuing.

The process used by MSDS Ltd. was vacuum deposition, which made film thickness and adhesion difficult to control, so the Agency undertook the development of an ion plating process at ESTL. (Ref.10, 11, 12).

The pre-eminent advantages of an ion plated film over a vacuum deposited one are excellent adhesion and close control of thickness, 0.35 ± 0.15μm at ESTL. At present only the races of the bearing are plated and the film is transferred to the balls during the running-in period in vacuum, which is an integral part of the process. Ion plating of the balls is now possible but, for most applications seems to offer little advantage.

Lead lubricated bearings are flying in the OTS and MARECS Solar Array Drives and will fly in the SADs of ECS, EXOSAT, L-SAT and the French SPOT. The bearings of the de-spin mechanism of GIOTTO, the European Halley's Comet probe, will also be lead lubricated. The selection of lead in this case was made to take advantage of its immunity from torque changes due to temperature. The torque at -40°C is virtually identical to that at room temperature in a well designed system; a performance which no liquid lubricated bearing can equal. Torque noise, however, will be higher than an equivalent liquid system which was the reason for the final rejection of lead, and all other dry film systems, for the Space Telescope Solar Array Drive, where smoothness of operation is a prime requirement.

Lead lubricated bearings are finding application in scientific instruments and lead lubricated ball screws are flying on Nimbus G. A similar ball screw has performed very well at 4°K and demonstrated the effectiveness of the process at cryogenic temperatures. Ion plated lead is now being applied to gears although it operates better in a rolling than in a sliding contact. Tests have shown it to be a very effective lubricant for gears of nitrided steel on nitrided 440C steel and it has been adopted for the gears of the L-SAT Solar Array Drive. Most importantly the process has been fully established, codified and documented at ESTL to ensure complete repeatability of film parameters and friction and wear characteristics, and it is now offered as a standard process routinely carried out.

The pros and cons for lead lubrication may be listed as:

Pro: - excellent adhesion
     - low, consistent and temperature-independent torque
     - very long life under vacuum
     - applicable to bearings, gears and ball screws
     - usable at cryogenic temperatures

Con: - flake wear debris from the lead bronze cage can cause torque noise problems. ESTL is investigating alternative materials for cages.

     - operation in air must be limited to fairly low speed and short duration e.g. 10^5 revs. at 100 rpm for a 20mm ID bearing.

4.4 Bearing Torque Characteristics of Solid Lubricant Films in Vacuo.

With the aim of comparing levels of torque and of torque noise from ball bearings under identical conditions, data from ESTL have been assembled to show what is to be expected of the three most common lubricants discussed above.

Table 2 shows the details of the method of lubricant application, the associated cage material, the type, size and preloading of the ball bearings used in this comparison. In all the tests an arbitrary level of torque ten times the initial DC level was chosen to represent torque "failure".

4.4.1 Torque Results

The histories of torque tests over 2 million revolutions (or until torque failure, as defined above) are collected in Table 3. As a comparison the result for an unlubricated, but degreased, bearing is included. Very rapid failure is encountered.

The MoS_2-sputtered bearings tended to start with a moderately high torque (20×10^{-4} Nm) but it soon fell to a $6 - 10^{-4}$ Nm (at 40N preload) where it remained until usually a sudden steep rise and torque failure. Films sputtered at ESTL and at LSRH Switzerland exhibited broadly similar behaviour but bearing lives varied from $0.7 - 3 \times 10^6$ revs for films sputtered at ESTL and from $2 - 3.66 \times 10^6$ revs for films sputtered at LSRH. The torque traces were some of the smoothest that we have observed and we would now recommend sputtered MoS_2 where low torque noise is important and the life requirement is not above 10^5 revolutions, or equivalent rolled distance.

The ion-plated, lead-filmed bearings exhibit reasonably steady DC (average) torque but there is considerable fluctuation in the peak-to peak torque. This latter is defined here as the maximum total swing of torque either side of zero during slow rotation of the bearings in two opposite directions. This bi-directional check of the torque was carried out at regular intervals in these tests. For the lead film there is no effect of a previous run at 100N preload on the torque. None of the lead bearings failed by the above criterion. The PTFE-composite caged bearings were considerably more erratic in their average torque than the lead filmed bearings and the torque was prone to rise quite suddenly over a few thousand revolutions and as quickly to subside. Nevertheless, the peak-to-peak amplitudes of torque were less than those of the ion-plated lead, as Table 3 shows. There was a

151

discernable effect of previous running at low load (i.e. below critical stress) upon the torque in the 250N run (above critical). The bearings run immediately at 250N showed high DC torque initially but this gradually reduced with time. Such initial high torque was absent if the bearings had been run at sub-critical stress beforehand but there were still periods of high torque during the run. For these bearings 100N preload caused the critical Hertzian stress.

It is appropriate to complete this section with the second precept of space tribology:

> Precept 2: In designing a space mechanism avoid making its operation dependent upon close control of the coefficient of friction.

5.   LIQUID LUBRICANTS

The use of liquid lubricants in space is not new and it is fair to say that all early American satellites relied upon it. Ball Bros. were the leaders in the field in the USA and the move to dry lubrication has been slow. In Europe we went strongly for dry lubricated systems, largely driven by the success of the lead system, but also because their advantages were clear.

In consequence liquid systems in Europe have received less attention but have not been ignored. A de-spin system which has been running under test in vacuum at ESTL for more than 7 years is liquid lubricated. The Solar Array Drive for Space Telescope, the Instrument Pointing System and the bearings of the Antenna Pointing Mechanism on L-SAT, all have liquid lubricated bearings.

5.1 Hydrocarbon Oils

In Europe BP Ltd (U.K) developed, at the request of RAE, two space oils with very low vapour pressure.

BP 110 is a very fine cut natural hydrocarbon with a claimed vapour pressure of 10^{-10} torr or lower at room temperature.

BP 135 is a synthetic tri-ester with a higher viscosity index than BP 110.

BP. Ltd. further developed two greases upon these oils but of quite different formulation from normal greases. (Ref.13).

The thickener is based on an oleophilic graphite-lead composite capable of forming stable semi-solid structures on dispersion in a suitable base fluid. The proportion of thickener is 17%, which is unusually high, but it also provides a marked improvement in the extreme pressure and boundary lubricating properties.

The use of graphite in a space environment is very unusual since it is well known that graphite in a wholly dry atmosphere acts as an abrasive. In the case of the BP greases the special method of preparing the graphite to render it oleophilic makes it operate in a vacuum environment very satisfactorily.

Both of these oils have been subjected to extensive testing over the past seven years and the BP 110 and its grease BP 2110, have been shown to give longer bearing lives in vacuum than competitive lubricants. In most tests BP 135 has not equalled the life achieved by BP 110 but has given good results for many applications, and is useful where its higher viscosity index may be an advantage.

At ESTL a de-spin mechanism has been running at 60 rp.m. under thermal vacuum conditions for seven years without any change in performance. Lubrication is achieved by a small quantity (about 5% fill) of BP 2110 grease in each bearing combined with Nylasint oil stores charged with BP 110 oil. The thermal conditions in the chamber are maintained at 20°C whilst the shaft of the mechanism is driven between -5°C and + 45°C. The performance of this mechanism has been excellent with no change in any operating parameter except a small rise in motor current attributable to magnet deterioration.

5.2 Synthetic Oils

In general, synthetic fluids usable as space lubricants fall into two broad categories:-

-       silicones
-       fluorinated polyethers.

In the early years of space many silicones were tried as space lubricants with widely different results. Their advantages are a high viscosity-index, good thermal stability and fair to good lubricating properties, but their volatility in space environment may not be significantly better than some hydrocarbon oils, and their surface creep characteristics are notoriously bad.

Silicones in space have now been displaced by perfluoralkylethers such as the Krytox and Fomblin fluids. However, the lubricating properties of these fluids are not general and only some of them have found application in space. In some applications such as sliprings they can lead to excessive wear rates and the formation of polymers. However used in the right application they can give excellent performance and their extremely low (10^{-11} - 10^{-13} torr) vapour pressures at room temperature makes them most attractive for mechanisms with critical cleanliness requirements.

5.2.1   Fomblin Z25

Probably the most significant addition to space lubricants is a polyfluoralkylether manufacturered by Montedison in Milan called Fomblin Z25. The Fomblin Y series of fluids has been well known in the vacuum industry for many years but the introduction of Fomblin Z, which has a different chemical structure, was something of a breakthrough for space lubrication. The fluid has the very high viscosity index of 345 and a vapour pressure that is certainly below 5×10^{-12} torr at room temperature. It is the only oil known to the authors which passes the Agency's material out-gassing test, giving TML/RML % 0.01/0.01 and cvcM% 0.00.

The Bray Oil Co. in the USA have for several years distilled the Italian raw stock to make Bray 815Z, which has been flown on a number of USA satellites mostly in the form of a PTFE thickened grease, Bray 3L38RP, and will fly in Space Telescope in the solar array development mechanism.

During the last two years the fluid has been subjected to numerous tests at ESTL.

Bearings of two different conformities are being run at speeds of 20, 100, 200 and 1400 rpm. Full elasto-hydrodynamic lubrication is achieved at about 250 rpm. In the case of low conformity bearings at 1400 rpm the torque reduces suddenly and torque noise increases after a short time indicating the onset of starvation conditions. A black deposit has been observed in these bearings which has been identified as a polymer. In initial tests the high conformity bearings at 1400 rpm were accidentally contaminated with a hydrocarbon oil which resulted in a good performance but a milky deposit. Repeat tests with rigorously pre-cleaned cages are showing similar deterioration in average torque and torque noise. This work will be fully reported when complete.

The load carrying capacity of the fluid tested on the Falex machine is very good, giving a failure load in excess of 15,500N (3500 lb).

In pin and disc tests to evaluate its boundary lubricating properties it has performed equally with KG 80 and BP 110 for wear rate and coefficient of friction. (Ref.10).

The conclusion to date is that its use is acceptable in low speed applications where boundary lubrication is required and in Europe it is currently being applied to the bearings of IPS, Space Telescope Solar Array Drive and the L-SAT Antenna Pointing Mechanism.

For the present the oil should not be used in sliprings where it may increase the wear rate and can lead to complete failure by the formation of debris.

The grease 3L38RP has been used in small gear boxes but data on permissable tooth loading, shear breakdown or polymer formation is very limited.

Future use of the oil under conditions of EHD lubrication is dependent upon the results of the ongoing test programme at ESTL, the objective of which is to determine the limits of its range of application. Its use in optical instruments is not excluded but has yet to be demonstrated by valid test.

Before any application of this oil to a space mechanism is contemplated it is strongly advised to seek the guidance of ESTL who has carried out most of the relevant work.

6.   THE CODIFICATION OF SPACE LUBRICANT SYSTEMS & PROCESSES

The behaviour of any tribological systems is governed by a large number of factors, many of which are difficult to control. Material properties, both macro and micro, surface condition, presence of micro quantities of contaminants, system geometry, speed, load, duty cycle are only some of the variables. In consequence any lubrication process to be acceptable for space must be subject to the following precepts:-

Precept 3: - Any lubrication process used for space application must be fully codified and documented to ensure consistent repeatability of performance.

Precept 4: - The lubricant used must be approved and validated for space to a recognised specification and must be source traceable.

Precept 5: - The test programme to determine the performance of the lubrication system must reproduce all the operational conditions of duty cycle, environment and life that it will experience in the application.

The use of a commercial system for a space application is acceptable only if it fulfils these three precepts. The Agency has issued a guide to the preparation of process procedures. (Ref.14).

7. CONCLUSION

In attempting to cover the whole field of space lubrication in a short paper the authors have set themselves an impossible task. Much relevant and important detail has had to be omitted. But the purpose of the paper is to provide a useful brief overall view which will, it is hoped, give the non-specialist a picture of a highly complex subject, and some guidance in the choice and application of a lubricant in spacecraft mechanisms.

TABLE 1.

APPLICATION	DUTY	LIFE	ENVIRONMENT	LUB. REQUIREMENTS
Momentum Wheels	Continuous Rotation at 3000-4500 rpm.	7-10 yrs	Closed low pressure He	Liquid: Maintenance of a controlled flow of lubricant into EHL zone to achieve low and consistent torque. Low wear rate.
Solar Array Drive	Continuous rotation at 1-16 revs. per day	7-10 yrs	Space exposed Thermal -40°C to +65°C	Solid or liquid: Boundary lubrication with consistent torque. No contamination, moderate load, corrosion protection.
Antenna Pointing Mechanism	Slow intermittent operation over small angle and occasional fast tracking	7-10 yrs	Space exposed Thermal -40°C to +65°C	Solid or liquid: Boundary lubrication with consistent torque. No contamination. Moderate to high load capacity, corrosion protection
Instrument Pointing Systems	Slow intermittent operation over small angle, very high precision	1-4 weeks 5 yrs storage	Space exposed Thermal -20°C to +60°C	Liquid: Boundary lubrication with low and consistent torque. No contamination. High load capacity, corrosion protection
De-Spin Mechanism	Continuous operation at 15-60 rp.m.	7-10 yrs	Space exposed Thermal -40°C to +65°C	Solid or liquid, controlled quantity of oil or grease. Low and consistent torque over temperature range. Low wear rate, corrosion protection.

157

TABLE 1 (cont.)

APPLICATION	DUTY	LIFE	ENVIRONMENT	LUB. REQUIREMENTS
Solar Array (Hold down points and latches)	Deploy and retract	50 oper- ations	Space exposed Thermal -60°C to -80°C Launch vibration loads	Dry. High load capability, good fretting resistance
Focussing Mechanism	Intermittant operation cycles 10000 to 100000	7-10 yrs	Space exposed Thermal 0°C to 20°C	Dry. Low and consistent friction. Absolute freedom from contamination. Low wear rate. Corrosion protected.
Filter wheels, shutters, beam splitters etc.	Intermittant operation 10.000 - 20.000 cycles	7-10 yrs	Space exposed Thermal 0°C to 20°C	Dry or liquid. Low and Abso- lute freedom from contamin- ation. Corrosion protection.
Slip rings and brushes	1) Low speed continuous	7-10 yrs	Space exposed	Dry. Low and consistent friction. Absolute freedom from contamination. Low and consistent electrical noise.
	2) High speed intermittent	7-10 yrs	Thermal -40°C to +65°C	Dry or liquid, low wear rate, low electrical noise
Gears	1) Intermittent operation 1000 Cycles	7-10 yrs in space	Space exposed Thermal -40°C to +65°C	Dry or liquid. Low and consistent-friction. No contamination
	2) Continuous operation	7-10 yrs	" "	Dry or liquid. Low wear and low friction. No contamination

158

TABLE 1. (cont.)

APPLICATION	DUTY	LIFE	ENVIRONMENT	LUB. REQUIREMENTS
Rotating Scanner	Continuous operation operation at 15-60 rpm.	7-10 yrs	Space exposed Thermal -0°C - 20°C	Solid or liquid, controlled quantity of oil or grease. Low and consistent torque. Absolute freedom from contamination.
Booms	1) Deploy only	50 operations	Space expoosed -40°C to +65°C	Dry: Consistent friction over temperature range
	2) Deploy and retract.	100 operations after long space stay	Space exposed Launch vibration loads	Dry: Consistent friction over temperature range and life. No contamination.
Solar Array (Hinges)	1) Deploy only	20 operations	Space exposed Thermal -60°C to +80C	Dry.Consistent friction over temp.range.
	2) Deploy and retract.	50 operations after long space stay	Space exposed Thermal -60°C to +80C	Dry or liquid. Consistent friction over temp.range for long life required. No contamination.
Antenna Deployment	Deploy only	20-50 operations	Space exposed Thermal -60°C to +80°C	Dry.Consistent friction over temperature range.

TABLE 2

LUBRICANTS AND BEARING DATA

LUBRICANTS

Solid Lubricant	Cage Material	Method of Application
PTFE film	PTFE/glass fibre/MoS$_2$ composite (commercially available).	Film formed by transfer from the cage during bearing rotation.
MoS$_2$	1% C/1% Cr steel (EN31), machined and ground	RF sputtered film approx. 0.5 micron thickness All bearing components sputter coated.
Pb	11% Pb tin bronze cast alloy (Commercially available)	Raceways ion-plated with lead to thickness between 0.2-0. 5 μm Balls not coated

BEARING DATA

Type of bearing	angular contact
Size	20mm ID, 42mm OD
Contact angle	15°
No. of balls	10
Ball diameter	7.14mm
Ball conformity	1.14
Precision	ABEC 7 or 9
Axial preloads used	40N, 100N and 250N

160

Table 3. Torque Results From Bearing Tests (All Bearings Completed 2×10^6 Revs at 100-600 RPM Unless Stated).

Lubricant	Axial Preload	Range of Average torque (Nmx10^{-4})	Range of Peak-to-peak torque (Nmx 10^{-4})	Remarks
NONE (DEGREASED)	40N	12 – 43	105 - 144	Torque failure after 1,340 revs
SPUTTERED MoS$_2$	40N	6 - 10 (20 initally)	50 - 100 (up to 220 initially)	Low DC torque noise,but torque failures between 0.7 and 3.6×10^6revs.
ION-PLATED LEAD	40N 100N 250N (After 100N) 250N (No run-in)	16 – 26 40 – 50 100 –180 80 –160	80 - 150 150 - 450 400 –1100 500 –1300	NO TORQUE FAILURES DC torque steadier than for PTFE but peak-to-peak torque (noise) greater
PTFE- COMPOSITE CAGE	40N 100N (After 40N) 250N (After 40/ 100N) 250N (No run-in)	6 – 50 10 – 60 50 –500 40 –500	34 - 160 75 - 250 250 –1050 150 - 700	NO TORQUE FAILURES -Stress below limit -Stress at limit (1×10^6 revs only) -Stress above limit.Periods of high DC torque. -Stress above limit. Initially high DC torque.

* Bearing lives with MoS$_2$ -sputtered film varied from 0.7 to 3×10^6 revs for films sputtered at ESTL to 3.6×10^6 revs for films from LSRH, Switzerland.

REFERENCES

1. Lansdown, A.R., "Molybdenum Disulphide Lubrication", ESRO Contractor Report CR(P)-402, (May 1974).

2. Lansdown, A.R, " Molybdenum Disulphide Lubrication, a Continuation Survey 1973-4," ESA Contractor Report CR(P)-764.

3. Lansdown, A.R., "Molybdenum Disulphide Lubrication, a Continuation Survey 1975-76", ESA Contractor Report CR(P)-1045.

4. Lansdown, A.R., "Molybdenum Disulphide Lubrication, a Continuation Survey 1977-78", ESA/TRIB/4, (February 1981).

5. Lansdown, A.R., "Molybdenum Disulphide Lubrication, a Continuation Survey 1979-80", ESA Contractor Report CR(P)-1518.

6. Spalvins, T. "Bearing Endurance Tests in Vacuum for Sputtered Molybdenum Disulphide Films". NASA TMX 3193 (1975).

7. Christy, R.I and Barnett, G.C. "Sputtered MoS_2 Lubrication System for Spacecraft Gimbal Bearings". ASLE 32nd Annual Meeting, Quebec, 1977.

8. Bergmann, E. et al., "Friction Properties of Sputtered Dichalcogenide Layers". Tribology International December 1981.

9. Stevens, K.T. and Todd, M.J. "Parametric Study of Solid Lubricant Composites as Ball Bearing Cages." Tribology Int. Oct. (1982).

10. Todd, M.J. and Robbins E.J., "Ion Plated Lead as Film Lubricant for Bearings in Vacuum". ESA Trib/1.

11. Thomas, A., Todd, M.J., Garnham, A.L., "Current Status of Lead Lubrication of Ball Bearings". Proceedings 2nd Space Tribology Workshop, ESA SP-158.

REFERENCES

12. Todd, M.J., "Lead Film Lubrication in Vacuum" ASLE Conf. on Solid Lubrication, Denver (1978).

13. Friend, G.C., Groszek, A.J., et al., "Development of New Space Lubricants". First European Space Tribology Symposium 1975. ESA SP 111.

14. "A Guide to the Writing of Process Procedures" ESA PSS 30.

Contact Heat Transfer—The Last Decade

C. V. Madhusudana
The University of New South Wales, New South Wales, Australia
and
L. S. Fletcher
Texas A&M University, College Station, Texas

Nomenclature

a = radius of contact spot, m
A = area, m^2
b = radius of heat flow channel, m
b_1 = radius of contour area, m
d = dimension defining specimen size, m
E = Young's modulus of elasticity, N/m^2
E' = effective elastic modulus,

$$= 2\left(\frac{1-\gamma_1^2}{E_1} + \frac{1-\gamma_2^2}{E_2}\right)^{-1}, \text{ N/m}^2$$

f = constriction alleviation factor
g = temperature-jump distance, m
h = thermal conductance, W/m^2-K
H = microhardness of softer material, N/m^2
k = harmonic mean of the thermal conductivities of the two solids in contact, W/m-K
m = moment of the power spectral density of the surface profile (units depend on the order of the moment)
n = number of contact spots
N = density of contact spots, 1/m^2
P = contact pressure, N/m^2
q = heat flux, W/m^2
Q = heat flow, W
r = radial coordinate, m
R = resistance, m$^2\cdot$K/W
S_u = ultimate strength of the softer material, N/m^2
t = thickness of surface film, m
T = temperature, K or °C
ΔT = temperature difference, K
W = total load on the surface, N

α = coefficient of linear thermal expansion, 1/K
δ = effective thickness of gas gap, m; also, separation between mean surface planes
λ = mean free path of gas molecules, m
ν = Poisson's ratio
ξ = waviness number, $= Wd/E'\sigma_1$
ρ = radius of curvature, m
σ = rms surface roughness, m
σ_1 = total rms surface roughness (dependent on σ and d), m

Subscripts

av = average
c = constriction
cd = disk constriction
f = surface film
g = gas
i = individual
n = nominal
T = total
1,2 = solids in contact

Introduction

THERMAL contact resistance is the resistance to heat flow offered by a joint because the area of actual contact is only a small fraction of the nominal area (Fig. 1). It is defined as the ratio of the temperature drop at the interface to the heat flux (Fig. 2),

$$R = \frac{\Delta T}{Q/A} \tag{1}$$

C. V. Madhusudana has been Senior Lecturer in Mechanical Engineering since 1973. He received his B.E. degree from Mypore University in 1963, M.E. from the Indian Institute of Science in 1965, and Ph.D. from Monash University in 1972. His fields of research, in which he has published numerous articles, include heat flow across metallic joints, the general problem of contact heat transfer, and the strength of vertically laminated timber beams. Dr. Madhusudana is a member of the American Institute of Aeronautics and Astronautics, the American Society of Mechanical Engineers, and the Institution of Engineers, Australia.

Leroy S. Fletcher is Professor of Mechanical Engineering and Associate Dean of the College of Engineering. He received his Ph.D. degree from Arizona State University in 1968. He served as Instructor and Research Associate at Stanford and Arizona State. He was a Research Scientist at NASA-Ames Research Center for six years, and has held full-time academic positions at Rutgers University and the University of Virginia. His major areas of specialization are heat transfer and fluid mechanics. Professor Fletcher has published several books and numerous papers and reports. His present research specialities include the thermal conductance characteristics of energy transfer across dissimilar metal interfaces and convective heat transfer coefficients of thin liquid films flowing over horizontal tubes. He is a Fellow of the American Institute of Aeronautics and Astronautics, the American Association for the Advancement of Science, the American Astronautical Society, the American Society for Engineering Education, the Institution of Mechanical Engineers, and the American Society of Mechanical Engineers, of which he is currently President.

The thermal contact conductance is then

$$h = \frac{Q/A}{\Delta T} \qquad (2)$$

The conductance should be high for applications such as nuclear reactors, gas turbines, aircraft structural joints, surface temperature measurements by thermocouples, cooking on a hot plate, and the setting process during a total hip prosthesis.[1-8] When mechanically strong insulation is required, however, such as in the storage of cryogenic liquids, thermal isolation of spacecraft components, mechanically strong insulating cylinders for internal combustion engines, and thermal insulation of high-temperature batteries,[9-12] the contact resistance must be as high as possible.

The heat transfer across a joint may take place by conduction through the actual contact spots, conduction through the interstitial material, and radiation across the gaps. Since radiation is significant only at high temperatures, this mode of heat transfer is usually neglected. This paper, therefore, considers only the theoretical and experimental studies of conduction through the actual contact spots and through the interstitial material. Special problems introduced because of the geometry of contact, cycling of load, and heat flux are also discussed. The review is restricted to steady-state problems.

It must be pointed out that bibliographies and views exist that cover the work done in the field of contact heat transfer into the late 1960's.[13-21] The present paper deals primarily with more recent work, i.e., since 1970. Where reference has been made to an earlier work, it is with a view to developing a particular topic systematically.

Resistance of a Single Constriction

Since thermal resistance results from most of the heat being constrained to flow through actual contact spots, the first logical step in determining the contact resistance would be to estimate the resistance associated with a single contact spot. The constriction resistance of such a spot is a measure of the additional temperature drop due to the presence of the constriction. The shape of a constriction and associated boundary conditions depend upon the nature of the problem being considered.

Disk Constriction in Vacuum

The solutions to the constriction resistance problem date back to 1949.[22-27] Gibson[28] attempted a direct solution to the mixed boundary value problem at the interface. His approach is very similar to that of Hunter and Williams,[26] and his result can be expressed as

$$f = 1 - 409183(a/b) + 0.338010(a/b)^3 + 0.06792(a/b)^5 + \dots \qquad (3)$$

where f is the nondimensional constriction resistance (R_c/R_{cd}); R_c is the constriction resistance based on total heat flow, and $R_{cd} = 1/(4ak)$, the constriction resistance of a disk of radius a in a half-space. Note that f is always less than 1 and is sometimes called the "constriction alleviation factor." It may also be pointed out that Gibson's solution differs very little from Roess' approximate solution[22] derived in 1949.

Yovanovich[29] obtained solutions for the preceding problem after replacing the constant temperature boundary condition on the contact area by the condition that the heat flux over the contact area is proportional to $(1 - (r/a)^2)\mu$. It may be noted that a similar approach was considered earlier by Roess[22] and Cooper et al.[25] Yovanovich developed relations for the nondimensional resistance almost identical to those developed by Cooper et al.[25]

Conical Constrictions

Since the asperities in contact can be considered to be cones of large semiangle (~80 deg), it is desirable to estimate the resistance offered to heat flow by a conical constriction. With such a model, it is a straightforward matter to include the effect of the fluid surrounding the constriction.

Theoretical and experimental studies of the thermal resistance of a conical frustum with a semiangle up to 80 deg in vacuum were studied by Williams,[30] who found that his theoretical predictions, in general, overestimated the thermal resistances. The difference was considered due to the contact areas being actually larger than those predicted using Meyer hardness.

The problem of conical constriction in vacuum was simulated in an electrolytic tank analog by Major and Williams.[31] The maximum value of the cone semiangle used in the experiments was 60 deg. The authors found that the resistance decreased as the cone semiangle increased.

A numerical and experimental analysis of the problem of conical constrictions has been considered by Madhusudana.[32] He considered cones of semiangles up to 85 deg surrounded by a vacuum or a conducting medium. He found that the presence of a conducting fluid significantly reduces the resistance, especially at the practically important low radius ratios. He also found that in a vacuum and for large values of the cone semiangle ($\approx$ 85 deg), there is very little difference between conical and disk constriction resistance. Therefore, for nominally flat, rough surfaces, whose asperities have small slopes with respect to the contact plane, the disk constriction resistance values could be used with confidence.

Major[33] has also presented a finite difference analysis of the problem of conical constriction in vacuum, although the configuration used is slightly different. Results were obtained for values of the cone semiangle up to 68 deg.

Constrictions of Other Shapes

In some applications, the constriction to heat flow may be other than a circular area. These studies[34-41] are presented in Table 1 and deal primarily with constrictions on the boundary of a half-space. Therefore, caution must be exercised when applying the results of these studies to specific problems.

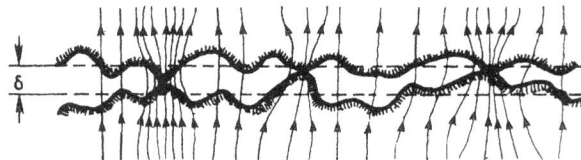

Fig. 1   Heat flow through a joint.

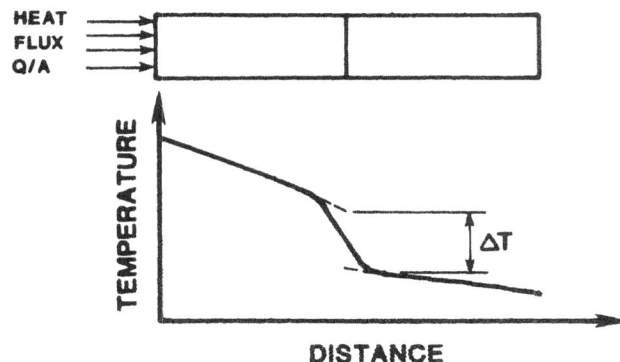

Fig. 2   Axial temperature distribution through a joint.

Table 1   Summary of studies dealing with constrictions on the boundary of a half-space

Author(s)	Types of constriction	Results
Veziroglu and Chandra[34]	Symmetrical two-dimensional and nonsymmetrical	Graphical
Veziroglu et al.[35]	Circular and rectangular	Graphical
Yovanovich[36]	Circular, rectangular, annular	Tabular form
Yovanovich and Schneider[37]	Circular annular	Closed form
Schneider[38]	Circular, rectangular, annular	Tabular form
Yovanovich et al.[39]	Annular, circular, rectangular, triangular	Correlations and tabular form
Yovanovich et al.[40]	Circular, square, astroidal	Tabular
Yovanovich and Burde[41]	Triangular, semicircular, L-shaped	Correlations

Thermal Resistance of Multiple Contact Spots in Vacuum

If R_i is the resistance associated with a single contact spot and the fluid surrounding it, the overall resistance of the joint of the heat flow is given by

$$\frac{1}{R_T} = \sum_{i=1}^{n} \frac{1}{R_i} \qquad (4)$$

where n is the number of contact spots in the joints.

In terms of conductances, one could write

$$h_T = \sum_{i=1}^{n} h_i \qquad (5)$$

Since each side of a constriction has a resistance equal to $f_i/(4ak)$, the resistance of a contact spot in vacuum is $f_i/(2ak)$. If the contact is between two different solids, then k would be the harmonic mean of the two conductivities. Thus, for joints in vacuum,

$$R = \sum_{i=1}^{n} \frac{f_i}{(2a_i k)} \qquad (6)$$

and

$$h = \sum_{i=1}^{n} \frac{(2a_i k)}{f_i} \qquad (7)$$

If $\bar{\alpha}$ is the average radius of the contact spot and the variation in the constriction alleviation factor f is neglected, then

$$R = f/2n\bar{a}k \qquad (8)$$

$$h = 2n\bar{a}k/f \qquad (9)$$

Thus the problem reduces to one of determining n and a. Now n and a both depend on δ, the separation between planes that defines the mean height of surface profiles (Fig. 1). Also, n depends on the profile height distribution, while $\bar{\alpha}$ may be expected to depend on the profile slope distribution. Thus, both deformation and surface analyses would be required in estimating the contact conductance.

First, it is necessary to know whether the deformation of the asperities is elastic or plastic. Greenwood[42] had shown that, unless the surfaces were carefully polished, the deformation would be plastic even at the lightest loads. His analysis was based on asperities of spherical shape.

Mikic[43] suggested the use of an index

$$\gamma = H/(E' |\tan \theta|) \qquad (10)$$

where H is the microhardness of the softer material, E' the effective elastic modulus, and $\tan \theta$ the mean absolute slope of the surface profiles. According to Mikic, the deformation is predominately plastic for $\gamma < 0.33$ (for most surfaces $\gamma < 0.1$).

Based on statistical geometry theory, Bush and Gibson[44] defined a new plasticity index that depends on m_0, m_2, and m_4 (the first three moments of the power spectral density of the surface profile), as well as on E and H. The asperities are expected to deform plastically if $\psi > 2$. It may be noted that m_0 and m_2 are, in fact, related to the standard deviation of profile height (i.e., the surface roughness, σ) and $\tan \theta$, respectively.

It is clear from the preceding discussion that the statistical nature of surface profiles controls not only the number and size of contact spots, but the mode of deformation as well. The statistical nature of contact conductance was demonstrated experimentally by Veziroglu,[45] who found variations of up to $\pm 60\%$ in the measured value of conductance of 15 pairs of apparently similarly finished surfaces.

Nominally Flat Rough Surfaces

If the average contact spot radius is assumed to be a constant, then the number of contact spots and, therefore, the conductance, are directly proportional to the actual contact area, A_{act}. Popov[46] empirically determined that

$$A_{act} = A_{nom}(PB/E)^{0.8} \qquad (11)$$

where A_{nom} is the nominal contact area, P the contact pressure, and B a function depending upon the sum of the average heights of asperities of the two surfaces in contact. Thus, according to Popov, the conductance is proportional to $P^{0.8}$. No mathematical expression for B was given in Popov's paper.

Novikov[47,48] derived the relationship between the thermal resistance and the load by performing a statistical analysis assuming spherical asperities whose heights obeyed a Gaussian distribution. He used Cetinkale and Fishenden's[23] expression for constriction resistance. For elastic deformations at low loads, the resistance was found to vary exponentially with the separation (and, therefore, the load) between the surfaces; at larger loads, the variation was found to be linear. For elastic-plastic contacts, the ratio of the number of plastically deformed contact spots to their total number was taken to be $\exp(-b^*/\sigma)$, where b^* is the critical deformation at which the projections begin to deform plastically.

In the statistical analysis of Tsukizoe and Hisakado,[49,50] it was assumed that the asperities were conical in shape, with normally distributed heights. The thermal resistance could then be expressed as a function of the dimensionless separation between the surfaces. Separation is a function of the applied load and, thus, a relationship between the thermal resistance and the load is established. The deformations were assumed to be plastic under a constant flow pressure. A noteworthy feature of this work is that a tentative correlation was proposed between the profile slope and the maximum height of asperities which took into account different types of surface finish.

Mikic's[43] analysis considered the Gaussian distribution of profile heights and slopes and the constriction alleviation factor given by Cooper et al.[51] Whether the deformation was plastic or elastic, Mikic found that the conductance was proportional to $P^{0.94}$. He also developed a theory for the increase in conductance, including the elastic deformation of the substrate.

Sayles and Thomas[52] considered elastic deformation of rough Gaussian surfaces and found that the number of contacts was proportional to load, while the mean contact spot size was almost independent of load. This resulted in the conductance being nearly proportional to the load and the mean slope, confirming Mikic's results. Sayles and Thomas also discussed how their method could be extended to anisotropic surfaces. No constriction alleviation factor was used in their analysis.

As mentioned earlier in this section, Bush and Gibson[44] considered statistical geometry theory to evaluate the conductance. The constriction alleviation factor used was that of Gibson.[28] Like Mikic, they found that whether the contacts were plastic or elastic, the thermal contact conductance was proportional to $P^{0.94}$. In view of the asperity interactions, the theory is not applicable for large loads.

The preceding discussion of nominally flat rough surfaces in a vacuum reveals that

1) whether the contact is elastic or plastic, the thermal conductance varies nearly linearly with the load; and 2) the thermal conductance is proportional to mean absolute slope of surface profiles. The first of these conclusions implies that the average contact spot size remains substantially constant with the load. If the number of contact spots remained constant however, the conductance would be proportional to the (load)$^{0.5}$.

Effects of Large-Scale Surface Irregularities

Most practical surfaces contain, in addition to roughness, large-scale errors of form such as waviness and deviation from flatness. These departures from an ideal flat surface will also affect the thermal resistance.

If two rough spherical surfaces (or the crests of two wavy and rough surfaces) are in contact, the configuration can be considered to be as shown in Fig. 3. The contact geometry would be similar if one of the surfaces were flat. In these cases, the resistance is obtained by adding the constriction resistance of the "contour" area of radius b to the constriction resistance of the multiple contact spots distributed within the contour area.[53]

In most of the work carried out before 1970, the size and distribution of the contact spots were assumed to be known. Actually, these have to be determined as functions of the mechanical loading, surface characteristics, and mechanical properties of the contacting solids. It is usually assumed that the roughness heights deform plastically while the large-scale irregularities deform elastically. The work of McMillian and Mikic[54] was based on the above considerations. One of the unexpected conclusions from their work was that the conductance of wavy surfaces can be increased by making the surfaces rough, thereby increasing the contour area.

Hsieh et al.[55] approximated the contact of two rough wavy surfaces by a model consisting of a smooth spherical surface in contact with a flat rough surface. The roughness asperities were assumed to be conical in shape and normally distributed. The results showed that conductance is strongly dependent on surface roughness, and the resistance of the macroscopic constriction, due to waviness, can be neglected. This conclusion is contrary to that of Clausing and Chao,[56] whose experimental work suggested that the microscopic constriction is of secondary importance for many surfaces. Yovanovich[57] considered the contour area to be a very long spiral when a soft turned surface is in contact with a harder, smooth flat surface. At relatively high loads ($P/H < 0.05$), the contact would be continuous along the spiral whose resistance alone would be the thermal resistance. At moderate and low loads ($P/H < 0.03$), however, a contribution due to roughness must be added.

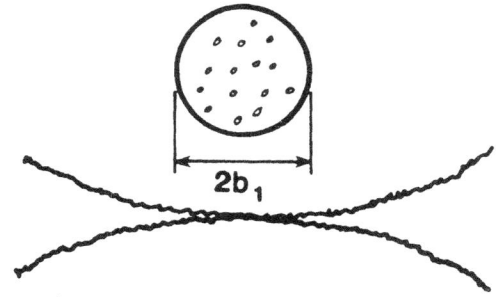

Fig. 3 Contact of two spherically convex surfaces.

All of the work discussed thus far considered the total resistance to be the sum of the resistance due to roughness and the resistance due to waviness or flatness deviation. A different approach was adopted by Popov and Yanin,[58] who considered the total resistance to be given by

$$R = R_r(A_n/A_c) \qquad (12)$$

in which R_r is the resistance due to roughness, A_n the nominal area, and A_c the contour area. The contour areas were determined assuming elastic deformation of the waves. For spherical waves, the contour area was found to be proportional to $P^{2/3}$, whereas for cylindrical waves, the contour area varied as $P^{1/2}$. The experiments clearly demonstrated that the resistance increased when waviness or flatness deviation was present. It was also noted that the resistance for a spherical surface was larger than the resistance for a corresponding cylindrical surface. However, it must be pointed out that all of the tests were conducted in air and, therefore, the resistance and its variation would be affected by the effective gap thickness.

A vertical section through a surface can be considered to contain a continuous spectrum of wavelengths extending down to atomic dimensions. Therefore, Thomas and Sayles[59] argued that waviness and roughness should be discussed in terms of the bandwidths of the spectrum rather than of fixed wavelengths. They found that the dimensionless contour radius b_1/b was given by

$$b_1/b = 0.44\xi^{1/3} \qquad (13)$$

where $\xi = WE'/\sigma_1 d$, the "waviness number"; W is the total load on the surface; and σ_1 the "total" rms roughness.

It was found that the effect of waviness could be neglected if $\xi > 1$. Since in all practical situations $\xi < 1$, the authors concluded that the waviness effect should never be neglected.

Dundurs and Panek[60] solved the elasticity and heat conduction equations simultaneously so that the contact area could be expressed as a function of both the applied pressure and the heat flux. They considered a two-dimensional wavy surface, completely ignoring the roughness and the statistical nature of the waviness. They found that, due to heat transmission, perfectly flat surfaces might become wavy.

The experiments of Edmonds et al.[61] also confirmed that, at low pressures, the effect of waviness predominates; this is indicated by a lower value ($\sim 2/3$) for the load exponent. At higher loads, the waviness undulations would have been mostly flattened and microconstriction due to roughness would become more important.

Correlations for Thermal Contact Conductance in a Vacuum

Several correlations between the conductance and the parameters affecting it have been developed over the past 20 years. However, in keeping with the constraints of the present paper, attention will be focused on the more recent ones.

A least-squares analysis of some 92 experimental data points led Mal'kov[62] to the following correlation:

$$h\bar{a}/k = 0.118 \, (PK_1/3S_u)^{0.66} \tag{14}$$

where S_u is the ultimate strength of the softer material and K_1 a constant depending on the average of the surface roughness heights. Apart from the use of $3S_u$ for the hardness, the major difference between this and the correlations proposed earlier by different authors is the rather low value (0.66) of the exponent.

Fletcher and Gyorog,[63] in their correlation for similar metals in contact, considered the following additional factors: 1) the mean junction temperature, 2) a gap dimension parameter accounting for the roughness as well as the flatness deviation, and 3) the variation of the above parameter with contact presure. They also made use of the Young's modulus E, rather than the microhardness, to nondimensionalize the pressure. The correlation fitted some 400 data points, representing the work of seven investigators, with an error of less than 24%.

The dimensional analysis of Thomas and Probert[64] considered that the nominal contact area did not play an important role in the variation of the conductance and therefore the total load, rather than the interface pressure, was taken to be one of the variables. Their analysis yielded two dimensionless groups, $C^* = C/(\sigma\kappa)$ and $W^* = W/\sigma^2 H$, where C is the total conductance of the contact. The data for stainless-steel/stainless-steel and aluminum/aluminum surfaces fell into two distinct groups, indicating that the dimensional analysis was incomplete.

The data for aluminum in the preceding work were combined with other data by O'Callaghan and Probert,[65] resulting in a total of 344 points, which followed the correlation

$$C^* = 3.73 \, (W^*)^{0.66} \tag{15}$$

However, when the stainless-steel data were also added, the scatter increased.

The experimental results of Al-Astrabadi et al.[66] were also found to belong to the correlation for stainless-steel surfaces proposed by Thomas and Probert.[64] The experiments of Edmonds et al.[67] on the contact of optically flat copper surfaces with stainless-steel surfaces of various degrees of surface finish, on the whole, obeyed the relationship

$$C^* = 0.26 \, (W^*)^{0.96} \tag{16}$$

However, when the data were grouped into two regions according to their roughnesses, the load term exponents were 0.60 and 0.61 for high and low roughness regions, respectively, indicating that both roughness and waviness contributed to the resistance.

Popov[68] proposed a correlation similar to that of Mal'kov,[62] except that the analysis was restricted to nominally flat rough surfaces. The pressure term exponent was found to be 0.956—the higher value indicating the absence of waviness.

Theoretical correlations assuming Gaussian distribution of asperity weights and elastic deformation of asperities were proposed by Blahey et al.[69] Their analysis showed that the pressure term exponent had a range of 0.93-0.95, depending on the asperity tip radius and surface roughness.

Effect of Surface Films

The existence of a surface film may be either intentional, as in electroplated surfaces, or unavoidable, as in oxidized surfaces. A comprehensive review of the literature, covering the work until the late 1960's on the effect of oxide films on thermal resistance, has been presented by Gale.[70] A generally accepted conclusion is that oxide films, unless sufficiently thick, do not appreciably increase the resistance although they may have a major effect in electrical contact resistance. Assuming one-dimensional flow through the film and no interaction between the solid spot and film resistances, Gale obtained the following approximate expression for the resistance of a single spot:

$$R = (\tfrac{1}{4}ak) + (t/\pi a^2 k_f) \tag{17}$$

where t is the thickness of the film and k_f its conductivity. Based on this equation, a constriction magnification factor C_m was defined as

$$C_m = 1 + (4tk/\pi a k_f) \tag{18}$$

which shows that the increase in resistance is proportional to the ratio of the thickness of film to contact spot radius and the ratio of the conductivity of the parent metal to that of the oxide film. The oxide film contributes to the total resistance for contacts having radii less than 10.

The experiments of Tsao and Heimburg[71] on aluminum 7075-T6 surfaces in dry air showed expected trends, namely, that the time of exposure increased the resistance, while degassing of the surfaces decreased resistance. However, exactly opposite trends were noted for the specimens aged in laboratory (humid) air. This anomalous behavior was thought to be due to the decrease of fracture stress of the aluminum oxide films in the presence of absorbed gases, especially moisture.

The analytic solution of Mikic and Carnascialli[72] is based on the premise that, for a fixed geometry, any increase in the thermal conductivity in the vicinity of contact points should reduce the value of the resistance. Their results showed that the increase in conductance due to plating is directly dependent on t/a and k_p/k, where k_p is the conductivity of the plating material. For wavy surfaces in contact, the contour area radius is so large that the plating must be very thick to achieve any significant reduction in resistance.

Based on experimental measurements, Tsukizoe and Hisakado[49,50] estimated that, for copper surfaces covered with oxide films, the ratio of the thermally conducting area to the apparent area of contact was larger than the corresponding ratio for the electrically conducting area. This again confirms that electrically insulating surface films may be thermally conducting since they permit a flow of phonons even though the motion of electrons is inhibited. It was also suggested that, for smooth surfaces, the slope of asperities is small and, therefore, the oxide films are less likely to break down. Thus, if the oxide films are present, the smoother surfaces may have higher resistance.

The theoretical work of Kharitonov et al.[73] considered the effects of both oxide films and coatings of higher conductivity metals, and led to the following conclusions:

1) For flat rough surfaces, the coating of a few-tens-of-microns thick will noticeably reduce the thermal contact resistance.

2) For wavy surfaces the contour radius might have a value of a fraction of 1 mm, and the coating thickness mentioned in point 1 would be ineffective.

3) Since the oxide layers usually have thicknesses of 1 μm and their conductivity is smaller than that of metals by factors of 3-30, the resistance depends only weakly on the oxide layer.

It may be added that the last conclusion may not be quite valid in view of the findings of Gale[70] previously mentioned.

Yip,[74] considering Gaussian surfaces and assuming uniform thickness of oxide films, demonstrated that the oxide films can cause a drastic increase in the resistance. His experiments on three pairs of aluminum alloy (6061-T6) surfaces clearly demonstrated that the effect of oxide film is more pronounced for smoother surfaces.

169

The results of previous investigators were confirmed in the experiments of Mian et al.[75] on steel-steel (EN3B) surfaces in a vacuum. They also found that the initial thickness of the oxide caused a considerable rise in the resistance, but additional thicknesses of oxide led to smaller increments of resistance.

Effect of Interstitial Materials

Whereas the solid spot constriction resistance is in series with the surface film, it is in parallel with the resistance of the interstitial material. In the following discussion, the interstitial media is broadly classified as gaseous and non-gaseous.

Heat Flow Through the Interstitial Gaseous Medium

Since the heat transfer across the gas filling the voids between the contacting surfaces is principally by conduction, then

$$h_g = k_g / \delta \qquad (19)$$

where h_g is the heat-transfer coefficient for the gas gap, k_g the thermal conductivity of the gas, and δ the effective thickness of the gas gap.

For normal engineering surfaces in contact, the effective gap thickness would be similar in magnitude to the mean free path of the gas molecules. Under these conditions the "temperature-jump" effect becomes important (see Fig. 4), so that Eq. (20) is modified to

$$h_g = k_g / (\delta + g_1 + g_2) \qquad (20)$$

Thus, the problem of determining the gap thermal conductance for a given gas (or gas mixture) reduces to one of determining the effective gap width and the temperature-jump distances.

Some investigators (for example, Cohen et al.[76]) had observed that the conductance between the fuel and the jacket in a nuclear reactor was independent of the gas composition. Kharitonov et al.[77] offered an explanation based on the fact that the accommodation coefficient α depends on the molecular mass of the gas. Thus, for helium with a low molecular mass, the accommodation coefficient will also be small. In such a case, the temperature-jump distance is approximately given by

$$g = \lambda / \alpha \qquad (21)$$

where λ is the mean free path of the gas molecules.

Now, if λ is large compared to δ, Eq. (21) may be written

$$h_g \simeq k_g / g \simeq k_g \alpha / \lambda \qquad (22)$$

Thus, although the conductivity of helium is very high, its accommodation coefficient is low, and, by Eq. (22), the gas conductance is only weakly dependent on the nature of the gas. For example, for pure surfaces of heavy metals, the gas conductance of xenon can be shown to be greater than that of helium although the conductivity of helium is 30 times that of xenon.

The experimental work of Madhusudana[78] indicated that the gas conductance increased with contact pressure due to the reduction in gap thickness with load. It was also found that, at any given contact pressure, the reduction in fluid conduction contribution was noticeable only at absolute pressures below 100 Torr (13 kPa).

The following expression for the gas gap conductance was proposed by Popov and Krasnoborod'ko.[79]

$$h_g = k_g Y_1 / [(h_{max1} + h_{max2})(1 - \epsilon)] \qquad (23)$$

where h_{max1} and h_{max2} are the maximum heights of roughness for surfaces 1 and 2, respectively; ϵ is the approach of sur-

faces under load; and Y_1 is a function of $(1 - \epsilon)$, the maximum thickness of the gas layer and the temperature-jump distances.

The temperature-jump distance g_m for a mixture of gases was determined by Vickerman and Harris[80] to be

$$g_m = \Sigma (x_i g_i / m_i^{1/2}) / \Sigma (x_i / m_i^{1/2}) \qquad (24)$$

where g_i is the temperature-jump distance of constituent gas i, x_i the mass fraction of constituent gas i, and m_i the molecular mass of constituent gas i. Their results for He-N$_2$ and He-Ar mixtures showed fair agreement with the data available in the literature.

The theoretical expressions such as Eq. (20) predict first an increase and then a decrease of gas conductance with temperature. However, the experiments of Garnier and Begej[81] indicated a continual increase of gas conductance with temperature, especially when the gap widths were comparatively small. The authors considered this to be due to the presence of free molecular conduction. Yovanovich et al.[82,83] have proposed correlations for the estimation of gas gap conductance. The correlations are somewhat similar to that of Dutkiewicz.[84]

A comprehensive review summarizing the current state of knowledge on gas gap conductance has been presented by Madhusudana and Fletcher.[85]

Effect of Nongaseous Interstitial Materials

The interstices may be filled with materials with a view to either decreasing or increasing resistance, thus providing a means for thermal control. Furthermore, a joint with a filler material is less sensitive to loads and surface conditions and thus offers the advantage of predictability of heat-transfer behavior. The interstitial materials may be metal foils, wire screens, greases, powders, or insulating sheets, depending on the application.

The use of interstitial materials as a means of thermal control, especially in spacecraft systems, has been discussed by Fletcher and co-workers.[86-92] These works also contain reviews of previous experimental work on low- and high-conductance filler materials. To classify various interstitial material/base metal combinations, Fletcher[13] proposed the use of a nondimensional conductance η,

$$\eta = (h_c t)_f / (h_c \delta_0)_b \qquad (25)$$

where h_c is the contact conductance, t the thickness of filler material, δ_0 the equivalent gap thickness, f the junction with filler material, and b the base metallic junction.

It was observed that combinations for which $1 \leq \eta \leq 10$ (e.g., carbon fiber paper) offer excellent thermal isolation qualities and light weight. But, if strength is also required, then the use of medium mesh screen wire of low thermal conductivity material was recommended. Thermal control materials, for which $100 \leq \eta \leq 1000$, enhance contact heat transfer. Indium foil and filled silicon grease appeared to be the most suitable materials in this category, although grease may not provide a good environmental seal.

Feldman et al.[93] investigated the thermal conductance of selected thermal joint compounds to ascertain their thermal conductivity, weight loss, and resistance to hardening. Results indicated that silicon base thermal joint compounds had reasonable thermal conductivity and moderate weight loss over the range of test conditions.

A theoretical model for the prediction of the conductance of a screen wire contacting two solids was proposed by Cividino et al.[94] The model assumed elastic deformation of smooth clean wires and equal loading of all modes. Probably because of these and other simplifying assumptions, the theory consistently overestimated the conductance when compared with measured values.

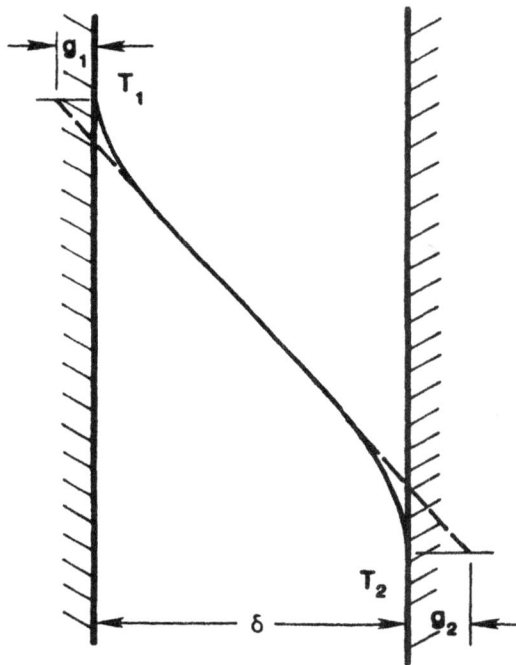

Fig. 4 Temperature-jump distance.

The experimental results of O'Callaghan et al.[95] indicated that the presence of a copper wire gauze between stainless-steel surfaces increased the conductance in a vacuum but decreased it in air. Further work by Al-Astrabadi et al.[96] showed that macroscopic constriction effects, due to either thermal distortion or badly mating surfaces, can be reduced or even eliminated by the insertion of such gauzes. They also noted that the method of weaving the screen wires results in the weft being a series of almost straight wires all in one plane with the warp interlaced. Therefore, contact occurs only between the warp and the solid surfaces, i.e., only at every other wire crossing. Another reason why Cividino et al.[94] overestimated the conductance is that they assumed contact to occur at every crossing instead of every other crossing.

Sauer et al.[97] found that with screens the conductance increased with increasing mesh size and the corresponding increase in the number of contacting regions. In another experimental work, Sauer et al.[98] found that films of lithium, graphite, silicon, and molykote lubricants improved the conductance of stainless-steel joints; such improvements being more noticeable (eight- to seventyfold) in a vacuum than in air (zero- to sixfold).

Effect of Heat Flow Direction

Some investigators[99,100] observed that the conductance is sensitive to direction of heat flow, especially if the joint is made up of dissimilar materials. This phenomenon, which may be called thermal rectification, could have applications in thermal control of systems; that is, suitable material combinations could be used as thermal switches.

Since the solid-state theory of Moon and Keeler[101] cannot explain the rectification behavior observed in joints of similar materials, it is now generally accepted that thermal rectification is caused by the distortion of the contact surface due to local temperature gradients.

The experimental studies of Williams[102] indicated, among other things, that the contact elements do not have to be dissimilar to exhibit rectification, and that the rectification effect decreases rapidly as the number of reversals increases. His experiments on a pair of Nilo 36 (an alloy of low thermal expansion) specimens revealed no effect on direction, confirming that the thermal distortion of contacting surfaces is necessary for rectification.

Veziroglu and Chandra[103] noted that the temperature gradients in the two specimens, causing differential radial expansions, result in the change of curvatures of the contacting surfaces. This theory, however, cannot explain the direction effect observed for similar materials.

A different approach to the problem was proposed by Barber,[104] who found the steady-state temperature distribution in two semi-infinite solids with spherical surfaces in contact, assuming no heat flow across the interface except through the circular contact area. This temperature distribution causes thermal strains through which one can find the load that must be applied to the solids to establish a contact area of the assumed size. Thus, the contact resistance, which depends on the contact area, was expressed as a function of the load as well as the heat flux (and its direction). Elastic deformation was assumed. Again, Barber's theory cannot explain the rectification behavior observed with similar materials.

Thomas and Probert,[105] in their experiments on stainless-steel/stainless-steel contacts, observed a large direction effect when one of the surfaces was bead-blasted and the other lapped; the conductance was higher when the heat flowed from the rougher to the smoother surface. When both surfaces were rough (bead-blasted), however, the rectification effect was not as significant. According to the models of Veziroglu and Chandra[103] or Clausing,[106] the conductance for heat flow in the SS→Al direction must be greater. Thomas and Probert concluded that the geometrical theories could not explain their results. Also, contrary to William's observations,[102] they noted that the directional effect increased with the number of reversals.

Using the principle of rectification, O'Callaghan et al.[107] constructed a thermal rectifier consisting of a multilayer stack of thin disks, each disk having one surface roughened, the other surface smoother, and so arranged that all of the rough surfaces were pointing in the same direction. For the stack consisting of brass disks, the conductance was higher when heat flowed from a rough to a smooth surface, while the opposite was true for stainless-steel stacks.

Further experiments of O'Callaghan and Probert[108] indicated that all contacts between dissimilar materials or dissimilar surfaces of the same material exhibited rectification with the exception of the contact between Fluorosint (reinforced PTFE) and Invar. Repeated thermal cycling reduced the direction effect to zero, endorsing the observation of Williams.[102]

Jones et al.[109,110] observed that, in the absence of radial heat losses, a constant axial temperature gradient causes the flat sections to bow with a radius of curvature, with the hotter end of each specimen becoming convex and the colder end concave. Thus if heat flows from material 1 to 2, where $\alpha_1/k_1 > \alpha_2/k_2$, the contact would be peripheral, whereas for flow direction 2→1, the contact would be central (Fig. 5). The experiments on stainless-steel/aluminum contacts in a vacuum confirmed the theory. It may be noted that the constriction of heat flow toward a central contact is greater than the constriction when the heat flow is peripheral. Therefore, the conductance for the SS→Al direction must be higher. This is in contrast to the results observed by previous investigators of flat contacts.[99,100,108,111]

The experiments of Al-Astrabadi et al.[112] indicated that the conductance values were always higher for the heat flow in the SS→Cu direction irrespective of whether one surface was initially flat and the other surface convex. Thus, their results for flat contacts again disagree with those of previous workers. A possible reason is that, the surfaces being optically flat, the microscopic resistance would be small compared to the macroscopic resistance. Also in Refs. 109, 110, and 112, the order in which the joints were tested is not specified. This factor could be important in the rectification behavior of dissimilar materials.[102] Their results for convex/flat surfaces, however, follow expected trends, since the resistance

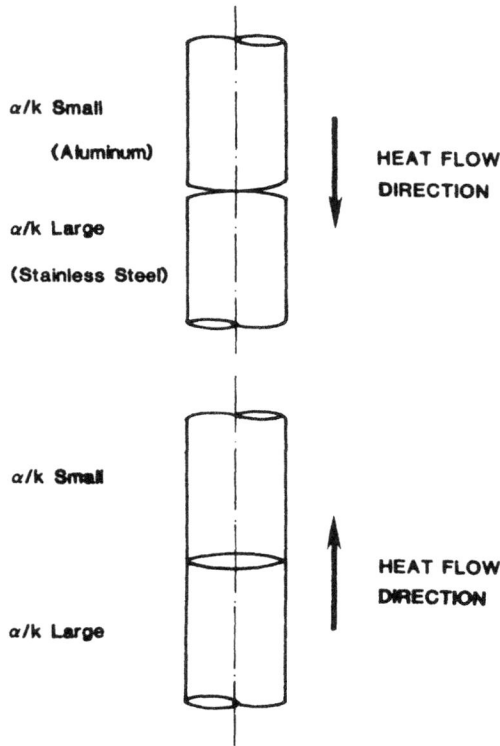

Fig. 5 Distortion of initially flat surfaces due to heat flow.[109,110]

in this case is obviously due to macroscopic constriction. Al-Astrabadi et al.[112] also observed that for stainless-steel (convex)/copper (optically flat) surfaces in contact, the conductance increased by a factor of 68 as the heat flux in the SS→Cu direction increased from 0 to 30 kW/m². This was due to a decrease in the disk-type constriction. For heat flow in the Cu→SS direction, the opposite was true, although the decrease in the conductance was only slight.

The theoretical study of Dundurs and Panek,[60] discussed earlier, also indicated that the conductance is higher for heat flowing from the material with the higher "distortivity." Thus, for example, the conductance for the SS→Cu direction would be higher than for the opposite direction. This is in agreement with the results of previous investigators of macroscopic contacts, except those of Thomas and Probert.[105]

The surfaces used by Somers et al.[113] in their experiments on conductance of dissimilar materials had substantial flatness deviations. Their results, therefore, confirmed the macroscopic theory, with the conductance for heat flow from the material of higher α/k being the greater in all cases except zircaloy/aluminum, for which no directional effect could be found.

A novel thermal rectifier making use of a stack of stainless-steel disks and an aluminum bar expansion element has been described by Jones et al.[114] In this device, the differences in resistance due to heat flow direction were on the order of 35-50%.

The preceding discussion shows that the following conclusions may be made regarding the thermal rectification effect, although some anomalies still remain.

1) For surfaces that are initially convex, the conductance is higher when the heat flows from the material with the higher α/k ratio. This conclusion is also generally true for flat-smooth surfaces.

2) For flat-rough surfaces, the conductance is higher when the heat flows from the material with the lower α/k ratio.

3) Similar materials with dissimilar surfaces may exhibit the rectification effect.

4) At present, no satisfactory theory exists to explain the rectification behavior observed for similar materials in contact.

5) The rectification effect decreases as the number of reversals increases; however, results opposite to this conclusion have been observed.

Some Special Topics in Contact Heat Transfer
Stacks of Laminations

Heat transfer takes place through stacks of thin laminations in electrical machinery such as transformers and generators. Such stacks have also been used as strong insulating structures for cryogenic tanks and in thermal isolation systems for spacecraft.[115]

The theoretical work of Williams[116] indicated that the effective thermal conductivity of a typical turbogenerator core when immersed in hydrogen would be about three times the value in normal atmospheric air. Williams' experiments[117] on stacks of steel laminations in different atmospheres showed that, for tests in air, the solid spot conductance was predominant. As the conductivity of the interface fluid was increased, however, the fluid conductance contribution became progressively more significant, and with helium as the interstitial fluid, the proportion of the solid spot conductance was quite negligible.

O'Callaghan et al.,[4] in their experiments on grain-oriented silicon-steel and nonoriented electric-steel laminations, found that the resistance per lamination-to-lamination was independent of the number of laminations. The resistance initially decreased with load, presumably due to the pressing out of any slight buckle present in the laminations, but it reached an asymptotic value at a contact pressure of about 1 kPa. At pressures up to this value, there was no electrical breakdown of the insulating surface films. Therefore, to reduce the equilibrium temperatures and, hence, increase the thermal efficiency, the authors recommended that the transformer lamination cores be prestressed to 1 kPa.

The theoretical work of Veziroglu et al.[35] considered the heat flow channel to be finite, a necessary requirement when the contacts are between thin sheets. The predicted conductances were consistently higher than the experimentally measured conductances of Williams.[117] A possible reason for the discrepancy is that the theoretical analysis considered bare steel sheets, whereas the experiments were performed on enameled sheets. The experimental results of Veziroglu et al.[35] on stainless-steel stacks in air, however, showed good agreement with their theory. A definite hysteresis behavior was observed for this stack of laminations, with the conductance values during unloading larger than the corresponding values during loading.

The experimental work of Sheffield et al.[118] on stacks of electrically insulated sheets in a vacuum resulted in a simple correlation

$$k^* = 1.24\sqrt{P^*} \qquad (26)$$

where $k^* = k_{eff}/k_s$; k_{eff} is the effective thermal conductivity of the stack and $P^* = P/H$.

Contrary to the observation of O'Callaghan et al.,[4] the conductance was found to increase continuously throughout the loading (10-45 MPa).

Al-Astrabadi et al.[119] proposed the following correlation for the conductance of stacks of thin layers in vacuum:

$$h^* = 3.025(P^*)^{0.58} \qquad (27)$$

where $h^* = h_{11}t/k_s$; h_{11} is the thermal conductance from layer to layer; and t the thickness of an individual layer.

The correlation was based on the results of 18 experiments obtained in 6 separate investigations and was deemed to be successful since the scatter band was relatively narrow.

172

Hysteresis

The hyteresis effect refers to the experimental observation that, when a joint is subjected to cyclic loading, the conductance values during unloading are usually higher than the corresponding values during the first loading. The work done prior to the period under review indicated the following as possible causes for hysteresis.

1) During first loading, the actual contact is formed by the elastoplastic deformation of asperities in contact, coupled with the elastic deformation of the underlying material. During unloading, the reduction of the contact area is due only to the elastic recovery of both the asperities and the bulk sublayers. Thus the formation of the contact area during loading is greater than the recovery during unloading for a given change in the load. This would result in the actual contact area and, therefore, the conductance being higher during unloading.

2) Cold welding occurs between perfectly clean surfaces immediately on the establishment of the metal-to-metal contact. This effect would be more significant for soft metals such as copper, rather than for harder materials such as tungsten. Since the spots where cold welding has taken place will remain in contact even when the load is removed, the contact area and therefore the conductance can be expected to be higher than the corresponding values during loading.

Since the hardness of metals decreases with the temperature and duration of stress (see, for example, Ref. 120), the conductance can also be expected to increase with the duration of the loading. Thus, the conductance values during subsequent unloading and reloading must be higher than those during first loading. This effort of contact duration, however, would be significant only at temperatures considerably higher than ambient temperatures.

McKenzie[121] observed that when the contact pressure was changed from 0.8 to 1.07 MPa, it took more than 50 h before equilibrium was restored. Therefore, he suggested that the hysteresis effects might have been caused by the unintentional use of nonsteady-state data.

In a theoretical study, Mikic[122] considered a Gaussian distribution of asperity heights and assumed that during first loading the asperities deformed plastically, while for a reduction and a subsequent increase (up to the maximum load achieved in the first loading) the asperities deformed elastically. This would result in the actual contact areas during unloading (and subsequent reloading) being greater than those during first loading. The analysis showed that the number of contact spots was also substantially higher in descending loading. As a result of both of these effects, the conductance would also be higher during unloading. It should be noted that Mikic's analysis applies to nominally flat, rough surfaces in a vacuum and does not consider the deformation of the bulk material.

In the experimental work of Madhusudana and Williams,[123] eight different pairs of surface combinations were tested. Hysteresis was observed in all cases, whether the materials were similar or dissimilar and whether the tests were conducted in a vacuum or in air. For specially prepared surface combinations, such as pyramids contacting flats, the hysteresis effect was more striking than for contacts between randomly rough flat surfaces. This was explained by the fact that, in the former case, all of the contact spots are nearly coplanar and break simultaneously on complete removal of loads; with the second type of surfaces, however, the contact spots will not be in the same place, and the area reduction on decreasing load takes place more gradually. It was also pointed out that the observed hysteresis effect was most probably not due to cold welding since the contacting surfaces were most likely contaminated.

Bolted Joints

Most contact heat-transfer studies assume that the contact pressure is uniform over the interface. However, the pressure distribution is nonuniform over the contact area formed between metal surfaces joined together by means of bolts, screws, or rivets. The usual contact heat-transfer theories, therefore, require some modification when applied to bolted joints.

Whitehurst and Durbin[124] considered the analytical solution for a bolted joint to be very complicated and, therefore, suggested an experimental method called the "effective fin method." Basically, this method would involve the measurement of temperature gradients on both sides of the joint, and an effective thermal conductivity of the joint based on the length of the overlap and the temperature difference between the two ends of the overlap. This method would be somewhat impractical when a large number of joints is involved, since this would mean that many models would need to be constructed and tested unless all of the joints were standardized.

In the theoretical study of Veilleux and Mark,[125] the physical contact between the metal sheets was assumed to occur only under the head of the bolts fastening the sheets together. The good agreement between theory and experiment confirmed this assumption.

Bradley et al.[126] used a stress-freezing technique to determined the pressure distribution over the interface of a bolted joint. It was found that the interface pressure distribution depended upon the radius of the bolt, the bolt-head dimension, and the plate thickness.

The finite element analysis of Gould and Mikic[127,128] also considered bolted joints of smooth surfaces. The area of contact was measured by both optical and autoradiograph methods. There was good agreement between theory and experiment. The results indicated, in conformity with those of Bradley et al.,[126] that the interface pressure reduced from a maximum value at the bolt axis to zero at a distance of about 2 bolt diameters from the axis.

Roca and Mikic[129] also took surface roughness into consideration in their analytical study of thermal conductance of bolted joints. One significant conclusion of this work was that, for a given pair of plates and loading conditions, the radius of contact increased with the surface roughness. This means that the macroscopic constriction decreases as the surface roughness increases, while the opposite is true of the microscopic constriction.

A general expression capable of handling nonuniform pressure distributions for the thermal resistance in vacuum was derived by Yip.[130] His calculations indicated that nonuniform stress distributions did not appreciably affect the microcontact resistance. It should be noted, however, that the extent of the contact zone and, therefore, the macroscopic resistance, will still depend upon the stress distribution.

The conductance of two aluminum plates bolted together was investigated by Oehler et al.[131] The conductance values in air were found to be about twice the values obtained in vacuum, but the joints loaded with grease in a vacuum were only about 10% more conducting than the dry joints in a vacuum. No information was given about the surface texture of the plates. It was also found that the contact conductances based on steady-state tests were about 50% higher than those based on transient tests. Since it was stated that transient measurements were subject to greater degrees of error, the suitability of the transient technique becomes questionable.

Cylindrical Joints

When the heat flow is radial across a joint formed by two concentric cylinders, the contact pressure is not explicitly known but depends on the

1) initial interference or clearance between the cylinders,
2) differential expansion of the cylinders due to the temperature gradient caused by the heat flow,

3) thermal and mechanical properties of the cylinder materials, and

4) differential expansion due to the temperature drop ΔT at the interface.

Therefore, for a given pair of cylinders, the contact pressure depends upon the heat flux. It is also to be noted that ΔT is itself, by definition, dependent upon the contact conductance and, therefore, the contact pressure. In other words, the contact pressure and the temperature drop are interdependent.

In the work prior to the period under review, the resistance of cylindrical joints was usually estimated indirectly. In such a procedure, the overall resistance to heat flow from the fluid heating the inner of the two concentric cylinders to the coolant surrounding the outer cylinder is measured. The individual component resistances, except the contact resistance, are then either measured or estimated separately and subtracted from the overall resistance to obtain the contact resistance. One serious drawback of such a method is that large errors are likely to occur when the contact resistance is relatively small and is determined as the difference between two comparatively large resistances.

In the direct method, the temperatures are measured at various radial locations in the cylinders, and the temperature distributions are then extrapolated to obtain the temperature drop at the interface directly. Thus, this method is similar to the usual method for flat surfaces, except that the heat flux, rather than the implicit contact pressure, would be the independent variable.

Williams and Madhusudana[132] described a simple theoretical analysis of heat flow across cylindrical joints of similar materials in a vacuum. The theory showed that the contact pressure and, therefore, the solid spot conductance increased directly with heat flux. For stainless-steel specimens with initial clearance, the experimental results obtained by the direct method confirmed the theoretical findings. The theory also showed that, for specimens with initial interference fit, the contact pressures are so high that the contact resistance becomes insignificant. This was verified by the experimental observation that no measurable interface temperature drop existed for stainless-steel specimens with an initial interference fit. However, their experimental results for specimens with initial zero clearance showed no definite trends and were therefore inconclusive.

Direct measurements of resistance in a compound cylinder have also been reported by Hsu and Tam.[133] In this case, the cylinder was of aluminum alloy on which a stainless-steel tube was shrunk. The heat flow was radially outward, and, since aluminum has a higher coefficient of thermal expansion, it is to be expected that the resistance would reduce with heat flux.[134] This fact was confirmed by experiments conducted in air.

More recently, Madhusudana[135] extended the theory presented in Ref. 132 to include dissimilar materials, also taking into account the more refined general contact heat-transfer theories that had become available since the first theory was published. The following additional conclusions were drawn:

1) For a given heat flux, the material combination in which the inner cylinder has the higher α/E ratio would yield higher values of the contact conductance if the heat flow is radially outward; the opposite would be true for inward flow.

2) For cylindrical joints, the contact pressure increases with the surface roughness. Therefore, the contact conductance would be expected to increase with the surface roughness. However, since the constriction resistance also increases with roughness, the relation between conductance and roughness in the case of cylindrical contacts is more complex than for flat contacts.

Nuclear Reactor Fuel Elements

A considerable amount of analytical and experimental work was carried out in the 1950's and 1960's on the conductance and strength of nuclear elements.

Based on the Mikic model,[51] Jacobs and Todreas[136] proposed a correlation for the solid spot conductance in reactor fuel elements. This correlation was tested against the data of Dean,[137] Ross and Stoute,[138] and Rapier et al.[139] Dean's tests were actually conducted in an argon atmosphere, and Jacobs and Todreas deduced the solid spot conductance values assuming that the fluid conductance contribution was 10% of the total. Since the information regarding $|\tan \theta|$ was not available in most of the tests, the data for one run of each surface was first normalized before a comparison was made between the proposed correlation and the experimental results. Thus the agreement between the correlation and the data was forced to a certain extent.

A model (GEGAP-III) for the calculation of pellet-cladding conductance in boiling water reactor fuel rods was presented in Ref. 140. Separate expressions were given for the solid spot, gas, and radiant heat-transfer coefficients. Comparison with experimental values showed that this model slightly underestimated the conductances.

A review of methods applicable to the calculation of gap conductance in zircaloy-clad uranium dioxide fuel rods was reported by Lanning and Hann.[141] They recommended the use of the Jacobs-Todreas model (reduced by a factor of 4) for the prediction of solid spot conductance until more data became available. Their expression for the gas conductance made allowance for the reduction of gap width with the contact pressure. The work of Lanning and Hann was extended by Beyer et al.,[142] whose correlation for the solid spot conductance also took into account the waviness of the surfaces, but not the profile slope.

As mentioned previously, Garnier and Begej[81] measured the gap conductance between uranium dioxide and zircaloy-4 surfaces in nominal contact. The experiments with helium, argon, and a 50:50 mixture of helium and argon indicated a continual increase of gas conductance with temperature. This is in contrast to the trend predicted by, for example, GAPCON-Thermal-2.[142]

Experimental results for the conductance of zircaloy-2/uranium dioxide interfaces have also been reported by Cross and Fletcher[143] and Madhusudana.[144] Tests were conducted in a vacuum, as well as in atmospheres of argon and helium. The improvement in conductance due to the presence of helium ranged from about 800 to 60% as the contact pressure increased from 0.89 to 20.13 MPa. Over the same range, the improvement due to the presence of argon ranged from about 400 to 10%.

A correlation for the solid spot thermal conductance of zircaloy-2/uranium dioxide interfaces has been proposed by Madhusudana and Fletcher.[145] This correlation considers all available data on direct measurements for this joint combination and thus includes data in Refs. 136, 137, 143, 145, and 146. This correlation takes into account the variation in the mean junction temperature obtained in the tests by different investigators, but does not include the effect of slope since, as pointed out earlier in this section, this information is not reported in most of the literature reviewed. The following correlation fitted 78 data points with remarkably small scatter:

$$h\sigma/k = 12.29 \times 10^{-3} \ (P/H)^{0.66} \tag{28}$$

It was suggested that further work on this type of contacts should include: 1) experimental investigation of the effect of varying the mean junction temperature; 2) quantitative evaluation of surface parameters, other than roughness, on the solid spot conductance; and 3) direct measurement (in a vacuum) of the solid spot conductance over the pressure range $0.001 < P/H < 0.005$.

Recommendations for Future Work

The preceding review of the research on contact heat transfer over the past decade identifies the following as some areas where need exists for further research.

1) Resistance of noncircular constrictions located at the ends of semi-infinite cylinders or prisms.

2) Conductance of rough wavy surfaces and comparison of the performance of these surfaces with that of rough flat surfaces.

3) Accommodation coefficient for practical gas/solid interfaces.

4) Determination of a reliable effective gap width for the calculation of conduction through the interstitial gas.

5) A theory for the explanation of the rectification effect observed in similar materials in contact.

6) Effect of mean-junction temperature on the conductance of nuclear fuel elements. (Experimental data over a significant range of contact pressures are also lacking for nuclear fuel elements.)

7) Resistance of a circular annulus and the pressure distribution in a bolt joint necessary to determine the conductance of such joints.

8) Direct experimental measurement of the thermal conductance of cylindrical joints and the extension of the present theory for such joints to include the effect of the interstitial medium on the heat transfer.

References

[1]Skipper, R. G. S. and Wootton, K. J., "Thermal Resistance between Uranium and Can," *Proceedings of the 2nd International Conference on the Peaceful Uses of Atomic Energy,* Geneva, Paper P/87, Sept. 1958.

[2]Il'Chenki, O. T. and Kapinos, V. M., "Thermal Resistance of Tail Connections of Turbine Blades," *Energomashinostroyeniye,* Vol. 5, No. 6, 1959, pp. 23-26 (in Russian). (Translation by Foreign Technology Div., Wright-Patterson AFB, OH, 1965.)

[3]Gardner, K. A. and Carnavos, T. C., "Thermal Contact Resistance in a Finned Tubing," *ASME Journal of Heat Transfer,* Vol. 82, No. 2, 1960, pp. 279-293.

[4]O'Callaghan, P. W., Jones, A. M., and Probert, S. D., "Effect of Thermal Contact Resistance on the Performance of Transformer Lamination Stacks," *Applied Energy,* Vol. 3, 1977, pp. 13-22.

[5]Barzelay, M. E., Tong, K. N., and Holio, G., "Thermal Conductance of Contacts in Aircraft Joints," NASA TN 3167, March 1954.

[6]Kaspareck, W. E. and Dailey, R. M., "Measurements of Thermal Conductance between Dissimilar Metals in Vacuum," ASME Paper 64-HT-38, Aug. 1964.

[7]Huiskes, R., "Some Fundamental Aspects of Human Joint Replacement," *Acta Orthopaedica Scandinavia,* Munksgaard, Copenhagen, 1980, Supp. 185.

[8]Cull, J. P., "Thermal Contact Resistance in Transient Conductivity Measurements," *Journal of Physics E: Scientific Instruments,* Vol. 11, 1978, pp. 323-326.

[9]Mikesell, R. P. and Scott, R. B., "Heat Conduction Through Insulating Supports in Very Low Temperature Equipment," *National Bureau of Standards Journal of Research,* Vol. 57, No. 6, 1956, pp. 371-378.

[10]Smuda, P. A. and Gyorog, D. A., "Thermal Isolation with Low Conductance Interstitial Materials Under Compressive Loads," AIAA Paper 69-25, Jan. 1969.

[11]Thomsen, D. M. and Zavoico, A. B., "Conductive Heat Transfer Resistance of Compound Barrel Interface," U. S. Army Weapons Command, Research Development and Engineering Directorate, Weapons Lab., Rock Island, RE TR 73-36, 1971.

[12]Reiss, H., "An Evacuated Powder Insulation for a High Temperatures Na/S Battery," AIAA Paper 81-1107, 1981.

[13]Fletcher, L. S., "A Review of Thermal Control Materials for Metallic Junctions," *Journal of Spacecraft and Rockets,* Vol. 9, Dec. 1972, pp. 849-850.

[14]Wong, H. Y., "A Survey of Thermal Conductance of Metallic Contacts," ARC CP 973, Her Majesty's Stationery Office, London, 1968.

[15]Minges, M. L., "Thermal Contact Resistance, Volume 1—A Review of the Literature," AFML-TR-65-375, April 1966.

[16]Williams, A., "Heat Transfer Across Metallic Joints," *Mechanical and Chemical Engineering Transactions,* Institution of Engineers, Australia, Paper 2305, Nov. 1968, pp. 247-254.

[17]Gex, R. C., "Thermal Resistance of Metal-to-Metal Contacts, an Annotated Bibliography," Lockheed Missiles and Space Div., Sunnyvale, CA, Rept. 58-61-30, July 1961.

[18]Atkins, H., "Bibliography on Thermal Metallic Contact Conductance," NASA TMS-53227, April 1965.

[19]Vidoni, G. M., "Thermal Resistance of Contacting Surfaces: Heat Transfer Bibliography," Lawrence Radiation Lab., Univ. of California, Livermore, CA, AEC Contract W-7405-ENG-48, June 1965.

[20]Fry, E. M., "Bibliography—Thermal Contact Conductance (to 1966)," Bell Telephone Labs., Whippany, NJ, Rept. 101, Dec. 1966.

[21]Moore, C. J., Atkins, H., and Blum, H. A., "Subject Classification Bibliography for Thermal Contact Resistance Studies," ASME Paper 68-WA/HT-18, New York, 1968.

[22]Roess, L. C., "Theory of Spreading Conductance," Beacon Labs. of Texas, Beacon, NY, unpublished report, 1949, App. A.

[23]Cetinkale, T. N. and Fishenden, M., "Thermal Conductance of Metallic Surfaces in Contact," *Proceedings of the General Discussion on Heat Transfer,* Institution of Mechanical Engineers, London, Sept. 1951, pp. 271-275.

[24]Fenech, H. and Rohsenow, W. M., "Prediction of Thermal Conductance of Metallic Surfaces in Contact," *ASME Journal of Heat Transfer,* Vol. 85, Feb. 1963, pp. 15-24.

[25]Cooper, M. G., Mikic, B. B., and Yovanovich, M. M., "Thermal Contact Conductance," *International Journal of Heat and Mass Transfer,* Vol. 12, 1969, pp. 279-300.

[26]Hunter, A. J. and Williams, A., "Heat Flow Across Metallic Joints—The Constriction Alleviation Factor," *International Journal of Heat and Mass Transfer,* Vol. 12, 1969, pp. 524-526.

[27]Sanokawa, K., "Heat Transfer Between Metallic Surfaces in Contact, 1st Report," *Bulletin of the Japan Society of Mechanical Engineers,* Vol. 11, No. 4, 1968, pp. 253-263.

[28]Gibson, R. D., "The Contact Resistance for a Semi-infinite Cylinder in Vacuum," *Applied Energy,* Vol. 2, 1976, pp. 57-65.

[29]Yovanovich, M. M., "General Expressions for Circular Constriction Resistance for Arbitrary Flux Distribution," AIAA Paper 75-188, Jan. 1975.

[30]Williams, A., "Heat Flow Through Single Spots of Metallic Contacts of Simple Shapes," AIAA Paper 74-692, July 1974.

[31]Major, S. J. and Williams, A., "The Solution of a Steady State Conduction Heat Transfer Problem Using and Electrolytic Tank Analogue," *Mechanical Engineering Transactions,* Institution of Engineers, Australia, 1977, pp. 7-11.

[32]Madhusudana, C. V., "Heat Flow Through Conical Constrictions in Vacuum and in Conducting Media," AIAA Paper 79-1071, June 1979.

[33]Major, S. J., "The Finite Difference Solution of Conduction Problems in Cylindrical Coordinates," *Mechanical Engineering Transactions,* Institution of Engineers, Australia, 1980, pp. 28-34.

[34]Veziroglu, T. N. and Chandra, S., "Thermal Conductance of Two Dimensional Constrictions," AIAA Paper 68-761, June 1968.

[35]Veziroglu, T. N., Williams, A., Kakac, S., and Nayak, P., "Prediction and Measurement of the Thermal Conductance of Laminated Stacks," *International Journal of Heat and Mass Transfer,* Vol. 22, No. 3, pp. 447-459.

[36]Yovanovich, M. M., "Thermal Constriction Parameters of Contacts on a Half Space: Integral Formulation," AIAA Paper 75-708, May 1975.

[37]Yovanovich, M. M., and Schneider, G. E., "Thermal Constriction Resistance due to a Circular Annular Contact," AIAA Paper 76-142, Jan. 1976.

[38]Schneider, G. E., "Thermal Resistance due to Arbitrary Dirichlet Contacts on a Half Space," AIAA Paper 78-870, May 1978.

[39]Yovanovich, M. M., Martin, K. A., and Schneider, G. E., "Constriction Resistance of Doubly Connected Areas Under Uniform Flux," AIAA Paper 79-1070, June 1979.

[40]Yovanovich, M. M., Burde, S. S., and Thompson, J. C., "Thermal Constriction Resistance of Arbitrary Planar Contact with Constant Flux," *Progress in Astronautics and Aeronautics, Thermophysics of Spacecraft and Outer Planet Entry Probes,* edited by A. M. Smith, Vol. 56, AIAA, New York, 1977, pp. 127-140.

[41]Yovanovich, M. M. and Burde, S. S., "Centroidal and Area Average Resistances Non-Symmetric, Singly Connected Contacts," *AIAA Journal,* Vol. 15, 1977, pp. 1523-1525.

[42]Greenwood, J. A., "The Area of Contact between Rough Surfaces and Flats," *ASME Journal of Lubrication Technology,* Vol.

89, Ser. F., Jan. 1967, pp. 81-91.

[43] Mikic, B. B., "Thermal Contact Conductance: Theoretical Considerations," *International Journal of Heat and Mass Transfer*, Vol. 17, 1974, pp. 205-214.

[44] Bush, A. W., and Gibson, R. D., "A Theoretical Investigation of Thermal Contact Conductance," *Applied Energy*, Vol. 5, 1979, pp. 11-22.

[45] Veziroglu, T. N., "Statistical Study of Thermal Contact Conductance," Univ. of Miami, Coral Gables, FL, NASA Grant NG 10-007-010 Report, June 1972.

[46] Popov, V. M., "Determination of the Thermal Contact Resistance of Plane-Rough Surfaces with the Roughnesses Deforming in Different Manners," *Heat Transfer—Soviet Research*, Vol. 2, Sept. 1970, pp. 26-31.

[47] Novikov, V. S., "The Thermal Contact Resistance as a Function of Compression of Rough Surfaces," *Heat Transfer—Soviet Research*, Vol. 2, Nov. 1970, pp. 160-165.

[48] Novikov, V. S., "Thermal Contact Resistance of Rough Surfaces under Compression," *Heat Transfer—Soviet Research*, Vol. 5, Sept.-Oct. 1973, pp. 151-159.

[49] Tsukizoe, T. and Hisakado, T., "On the Mechanism of Heat Transfer between Metal Surfaces in Cotnact, Part 1," *Heat Transfer—Japanese Research*, Vol. 1, Jan.-March 1972, pp. 104-112.

[50] Tsukiizoe, T. and Hisakado, T., "On the Mechanism of Heat Transfer between Metal Surfaces in Contact, 2nd Report," *Heat Transfer—Japanese Research*, Vol. 1, April-June 1972, pp. 23-31.

[51] Cooper, M. G., Mikic, B. B., and Yovanovich, M. M., "Thermal Contact Conductance," *International Journal of Heat and Mass Transfer*, Vol. 12, 1969, pp. 279-300.

[52] Sayles, R. S. and Thomas, T. R., "Thermal Conductance of a Rough Elastic Contact," *Applied Energy*, Vol. 2, 1976, pp. 249-267.

[53] Holm, R., *Electric Contacts—Theory and Application*, 4th Ed., Springer-Verlag, New York, 1967, p. 16.

[54] McMillan, R. Jr., and Mikic, B. B., "Thermal Contact Resistance with Non-uniform Interface Pressures," Dept. of Mechanical Engineering, Massachusetts Institute of Technology, Cambridge, MA, Rept. DSR 72105-70, Nov. 1970.

[55] Hsieh, C. K., Yeddanapudi, K. M., and Touloukian, Y. S., "An Analytical Study of Thermal Contact Conductance for Two Rough and Wavy Surfaces Under a Pressure Contact," *Proceedings of the Ninth Conference on Thermal Conductivity*, Ames, IA, April 1970, pp. 554-570.

[56] Clausing, A. M. and Chao, B. T., "Thermal Contact Resistance in a Vacuum Environment," *ASME Journal of Heat Transfer*, Vol. 93, 1964.

[57] Yovanovich, M. M., "Thermal Contact Conductance of Turned Surfaces," AIAA Paper 71-80, Jan. 1971.

[58] Popov, V. M. and Yanin, L. F., "Heat Transfer during Contact of Machined Surfaces with Waviness and Microroughnesses," *Heat Transfer—Soviet Research*, Vol. 4, 1972, pp. 162-167.

[59] Thomas, T. R. and Sayles, R. S., "Random-Process Analysis of the Effect of Waviness on Thermal Contact Resistance," AIAA Paper 74-691, July 1974.

[60] Dundurs, J. and Panek, C., "Heat Conduction between Bodies with Wavy Surfaces," *International Journal of Heat and Mass Transfer*, Vol. 19, 1976, pp. 731-736.

[61] Edmonds, M. J., Jones, A. M., and Probert, S. D., "Thermal Contact Resistance of Hard Machined Surfaces Pressed Against Relatively Soft Optical Flats," *Applied Energy*, Vol. 6, 1980, pp. 405-427.

[62] Mal'kov, V. A., "Thermal Contact Resistance of Machined Metal Surfaces in a Vacuum Environment," *Heat Transfer—Soviet Research*, Vol. 2, 1970, pp. 24-33.

[63] Fletcher, L. S., and Gyorog, D. A., "Prediction of Thermal Contact Conductance between Similar Metal Surfaces," AIAA Paper 70-852, 1970.

[64] Thomas, T. R. and Probert, S. D., "Correlations for Thermal Contact Conductance in Vacuo," *ASME Journal of Heat Transfer*, Vol. 94, Aug. 1972, pp. 276-281.

[65] O'Callaghan, P. W., and Probert, S. A., "Thermal Resistance and Directional Index of Pressed Contacts between Smooth Non-Wavy Surfaces," *Journal of Mechanical Engineering Science*, Vol. 16, 1974, pp. 41-55.

[66] Al-Astrabadi, F. R., O'Callaghan, P. W., and Probert, S. D., "Thermal Contact Resistance Dependence on Surface Topography," AIAA Paper 79-1065, July 1979.

[67] Edmonds, J. J., Jones, A. M., and Probert, S. D., "Thermal Contact Resistances for Hard Machined Surfaces Pressed Against Relatively Soft Optical-Flats," *Applied Energy*, Vol. 6, 1980, pp. 405-427.

[68] Popov, V. M., "Concerning the Problem of Investigating Thermal Contact Resistance," *Power Engineering* (NY) Vol. 14, 1976, pp. 158-163.

[69] Blahey, A., Tevaarwerk, J. L., and Yovanovich, M. M., "Contact Conductance Correlations of Elastically Deforming Flat Rough Surfaces," AIAA Paper 80-1470, July 1980.

[70] Gale, E. H. Jr., "Effect of Surface Films on Thermal Contact Conductance: Part 1—Microscopic Experiments" ASME Paper 70-HT/SpT-26, June 1970.

[71] Tsao, Y. H. and Heimburg, R. W., "Effects of Surface Films on Thermal Contact Conductance: Part 2—Macroscopic Experiments," ASME Paper 70-HT/SpT-27, June 1970.

[72] Mikic, B. and Carnasciali, G., "The Effect of Thermal Conductivity of Plating Material on Thermal Contact Resistance," *ASME Journal of Heat Transfer*, Aug. 1970, pp. 475-482.

[73] Kharitonov, V. V., Kokorev, L. S., and Tyurin, Yu. A., "Effect of Thermal Conductivity of Surface Layer on Contact Thermal Resistance," *Soviet Atomic Energy*, Vol. 36, April 1974, pp. 385-387.

[74] Yip, F. C., "The Effect of Oxide Films on Thermal Contact Resistance," AIAA Paper 74-693, July 1974.

[75] Mian, M. N., Al-Astrabadi, F. R., O'Callaghan, P. W., and Probert, S. D., "Thermal Resistance of Pressed Contacts between Steel Surfaces: Influence of Oxide Films," *Journal of Mechanical Engineering Science*, Vol. 21, 1979, pp. 159-166.

[76] Cohen, I., Lustman, B., and Eichenberg, J. D., "Measurement of Thermal Conductivity of Metal Clad Uranium Oxide Rods during Irradiation," U. S. Atomic Energy Commission, Westinghouse Electric Corporation, Pittsburgh, PA, Rept. WAPD-228, 1960.

[77] Kharitonov, V. V., Kokorev, L. S., and Del'vin, N. N., "The Role of the Accommodation Coefficient in Contact Heat Exchanger," *Soviet Atomic Energy*, Vol. 35, 1973, pp. 1050-1051.

[78] Madhusudana, C. V., "The Effect of Interface Fluid on Thermal Contact Conductance," *International Journal of Heat and Mass Transfer*, Vol. 18, 1975, pp. 989-991.

[79] Popov, V. M., and Krasnoborod'ko, A. I., "Thermal Contact Resistance in a Gaseous Medium," *Journal of Engineering Physics*, Vol. 28, 1975, pp. 633-638.

[80] Vickerman, R. H. and Harris, R., "The Thermal Conductivity and Temperature Jump Distance of Gas Mixtures," paper presented at the American Nuclear Society Winter Meeting, San Francisco, CA, Nov. 1975.

[81] Garnier, J. E. and Begej, S., "Ex-Reactor Determination of Thermal Gap Conductance between Uranium-Dioxide: Zircaloy-4 Interfaces," *Thermal Conductivity*, Vol. 15, Plenum Press, New York, 1978, pp. 115-123.

[82] Yovanovich, M. M., Hegazy, A. H., and DeVaal, J., "Surface Hardness Distribution Effects Upon Contact, Gap and Joint Conductance," AIAA Paper 82-0887, June 1982.

[83] Yovanovich, M. M., DeVaal, J., and Hegazy, A. H., "A Statistical Model to Predict Thermal Gap Conductance between Conforming Rough Surfaces," AIAA Paper 82-0888, June 1982.

[84] Dutkiewicz, R. K., "Interfacial Gas Gap for Heat Transfer between Two Randomly Rough Surfaces," *Proceeding of the 3rd International Heat Transfer Conference*, Vol. IV, 1966, pp. 118-126.

[85] Madhusudana, C. V. and Fletcher, L. S., "Gas Conductance Contribution to Contact Heat Transfer," AIAA Paper 81-1163, June 1981.

[86] Fletcher, L. S., "Thermal Control in Space Shuttle Systems," *Proceedings of the First Western Space Congress*, Santa Maria, CA, Oct. 1970, pp. 856-865.

[87] Miller, R. G. and Fletcher, L. S., "Thermal Conductance of Gasket Materials for Spacecrafts Joints," *Progress in Astronautics and Aeronautics, Thermophysics and Spacecraft Thermal Contol*, Vol. 35, edited by R. G. Hering, AIAA, New York, 1974, pp. 335-339.

[88] Miller, R. G. and Fletcher, L. S., "Thermal Contact Conductance of Porous Materials in a Vacuum Environment," *Progress in Astronautics and Aeronautics, Thermophysics and Spacecraft Thermal Control*, Vol. 35, edited by R. G. Hering, AIAA, New York, 1975, pp. 321-334.

[89] Miller, R. G. and Fletcher, L. S., "A Thermal Contact Conductance Correlation for Porous Metals," *Progress in Astronautics and Aeronautics, Heat Transfer with Thermal Control Applications*, Vol. 39, edited by M. M. Yovanovich, AIAA, New York, 1975, pp. 81-92.

[90] Fletcher, L. S., and Ott, W. R., "Thermal Conductance of Lead Ferrite and Boron Nitride," *ASME Journal of Heat Transfer*, Vol. 98, May 1976, pp. 331-332.

[91]Fletcher, L. S., "Thermal Control Materials for Spacecraft Systems," *Proceedings of the Tenth International Symposium on Space Technology and Science,* Tokyo, Sept. 1973, pp. 579-586.

[92]Fletcher, L. S., Cerza, M. R., and Boysen, R. L., "Thermal Conductance and Thermal Conductivity of Selected Polyethylene Insulation Materials," *Progress in Astronautics and Aeronautics, Radiative Transfer and Thermal Control,* Vol. 49, edited by A. M. Smith, AIAA, New York, Dec. 1976, pp. 371-380.

[93]Feldman, K., Hong, Y., and Marjon, P., "Test of Thermal Joint Compounds to 200°C," AIAA Paper 80-1466, July 1980.

[94]Cividino, S., Yovanovich, M. M., and Fletcher, L. S., "A Model for Predicting the Joint Conductance and Woven Wire Screen Contacting Solids," *Progress in Astronautics and Aeronautics, Heat Transfer with Thermal Control Applications,* Vol. 39, AIAA, New York, 1975, pp. 111-128.

[95]O'Callaghan, P. W., Jones, A. M., and Probert, S. D., "Research Note: The Thermal Behavior of Gauzes as Interfacial Inserts between Solids," *Journal of Mechanical Engineering Science,* Vol. 17, 1975, pp. 233-236.

[96]Al-Astrabadi, F. R., Probert, S. D., O'Callaghan, P. W., and Jones, A. M., "Reduction of Energy Dissipations at Thermally Distorted Pressed Contacts," *Applied Energy,* Vol. 5, 1979, pp. 23-51.

[97]Sauer, Jr., H. J., Remington, C. R., Stewart, W. E., Jr., and Lin, J. T., "Thermal Contact Conductance with Several Interstitial Materials," *Proceedings of the 11th International Conference on Thermal Conductivity,* Albuquerque, NM, Sept.-Oct. 1971, pp. 22-23.

[98]Sauer, H. J., Jr., Remington, C. R., and Heizer, G. A., "Thermal Contact Conductance of Lubricant Films," *Proceedings of the 11th International Conference on Thermal Conductivity,* Albuquerque, NM, Sept.-Oct. 1971, pp. 24-25.

[99]Barzelay, M. E., Tong, K. N., and Holloway, G. F., "Effect of Pressure on Thermal Conductance of Joints," NACA TN-3295, 1955.

[100]Rogers, G. F. C., "Heat Transfer at the Interface of Dissimilar Metals," *International Journal of Heat and Mass Transfer,* Vol. 2, 1961, pp. 150-154.

[101]Moon, J. S., and Keeler, R. N., "A Theoretical Consideration of Directional Effects of Heat Flow at the Interface of Dissimilar Metals," *International Journal of Heat and Mass Transfer,* Vol. 5, 1962, pp. 967-971.

[102]Williams, A., "Directional Effects of Heat Flow Across Metallic Joints," *Mechanical Engineering Transactions,* Institution of Engineers, Australia, Paper 3448, 1976.

[103]Veziroglu, T. N. and Chandra, A., "Direction Effect in Thermal Contact Resistance," *Heat Transfer 1970,* Vol. I, Paper Cu 3-5, Elsevier, Amsterdam, 1970. (See also discussion in Vol. X.)

[104]Barber, J. R., "The Effect of Thermal Distortion on Constriction Resistance," *International Journal of Heat and Mass Transfer,* Vol. 14, 1971, pp. 751-766.

[105]Thomas, T. R. and Probert, S. D., "Thermal Contact Resistance: The Directional Effect and Other Problems," *International Journal of Heat and Mass Transfer,* Vol. 13, 1970, pp. 789-807.

[106]Clausing, A. M., "Heat Transfer at the Interface of Dissimilar Metals—The Influence of Thermal Strain," *International Journal of Heat and Mass Transfer,* Vol. 19, 1966, pp. 791-801.

[107]O'Callaghan, P. W., Probert, S. D., and Jones, A., "A Thermal Rectifier," *Journal of Physics D: Applied Physics,* Vol. 3, 1970, pp. 1352-1358.

[108]O'Callaghan, P. W. and Probert, S. D., "Thermal Resistance and Directional Index of Pressed Contacts between Smooth Non-Wavy Surfaces," *Journal of Mechanical Engineering Science,* Vol. 16, 1974, pp. 41-55.

[109]Jones, A. M., O'Callaghan, P. W., and Probert, S. D., "Effect of Interfacial Distortions on Thermal Contact Resistance of Coaxial Cylinders," AIAA Paper 74-689, July 1974.

[110]Jones, A. M., O'Callaghan, P. W., and Probert, S. D., "Thermal Rectification Due to Distortions Induced by Heat Fluxes across Contact between Smooth Surfaces," *Journal of Mechanical Engineering Science,* Vol. 17, 1975, pp. 252-261.

[111]Lewis, D. W., and Perkins, H. C., "Heat Transfer at the Interface of Stainless Steel and Aluminum—The Influence of Surface Conditions on the Directional Effect," *International Journal of Heat and Mass Transfer,* Vol. 12, 1968, pp. 1371-1383.

[112]Al-Astrabadi, F. R., Jones, A. M., Probert, S. D., and O'Callaghan, P. W., "Effect of Surface Distortions on the Thermal Resistance of Pressed Contacts," *Journal of Mechanical Engineering Science,* Vol. 21, 1979, pp. 317-322.

[113]Somers, R. R., II, Miller, J. W., and Fletcher, L. S., "Thermal Contact Conductance of Dissimilar Metals," *Progress in Astronautics and Aeronautics, Thermophysics and Thermal Control,* Vol. 65, edited by R. Viskanta, AIAA, New York, March 1979, pp. 149-175.

[114]Jones, A., O'Callaghan, P. W., and Probert, S. D., "Differential Expansion Thermal Rectifier," *Journal of Physics E: Scientific Instruments,* Vol. 4, 1971, pp. 438-440.

[115]Androulakis, J. G., "Effective Thermal Conductivity Parallel to the Laminations of Multilayer Insulation," AIAA Paper 70-846, June-July 1970.

[116]Williams, A., "Heat Flow Across Stacks of Steel Laminations," *Journal of Mechanical Engineering Science,* Vol. 13, 1971, pp. 217-233.

[117]Williams, A., "Experiments on the Flow of Heat Across Stacks of Steel Laminations," *Journal of Mechanical Engineering Science,* Vol. 14, 1972 pp. 151-154.

[118]Sheffield, J. W., Veziroglu, T. N., and Williams, A., "An Experimental Investigation of Thermal Contact Conductance of Multilayered Electrically Insulated Sheets," AIAA Paper 79-1067, June 1979.

[119]Al-Astrabadi, F. R., O'Callaghan, P. W., Probert, S. E., and Jones, A. M., "Thermal Contact Conductance Correlation for Stacks of Thin Layers in High Vacuums," *ASME Journal of Heat Transfer,* Vol. 99, Feb. 1977, pp. 139-142.

[120]Bordzdyka, A. M., "Elevated Temperature Testing of Metals," Israel Program for Scientific Transactions, Jerusalem, 1965.

[121]McKenzie, D. J. Jr., "Experimental Confirmation of Cyclic Thermal Join Conductance," AIAA Paper 70-853, June-July 1970.

[122]Mikic, B., "Analytical Studies of Contact of Nominally Flat Surfaces; Effect of Previous Loading," *ASME Journal of Lubrication Technology,* Vol. XX, Oct. 1971, pp. 451-456.

[123]Madhusudana, C. V. and Williams, A., "Heat Flow Through Metallic Contacts—The Influence of Cycling the Contact Pressure," *Proceedings of the First Australia Conference on Heat and Mass Transfer,* Melbourne, Australia, 1973, Sec. 4.1, pp. 33-40.

[124]Whitehurst, C. A. and Durbin, W. T., "A Study of the Thermal Conductance of Bolted Joints," NASA-CR-102639, 1970.

[125]Veilleux, E. and Mark, M., "Thermal Resistance of Bolted or Screened Sheet Metal Joints in a Vacuum," *Journal of Spacecraft and Rockets,* Vol. 6, 1969, pp. 339-342.

[126]Bradley, T. L., Lardner, T. J., and Mikic, B. B., "Bolted Joint Interface Pressure for Thermal Contact Resistance," *ASME Journal of Applied Mechanics,* June 1971, pp. 542-545.

[127]Gould, H. H., and Mikic, B. B., "Areas of Contact and Pressure Distribution in Bolted Joints," Engineering Projects Laboratory, Department of Mechanical Engineering, Massachusetts Institute of Technology, Cambridge, MA, Rept. OSR 71821-68, 1970.

[128]Gould, H. H., and Mikic, B. B., "Areas of Contact and Pressure Distribution in Bolted Joints," ASME Paper 71-WA/DE-3, Nov.-Dec. 1971.

[129]Roca, R. T. and Mikic, B. B., "Thermal Conductance in a Bolted Joint," AIAA Paper 72-282, April 1972.

[130]Yip, F. C., "Theory of Thermal Contact Resistance in Vacuum with an Application to Bolted Joints," AIAA Paper 72-281, April 1972.

[131]Oehler, S. A., McMordie, R. K., and Allerton, A. B., "Thermal Contact Conductance Across a Bolted Joint in a Vacuum," AIAA Paper 79-1068, June 1979.

[132]Williams, A. and Madhusudana, C. V., "Heat Flow Across Cylindrical Metallic Joints," *Heat Transfer 1970,* Paper Cu 3-6, Elsevier, Amsterdam, 1970.

[133]Hsu, T. R. and Tam, W. K., "On Thermal Contact Resistance in Compound Cylinders," AIAA Paper 79-1069, June 1979.

[134]Novikov, I. I., Kokorev, L. S., and Del'vin, N. N., "Experimental Investigation of Contact Heat Exchange between Coaxial Cylindrical Casings in Vacuum," *Soviet Atomic Energy,* Vol. 32, June 1972, pp. 474-475.

[135]Madhusudana, C. V., "Contact Heat Transfer between Coaxial Cylinders of Similar of Dissimilar Materials," *Proceedings of the ASME-JSME Thermal Engineering Joint Conference,* Vol. III, Honolulu, HI, March 1983, pp. 317-322.

[136]Jacobs, G. and Todreas, N., "Thermal Contact Conductance in Reactor Fuel Elements," *Nuclear Science and Engineering,* Vol. 50, 1973, pp. 283-290.

[137]Dean, R. A., "Thermal Contact Conductance between UO and Zircaloy-2," Atomic Power, Div., Westinghouse Electric Corp., Pittsburgh, PA, Rept. CVNA-127, May 1962.

[138] Ross, A. M. and Stoute, R. L., "Heat Transfer Coefficient between UO and Zircaloy-2," Atomic Energy of Canada Ltd., Chalk River, Ontario, Rept. CRFD-1075, 1962.

[139] Rapier, A. C., Jones, T. M., and McIntosh, J. E., "The Thermal Conductance of Uranium Dioxide/Stainless Steel Interface," *International Journal of Heat and Mass Transfer,* Vol. 6, 1963, pp. 397-416.

[140] "A Model for the Prediction of Pellet-Cladding Thermal Conductance in BWR Fuel Rods," Boiling Water Reactors Dept., General Electric Co., San Jose, CA, Rept. NEDO-20181, Nov. 1973.

[141] Lanning, D. D. and Hann, C. R., "Review of Methods Applicable to the Calculation of Gap Conductance in Zircaloy-Clad UO Fuel Rods," Batelle Pacific Northwest Labs., Richland, WA, Rept. BNWL-1894, April 1975.

[142] Beyer, C. E., Hann, C. R., Lanning, D. D., Panisko, F. E., and Pachen, L. J., "GAPCON-Thermal-2: A Computer Program for Calculating the Thermal Behavior of an Oxide Fuel Rod," Batelle Pacific Northwest Labs. Richland, WA, Rept. BNWL-1898, Nov. 1975.

[143] Cross, R. W. and Fletcher, L. S., "Thermal Contact Conductance of Uranium Dioxide-Zircaloy Interfaces," AIAA Paper 78-85, Jan. 1978.

[144] Madhusudana, C. V., "Experiments on Heat Flow Through Zircaloy-2/Uranium Dioxide Surfaces in Contact," *Journal of Nuclear Materials,* Vol. 92, 1980, pp. 345-348.

[145] Madhusudana, C. V. and Fletcher, L. S., "Solid Spot Thermal Conductance of Zircaloy-2/Uranium Dioxide Interfaces," *Nuclear Science and Engineering,* Vol. 83, 1983, pp. 327-332.

[146] Garnier, J. S. and Begej, S., "Ex-Reactor Determination of Thermal Gap and Contact Conductance between Uranium Dioxide: Ziracloy-4 Interfaces Stage I: Low Gas Pressure," Pacific Northwest Lab., Richland, WA, Rept PNL-2696, April 1979.

Chapter 4

Vacuum Mechatronics Components

Majid Shirazi
Lakshmanan Karuppiah
Degang Chen

4.1 Actuators

In many cases the vacuum environment is utilized effectively only when it is possible to input the power to actuate mechanisms which can do useful work. This can be accomplished by mechanical feedthroughs or by electrical feedthroughs beyond the vacuum boundary. Power can either do the work directly, or can be converted by an actuator into the necessary form, e.g., electrical to mechanical by motor, shape memory alloy, or piezoelectric actuators. During the conversion and transfer, a certain amount of heat is generated and the resulting mechanism is also a potential source of particle contamination. Thus, the transduction mechanisms must be very efficient in order to minimize heat generation and mechanical contact. In this section, various types of actuators for the vacuum environment will be discussed.

Rotary and Linear Feedthroughs

Mechanical feedthroughs are divided into two classes: rotary and linear. For applications at pressures down to 10^{-6} Torr, O-ring seals (e.g., "Viton A" O-rings lubricated with graphite) are often used to seal the transmission of rotational motion. Figure 4.1 shows two typical assemblies of low-speed rotary feedthroughs. Multiple seals are sometimes used to reduce the leakage rate, and pumping of the cavities formed by consecutive seals is even more effective. The seals are not bakeable, however, and thus are not suitable for ultra-high vacuum applications.

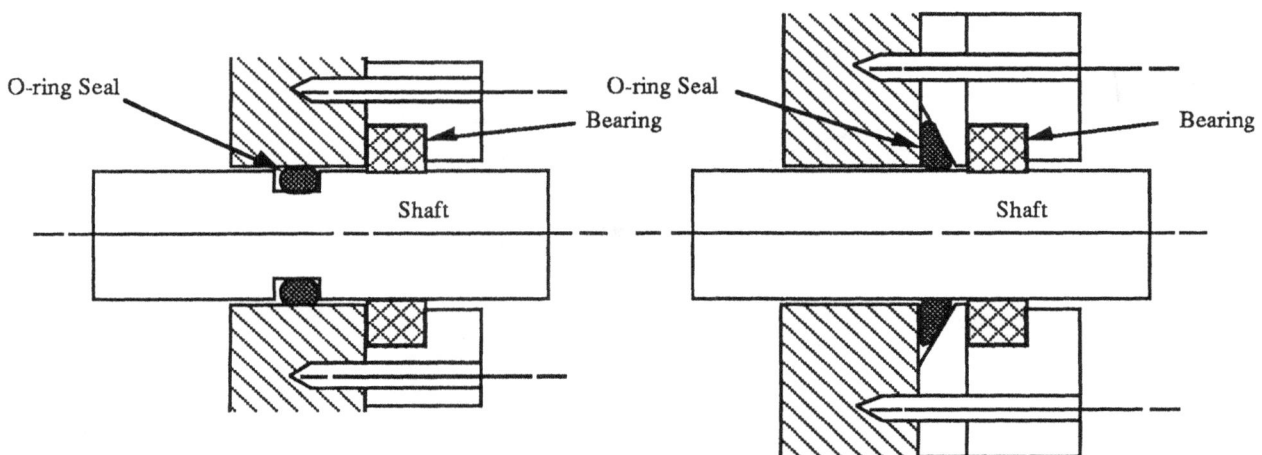

Fig. 4.1 Rotary feedthroughs for low speed.

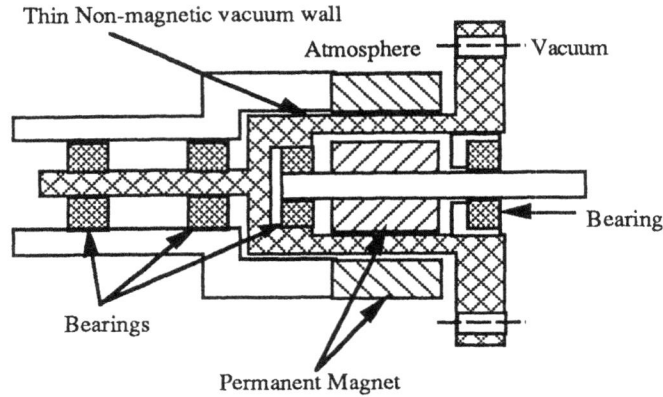

Fig.4.2 Rotary motion feedthrough using magnetic coupling.

A magnetic coupling effect can be used to implement a low-torque feedthrough, as shown in Figure 4.2. Opposite poles of permanent magnets are lined up against each other to produce the magnetic coupling. The rotational movement of permanent magnets on one side forces the shaft on the other side to move in the same direction. The transmitted torque is limited by the strength of the magnetic seal, which is inversely proportional to the distance between the two permanent magnets. There is a one-to-one ratio between the input and output shaft positions when the feedthroughs are operated below the saturation torque.

Sealing of rotary motion can also be achieved using a magnetic liquid suspension as a vacuum seal (Figure 4.3). Multiple seals are formed by a magnetic liquid material and magnetic field which is provided by the permanent magnet in the housing. This feedthrough allows transmission of high torques and high rotational speeds.

O-ring seals have a tendency to roll and twist in the gland under relative linear motion, and so are not suitable for use in linear feedthroughs. Linear motion in a vacuum can be achieved by a rotary feedthrough in concert with a rack-and-pinion at one end. This arrangement requires lubrication of mechanical components with relative motion in vacuum. This mechanical contact will introduce contaminants into the vacuum environment.

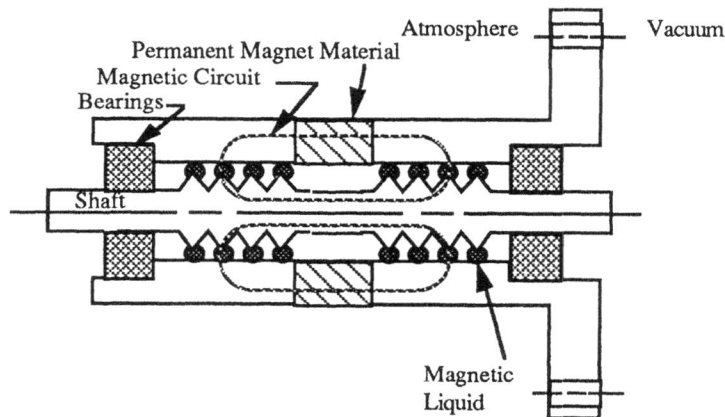

Fig. 4.3 Rotary motion feedthrough with magnetic liquid sealing.

180

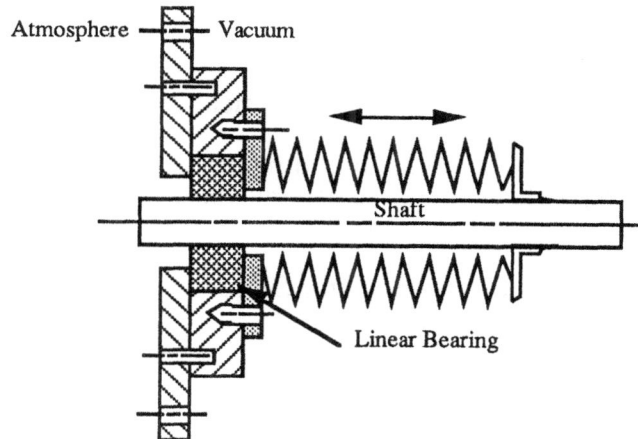

Fig. 4.4 Bellows-sealed linear motion feedthrough

A satisfactory linear motion feedthrough is achieved by using bellows, as shown in Figure 4.4. Two types of bellows are available: hydraulically-formed bellows and edge-welded bellows. The formed bellows are generally less costly, but the welded bellows give a superior performance in terms of flexibility and linear motion along the axis per unit length of bellows. Welded bellows are usually made of 18/8 stainless steel and thus can be baked to 450°C. The bellow system is suitable for low linear speeds and small strokes. Miniature metal bellows with low stiffness have been manufactured by an electro-deposit method. These bellows are made by forming a mandrel to the shape of the inside of the bellows and depositing metal with the desired stiffness on the mandrel. The ends are trimmed and the mandrel is dissolved. These bellows have been made with outside diameters as small as 0.035", and are extremely flexible.

Vacuum Motors

Motion inside a vacuum can also be achieved without a mechanical feedthrough, using motors that are specially designed to operate in a vacuum. The design constraints include: minimum power dissipation, a heat conduction path to the atmosphere, vacuum-compatible lubricants to avoid bearing seizure, and winding insulation with a high temperature rating (>150°C).

Stepper motors are commonly used for vacuum applications. The motor mount is used to conduct heat through the vacuum wall to an external heat sink. For continuous heat dissipation, water cooling may be provided at the heat sink. Temperature sensing is essential and is achieved by attaching a thermistor to the stator or to the housing furthest from the mounting. There is a time-delay on the order of 20 minutes before the casing reaches the same temperature as the inside winding insulation. When operating a motor inside a vacuum, the holding current should be reduced or eliminated when the mechanism is in a holding mode, and the power should be reduced after the motor's initial movement.

Motors operating in vacuum will outgas with rising temperature. For good vacuum performance, the motor temperature must usually be kept below 100°C. Motors can be baked prior to the vacuum application in order to minimize the initial outgassing. Commercially available vacuum-compatible motors and encoders are discussed in chapter 6. Appendix A lists company names.

Piezoelectric Actuators

Certain types of crystals, when mechanically deformed, produce an electric field; conversely, when an electric field is applied the crystals undergo a physical deformation. This phenomenon is the basis of the piezoelectric actuator. Single crystals of many compounds show piezoelectric properties, e.g. quartz, rochelle salt, ammonium dihydrogen phosphate, and tourmaline. Recent advances in ceramic technology have produced ceramic materials, such as barium titanate and lead zirconate titanate, with piezoelectric properties.

The general expression relating the strain vector S produced in the piezoelectric block by the electric field E is:

$$S = d \times E \qquad \qquad \text{Eq. (4.1)}$$

where d is the piezoelectric strain coefficient. By measuring the strain $\Delta x/x$, for certain applied voltages at a no-load condition, and then adding a load that returns the dimensions to the original length, a stress-strain characteristics diagram can be generated. Repeating the experiment for different voltages produces a family of graphs, as shown in Figure 4.5.

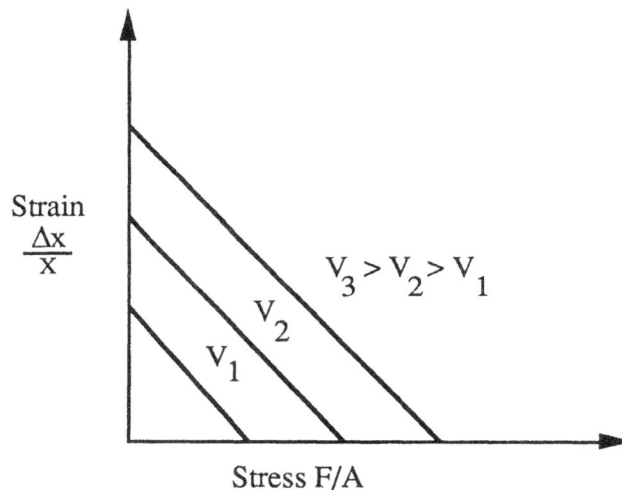

Fig. 4.5 Stress-strain curve for a piezoelectric block [4.1]

Piezoelectric elements are essentially capacitors. During static operation, current is not drawn and power is not consumed to maintain a state of actuation. Power is required only to change the voltage. Piezoelectric ceramics dissipate power in proportion to dielectric loss in the crystal, which is defined as the ratio of the effective series resistance to the effective series reactance. This factor varies from 0.02 for soft ceramics to 0.0005 for hard ceramics at 1000 Hz. At low frequency, this factor is negligible, making these materials suitable for vacuum applications.

Piezoelectric actuator designs vary depending on the application (stroke length, linear motion, rotary motion, etc.). A common design is a linear actuator with a stroke of 6 to 8 inches. Figure 4.6 shows the principle of operation of one such actuator. When voltage is applied to the first element, it clamps the shaft. A variable-rate voltage is then applied to element 2, causing it to change length in discrete steps at a few nanometers per step. Voltage is then applied to element 3, causing it to grip the shaft. If necessary, 1 may be unclamped, 2 contracted, 1 clamped and 3 unclamped to reach the initial configuration. This sequence can be repeated until the desired stroke

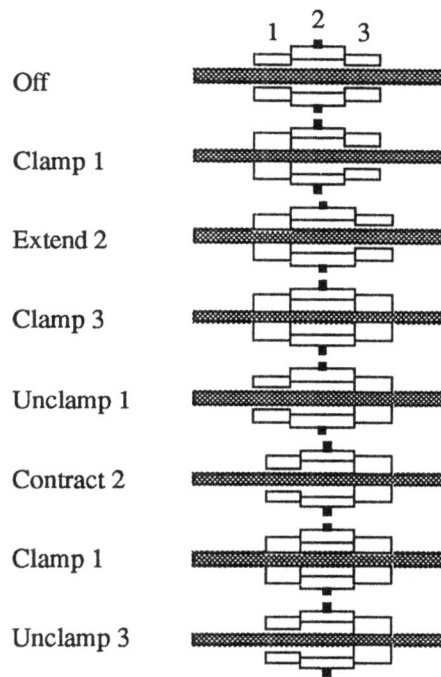

Fig. 4.6 Operation of a Piezoelectric Linear Actuator
(The Inchworm™ from Burleigh Instruments, Inc. [4.1])

is reached. The actuator is mounted at element 2. For a clamp frequency of 400 Hz, speeds exceeding 1 mm/s can be achieved.

Piezoelectric devices are inherently simple and can be made with good outgassing properties. This factor, coupled with their high heat dissipation, makes them excellent actuators in vacuum at low speeds and torques. Piezoelectric actuators for high resolution positioning normally use a high driving voltage (approximately 1000 volts) which makes them less sensitive to electrical noise, resulting in less jitter. When operating these devices in vacuum, the voltage must be turned off when the artificial vacuum chamber is being evacuated because there is a danger of electrical arcing as the pressure goes through the corona region (0.1 to 100 Torr). [Note: arcing leaves carbon deposits, and therefore creates the possibility of permanently shorting the actuator.] One disadvantage in employing such actuators is that the mechanical contact produces contamination undesirable in a vacuum environment.

Shape Memory Alloy (SMA) Actuator

After an apparent plastic deformation, certain metallic materials will return to their original shape when heated. This phenomenon is known as the shape memory effect. A part which is deformed or stretched at one temperature will completely recover its original shape when heated to a second temperature. In the process, the moving alloy delivers a force.

Shape memory is a behavior unique to alloys which undergo martensitic transformation. The martensite phase in these alloys, unlike martensite phase in steel, is thermoelastic; i.e., it continually appears and disappears with falling and rising temperatures. Figure 4.7 shows the stages for producing the shape memory effect in a copper alloy. The temperature at which martensite starts to form, after cooling from the parent phase temperature, is called M_S. The temperature at which the bulk transforms is M_f. When the alloy is heated, the initial parent phase reforms at A_S and is completed at A_f. Some common alloys are Ni-Ti, Cu-Zn-Al and Cu-Ni-Al.

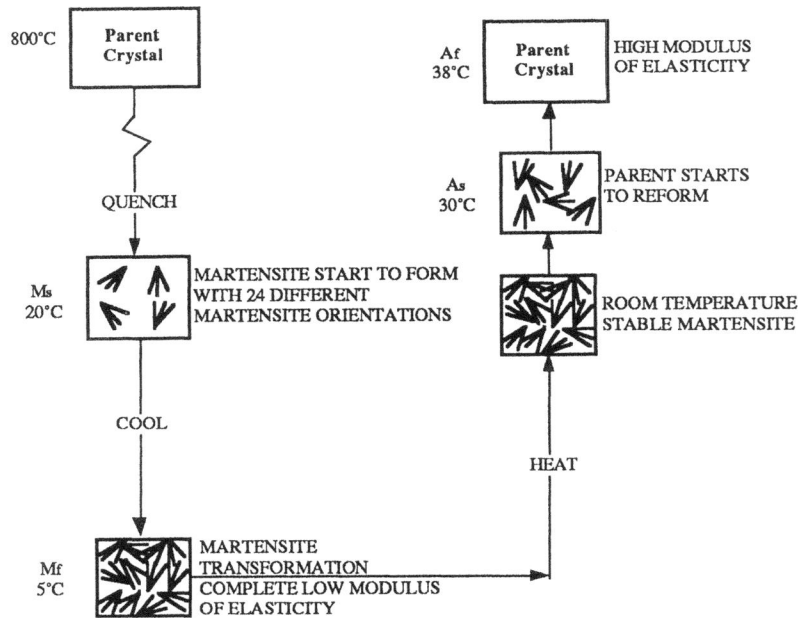

Figure 4.7 Stages for producing the shape memory effect in a copper alloy (From [4.3]).

The M_S of these alloys can be varied from -105°C to 170°C. Alloy heating can be accomplished by contact with a heating element or by passing current through the shape memory alloy itself. Cooling is accomplished naturally or with the aid of convective devices (e.g fans), or by contact.

Figure 4.8 shows two manifestations of the shape memory effect: one-way memory and two-way memory. In one-way memory, the deformed SMA can be heated to its original shape. In two-way memory, the deformed SMA cycles between the original shape and the deformed shape, depending on whether it is heated or cooled. The latter can be used for two-state applications.

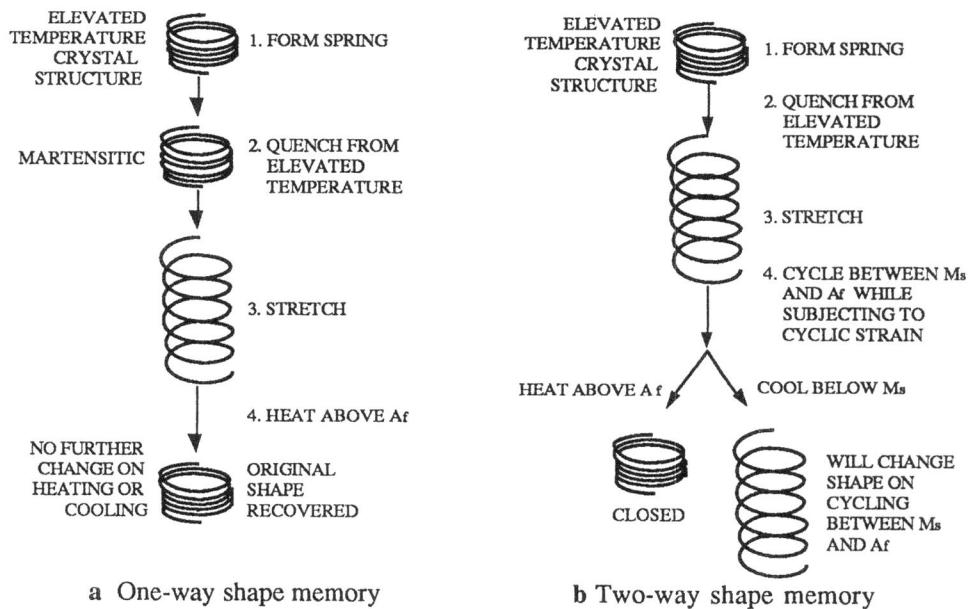

a One-way shape memory   **b** Two-way shape memory

Fig. 4.8 Shape memory (From [4.3])

184

The Ni-Ti alloy has been found to be suitable for vacuum applications. It has a low vapor pressure, a low rate of outgassing, and a relatively high transformation temperature. It also has a cleanliness rating equivalent to class 10 or better. This alloy has been used in ultra-high vacuum to open and close a shutter mechanism, as discussed in the reprint by Jardine *et al.* [4.2] included at the end of this chapter. For repeated operations, the alloy must be cooled and heated. If the cycle time is long (>30 mins), radiation effects are sufficient to cool the alloy. For shorter cycle times, forced convection is necessary. This poses several design problems because the coolant must be passed through the vacuum envelope in a vacuum-compatible flexible tube. Also, a conduction path must be established between the SMA and the coolant while simultaneously preventing an electrical short circuit to the cooling tube. The tube may be copper or another conductive material.

4.2 Sensors

This section will discuss sensors other than the ones used for monitoring the vacuum and measuring particle contamination. Sensors are one of the means of monitoring the environment. They may be used to simply inform the human operator of the status of the system. Now, however, they are being used more and more to allow the system to behave intelligently.

Early automation attempts were repetitive and allowed operation only in a known, well-structured environment with little change in the process or handling mechanism. This level of performance was adequate in many situations, yet the automation equipment, e.g. robots, required a vast amount of programming and information about the environment. This was true, for example, in the spot welding of automobile bodies where each car that came along was secured to the same point on the floor. The robot did not need any sensory capabilities in order to locate welding spots.

Today's stringent requirements on the quality of industrial products, especially in microelectronics and other high-precision technologies, demands a more intelligent, flexible machine. Such a programmable machine allows the system to adapt itself to the environment by using information from sensors, thus interpreting the given commands to suit the situation. In a remotely operated system, such as an in-vacuum manipulator, or a telerobot, the operator or controller uses information from many sensors to control the robot manipulator in highly variable and essentially unpredictable situations (See Fig. 4.9). Position sensors identify the end-effector position, vision sensors recognize the position and orientation of the object to be handled, and other sensors can then be used to ensure proper control and stability of the system [4.5].

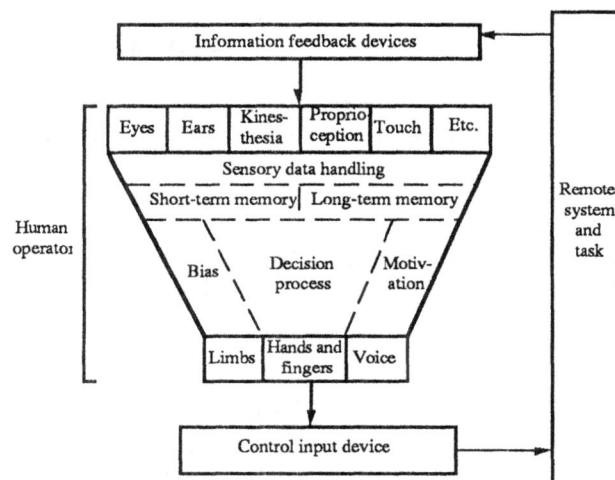

Fig. 4.9 Operator-centered view of teleoperation (from [4.4], p. 321)

The mechanism's performance and reliability are a function of the proper use of its components within specification. Many components are sensitive to temperature variations and tend to lose their effectiveness at extreme temperatures. An electric motor is specified to operate below a maximum temperature, and its efficiency decreases at high temperatures. In such situations, temperature feedback via sensors can be used to shut the system down. This is especially critical for vacuum-compatible mechanisms since air convection does not exist.

Sensor Applications for Vacuum Mechatronics

In controlling a mechanism in vacuum, a number of control variables must be monitored. These include geometric variables such as position and orientation, and fluid and thermal quantities, such as temperature and pressure. This section identifies the sensory systems which are the most appropriate for each parameter.

Position Sensors

The control of a robot manipulator and other mechatronic devices in a non-accessible environment requires position sensing and feedback. This information is used in determining the appropriate torque to apply to the manipulator joint in order to position the device. Angular position sensors are used at the joints of the robot, providing information to solve the forward kinematic relationship. Position sensors can also be used to identify the end-effector orientation relative to a reference frame, which is particularly important when an object must be grasped and manipulated.

A robot mechanism also possesses "robot singular configurations" in which the arm movement is constrained in particular directions where the load on the joints is very high. Angular position information can be used to prevent the robot arm from exceeding its singularity limits.

Other applications of robot position sensing involve obstacle avoidance. In these cases, the position of each object in the work area is established relative to a particular frame of reference. By using remote position sensors, which are either attached to the robot arm or stationed in the environment, the controller uses the sensor information to plan movements that will guide the robot to the desired position/configuration without collisions.

Vision Sensors

In robotics, computer vision is used for locating the position of objects in the environment, including the manipulator, and for inspecting objects for errors or defects. In manipulator control. the vision system is able to identify the workpiece position, orientation and condition. This enables the robot to handle inaccurately positioned objects with variable orientation. Accuracy of the system depends on the resolution of its vision; resolution improves by reducing the field of view through the use of magnifying lenses, or by using vision components with higher resolution. The latter is more expensive.

A vision inspection system helps in identifying surface defects such as cracks, voids, and stains. It can also be used for measuring the critical dimensions of mechanical parts and even for gauging "wear and tear." The data on the various defects is categorized by type, size, location, etc., and is readily stored in databases for analysis.

It is important to have a remote global sensor such as a camera in the vacuum chamber in order to monitor the movements of all robots and/or objects. Other applications include end-point positioning, where the workpiece must be positioned accurately relative to a reference frame which is attached to the chamber. *In-Situ* inspection during in-vacuum manufacturing is another application which may include vision sensors.

Temperature Sensors

One purpose of in-vacuum manufacturing is to assemble clean parts with close tolerances. High accuracy and repeatability (typically less than 10 μm) are required for many of these tasks. Energy waste, resulting from friction and driving mechanism losses, raises the temperature of the components and may affect their physical stability. For example, when using aluminum 6061-T6 or stainless steel 304, a linear expansion of manipulator links on the order of 20μm/m per °C occurs. Accounting for this expansion during operation in vacuum is therefore critical. Other material properties that are sensitive to temperature variation, such as the tensile and torsional modulus of elasticity, must be accounted for when designing mechanisms for the space environment.

Heat generated during operation in vacuum is primarily dissipated by conduction. Some heat is carried out through leads, increasing their temperature. This heat, if excessive, can melt solder from joints and loosen connections. The temperature can be controlled by cutting power to idle circuits. It is recommended that thermocouples be incorporated into the mechanism's design.

Sensor Types

The vacuum chamber interior is considered an "inaccessible" environment, and thus requires a large number of vacuum-compatible sensors in order to perform an intelligent task. Most sensors used in vacuum have the same operating principles as those used under normal atmospheric conditions. The key parameters in the design of sensors for vacuum are proper surface materials and an adequate cooling system for the electrical components. Various low vapor pressure coatings must protect the surfaces that are exposed to vacuum. For example, soldered joints are coated to reduce outgassing. Vacuum-compatible components must be highly efficient in order to prevent the heat generation. Forced convection methods are normally used to cool electro-mechanical components with a large power consumption. Such methods are discussed in section 3.3.

Robotic sensors may be functionally divided into two kinds: contact and non-contact. Contact sensors include touch, tactile, and force/torque sensors. Non-contact sensors include proximity, range, and vision sensors. These sensors might be used separately or in various combinations.

Contact

Contact sensors are responsive to actual physical contact. In robotics, touch sensors are used partly in determining the relationship between the object and the manipulator "hand", thereby allowing control of the exerted force on the object. They are compliant devices, and transform the strain in a material to electrical signals. For example, the UCSB Hand uses an advanced form of force-torque sensor with strain-gauge bridge circuits to measure contact pressure [4.6]. Each finger of the three-fingered "hand" can sense and measure force and moments in three dimensions. Other methods of measuring strain include optical systems and lasers. Optical systems have a sensitivity threshold of 0.04 grams [4.7, 4.8]. Laser devices possess higher resolution but are more expensive [4.9].

Slip sensors measure the distribution of pressure between a robot hand and grasped objects. They are of two types: single point and array. Transducers used in this type of sensor are strain gauges, pressure-conductive elastomers, capacitive and piezoelectric elements [4.4].

Force and torque sensors are used in measuring the reaction forces at robot wrists and joints. Deflection is one possible result of applying external forces to a robot. The wrist sensor is positioned between the tip of the robot arm and the end-effector, and consists of various strain-gauge bridge circuits that measure deflection of the sensor's structure. This type of sensor normally has a stiffer structure and is used to measure larger forces than the touch sensor. Joint sensors estimate the total actuator force supplied to any joint. When a DC motor is driving the joint, the joint sensor measures the motor's current as well. When hydraulics or pneumatic actuators are used, the back pressure is measured.

Non-Contact

A non-contact sensor outputs a signal proportional to the electromagnetic or acoustic radiation in the environment. Many non-contact sensing devices have been designed for aerospace and vacuum environment applications, including electromagnetic, microwave, infrared (IR), laser, imaging, and particle sensors. New sensors are continually being designed, while simultaneously the older models are being improved. Hord [4.10] lists many of the sensors used for space technology and makes predictions on future sensor developments.

Microwave systems are used in sensing environmental conditions such as temperature, windspeed, humidity and precipitation. IR sensors are used for remote sensing, such as in reading planetary surfaces and industrial furnace temperatures. Laser systems are used primarily for the range detection of atmospheric and chemical species, and for their profile measurements. Charge-Coupled Device (CCD) imaging has been used for remote sensing, attitude and orbit control, and target detection [4.11]. Charged particles traversing a magnetic field are measured by a magnetic spectrometer particle detector.

Proximity, range, and vision sensors are also used in vacuum mechatronics, enabling mechanisms to interact in vacuum, by indicating the presence or absence of objects. Proximity sensors generally have binary output and utilize either electro-optical or electromagnetic measurement principles. Electro-optical proximity sensing consists of a light emitter (typically a light emitting diode or a low-power laser) which acts as a transmitter and a photodetector receiver. Electromagnetic proximity sensing devices can be in the form of inductive, Hall-effect, and capacitive sensing units. The inductive and Hall-effect sensors detect only ferromagnetic materials, but capacitive sensors are capable of detecting all solid and liquid materials. Each type may be used in vacuum, assuming proper material and thermal design.

Range sensors are used for measuring the distance from reference points to objects. Triangulation and time-of-flight techniques are used, along with electro-optical and acoustic principles, to estimate the desired range. Refer to Nof [4.4] for a description of these techniques.

Cameras are visual sensors. Vidicon and solid-state array are the two types of cameras that are used in robotics. Solid-state array cameras are used most often, even though they are slightly more expensive than Vidicons. Advantages of the solid state imaging devices include: lighter weight, smaller size, longer life, and lower power consumption. However, some Vidicon cameras have resolutions beyond the capabilities of solid state cameras. The resolution of CCD cameras ranges from 32x32 to 1024x1024. Reference [4.12] describes the principles of operation of both cameras. For black and white cameras, a video signal proportional to the intensity of input image is produced, and for color cameras a composite color image with red, green and blue components is output.

In addition to weight and resolution, vacuum-compatible cameras have the following requirements:

- vacuum-compatible housing,
- low power consumption, and
- cooling.

The power consumed by the electronics and CCD array typically does not exceed 7 watts.

a. Flanged face coupling   b. Keyless compression coupling

Fig. 4.10 Rigid couplings (From [4.13])

4.3 Energy Transmission

Many types of transmission elements have been used in robotics to transmit mechanical power from a source to a load. The type of element used depends on the nature of the motion, power requirements and the geometry and configuration of the system. The factors involved in the transmission design are efficiency, stiffness, and cost [4.13]. In this section a number of transmission elements are presented.

Couplings

One method of transmitting the motion from one shaft to another is by means of a coupling, which may be rigid or flexible. A rigid coupling is used to connect the shafts with perfect alignment and is rigid in axial, radial and tangential directions. They possess different shapes depending on the configuration of the shafts' end-faces and the alignment method used. Figure 4.10 shows two types of rigid couplings. The flanges are either keyed to the shafts or to a tapered cone-shaped sleeve for automatic centering of the shafts. This type of coupling introduces an additional dynamic load and moment, which are functions of the alignment tolerances on the shaft support bearings.

The flexible coupling employs a spring or rubber material which provides flexibility in the axial and radial directions but has high rigidity in the torsional direction. This type of coupling is useful for misaligned shafts in the lateral and/or angular direction. A helical flexible coupling is shown in Fig. 4.11. This type of coupling eliminates the dynamic loads and moments produced by misaligned shafts, but may fail with excessive vibration, acceleration or braking loads.

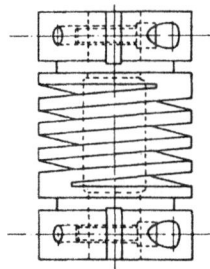

Fig. 4.11 Helical coupling (From [4.4, p. 60])

Brakes/Clutches

Brakes are used to convert the kinetic energy of a moving mass to heat or other forms of energy. Mechanical and electromagnetic principles are used to slow down, or in emergency situations, stop the motion of the motor shaft. Mechanical brakes are commercially available in different types: block, cone, disk, and spring. They are generally constructed with a stationary part which is clamped to the mechanism's body and a free part which is attached to the moving mass. The friction between the two parts creates the braking action.

Electromagnetic brakes are more widely used in computer-controlled mechanisms. Commercially available electrical brakes include eddy-current and magnetically actuated brakes. The eddy-current brake is used with flywheels where high braking power is necessary to reduce kinetic energy. Electrically actuated magnetic poles create a magnetic flux that permeates the gap and the iron of the wheel rim. The kinetic energy of the fly wheel is converted to heat in this manner. The magnetic type of brake includes an electric coil, springs, a rotating member attached to the shaft, and a stationary plate attached to the housing. The brake force is generated by the springs and counteracted by the magnetic force of the coil. This is mostly used in fail-safe brake designs. These types of brakes/clutches are important for use in vacuum-compatible mechanisms in order to reduce the heat generation at joints during a holding mode.

Mechanical motion is sometimes transmitted from one component to another by means of a clutch. When both members must be connected and disconnected frequently, a clutch is the most useful device. The principal of operation is the same as a brake, but with no stationary parts involved.

It is important to note that the effectiveness of a brake or clutch is dependent on its ability to convert power to thermal energy and dispose of the frictional heat. The amount of energy to be dissipated is proportional to the frictional force and the speed. Special cooling methods may be necessary to remove the heat. Particles are generated from the frictional effects which must be trapped in a housing enclosure around the device.

Gears

Motion can be transferred from one shaft to another and a definite ratio between the velocities of the shafts can be maintained if a gear assembly is used. Gears are commercially available in many forms and are grouped according to the tooth forms, shaft arrangement, pitch and quality. The following summarizes these properties:

Tooth form	*Shaft arrangement*
Spur	Parallel
Helical	Parallel or skew
Worm	Skew
Bevel	Intersecting
Hypoid	Skew

Available types: Coarse (for $P_d \leq 20$),
Fine (for $P_d > 20$),
where P_d is the diametral pitch, defined as the ratio of the
number of teeth in the gear to the pitch circle diameter.

Available classes: Commercial,
Precision, and
Ultra-precision.

Material and Designation	Tensile strength (psi)	Yield strength (psi)	Hardness (BHN)	Condition
Cast irons:				
ASTM 20	22,000	. . . . .	156	As cast
30	31,000	. . . . .	201	As cast
60	62,5000	. . . . .	262	As cast
Plain carbon steels:				
AISI 1020	55,000	30,000	110	Hot-rolled
1020	78,000	66,000	155	Cold-worked
1040	76,000	42,000	150	Hot-rolled
1040	123,000	93,000	350	Cold-worked
1080	112,000	61,000	230	Hot-rolled
1080	189,000	142,000	385	Cold-rolled
1117	62,000	34,000	120	Hot-rolled
1117	80,000	68,000	163	Cold-worked
Alloy steels:				
AISI 3140	105,000	90,000	280	Heat-treated
3140	228,000	209,000	450	Heat-treated
4140	145,000	120,000	290	Normalized
4140	215,000	190,000	440	Heat-treated
4820	150,000	125,000	325	Heat-treated
4820	206,000	166,000	415	Heat-treated
6120	125,000	94,000	. . . .	Heat-treated
8620	122,000	98,000	245	Normalized
8620	173,000	142,000	375	Heat-treated
9310	152,000	120,000	350	Heat-treated
9310	180,000	140,000	375	Heat-treated
Stainless steels:				
AISI 303	90,000	35,000	160	Annealed
303	110,000	75,000	240	Cold-worked
416	75,000	40,000	155	Annealed
416	160,000	140,000	350	Heat-treated
Bronzes:				
Aluminum bronze ASTMB139	105,000	60,000	B100	
Phosphor bronze ASTMB1397	60,000	45,000	B70	
Silicon bronze ASTMB 99	58,000	25,000	B100	
Aluminum alloys:				
2024-T4	68,000	47,000	120	Heat-treated 1/2 hard
7075-T6	83,000	73,000	150	Heat-treated 3/4 hard
Non-metallics:				
Phenolic laminate				
NEMA, Grade C	11,000	. . . . .	M-103*	
NEMA, Grade L	14,000	. . . . .	M-105*	
Nylon				
ASTM6	8,700	6,000	M-100	2.5% moisture
ASTM 66	11,000	8,500	M-108	2.5% moisture

*Rockwell

Table 4.1 Typical Gear Materials (From [4.13], p. 8-117)

Considerations in designing geared transmissions include gear ratio, type of gear, gear shaft support, backlash, and lubrication. Backlash is important for precision machines. It normally arises due to the fabrication tolerances of the tooth thickness, and the clearance between meshing teeth allowed for lubricant and thermal expansion. Backlash results in the loss of motion when reversing the gear rotation. Backlash is also a function of the change in distance between the centers of the mating gears and the pressure angle:

$$B = 2(\Delta C) \cdot \tan \Phi$$

Where ΔC = Small change in center distance, and

         Φ = Pressure angle.

Typical gear materials and their physical properties are listed in Table 4.1.

A harmonic drive assembly is a transmission that has a high gear ratio and near zero backlash, but has high static friction. It consists of three components: 1) a wave generator, 2) a flexible spline, and 3) a circular spline which has teeth of the same pitch as the flexible spline (see Fig. 4.12). The wave generator includes an elliptical cam and a ball bearing with a flexible outer ring. A motor shaft drives the wave generator. The cam motion deforms the flexible spline and causes meshing of some involute teeth on the flexible spline with those on the circular spline. The circular spline is attached to the housing and is stationary. The difference in the number of teeth on the splines creates a small motion in the flexible spline of one revolution per cam rotation. This motion is transmitted to the output by the flexible spline body. The harmonic drive is used with in-line parallel shafts. It is commercially available in compact form and is used as a speed reducer for precision machines. It is usually immersed in oil while operating.

Several factors must be considered when using gears in vacuum. Material selection and lubrication are the most important. Significant mechanical and physical properties of the material include:

- yield strength
- surface hardness
- contact pressure fatigue resistance
- damping capabilities
- wear resistance
- machinability.

The first three characteristics determine the load-carrying capacity of the gear. The vibration of

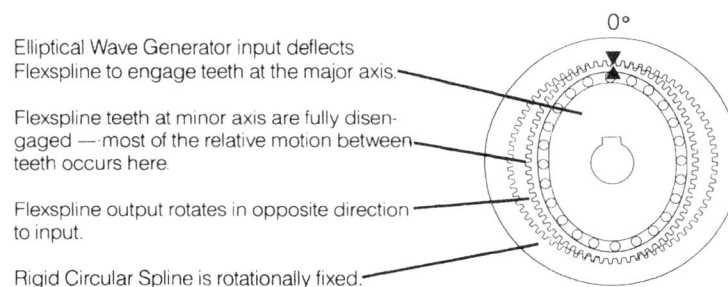

Elliptical Wave Generator input deflects Flexspline to engage teeth at the major axis.

Flexspline teeth at minor axis are fully disengaged — most of the relative motion between teeth occurs here.

Flexspline output rotates in opposite direction to input.

Rigid Circular Spline is rotationally fixed.

0°

Fig. 4.12 Harmonic drive assembly (From [4.14])

gears is also material dependent. Noise is reduced and damping is increased by using non-metallic gears. Wear-resistant material must be used to increase the life of the gear and reduce contamination. Good machinability is required in order to obtain good geometric tolerances.

A lubricating mechanism must be provided for the gear teeth in order to decrease friction and wear. The choice of a lubricant is a function of the gear load, speed, temperature, and ease of implementation. The lubricant is either wet or dry coated on the gear. The selection criteria for coatings includes:

- absence of scuffing in vacuum
- low wear rate
- low coefficient of friction
- lack of geometric modification during and after coating
- good adhesion between coating and substrate. [4.15]

Teflon gears do not require lubricants, and are useful for designing lubricant-free gear trains. They can be used as idler gears between stainless steel drivers and other driven components. The drawback of this method is that the power transmission is a function of the properties of Teflon, which is weaker and more temperature-sensitive than stainless steel.

Chain Drives

Motion can also be transmitted via chain drives, which can have high efficiency (98 to 99%) and no slippage. Two sprocket wheels (driver and follower) are used with the chain. The power rating of the chain depends on the number of teeth and rotary speed of the smaller sprocket. Multiple strand chains can be used for applications requiring high power. It is important to pre-load the chain properly to avoid drooping or breakage. For vacuum applications the chain drive assembly must be lubricated, and, in order to reduce particle contamination, cannot have direct exposure to the surroundings. An appropriate vacuum grease is the most widely used lubricant for this purpose.

4.4 Machine Elements

The design of vacuum-compatible mechanisms involves the use of machine elements that are suitable for that environment. This section identifies seals, bearings, linkages and joints with their vacuum properties.

Seals

In vacuum systems, static seals are needed where flanges are joined. Dynamic seals are used to transmit motion through the vacuum envelope. Characteristics of current dynamic sealing methods used in mechanical feedthroughs have been covered in section 4.1.

Static seals include elastic and metal gaskets. The former are used for pressures down to 7.5×10^{-9} Torr (10^{-7} Pa) and temperatures up to 300 °C, making them suitable for the baking cycle of the flange. In addition to these characteristics, the gasket material is chosen according to the outgassing rate and gas permeability. Table 1 in the reprint by Weston shows the properties of some synthetic materials used for vacuum seals.

The leak rate of an elastic gasket can be reduced by applying a high vacuum grease (e.g. silicone grease) to the gasket surface. Figure 4.13 shows the effect that the amount of squeeze has on the

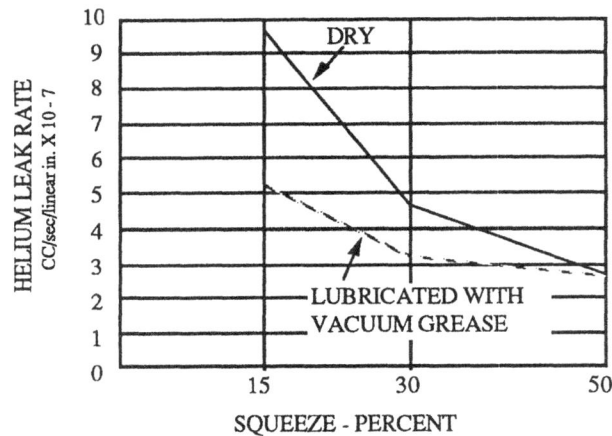

Fig. 4.13 Percent squeeze on the O-ring vs. leak rate (Adapted from [4.17])

leak rate. The groove design for this type of gasket requires a surface finish smoother than 16 μm to reduce the leak rate significantly.

Metal seals are mainly used for lower pressures and higher temperatures (up to 500°C). Gold, copper, indium, aluminium, and silver are sometimes used as metallic gaskets. A sharp knife-edge or other irregularity on the flange face is used to deform the soft gasket material beyond its elastic limit into the plastic region. Care must be taken to avoid damaging the high surface finish of the flange face. This type of seal is not re-usable.

Another type of metal gasket uses the springy characteristics of some metals, e.g., monel and stainless steel. These materials are coated with softer materials such as indium or copper and are compressed against the flange faces in the groove. This type of seal is re-usable but more expensive. The surface cleanliness, flatness, and finish are even more important than that for elastic gaskets. Weissler and Carlson [4.16] present a thorough explanation of vacuum flanges.

Bearings

There are three general types of bearings, classified according to the type of support they provide:

    1) Journal bearings, which support a rotating shaft in the radial direction,
    2) Thrust bearings, which support a rotating shaft in the axial direction, and
    3) Guide bearings, which support a non-rotating shaft in the radial direction.

The following parameters should be considered when selecting bearings for vacuum-compatible mechanisms:

    • frictional torque,
    • thermal conductance across the races,
    • bearing stiffness,
    • material, and
    • contamination.

Ideally, bearings do not show friction in the direction of motion. This can be achieved by using a non-contact levitating system (see section 4.5). The design and operation of roller and slider bearings are explained in the text by Shigley and Mitchell [4.18]. The contact friction in these types of bearings converts power to heat, which is especially undesirable in vacuum. The generated heat

194

is conducted to the sink through the bearing. If thermal conductance of the bearing is low, the temperature rise will cause thermal expansion and an additional load on the bearing mount. Experiments have shown that the frictional torque and the thermal conductance is a function of the type and amount of lubricant [4.19]. Methods of reducing friction and increasing thermal conductance are discussed in chapter 3. Dry lubrication methods for different types of bearings have been studied, and some examples are shown in Table 4.2.

Classification	Dry Lubrication Method	Configuration
Cross Roller	Outer ring, Inner ring, Roller WS_2. Surface process by Dicronite inc. of U.S.A. Roller and ball : Sintered alloy	
Thrust	Outer ring, Inner ring, Ball Sputtered MoS_2. Retainer : Sintered alloy	
Ball	Outer ring, Inner ring, Ball Sputtered MoS_2. Retainer : Sintered alloy	
Linear guide	Truck rail, Casing, Ball Burnished MoS_2	
Ball Screw	Screw, Nut, Ball Burnished MoS_2	
V roller	Outer ring: Sputtered MoS_2, Mover: PTFE	
Support roller	Same as the thrust bearing outer ring : Burnished MoS_2	

Table 4.2. Possible dry lubrication methods for various bearing types (adapted from [4.21])

The bearing stiffness is determined by measuring the bearing moment compliance. It is defined as the amount of torque exerted on the bearing per deflection angle of the bearing's inner race relative to the outer race. The bearing's frictional torque, thermal conductance, and stiffness are also a function of the preloading effects provided by the cage and external fit. The use of clearance fit for a four point ball bearing with adjustable clamps has been recommended to be able to control these parameters more closely [4.19].

The bearing material is typically stainless steel or other hard material with a high compressive and fatigue strength. Contamination results from the debris generated by the wear of moving components and absorbed foreign particles during assembly. Bearing seals are used primarily to reduce contamination due to lubricants and component wear.

A number of bearing types are available for vacuum, e.g., slider and sealed ball bearings. Slider bearings use fluorocarbon plastics, typically PTFE (Teflon), with a metal sleeve to reduce friction and provide support for the shaft. The metal sleeve also minimizes the effects of thermal expansion by the slider and increases the heat transfer rate to the housing. This type of bearing is typically used over a temperature range of -40 °F to 250 °F. The plastic slider will stick to the shaft at high temperatures. Some experiments have shown that the PTFE might rotate with the sleeve [4.20]. The PTFE can be locked to a perforated metallic sleeve because PTFE can expand through the holes when compressed. This type of bearing is maintenance-free for repeated cyclic temperature variation.

Stainless steel and ball and roller bearings for use in vacuum mechanisms are typically lubricated with vacuum grease. They are mainly used for high temperature applications or where a large variation in the temperature variations are expected. Ball and roller bearings are also more rigid and offer a higher stiffness than slider bearings.

Linkages/Joints

A mechanism is made up of a number of linkages which are connected together through joints. These links can be configured in different shapes to produce the desired outputs. They can be considered as transmission elements with a variable transmission ratio throughout the workspace. An important property of a link is its stiffness, which must be quite high for precise positioning and handling. This is specifically true when long links are used for moving loads. The space shuttle arm uses thin-walled multiple graphite-epoxy tubes with aluminum joint interfaces which provide an overall stiffness of 10 lb/in. (1760 N/m) [4.22]. Other desirable properties include low mass and thermal expansion, and oxidation resistance. Glass-ceramic matrix composites have been developed which exhibit these properties, but they are fragile and their manufacturing costs are high.

Revolute and prismatic joints are the types used most for industrial robots. Rotation about an axis is performed by the revolute joint and translation along an axis is allowed by the prismatic joint. Each has one degree of freedom. The joint design for vacuum environment includes the choice of materials, seals and geometry. Seals are necessary to reduce the contamination. The shape of the joint must be such that the seals are most effective.

4.5 Magnetic Levitation

Magnetic suspension/levitation maintains the position of a body along or about one or more axes attached to a certain reference frame. The levitation is a result of the force actions of various magnetic fields, and achieves the completely contactless support of its critical components, thereby eliminating mechanical friction [4.23].

Ever since Holmes [4.24] showed that a vertical ferromagnetic needle could be supported in macroscopic equilibrium by magnetic force alone (using a variable magnetic field) many theories and experiments have come about in an effort to broaden the "state of the art" of magnetic levitation systems. Among these, the magnetic suspension and balance system for wind tunnel models received early and extensive attention [4.25,4.26]. The different applications of contactless suspension and levitation to high-speed ground transportation, however, continue to be most compelling to scientists and the media [4.27].

The most recent application research trend in engineering has been the development of active mechanisms for critical applications in superclean, vacuum, and space environments. Motivation for this practice is the predominating need to meet the rapidly increasing demand for high cleanliness levels and flexible automation. As these demands are met, the more general intrinsic assets of magnetic levitation are also discovered and exploited.

Features of Magnetic Levitation

Because of their completely contactless support of critical components, magnetic levitation systems are considered important mechanisms for flexible and precise positioning and force control, especially in superclean and vacuum environments. As they are also contactless, they have no mechanical friction, and thus are useful for high-speed rotating axes. Levitation systems are wear-free and maintenance-free, and have no backlash. They thereby lend themselves to potential use in contamination-free and high-precision mechanisms. Multi-degree-of-freedom joints can be built using magnetic levitation.

Magnetic levitation systems generate no particle contaminants, and thus are highly suitable for operation in superclean environments, such as VLSI manufacturing and biotechnology experiments. Because no lubrication is necessary, and since vacuum-compatible materials may be employed, magnetic levitation systems are certain to become indispensable in vacuum applications.

On a commercial scale, magnetically levitated passenger trains are fast, power-conserving, comfortable, and quiet [4.28], but they are also more expensive to build, operate and maintain than conventional trains. Research is continually under way to make these magnetically levitated trains more cost-effective. Although various principles have been studied and a variety of prototypes designed, only two types of magnetically levitated vehicles have been built and presently operate: electromagnetic systems (in both Germany and Japan) and superconductivity (in Japan). The trains in Europe run on magnets that attract each other, and the trains in the Orient employ magnets that mutually repel.

Any magnetic levitation system is inherently unstable without proper control, so either analog control or fast digital control is imperative in making such a system stable. On the other hand, the dynamics of the levitation system are highly nonlinear and coupled, placing a heavy burden on controller speed. This, of course, contributes to the high cost.

Classifications of Magnetic Levitation

Magnetic levitation is classified into three types, depending on the type of magnet used. They may be electromagnets or permanent magnets or utilize super-conductivity/diamagnetism. Permanent magnet systems are the most economical, while superconductivity is the most expensive but the most energy-conserving. Electromagnets are the most flexible and currently the easiest to control; they are therefore the most widely used.

Magnetic levitation is further classified according to the types of magnetic fields employed:

1. Magnetic levitation with a constant field utilizes repulsive forces from either permanent magnets or super-conductivity/diamagnetism. The system can be stable without active control, but the control can be very difficult to achieve.

2. Magnetic levitation in a passive-variable field is achieved either by restoring forces of DC electromagnets, with eddy current repulsion of AC electromagnets or by using the resonant circuit of AC electromagnets.

3. Magnetic levitation with an active variable field utilizes attractive forces from actively controlled DC electromagnets. This type allows the most flexibility and controllability, and presently is very widely used.

Current Advances in Magnetic Levitation

With the recent rapid advances in solid-state electronics and in analog and digital control technology, many applications of magnetic levitation have become viable. Many published papers report a wide range of results from recent application research, as discussed in this section.

Using magnetic levitation to develop new types of actuators and sensors is a very attractive research topic. These types of actuators have all the good features of magnetic levitation, including cleanliness, and the lack of contact, friction and wear. Additionally, they have no lubrication requirements, and can be either linear or rotational. In [4.29] Maresca realized an integrated magnetic bearing actuator and sensor using a specially designed ferrite pole piece. The sensors have < 0.25 μm resolution and negligible temperature drift, each actuator providing a controllable force of 0-10 N.

Matsumura, *et al.* [4.30] designed a moving-magnet type of linear DC motor, which is completely levitated by the moving magnets. The chief operating characteristics of this motor are: short stroke, high accuracy, high response-speed, and large damping-force.

Higuchi, *et al.*[4.31] developed a new type of stepping motor with the rotor suspended, actuated and positioned without contact. They accomplished this by employing feedback-controlled magnetic forces. The researchers project that these motors will be useable as direct-drive motors for ultra-clean and vacuum environments.

Because of the absence of friction, magnetic bearings have been used in mechanisms with high-speed rotating axes [4.32] such as turbo-molecular pumps and centrifugal separators. Ulbrich [4.33] studied the theory and application of magnetically-supported rotors and subsequently discussed their modeling, measurement and control.

Richards, *et al.* [4.34] described the self-generated rotation of magnetic levitators using a proximity height detector, in which the supporting DC electromagnet provides a torque about the vertical axis, so that the floating object rotates continuously. In a more recent paper Ciric, *et al.* [4.35] demonstrated theoretically the stable electromagnetic levitation of spinning conducting cylindrical rotors and analyzed quantitatively the stability conditions, losses, torque/speed characteristics and high-speed rotation.

A much smaller mechanism, but equally complicated to design, would be a magnetically levitated multi-degree-of-freedom robotic hand. By employing magnetic levitation, a robot hand could have remote center compliance, fine position control, fine force control, variable damping and stiffness for several degrees of freedom at the same time. A magnetically supported intelligent hand (MSIH) as proposed in [4.36] has five degrees of freedom (three translational and two rotational) and is electromagnetically feedback-controlled, using magnetic gap sensors for both position-sensing and force-sensing.

A similar example is the IBM "magic wrist" [4.37] a six degree-of-freedom magnetically levitated fine-motion wrist. The wrist's floater is a hexagonal aluminum structure with embedded coils, and a stator with permanent magnets holds and controls the floater by passing and changing current through the coils. The positioning and orientation of the wrist are achieved with an optical sensing system that employs projecting co-planar beams from three LEDs attached to the wrist 120° apart. By changing parameters, the wrist can serve as either an RCC (remote compliance center) device, translator, rotator, slider, or plunger.

To meet the various requirements of flexible automation, it is necessary to integrate the above components into a reliable, flexible, intelligent robot. From the proper utilization of magnetic levitation, robots can exhibit greatly improved abilities and will be able to do precise manipulation, especially in superclean and vacuum environments. Higuchi, *et al.* [4.38, 4.39] have designed and constructed an experimental closed-loop five link robot with the two base joints direct-driven by two conventional step motors and the two upper joints composed of magnetic bearings capable of small-range, fine positioning. More practical magnetically levitated robots will most likely appear, and will play a pivotal role in manufacturing within the vacuum environment.

Considerations for Vacuum Applications

Being contactless, frictionless, and requiring no lubrication, magnetic levitation mechanisms have a promising future in terms of their applications to vacuum environments. In these applications, the working temperature range, efficiency of energy transfer, and the choice of materials all have to be considered. In a magnetic levitation system, most of the parts are steel, but the magnetic materials can be coated for high-vacuum applications without loss of magnetic performance. The insulation of windings, a concern in the operation of vacuum motors, warrants particularly close consideration.

The working temperature and power efficiency depend mainly on the structure and the control. In designing the structure, heat generation and dissipation must be considered and minimized. Once the choice of hardware is determined, the main factor influencing both temperature and efficiency is the control scheme employed, which can make a considerable difference. For example, if the local linearization technique is used, large operating currents are required, and many of the forces from these currents are cancelled in pairs, without any work output resulting. The wiring resistance, however, will then consume a large part of the input energy, reducing the efficiency and unnecessarily generating heat. Employing non-linear control reduces this energy consumption dramatically, and will usually bring it to a minimum.

References

[4.1]   Burleigh Instruments, Inc.,*The Piezo Book*, 1988. May be obtained from Burleigh Instruments, Burleigh Park, Fishers, NY, 14453.

[4.2]   Jardine, A.P., M. Ahmad, R.J. McClelland and J.M. Blakely, "A simple ultra-high vacuum shape memory effect shutter mechanism", *J. Vac. Sci. Technology*, A6 (5), Sept/Oct 1988.

[4.3]   Schetky, L., "Shape Memory Effect Alloys for Robotic Devices", *Robotics Age,* 13, July, 1984.
Figures 4.7 and 4.8 reprinted with permission from Robotics Age.
Copyright 1984 Helmers Publishing, Inc., Petersborough, NH 03458

[4.4]   Nof, S.Y., *Handbook of Industrial Robotics*, New York, New York, John Wiley & Sons, Inc., 1985.
Figures 4.9 and 4.11 reprinted by permission of John Wiley & Sons, Copyright © 1985.

[4.5]   Lindsey, J.S., L.A. Corkan, and D. Erb, "Robotic work station for synthetic chemistry: on-line absorption spectroscopy, quantitative automated thin-layer chromatograghy, and multiple reactions in parallel", *Review of Science and Instrumentation*, 59 (6), pp. 940-950, June, 1988.

[4.6]   Nakamura, Y. and T. Yoshikawa, "Design and Signal Processing of Six-Axis Force Sensors," in *4th International Symposium of Robotics Research*, Santa Cruz, California, August, 1987.

[4.7]   McAlpine, G.A., "Tactile Sensing", *Sensors*, Vol. 3, No. 4, pp. 7-16, April, 1986.

[4.8]   Prahl, B.W. and P.M. Tracey, "Pressure/Tactile Sensing With Intrinsic Fiber-Optic Sensors", *Sensors*, Vol. 3, No. 8, pp. 48-52, August, 1986.

[4.9]   Creighton, A. and M. Hercher, "The Laser Extensometer", *Sensors*, Vol. 3, No. 8, pp. 43-47, August, 1986.

[4.10] Hord, R.M., *Handbook of Space Technology*, Boca Raton, Florida, CRC Press, Inc., 1985.

[4.11] Bailly, M., M. Tulet, and S. Flamenbaum, "CCD Imaging Sensor," in *Proceedings of the Tenth IFAC Symposium*, Toulouse, France, June, 1985.

[4.12] Fu, K.S., R.C. Gonzalez, and C.S.G. Lee, *Robotics Control, Sensing, Vision, and Intelligence*, New York, New York, McGraw-Hill Book Company, 1987.

[4.13] Baumeister, T., E.A. Avallone, T. Baumeister III, *Mark's Standard Handbook for Mechanical Engineers*, New York: Mac Graw-Hill Co., 1978. Figure 4.10 and Table 4.1 reproduced with permission.

[4.14] "The Designers Drive", brochure from Harmonic Drive, A Division of Quinay Technologies, 51 Armory Street, Wakefield, MA, 01880. Tel: (617) 245-7802, Jan, 1984.

[4.15] Borrien, A., L. Petitjean, "Robotic Joint Experiments Under Ultravacuum", in *22nd Aerospace Mechanisms Symposium Proceedings*, Hampton, VA., May, 1988.

[4.16] Weissler, G.L., R.W. Carlson, "Vacuum Physics and Technology",*Methods of Experimental Physics*, Vol. 14., New York: AcademicPress, Inc., 1979.

[4.17] Parker O-ring Handbook, Catalogue number ORD 5700, Irvine, CA: Parker Seal Group, pp. A2-15, A2-16, 1982.

[4.18] Shigley, J.E. and L.D. Mitchell, *Mechanical Engineering Design, Fourth Ed.*, New York: McGraw Hill, 1983.

[4.19] Rowntree, R.A., "The Properties of Thin-Section, Four-Point-Contact Ball Bearings in Space", *19th Aerospace Mechanism Symposium*, NASA CP-2371, pp.141-166, May 1985.

[4.20] Kubiak, R.A.A., P. Driscoll, V. Manning, and R. Houghton,"The Use of Polytetrafluoroethyene Bearings in Ultrahigh Vacuum", *J. Vac. Sci. Technology*, A, 4(4), pp. 1951-1952, July/August 1986.

[4.21] Nio, S., T. Suzuki, H., Zenpo, K. Yokoyama, H. Wakizako and S. Belinski, "Vacuum Compatible Robot for Self-contained Manufacturing", *Proceedings of the First International Workshop on Vacuum Mechatronics*, University of California, Santa Barbara, Feb. 2-3, 1989.

[4.22] Aikenhead, B.A., "Canadarm and the Space Shuttle", *J. Vac. Sci. Technology*, A 1(2), pp. 126-132, April-June 1983.

[4.23] Jayawant, B.V., *Electromagnetic Levitation and Suspension Techniques*, London: Edward Arnold, 1981.

[4.24] Holmes, F.T., "Axial Magnetic Suspension", *Review of Scientific Instruments*, Vol. 8, p. 444, Nov, 1987.

[4.25] Tilton, E.L., "Dynamic Stability Testing with a Wind Tunnel Magnetic Model Suspension System", M.S. Thesis, MIT, Jan, 1963.

[4.26] Clemens, P.L. and A. H. Cortner, "Bibliography: The Magnetic Suspension of Wind Tunnel Models", Rep. no. AEDC-TDR-63-20, Feb, 1963.

[4.27] Rogg, D., "General Survey of the Possible Applications and Development Tendencies of Magnetic Levitation Technology", *IEEE Transactions on Magnetics*, Vol. Mag-20, no. 5, p. 1696, Sept, 1984.

[4.28] Ohtsuka, T., "Japanese National Railway System Using Electromagnetic Repulsive Force between Normal Track and on-board Superconducting Magnets", *IEEE Tansactions on Magnetics*, p. 1982, 1981.

[4.29] Maresca, R.L., "An Integrated Magnetic Actuator and Sensor for use in Linear or Rotary Magnetic Bearings", *IEEE Transactions on Magnetics*, Vol. Mag-19, no. 5, p. 2094, Sept. 1983.

[4.30] Matsumura, F., S. Maeda, and M. Fujita, "Completely Contactless Linear DC Motor Using Magnetic Suspension", *Electrical Engineering in Japan*, Vol. 107, no. 1, p. 95, 1987.

[4.31] Higuchi, T. and H. Kawakatu, "Super-clean Actuators for Machines and Robots", *Proceedings of IECON '87*, p. 303, 1987.

[4.32] Kant, M., "General Study of Electromagnetic Bearings", *IEEE Transactions on Magnetics*, Vol. Mag-11, no. 5, p. 1511, Sept, 1975.

[4.33] Ulbrich, H., G. Schweitzer, and E. Bausar, "A Rotor Supported without Contact: Theory and Application", *Proceedings of the Fifth World Congress on Theory of Machines and Mechanisms*, p. 181, 1979.

[4.34] Richards, A.H., J.G. Magondu, R.N.W. Laithwaite, and P.N. Murgatroyd, "Self-generated Rotation in a Magnetic Levitator", *IEEE Proceedings*, Vol. 128, pt. A, no. 6, p. 449, Sept, 1981.

[4.35] Ciric, I.R.and R.M. Mathur, "Electromagnetic Levitation of Rotating Cylinders", *IEEE Proceedings*, Vol. 132, pt. A, no. 1, p. 21, Jan, 1985.

[4.36] Higuchi, T., M. Tsuda, and S. Fujiwara, "Magnetically Supported Intelligent Hand for Automated Precise Assembly", *Proceedings of IECON'87*, p. 926, 1987.

[4.37] Hollis, R.L., A.P. Allan, and S. Salcudean, "A Six Degree-of-Freedom Magnetically Levitated Variable Compliance Fine Motion Wrist", *presented at* the Fourth International Symposium on Robotics Research, Santa Cruz, Aug, 1987.

[4.38] Higuchi, T., K. Oka, and H. Sugawara, "Development of Clean Room Robot with Contactless Joints Using Magnetic Bearings", *Proceedings of USA–Japan Symposium on Flexible Automation*, 1988.

[4.39] Higuchi,T., "Applications of Magnetic Bearings to Robotics", *Proceedings of First International Symposium on Magnetic Bearings*, Zurich, 1988.

Bibliography

Frazier, R.H., P.J. Gilinson, and G.A. Oberbeck, *Magnetic and Electric Suspensions*, MIT Press, 1974.

Matsuda, R., M. Nakagawa, and I. Yamada, "Multi Input–Output Control of Magnetically Suspended Linear Guide", *IEEE Transactions on Magnetics*, Vol. Mag-20, no. 5, p. 1690, Sept, 1984.

Matsumura, F. and S. Tachimori, "Magnetic Suspension System Suitable for Wide Range Operation", *Electrical Engineering in Japan,* Vol. 99, no. 1, p. 29, 1979.

Matsumura, F., and T. Yoshimoto, "System Modeling and Control Design of a Horizontal-Shaft Magnetic-Bearing System", *IEEE Transactions on Magnetics,* Vol. Mag-22, no. 3, p. 196, May, 1986.

Mizuno, T. and T. Higuchi, "Design of the Control System of Totally Active Magnetic Bearings: Structures of the Optimal Regulator", *Proceedings of International Symposium on Design and Synthesis,* Tokyo, July 11-13, p. 534, 1984.

Morishita, M., and T. Ida, "Constant Gap Width Control of Magnetic Levitation Systems by Attractive Force", *Electrical Engineering in Japan,* Vol. 103-B, no. 6, p. 95, 1983.

Nakamura, Y., M. Tsuda, and D. Chen, "Magnetic Servo Levitation with Large Air Gap", *a proposal submitted* to the National Science Foundation, Nov, 1988. For full details on the subject, contact the Center for Robotic Systems in Microelectronics, University of California, Santa Barbara.

Salcudean, S., and R.L. Hollis, "A Magnetically Levitated Fine Motion Wrist: Kinematics, Dynamics and Control", *Proceedings of IEEE International Conference on Robotics and Automation,* Philadelphia, p. 261, April, 1988.

Weissler, G.L., R.W. Carlson, *Methods of Experimental Physics*, New York: Academic Press, Inc. 1979.

Weston, G.F., *Ultra High Vacuum Practice*, London: Butterworths Publishers, 1985.

Weston, G.F., "Ultra-High Vacuum Line Components", *Vacuum*, 34(4), pp. 619-629, 1984.

Wong,T.H., "Design of a Magnetic Levitation Control System: An Undergraduate Project", *IEEE Transactions on Education,* Vol. E-29, no. 4, p. 196, Nov, 1986.

Vacuum/volume 34/number 6/pages 619 to 629/1984
Printed in Great Britain

0042-207X/84$3.00 + .00

Ultra-high vacuum line components*

G F Weston, *Philips Research Laboratories, Redhill, Surrey RH1 5HA, England*

received 29 April 1983; in revised form 2 August 1983

Vacuum line components such as demountable seals, mechanical feed-throughs and valves suitable for ultra-high vacuum application are reviewed, with emphasis on the most recent developments

1. Introduction

In any vacuum system there is usually a requirement for valves, demountable seals, electrical and mechanical feed-throughs, etc. which are classed together here as vacuum line components. Most vacuum equipment manufacturers include a range of such components amongst their products, with compatible interconnections which can be assembled with pumps and gauges to form the complete vacuum system.

For ultra-high vacuum applications such components must conform to the rigid material requirements previously discussed in an earlier article[1] so as to present the minimum source of gas to the vacuum system. In particular, they should be capable of being baked to at least 250°C, and give a low outgassing rate and negligible gas permeability when forming part of the chamber walls. In general this precludes synthetic materials such as elastomers, epoxy resins or plastics, which are employed extensively at higher pressures, above 10^{-5} Pa. The elastomers for example, because of their high elasticity are especially useful as a gasket material in demountable seals and valves where repeated opening and closing is required. Harder materials such as PTFE have properties which make them suitable as seatings or bushes for mechanical movements. Although there are some synthetics which can be baked above 200°C, in general there are no materials with similar mechanical properties which fulfil all the requirements of ultra-high vacuum applications. The vacuum engineer has to resort to rather different approaches to arrive at equivalent components for pressures below 10^{-6} Pa. The design and development of such components has been an integral part of the ultra-high vacuum studies since the 1950s, and today most manufacturers produce a range of ultra-high vacuum line components, which, although less convenient and more expensive than their higher pressure counterparts, provide a practical answer to the stringent environmental conditions imposed in attaining ultimate pressures below 10^{-6} Pa.

In this article the general design of such components is discussed with emphasis on the most recent developments.

2. Demountable seals

Demountable seals can be defined as the static seals made between the various vacuum components which can be broken and resealed to facilitate dismantling for system changes or maintenance. For ultra-high vacuum application such seals are almost entirely in the form of stainless steel flanges clamped or bolted together with some form of gasket compressed in between. The properties of the gasket are such that it must be deformable under the minimum pressure so that the gasket material can flow and fill surface irregularities on the flanges. On the other hand it should be resilient enough to maintain contact over a range of compressions to allow for expansion and possibly flexing. Natural or synthetic rubber is an almost ideal gasket material, giving leak tight joints with low loads on the flange bolts and being resilient enough to be re-usable several times.

Because in general the gasket surface in contact with the vacuum is minimal and the gas diffusion path length through the gasket is relatively long, the demands on degassing rate and gas permeability are less stringent than for the rest of the vacuum envelope. Synthetic rubbers are therefore suitable for vacuum equipment. However, for ultra-high vacuum systems it is essential that the material used for the gasket will withstand temperatures up to say 250°C in order that the flanges can be baked with the rest of the system to reduce the outgassing rates. Few rubber-type materials can be raised to such temperatures without changing their physical properties or decomposing. Nevertheless there are one or two which can withstand such temperatures and have a low enough outgassing rate after baking to be of interest, especially if the pressure requirements are at the upper end of the ultra-high vacuum range, $>10^{-7}$ Pa, and if the baking temperature can be lowered slightly. In Table 1, the outgassing rate and gas permeation constant of suitable materials are listed. The values depend on the history of the sample and are likely to vary from sample to sample so that they should only be taken as indicative. The values for the glass are listed for comparison. The high gas permeation rate through the synthetic materials especially for hydrogen and helium not only represents a limitation to the ultimate pressure that can be attained but can cause serious interference with helium leak detection.

The two best known synthetic materials which have been used for ultra-high vacuum systems are Viton-A, a product of du Pont,

---

* A paper in the Education Series: The Theory and Practice of Vacuum Science and Technology in Schools and Colleges.

Table 1. Outgassing rates and gas permeability constants for synthetic materials used for high vacuum seals

Material	Baking schedule	Outgassing rate Pa m³ s⁻¹ m⁻²	Gas permeability constant m² s⁻¹ at 23°C Nitrogen	Hydrogen	Helium
Viton-A	Unbaked	10^{-4}	—	2×10^{-12}	8×10^{-12}
	24 h at 200°C	3×10^{-6}			
Polyimide	Unbaked	5×10^{-5}	10^{-13}	10^{-12}	2×10^{-12}
	22 h at 300°C	4×10^{-8}			
Kalrez	300°C	4×10^{-8}			
Viton E60C	300°C	3×10^{-8} ($\sim10^{-6}$ at 150°C)			
Fernico sealing glass	Unbaked	10^{-5}	Not measurable		5×10^{-16}
	24 h at 300°C	10^{-12}			

and the polyimides, the best known of which is Kapton-H. Both have been used successfully on seals in vacuum systems pumping down to 10^{-7} Pa[2,3].

Viton-A is an elastomer which can be baked up to 200°C, after which its degassing rate is around 10^{-6} Pa m³ s⁻¹ m⁻² (10^{-9} torr 1 cm⁻² s⁻¹). It can be used in the form of a toroid ('O'-ring) in a conventional vacuum flange seal, i.e. compressed between a flat flange and one having a locating groove of trapezium cross-section or in the small KF type seals. Details of such seals are given in most text books on vacuum technology and will not be described here.

Kapton-H will withstand higher temperatures and after baking at say 300°C has a much lower outgassing rate than Viton-A. It is a much harder material, a bit like nylon, and cannot be satisfactorily used in a normal 'O'-ring seal configuration. It needs a higher compression force and will suffer permanent distortion if compressed more than 20%. Since it also has a high expansion coefficient the seal has to be very carefully designed. One system which appears to be satisfactory uses a thin sheet of Kapton-H compressed between a groove and a knife edge[3]. Another disadvantage is that Kapton-H is to some extent hygroscopic, so that on exposure to air, water vapour is readily adsorbed. The unbaked degassing rate can therefore be fairly high.

More recently de Chernatony[4] has reported on the vacuum properties of a new elastomer material introduced by du Pont under the trade name of Kalrez, which appears to have rather better properties. For example, its degassing rate after baking to 300°C is comparable with that of the polyimides. It is now available as a formed gasket for flange sealing but it is expensive. Du Pont have also introduced another variant of the Viton family under the code name of E60C. It differs from other Vitons mainly in the improved processing and although the vacuum properties have not been proven as better than for Viton-A, it has a lower outgassing rate at room temperature[4].

For lower pressures synthetic gaskets are not suitable and the vacuum engineer must resort to all metal seals, i.e. the use of metal gaskets. Metal gaskets are available in a number of forms with material ranging from soft metals such an indium, aluminium and gold to harder metals such as silver, copper, monel and even steel. They require considerably greater pressure than elastomers to ensure flow of the gasket metal into the irregularities of the flange faces and they lack the resilience, so that once the seal has been broken the gasket is rarely re-usable. Indeed Buchter[5] showed that leak tight metal seals could not be made unless the metal in the contact area was permanently deformed by compressing beyond the elastic limit into the plastic deformation range.

Numerous designs of ultra-high vacuum metal seals have been described in the literature and since there does not appear to be any really scientific foundation for seal design analysis, most depend on empirical evaluation. In general, the very soft metals, such as indium and gold, are used as unconfined gaskets between two flat flanges so that they are free to deform inwards and outwards along the flange surfaces. The harder metals are usually partly confined by contoured flanges so that the gasket is only free to deform in one direction.

The unconfined gasket normally consists of a wire ring fabricated by fusing the wire ends together. It is then mounted between two stainless steel flanges which are machined accurately flat and to a fine finish. Care must be taken to avoid any radial scratches on the flange surfaces. The ring may be held in position with a spider as in Figure 1, or just positioned with a suitable tool

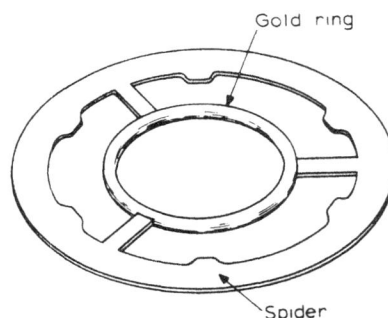

Figure 1. Gold wire gasket with positioning spider.

whilst the bolts are tightened. The main criterion for the design is that the wire should be of fairly uniform thickness. The diameter must be large enough to form a reasonable contact area along the flange surface when compressed but not so large that there is a danger that the atmospheric pressure will push it inward and burst through. 1.0 mm dia wire is about the maximum size, but 0.5 mm dia is normally employed compressed to about 0.2 mm, i.e. the spider in Figure 1 is 0.2 mm thick.

An alternative arrangement is offered by Leybold–Heraeus who market an aluminium disk with a diamond shaped raised sealing rim. The rim acts in the same way as the wire whilst the disk locates the sealing rim and limits the compression. The advantage of the ring type of seal is that the flanges are fairly easily machined and the wire rings can be made up by the user. The disadvantage is that in making the seal and· in subsequent temperature cycling the wire tends to weld to the flanges which may necessitate resurfacing the flanges before they can be re-used. They are therefore most suitable for semipermanent seals where long periods elapse before they are required to be opened. Another problem is that the flange surfaces are vulnerable to damage when exposed. Quite small scratches can degrade the vacuum performance.

Gold is the most satisfactory metal for such seals. It is not affected by contaminants such as oil vapours and it can be baked up to 500°C. Also the value of the gold can be reclaimed. Gold wire seals are sold commercially particularly for large diameter flanges.

For smaller flanges the trend has been to move to copper gasket seals using designs which have become standardized. Copper, being a harder metal than gold, requires a larger force to cause deformation. It is unsatisfactory as a wire seal and although 'diamond section' ring gaskets can be used most seals employ relatively 'heavy section' gaskets which are partially confined.

A number of configurations for copper gasket seals have been proposed in the literature and a selection are shown diagrammatically in Figure 2. The designs are self-explanatory and

Table 2. ISO metal flange recommended dimensions

Nominal tube i.d. (mm)	Outside diameter of flange (mm)	Diameter of bolt hole centres (mm)	No. of bolt holes	Diameter of bolt holes (mm)
16 (¾ in.)	34	27	6	4.3
35 (1½ in.)	69.5	58.7	6	6.6
63 (2½ in.)	113.5	92.1	8	8.4
100 (4 in.)	152	130.2	16	8.4
150 (6 in.)	202.5	181	20	8.4
200 (8 in.)	253	231.8	24	8.4
250 (10 in.)	306	283.5	32	8.4

common usage. Although the sealing method was not specified it was clear that, since contoured flanges were required, the sealing configuration had also to be agreed. Most manufacturers have opted for a Varian design[6] marketed under the trade name of ConFlat® and this has become the accepted standard.

The ConFlat flange seal is illustrated in Figure 3. It consists of two symmetrical flanges having an outside rim to confine the flat copper gasket and a concentric knife-edge of a particular cross

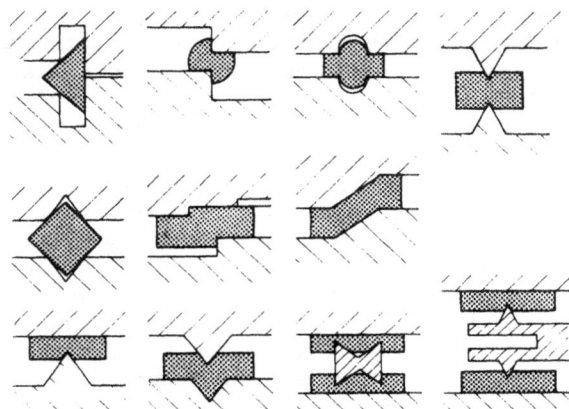

Figure 2. Some copper gasket seal configurations.

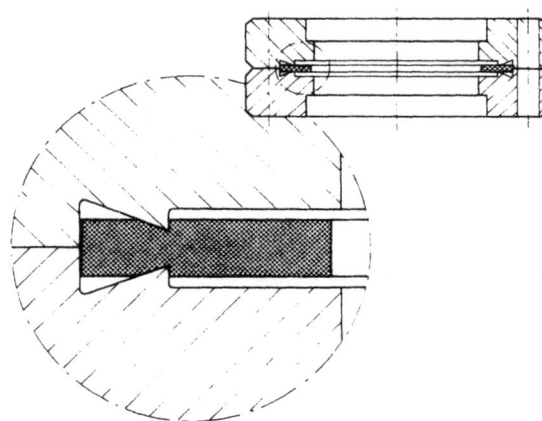

Figure 3. Basic design of the ConFlat® flange seal from Varian associates.

generally rely on a sharp-edge biting into the gasket. This gives the maximum specific force for deformation of the copper from the minimum force on the flange bolts. The relative merits of the various designs are difficult to assess and in the early days of ultra-high vacuum one or two of these designs were exploited commercially. The problem was that the different flange designs were not compatible and equipment from different manufacturers had to be modified to allow interconnection. A similar problem had arisen for the elastomer 'O'-ring seals and in the early 1960s the International Standards Organisation (ISO) in co-ordination with Pneurop (an assembly of European manufacturers of compressors, vacuum pumps and pneumatic tools) set about drawing up international standards for vacuum flanges as part of a wider study of vacuum equipment. In the case of the UHV metal gasket seals, recommended specifications have been arrived at for a range of flanges to cover the interconnection of different size tubing. The specification details the flange diameter, number of bolt holes and their position etc. but does not specify the actual sealing mechanism. Table 2 gives the dimensions of those in

section. When bolted together the knife-edges bite into the copper gasket causing lateral flow. The outer rim of the flange however restricts the flow increasing the interface pressure to ensure that the copper fills any imperfections in the flange surface. Properly employed, this arrangement gives virtually 'leak tight' seals in that the leak rate is undetectable with a helium leak detector (better than 10^{-12} Pa m^3 s^{-1}). Another advantage of the ConFlat flange is that the design allows the knife edge to be rigidly attached to the component whilst the outside ring with the bolt holes can be free to rotate, an important consideration in suitably positioning components such as valves. The outside rim also protects the knife-edge from mechanical damage. The flanges are bakeable up to 450°C. If long periods of baking or baking in an oxygen environment are required, the copper gasket can be silver plated to prevent excessive oxidation. Because of the plastic deformation the gasket is not re-usable.

Metals such as monel and stainless steel can also be used as gaskets by exploiting their springy characteristics. Various shaped cross-sectioned gaskets, such as X- and V-shaped, used in other applications have been adapted for vacuum use. Of particular interest is the use of metal tubular 'O'-rings. 'O'-rings of stainless steel tubing are manufactured for seals and can be used in similar systems to the elastomer 'O'-ring. The tubes may be filled to atmospheric pressure or may be pressurized. For ultra-high vacuum, the seal is not sufficiently leak tight using the steel ring alone. However, it can be coated with a softer metal, for example copper, or with Teflon. Teflon can be baked to 250°C and has a low outgassing rate, although its gas permeability is relatively high. Teflon coated rings have been used at the Cavendish Laboratory for vacuum systems pumping down to 5×10^{-8} Pa[7]. A variation on this system has recently been described[8] where the 'O'-ring consists of a helical spring core with a soft metal sheet wrapped around it. There is also a commercial design using a C-shaped cross sectioned ring, Figure 4. Although they are comparable in price to the copper gasket they are less critical on the flange contour and do not need such a high compression force. They can be re-used providing they have not been distorted or damaged by the previous sealing and/or temperature cycling.

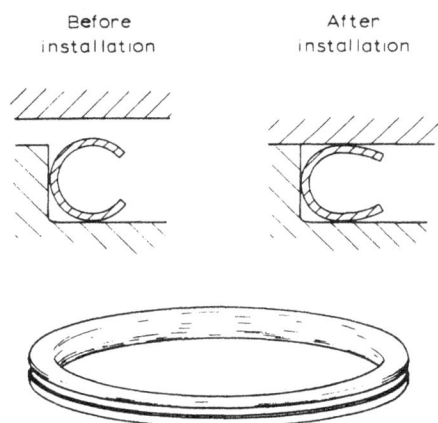

Figure 4. C-shaped ring gasket from Avica Equipment Ltd.

Whatever metal gasket sealing mechanism is employed there are certain essential design features common to them all that should be observed for ultra-high vacuum application. Firstly the material used must be of UHV quality. The stainless steel must not be porous or contain inclusions likely to affect the knife-edge or surface finish. The gasket should be chemically pure and in the case of copper, OFHC (oxygen free high conductivity) copper should be used. Secondly the surfaces should be free of scratches and be clean. Especially they should be free from dust and grease and should not be handled after cleaning without gloves. Thirdly when the flanges are connected to the tubing they should be welded on the inside with an argon arc to ensure there are no voids which could represent a virtual leak. Care must be taken not to distort the flange during welding; a single pass with a penetration depth of about 1.5 mm is suitable. The flanges should be bolted together with high tensile strength stainless steel bolts with a thermal expansion matched to that of the flanges. To ensure even pressure on making the seal the bolts should be tightened in a suitable sequence. The actual torque required need not be measured as the flanges are normally tightened until either they touch or a fixed gap, measured with a feeler gauge, is reached. Finally the gaskets should not be re-used and it is false economy to attempt to do so. The exception is the hard springy metal seals,

such as the 'O'-rings of stainless steel, which can be used more than once in some circumstances. Also Balzers have recently announced a re-usable aluminium seal which can be used in ISO clamped flange designs. The seal consists of two metal gaskets and an X-shaped cross-sectioned support ring placed between them. On sealing, the gaskets are forced into the cavity of the X-cross-section. To re-use, the gaskets are turned over and again pressed into the grooves. It can only be re-used if low baking temperatures, < 150°C, are employed.

3. Mechanical feed-throughs and dynamic seals

In many vacuum systems there is a need to transmit mechanical movement through the vacuum envelope. This may be to position a component or to provide a continuously moving part within the vacuum. In this section the various methods of achieving mechanical transmission whilst maintaining ultra-high vacuum conditions are considered. The applications can be conveniently divided into two categories, those in which manipulation of components over relatively small distances or rotation at low speeds, up to say 1000 rpm are required and those requiring large movements or rotation at high speed and/or with a high torque.

For vacuum equipment operating at pressures above 10^{-4} Pa, the first category is served by 'O'-ring seals or grease-packed lip washer seals using vacuum grease or oil as the lubricant. Although the leak rate for such seals can be low, they are not bakeable and therefore unsuitable for ultra-high vacuum. Such seals can be improved by using Viton-A 'O'-rings lubricated with graphite, but in general these are only satisfactory at the upper end of the ultra-high vacuum pressure range. Fortunately there is a satisfactory solution which meets the stringent vacuum requirement and which is not excessively expensive. The method uses flexible bellows as part of the vacuum envelope, to manipulate the component from outside. This is illustrated in Figure 5 which shows a simple 'wobble stick' from Vacuum Generators Ltd.

Two types of bellows are used for such manipulators, hydraulically formed bellows and edge welded bellows, illustrated in Figure 6. The formed bellows, as the name implies, are formed from tubing by forcing the convolutions into a collapsible die using hydraulic pressure. The edge welded bellows consist of preformed diaphragms, argon arc welded together along alternate inner and outer rims. Various profiles can be used for the diaphragms. The formed bellows are considerably cheaper than the welded bellows but the latter give a superior performance in terms of flexibility and in particular a greater linear motion along the axis per unit length of bellows. The vacuum performances are similar. Both can be made of 18/8 stainless steel and both can be baked up to 450°C. The bellows are used under compression and preferably should not be extended much beyond their free length.

The manner in which linear motion in any direction can be transmitted via the bellows is fairly obvious from Figure 5. What may not be so clear is how continuous rotation of a component may be achieved. Figure 7 shows one possible arrangement using ball and socket type joints. Other configurations have been employed based on a similar principle. Rotational speeds up to 2000 rpm have been achieved, but mainly they are designed to operate at much lower speeds or under manual control.

Linear and rotational motion feedthroughs based on the bellows system are marketed by most vacuum component manufacturers. They vary from simple 'wobble-sticks' to precision manipulators with vernier and adjustments and are usually fitted

Figure 5. 'Wobble stick' using bellows, courtesy Vacuum Generators Ltd.

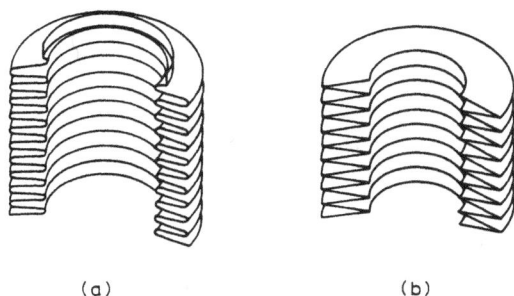

(a)   (b)

Figure 6. Types of metal bellows (a) hydraulically formed, (b) edge welded diaphragms.

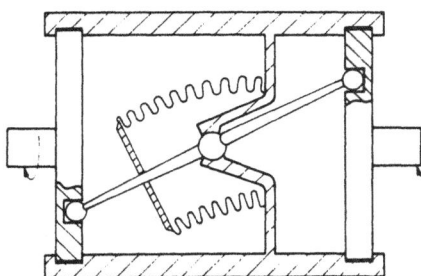

Figure 7. Schematic diagram of a rotating drive with metal bellows.

with flanges for connection onto the vacuum system. An example of a more complex manipulator is shown in Figure 8.

For the second category of applications requiring greater movement or faster rotational speeds the bellows system is less satisfactory. One convenient method, which also obviates the need for the shaft to pass through the vacuum envelope, is to use magnetic or electromagnetic coupling. For example the coil of an electric motor can be placed outside the vacuum envelope whilst the rotor is mounted inside to obtain rotary motion. Since the rotor must not be a source of gas it should comprise of the minimum bulk and be of suitable material which can be baked. A wound coiled rotor, unless encased in a sealed envelope, would be unsatisfactory and therefore ac induction motors with cage type

rotors or reluctance motors with magnetic rotors are employed. The problem with such arrangements is that the air gap between rotor and stator has to be sufficient to accommodate the vacuum envelope. This results in a reduction in the torque and in the efficiency. In general such motors are restricted to low torque requirements such as rotating small piece parts, although high speeds can be obtained. An alternative method is to produce a rotating magnetic field by rotating magnets outside the vacuum. This type of magnetic drive allows a larger air gap between the magnetic drive and magnetic rotor and thus a thicker vacuum envelope. Several designs were used in the 1960s, e.g. Coenraads and Lavelle[9], and recently Budgen[10] described the drive for a mechanical booster pump using an arrangement with three rotating magnets.

The movement of a magnet or ferromagnetic component inside the vacuum by a magnet outside is also an obvious method of obtaining extensive linear movement of components. The external magnet can either be moved over the required distance in parallel with the internal component or it can be rotated to drive a leadscrew. The only requirement here is that the vacuum envelope should not interfere with the magnetic coupling.

Thus for the majority of applications mechanical movements within an ultra-high vacuum system can be effected by one of the above methods without the need for the drive shaft to pass directly through the vacuum envelope. There are, however, some applications where a high torque is required at relatively high speeds and the vacuum engineer needs to resort to a dynamic seal on the drive shaft at the vacuum-atmospheric interface. Normal dynamic seals used on rotating shafts exploit mating surfaces with a thin oil film between them and are similar to bush bearings. Using low vapour pressure grease or oil, such seals can be used down to pressures of the order of 10^{-4} Pa. Oil can be eliminated by using PTFE bushes which can then be baked up to 200°C. A suitable shaft seal could be the BAL-seal manufactured by the BAL-seal Engineering Corp. The seal is a U-shaped PTFE ring with the opening directed towards the atmospheric pressure side with a special spring imbedded inside the U-cavity as illustrated in Figure 9. It is applied like an 'O'-ring and is suitable for rotating or reciprocating shafts. By using two seals and evacuating the space in between with a backing pump the leak rate is minimal. Similar

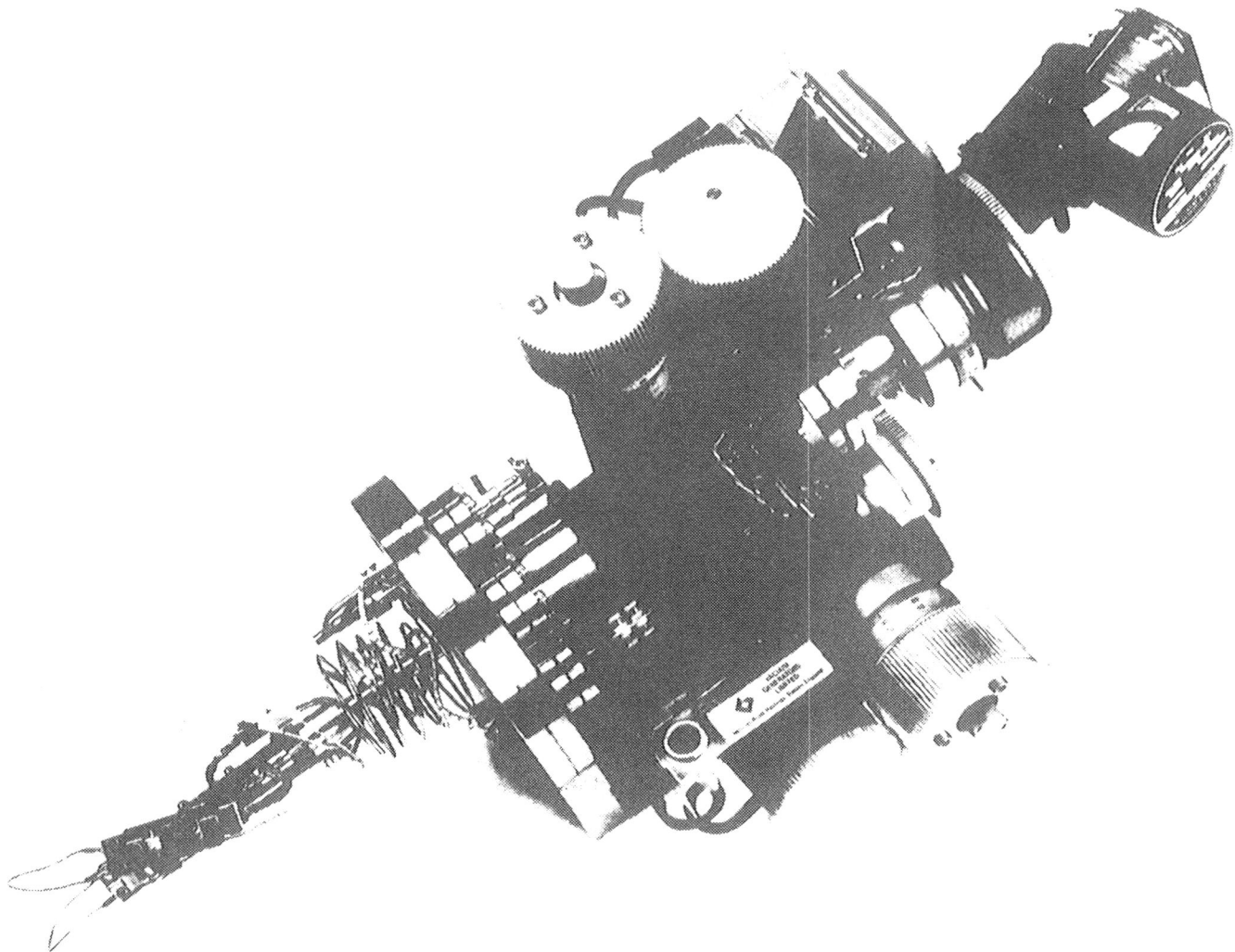

Figure 8. Ultra-high vacuum precision manipulator, courtesy Vacuum Generators Ltd.

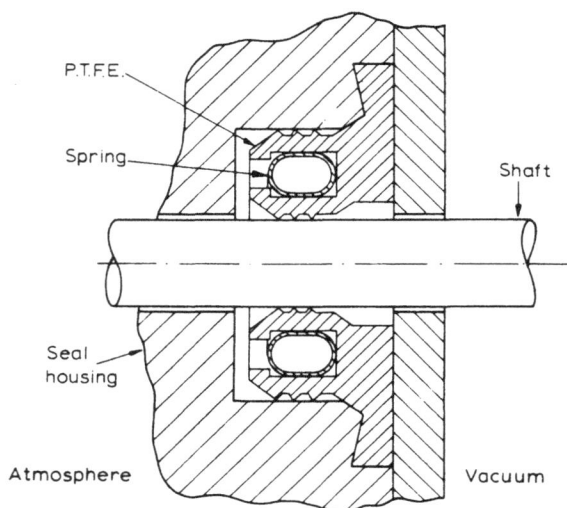

Figure 9. Basic design of PTFE dynamic seal from BAL-seal Engineering Corp.

PTFE dynamic seals are available from vacuum component manufacturers fitted with suitable vacuum flanges.

A new type of seal has been introduced recently for high vacuum applications (pressures down to 10^{-6} Pa) which is becoming more popular. It is based on the use of a magnetic fluid and is marketed by the Ferrofluidic Corporation[11]. The magnetic fluid consists of submicroscopic magnetic particles colloidally suspended in a carrier liquid which, for vacuum application, is a low vapour pressure oil. In the absence of a magnetic field the liquid behaves in its normal fashion. On application of the magnetic field the motion of the suspended particles along the applied field transports the oil by osmotic forces, causing the suspension as a whole to move. The liquid properties are retained but the liquid can now be contained within the magnetic field to build up an oil barrier which will stand up to a differential pressure. An illustration of such a seal for a rotating shaft is given in Figure 10. It uses a permanent magnet and multistage seal to stand up to the atmospheric pressure difference. Since there is little or no friction the seal has a long life and allows a high rotational speed of the shaft; many are rated above 5000 rpm and special

208

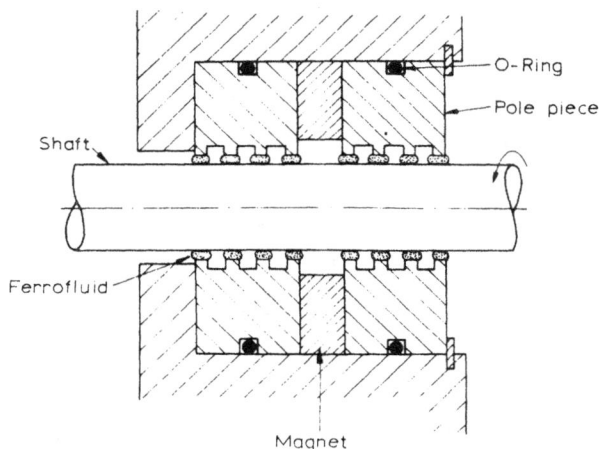

Figure 10. Ferrofluidic rotating shaft seal from Ferrofluidics Corp.

designs have performed above 50,000 rpm. They are relatively cheap compared with alternative systems and are conveniently designed with standard vacuum flange connections. Unfortunately there are no dynamic seals, known to the author, which are completely satisfactory for pressures below 10^{-7} Pa.

Whatever drive mechanism is used there will be a need to lubricate the moving components which are within the vacuum envelope, otherwise binding and even cold welding can occur. This poses the problem of finding a suitable lubricant which will not constitute a contamination and which can withstand the baking temperature, criteria which rule out most oils and greases. It is a problem which is also significant in the space programme and as a consequence it has been extensively studied. Solutions vary from soft metal coatings on ball bearings such as lead and silver to dry films which combine molybdenum disulphide with graphite. PTFE bushes or coatings are also very suitable if the baking temperature can be limited. Magnetic bearings which hold the component in levitation have also been described. For most purposes metal coatings probably offer the best solution where ultra-high vacuum pressures are to be attained.

4. Valves

Valves are probably the most important vacuum line component, allowing isolation of the various regions of the vacuum system essential for pumping down, processing and maintenance. Because of the number of roles they play in the vacuum system there is a wide variety of designs from the small air inlet valves to the large open-conductance gate valves used for example in the beam line of particle accelerators. There are, however, some basic design principles on which most of the commercial valves depend and by describing examples, the overall picture of ultra-high vacuum valve design for the range can be presented. The basic designs depend on the application. They depend on the required open conductance or tube bore in which the valve is to be connected, whether the valve is to be an on–off device or used to control the gas flow rate, and how stringent the vacuum conditions have to be. For some applications a straight through path may be required to pass a charged particle or radiation beam in the open position. The two main design types are the right angle valve so called because of the position of the vacuum tube connections, and the straight through valve where the tubes are in

line. In the latter case, a common design is the gate valve which facilitates a large unimpeded conductance.

There are certain criteria that all ultra-high vacuum valves have in common. They should have a minimum leak rate in the closed position (e.g. $< 10^{-11}$ Pa m^3 s^{-1}), have a maximum conductance in the open position and not be a source in themselves of gas contamination. The last criterion implies using materials compatible with the low pressure requirements with negligible gas permeability and which can be outgassed by baking to a temperature of at least 200°C but preferably to 450°C. The leak rate and baking criteria are the same as those applying to demountable seals and it is not surprising therefore that many of the valve seating designs resemble the configurations used in the flange seals. Similarly, operating the opening and closing mechanism of the valves requires a mechanical feed-through to the same specification as those described in the previous section.

At the upper end of the pressure range, $> 10^{-6}$ Pa, elastomer sealing techniques can be adapted for valve application. It is essential that the synthetic material used can be baked up to say 200°C, for example Viton 'A', and that the mechanical feed-through is also compatible with the vacuum requirements. A number of designs for this pressure region have been proposed and several alternative designs are sold on the market. For relatively small bore tubing right angle valves are normally employed, but this is for convenience in the design rather than any vacuum consideration. Probably the commonest design for this type of valve exploits the ubiquitous 'O'-ring seal; an example is illustrated in Figure 11. The body of the valve is stainless steel and

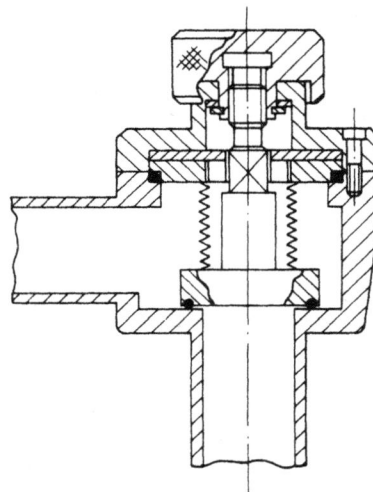

Figure 11. 'O'-ring elastomer sealed right angle valve from Leybold–Heraeus.

the drive mechanism is operated through a bellows seal. Using Viton-A 'O'-rings the valve is bakeable up to 200°C. Valves of this type are normally used for isolation in vacuum lines of 15–35 mm dia. For the large diameter bores, however, the gate valve offers advantages. It takes up less vacuum line space and allows a straight-through path.

The basic design of a gate valve is shown diagrammatically in Figure 12. The shaft first moves the seal plate into the gate and then applies the transverse motion to close the valve by an 'over-shoot' mechanism, shown here as ball bearings being pushed out of indentations, a method exploited by VAT who specialize in

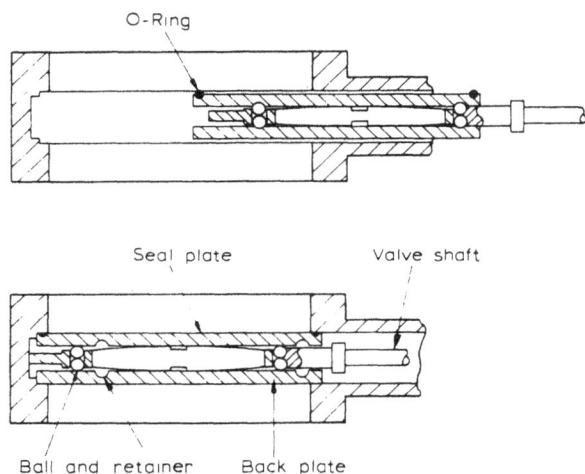

Figure 12. Elastomer sealed gate valve, basic design of VAT.

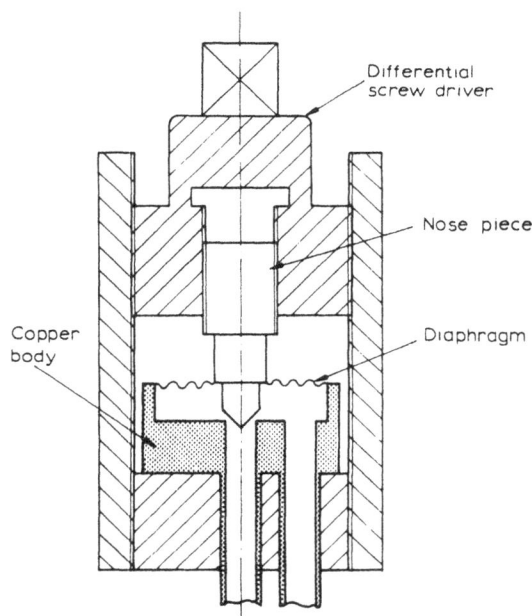

Figure 13. Schematic diagram of the Alpert Valve.

such valves. In this example the transverse force is applied between the seal plate and a backing plate, but in some designs the seal plate bears against the sliding carriage. The seal is made with a Viton-A 'O'-ring and for the bakeable version a bellows seal is employed on the shaft.

Valves required to control the flow of gas in this pressure region usually are based on the needle–valve principle. The diameter and taper of the needle determines the flow rate range. The needle and seating are of hard metal often stainless steel and in some designs a soft metal is deposited on the seating to give a better vacuum seal when closed. Some are provided with a separate seal in the closed position using a Viton 'O'-ring. Few of the designs can be baked, although they may be fitted with a bellows seal on the drive shaft and use materials compatible with high vacuum requirements. The control is normally via a micrometer type head acting on a differential screw which can be calibrated.

For lower pressures, below 10^{-6} Pa, 'all-metal' valves are required. The first all-metal valve was developed by Alpert[12] at Westinghouse at the inception of ultra-high vacuum techniques in the early 1950s. It was constructed in two parts, the main valve body and the drive mechanism which was detached for baking. A schematic diagram of the design is given in Figure 13. The valve body consisted of a copper cup with two orifices into which the copper tubes of approx. 6 mm bore were sealed. The front of the cup was sealed with a Kovar diaphragm. A nose was mounted at the centre of the diaphragm and had a polished 45° cone which came into contact with the edge of the central orifice when the valve was closed. The diaphragm, nose and tubes were all brazed onto the copper cup in one operation in a hydrogen furnace to prevent any oxidation. When the valve is first closed the copper flows sufficiently to give a reasonable surface area in contact with the nose cone. A force of about 10 kN is required. The driving mechanism is a differential screw which allows precise positioning of the cone relative to the seating. Because of this, the valve can be used as a gas flow control valve. The diaphragm allows a cone movement of *ca* 3 mm. The design or modifications of it were taken up by several companies and the same basic design for connecting to glass systems is available today, although inconel may be used instead of the copper with a stainless steel diaphragm. Such valves were bakeable up to 450°C in the open position whilst the later designs can also be baked closed.

Such valves were limited in size giving an open conductance of

around 5×10^{-4} m^{-3} s^{-1}. For larger valves, bellows are used instead of the diaphragm to give a larger movement and therefore a higher open conductance. The cone and orifice system was found unsuitable for larger diameters and the popular choice has been the knife-edge seal. Most manufacturers offer such valves in standard sizes to connect to ISO flanges, i.e. nominal vacuum tube diameters of 16, 35, 63 mm, etc. (see Table 2). The knife-edge is normally part of the main body of the valve and made of stainless steel. The valve plate has a flat surface which presses on to the knife-edge. Since the knife-edge will normally cause an indentation on the plate it is essential that the actuating mechanism precisely positions the plate to ensure that the indentation is aligned with the knife-edge at each closure. Two types of valve plate are common, in one a copper pad is fitted whilst in the other a hard material such as stainless steel or sapphire is used. In the latter case one of the surfaces, the knife edge or the plate is usually plated with a thin layer of gold. A copper pad design is illustrated in Figure 14. Most of the designs have a metal bonnet seal which allows the sealing pad on the valve-plate to be replaced when worn. Some have a seal arrangement which also allows replacement or re-machining of the knife-edge. The pros and cons of the two types have been expressed in the literature and data sheets, but in practice there is little to choose between them. Both can be baked in the open position to 450°C. Most of the designs can also be baked in the closed position, some up to 450°C. In general the hard plate type can be baked to the highest temperature in the closed position whilst rather lower temperatures 300–400°C are recommended for the copper pad designs. This is mainly due to the fact that the copper pad tends to soften by annealing. If there is oxygen likely to be around whilst baking, a situation that would arise with an air-inlet valve for example, the copper pad could become oxidized. However, a high temperature copper alloy with similar properties can be used as an alternative. The closing torque required is high, particularly for the hard seal version, several Nm, and there can be difficulties in lubricating the actuating mechanism to prevent binding when high temperature cycling is involved. When the

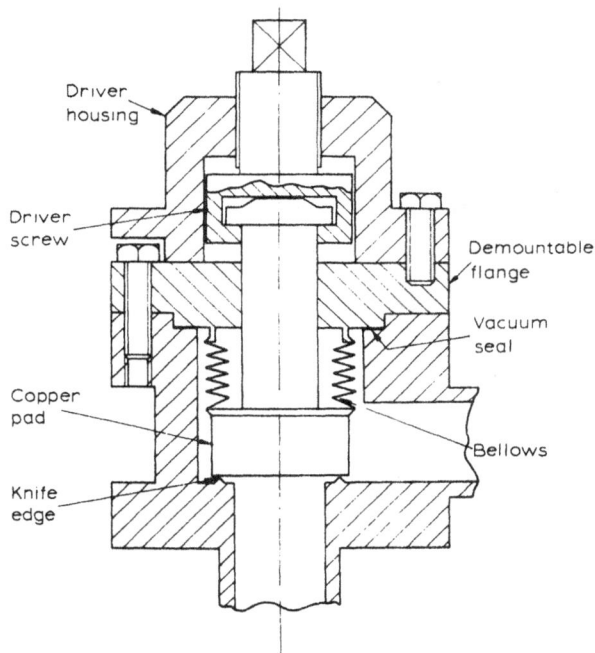

Figure 14. Schematic diagram of a knife-edge sealed right angle valve.

valve is to be baked in the open position it is often expedient to remove the actuating mechanism during baking and some designs are constructed to facilitate this.

Modification of the design as shown in Figure 15 will allow the same sealing mechanism to be adapted for a straight-through valve. The orifice in the open position will be rather limited, however, and for an unimpeded path through the valve a gate valve or similar design of valve is required. The design of an all-metal gate valve involves special technical problems because of the force required to make the gas tight seal in the closed position, which is at least 10 times that required for an 'O'-ring seal. There is also the need to attain exact positioning of the seal plate relative to the seating for repeated operation. Because of these factors, the designs of all metal gate valves have had varying degrees of success. It has been found to be very difficult to obtain operational lives comparable to those attained with the right angle valves, especially where frequent temperature cycling is imposed. The

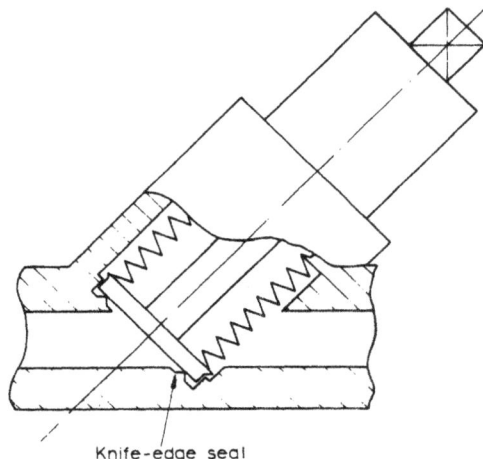

Figure 15. Straight through valve using a knife-edge seal.

larger the valve the more difficult it is to fulfil the requirements. Most of the large valves, with ports of 150 mm or more diameter, have been commissioned for specific particle accelerators and although they are available on the market they are very expensive; several thousand pounds.

As with the right angle valve the common main seal is the knife-edge. If it is used with a copper pad then there is the problem of ensuring alignment of the knife-edge with the indentation that is impressed on the copper. The hard metal seal with a thin plating of gold or silver is better in this respect. If the plating is applied to the knife edge, positioning is non-critical. On the other hand a larger sealing force may be required. The actuating mechanism can be similar to that used with the 'O'-ring design. However because of the large force required to close the valve, there is difficulty in providing a mechanical drive with a low enough friction level, bearing in mind that the use of grease or oil lubricants would be unacceptable. In some designs this problem has been overcome by providing a separate pneumatically operated sealing system. An example of this is a design by Granville Phillips Corp, where the seal plate is relatively thin and easily positioned and is pressed into position by an annular bellows welded to the port opposite the seat; the basic principle is illustrated in Figure 16. A more

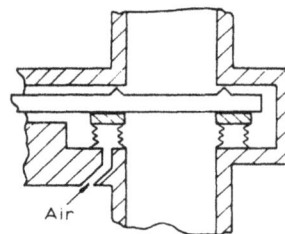

Figure 16. Basic principle of gate valve with pneumatic sealing mechanism from Granville Phillips Corp.

complex arrangement designed by CERN[13] involved a flat hard seal at each port and a bellows between the two seal plates with deformable knife-edges. The whole system of seal plates and pneumatically operated bellows were swung into position on a short shaft in a 'pendulum' motion, and the space between the two seals was separately evacuated to give a low leak rate. An added advantage of the pneumatically operated seal is that it maintains a constant sealing force during the life of the valve.

A novel gate valve design from Varian Associates[14] is claimed to overcome most of the problems. The design is illustrated in Figure 17. The seal plate is a conical shape of thin metal which is deformed under compression to form a cylindrical seal around its periphery (see inset). The sealing ridge is gold plated. The gate is swung into position in a pendulum motion and the pressure is applied by driving a wedge into the carriage mechanism. The shape of the cone and the fact that the seal is round the edge makes it less sensitive to positioning than the knife-edge seal. It is also much lighter in weight.

In general the baking temperatures for gate valves are more limited than for the right angle valves and most manufacturers do not recommend baking above 300°C.

Because of the precision of alignment between the valve plate and its seating and the exact point of closure, almost any of the all metal right angle valves can be adapted with a fine screw adjustment on the actuator to function as a gas flow control valve. Since a fairly restricted movement is all that is required the use of a

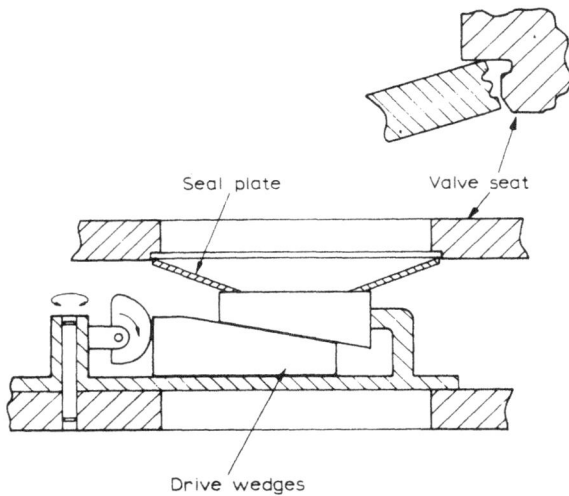

Figure 17. Gate valve design from Varian Associates.

diaphragm rather than bellows on the drive mechanism offers the minimum dead volume. To give a fine adjustment the control screw is often applied via a lever. Accurate and reproducible control is claimed for such valves, although few manufacturers produce any figures for the tolerances. For most applications the requirements are defined by the maintenance of a constant pressure drop across the valve and for this purpose most of the control valves are adequate.

5. Other vacuum line components

Apart from the need for pipeline components such as flexible connections, T-junctions, side tubes to vacuum chambers, etc., most ultra-high vacuum systems will also require electrical feed-throughs and viewing ports. There may also be a need for a liquid feed-through which will carry cooling fluid for example.

For the pipeline components, the main criterion is that the material, in most cases an austenitic stainless steel, is of high quality free from inclusions or cavities which could give rise to gas permeation or affect the outgassing rate. The welding of such components should be made as far as possible along the inside surfaces to ensure that no gas is trapped in the join which could leak into the vacuum system. Figure 18 gives some examples of good and bad welding practice for vacuum systems.

For electrical feed-throughs and viewing ports the controlling design factor is the sealing bond between the ceramic or glass and the metal, which has to withstand temperature cycling during manufacture and use. Because of the brittleness of glass and to some extent also of ceramics, generally such seals can only be made if the temperature coefficient of expansion of the insulator and of the metal are closely matched. This more or less precludes direct sealing to stainless steel where the expansion coefficient is around $17 \times 10^{-6}/°C$ as compared to the values for glass and ceramics which vary from around 3 to $9 \times 10^{-6}/°C$. Fortunately the need in the electronic industry for glass and ceramic to metal seals for thermionic valves has prompted the development of a number of metal alloys which closely match the expansion coefficient of certain glasses. Thus a 50% Ni–50% Fe alloy matches the soft glasses whereas a NiFeCo alloy in the ratio of 54:29:17 originally known under the trade name of Kovar matches the harder borosilicate glasses. The latter alloy also is a

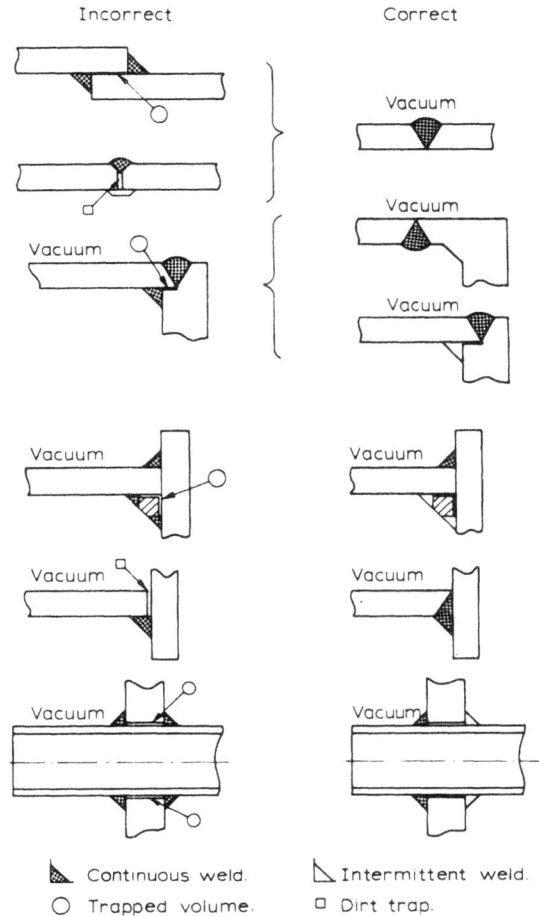

Figure 18. Examples of good and bad welding practice according to Kronberger[15].

reasonable match to the high strength alumina ceramics. The glass to metal seals are made by directly heating the glass and metal in contact to well above the glass softening point. Ceramic seals are usually made by first metallizing the ceramic by a sintering and plating technique and then brazing it to the metal. The metal alloys used can then be directly welded to the stainless steel providing the design allows the metal to take up the strain introduced by their differences in expansion coefficients. Unmatched seals between glass and metal can be made if a ductile metal such as indium or gold is introduced between the surfaces to take up the thermal expansion differences. The bond is made under pressure at a temperature below the melting point. The method is useful for sealing special windows such as sapphire or quartz. Sapphire and quartz crystal windows can also be sealed directly to glass providing a glass is chosen with a matching expansion coefficient. The glass will wet the surface without the crystals being deformed. The glazing can also be used to seal such windows into a metal holder. Sapphire crystals can also be metallized and sealed directly to metal in the same way as alumina ceramics.

Although most ceramics are best sealed to metals with a matched expansion coefficient, the high density alumina ceramic is strong enough to withstand the strain of sealing to unmatched metals providing the metal can flex sufficiently to compensate for the expansion differences. Thus copper rods can be brazed into alumina ceramics for current leads and direct sealing to stainless steel can be effected if the metal is thin enough at the braze. For more details on glass and ceramic to metal seals the readers are referred to Espe's book on the subject[16]. The techniques are very specialized and require considerable skill to perfect. It is therefore fortunate for the vacuum engineer that there is a wide range of electrical feed-throughs available from vacuum component manufacturers which will meet the majority of applications.

6. Conclusions

In conclusion a range of ultra-high vacuum line components are now available from vacuum equipment manufacturers which will fulfil most applicational requirements. They tend to be rather more expensive than components aimed at a higher pressure range, above 10^{-6} Pa, and in general the better the performance the higher the cost. For example a Viton sealed valve with a stainless steel body which can be baked to 200°C will cost two or three times its non-bakeable counterpart, whilst a similar sized all-metal value will cost twice as much again. Thus the vacuum engineer will often be faced with a trade-off between performance and cost, and as a consequence it is important for him to assess his requirements as accurately as possible before ordering the

equipment. Although he does not want to have to pump down for weeks on end to attain his specified ultimate pressure, it does not make economic sense to buy equipment whose performance is far better than the requirements.

References

[1] G F Weston, *Vacuum,* **25,** 469 (1975).
[2] P W Hait, *Vacuum,* **17,** 547 (1967).
[3] T J Edwards, J R Budge and W Hauptli, *J Vac Technol,* **14,** 740 (1977).
[4] L de Chernatony, *Vacuum,* **27,** 605 (1977).
[5] H H Buchter, *Industrial Sealing Technology,* Wiley, New York (1979).
[6] W R Wheeler and M Carlson, *Trans 8th Vac Symp and 2nd Int Congr 1961,* p 1309, Pergamon Press, Oxford (1982).
[7] P V Head, D M Martin, W Allison and R F Willis, *Vacuum,* **32,** 639 (1982).
[8] I Sakai, H Ishimaru and G Houkoshi, *Vacuum,* **32,** 33 (1982).
[9] C N Coenraads and J E Lavelle, *Rev Sci Instrum,* **33,** 879 (1962).
[10] L J Budgen, *Vacuum,* **32,** 627 (1982).
[11] K Raj and C Reiser, *Laser Focus 15th April,* 56 (1979).
[12] D Alpert, *J Appl Phys,* **24,** 860 (1953).
[13] W Bachler and T Wikberg, *Vacuum,* **21,** 457 (1971).
[14] W R Wheeler, *J Vac Sci Technol,* **13,** 503 (1976).
[15] H Kronberger, *Proc IME,* **172,** 113 (1958).
[16] W Espe, *Materials of High Vacuum Technology,* Vol. 2, 'Silicates'. Pergamon Press, New York (1968).

Ultrahigh vacuum leak sealing with a silicon resin product

William F. Egelhoff, Jr.

Surface Science Division, National Bureau of Standards, Gaithersburg, Maryland 20899

(Received 3 August 1987; accepted 27 December 1987)

During the course of the past eight years a large number of opportunities have arisen in the author's laboratory to seal a variety of leaks in an ultrahigh vacuum (UHV) system consisting of an x-ray photoelectron spectrometer and a sample preparation chamber. In many cases it was of great importance to seal a leak immediately so that a sequence of experiments could proceed uninterrupted. It seems likely that the lessons of these experiences could be useful to others in the UHV field in avoiding costly and unnecessary repairs and instrument downtime.

The silicon resin used in this work, SR-882M made by General Electric,[1] was briefly marketed as a leak sealer in the 1960s. It is available currently from local chemical distributors (see the Yellow Pages of the telephone directory) for approximately \$50/gal (which is essentially a lifetime supply). Apparently its use as a leak sealer has been the subject of only one short technical note.[2]

This silicon resin is a long-chain silicon–oxygen polymer with an abundance of hydroxyl groups.[3] It is supplied as a viscous liquid, having been diluted somewhat with toluene. After the volatile toluene solvent has evaporated, or been baked off, the resin remains behind and has an immeasurably small vapor pressure under UHV conditions ($< 10^{-11}$ Torr).[4] This immeasurably small vapor pressure is important since silicon-based diffusion pump oils are well known to decompose under electron bombardment. The decomposition products form insulating films that can cause problems in UHV equipment. With this leak sealer no volatile silicon compounds have been detected to outgas from this resin (even during 200 °C bakeout) using a U.T.I.-100C quadrupole mass spectrometer,[1] nor have any of the decomposition problems been observed in the present equipment. Therefore, it appears this resin is entirely safe to use. The fact that toluene, a volatile and commonly available chemical,[5] is the solvent is a major advantage since one can easily adjust the consistency of the resin either by adding more toluene or by allowing it to evaporate.

A good example of this advantage is in sealing leaks that are hard to reach. When the leak is located at the end of a narrow channel or out of sight around a corner (e.g., at a damaged conflat knife edge), leak sealing compounds have difficulty reaching the leak. However, diluting the silicon resin with an equal volume portion of toluene reduces the viscosity, allowing the solution to flow easily into very tight places. This diluted solution can also be useful for spray applications. An example of this occurred in the author's laboratory when a leak was located inside a liquid-nitrogen trap. It was impossible to locate the specific site of the leak; however, the leak was easily sealed (and it has held for years) by placing a $\frac{1}{4}$-in.-i.d. rubber hose in the entrance to the trap, loading the other end of the hose with a few ml of the solution

and connecting the hose to a compressed air source. The mist blown into the trap sealed the leak on the second attempt.[5] This general procedure is also very effective in sealing leaks in stainless-steel tubes inside the vacuum system, e.g., lines carrying liquid nitrogen.

The other extreme of viscosity, in which toluene is allowed to evaporate to leave the highly viscous silicon resin, is useful for sealing very large leaks. For example, an electrical feedthrough (8-pin miniconflat type) with a completely broken seal between the ceramic and the 1-mm rod was sealed by coating the ceramic–rod joint with the viscous resin and baking in air with a 200 °C heat gun for 30 min. The only shortcoming noted in sealing this obviously huge leak was that the resin became soft and the leak threatened to reopen if the feedthrough was heated above 50 °C, indicating that baking would not be advisable.

Fortunately, most leaks in UHV equipment are far smaller than the one in the above feedthrough and can often be sealed *in situ*. If the pressure does not rise above $\sim 10^{-6}$ Torr,[4] rebaking the system may not be necessary.

Accessible leaks that do not cause pressure rises above $\sim 10^{-8}$ Torr,[4] can generally be sealed immediately by squirting them with diluted resin from a wash bottle. For leaks of this size, the resulting partial pressure of toluene in the vacuum is almost undetectable on a quadrupole mass spectrometer, and a system with a base pressure of 2×10^{-10} Torr will generally return to that value within a few minutes of sealing the leak. Evaporation of the toluene may be hastened by warming the leak site with a heat gun, but most of the pressure rise will be found to be water from the chamber walls.

Sealing leaks in the 10^{-6} Torr range,[4] requires slightly more care. In these cases it is an advantage to evaporate the toluene promptly to minimize the amount of toluene outgassing from the sealed leak into the chamber. This is best done by preheating the site of the leak with a heat gun until it is too hot to touch. The toluene then boils off almost immediately when the solution is squirted on, and a base pressure of 2×10^{-10} Torr is often regained in a few hours or less.

If leaks in the 10^{-5} Torr range,[4] or above, are sealed with this resin an interesting phenomenon is occasionally observed during subsequent bakeouts. Tiny air bubbles travel slowly through the softened resin producing small pressure bursts at regular intervals (e.g., $\sim 10^{-7}$ Torr for ~ 1 s at 1-min intervals). This interval rapidly lengthens (approaching infinity) upon cooling the chamber after bakeout. The ultimate vacuum achieved is unaffected by this phenomenon.

Among the worst places for a leak to occur in a UHV system is in moving parts which include welded-diaphragm metal bellows (e.g., manipulators, transfer rods, wobble sticks). These are often an integral part of the system and the common remedy, replacement, often requires a very inconveniently long, and ill-timed, instrument downtime. Conventional leak sealing is generally unsuccessful in such movable bellows, however, such leaks often can easily be sealed using the procedures described above and immobilizing the leaking section with clamps. In welded-diaphragm metal bellows enough segments are usually present to allow the leaking segment to be clamped mechanically. Convenient clamps can be made by folding 1 cm^2 squares of 0.5-mm-thick stainless-steel sheet in half to be a tight fit over the folds of the bellows. Leaks in welded-diaphragm metal bellows occur more often than not at the inner (or i.d.) weld seam.[6] It may be helpful to point out that such leaks are often caused by fully and forcibly compressing the bellows (i.e., until the segments touch) a point not sufficiently emphasized in most catalogs.

When metal bellows approach the end of their normal, useful life of several thousand strokes, the appearance of a leak can easily put an instrument out of action for the typical eight-week delivery time of new, custom bellows. For many experimentalists such a delay is unacceptable. Fortunately, in many cases the leak can be repaired in a matter of minutes by the above procedures. Since one leak in an old bellows is generally a signal of more to come a short-term extention of bellows life (until delivery of the replacement) can be very valuable.

[1] This commercial product is specified only to identify the experimental conditions and does not signify any endorsement by the National Bureau of Standards.
[2] J. R. Young, Rev. Sci. Instrum. **35**, 116 (1964); note the product was formerly called SR-82 and the solvent used to be xylene.
[3] General Electric Data Sheet on SR882.
[4] For the reader's convenience this value is presented as a pressure. It assumes a typical pumping speed of 100 l/s in UHV systems. To convert it to a leak rate simply multiply by 100 l/s.
[5] Since exposure to high concentrations of toluene vapor can have adverse health effects (e.g., in chronic "glue sniffers"), good laboratory ventilation is recommended. See product safety label for further details.
[6] J. Chapman, Standard Welded Bellows Co. (private communication).

J. Vac. Sci. Technol. A, Vol. 6, No. 4, Jul/Aug 1988

216

A simple ultrahigh vacuum shape memory effect shutter mechanism

A. P. Jardine,[a] M. Ahmad,[b] R. J. McClelland, and J. M. Blakely

Department of Material Science, Cornell University, Ithaca, New York 14853

(Received 25 January 1988; accepted 5 April 1988)

The need for a simple, inexpensive method of opening and closing a shutter in an UHV environment prompted an examination of the shape memory effect (SME) as a possible shutter mechanism. We wished to use a shutter to control the dosing of a surface from a sulfur cell source of the type described by Shrott et al.,[1] and to examine the dependence of overlayer structures on submonolayer S coverages using low-energy electron diffraction (LEED). The shape memory effect exhibited by NiTi was utilized to provide an inexpensive but quite elegant alternative to electromagnetic shutters (where magnetic fields can be disruptive to LEED) or to mechanical shutters which require relatively expensive rotary feedthroughs and usually need a separate flange. In the apparatus described below, the sulfur cell and shutter mechanism were mounted on one $2\frac{3}{4}$-in. flange.

The shape memory effect refers to the macroscopic change in shape of a material from a deformed martensitic "cold" shape to a preferred high-temperature "hot" shape, when heated through a martensitic to austenitic transformation. (Brief descriptive articles can be found in Refs. 2 and 3; more extensive reviews can be found in Ref. 4.) The restoring force in the transformation is surprisingly large, $\sim 10^6$ kg/m^2. In its cold state, the microstructure is composed of twinned martensitic variants; one of which grows at the expense of the other under an applied stress. Thus the Youngs modulus of the martensitic material is considerably less than that of the fully austenitic material, and so is more easily deformed.

There are numerous alloys which exhibit the SME. NiTi was chosen as it is an UHV compatible material, is easily formed into a hot shape, and has a relatively high transformation temperature. The martensite-to-austenite transformation temperature is ~ 60 °C and the austenite-to-martensite transformation temperature is ~ 30 °C, so that the martensitic state of the material occurs at room temperature. Deformations producing $< 2\%$ local strain in the martensitic wires can be recovered by heating through the transformation. By setting the hot shape of the wires in a helical geometry, it is possible to get extensions of several hundred percent while keeping the local strain in an element of the wire less than 2%.[5] The hot shape of the NiTi helices was set by first wrapping the wire around a 4-40 stainless-steel threaded rod and then vacuum annealing at 500 °C for 1 h followed by furnace cooling. Vacuum annealing prevents the growth of oxide on the surface. If the helical hot shape becomes permanently deformed or requires resetting, reannealing either *in situ* or in a furnace will reset the hot shape.

The small diameter (0.45 mm) of the NiTi wire[6] used has, in its cold state, a relatively large resistance (6.2 Ω/m), allowing resistive heating of the wire through the transformation temperature of ~ 60 °C with 2.5 A of current in ~ 0.2 s.

With this diameter wire, the force of transformation was ~ 26 kg.

To utilize the SME, a Ta foil shutter at ground potential was attached to the middle of a NiTi helix; the free ends of

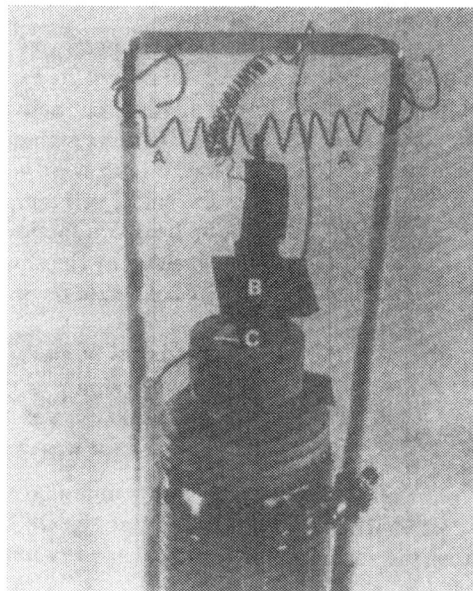

(a)

(b)

FIG. 1. (a) NiTi SME helices (A) were stretched from stainless-steel rods to a Ta shutter (B), so that the equilibrium position is over the sulfur cell aperture (C): (b) By applying 2.5 A to resistively heat the left-hand SME helix, the helix contraction moves the shutter to the left of the aperture. To move the shutter over the aperture, current is applied to the right-hand SME helix.

the helix were clamped to stainless-steel rods so that the helix was stretched from its preferred hot shape (see Fig. 1). A thin, flexible Ta wire connected the shutter to the chassis ground. In the cold state of the NiTi the shutter is in its equilibrium position over the aperture of the sulfur cell. On heating one segment of the helix, it contracts to its hot shape thereby pulling the shutter to one side of the aperture, allowing dosing of the sample. During dosing, it was not necessary to keep this helix hot. It was observed that although the shutter relaxed back towards the aperture, it did not block the aperture as the other cold helix became plastically deformed, and so has little restoring force. The second segment of the NiTi helix was heated to move the shutter over to its equilibrium position over the aperture.

The relatively low transformation temperatures of NiTi require that some care should be taken not to expose the NiTi to high temperatures or else inadvertent activation of the element could occur. In our case, the helix segments were balanced against each other so that, if both were activated, the shutter would not move. As the sulfur cell temperature does not exceed 160 °C with poor heat conduction to the NiTi wire which is also partially shielded by the shutter, we are able to maintain the NiTi in its cold state during sulfur cell operations.

Finally, we tested the effect of bakeouts at elevated temperatures on the hot shape by constraining a severely stretched (300%) helix and annealing at typical bakeout temperatures. It was found that the hot shape was unaffected by 215 °C, 24-h anneals, and so will be unaffected by typical bakeouts.

With this shutter we are able to control the dosing time to within 0.5 s. The arrangement is inexpensive, simple to manufacture, and requires only two electrical feedthrough pins and a modest current supply. Finally, we note that in some applications where the shutter is not required to switch quickly, a stainless-steel spring of similar spring constant as the hot state NiTi helix can be substituted for one SME helix, reducing the number of electrical feedthrough pins required to one.

Acknowledgments: This work was supported by NSF Grant No. DMR 8403444 and by the Materials Science Center of Cornell University.

a) Present address: Departments of Materials Science and Engineering, SUNY at Stony Brook, Stony Brook, NY 11794.

b) Present address: Department of Physics, Wayne State University, Detroit, MI 48202.

[1] A. Schrott, R. J. Lad, and J. M. Blakely, Materials Science Center Report No. 4790, Cornell University, 1982.

[2] L. M. Schetky, Sci. Am. **241**, 68 (1979).

[3] A. Golstaneh, Phys. Today **37**, 62 (1984).

[4] *Shape Memory Effects in Alloys*, edited by J. Perkins (Plenum, New York, 1975).

[5] A. P. Jardine, Ph. D thesis, University of Bristol, 1984.

[6] The NiTi used was supplied indirectly from Raychem Inc. of Menlo Park, CA, through Cory Laboratories of Escondido, CA.

ROBOTIC JOINT EXPERIMENTS UNDER ULTRAVACUUM

A. Borrien and L. Petitjean*

ABSTRACT

In the first part of the paper, various aspects of a robotic joint development program, including gearbox technology, electromechanical components, lubrication, and test results, are discussed.

A test prototype of the joint allowing simulation of robotic arm dynamic effects is presented in the second part of the paper. This prototype is tested under vacuum with different types of motors and sensors to characterize the functional parameters: angular position error, mechanical backlash, gearbox efficiency, and lifetime.

1. INTRODUCTION

The objectives of this paper are to present the gear technology, lubrication, and ultravacuum test results of a robotic joint with high Hertzian pressure on the gear teeth. Dynamic operation of the joint is studied to determine the influence of the control feedback law on significant parameters such as angular precision, speed variation, and damping rate of the joint. Finally, simulated lifetime results of the joint are discussed. After the ultravacuum lifetime test, mechanical and tribological reducer effects in the gearbox are examined to quantify wear level versus number of cycles.

2. JOINT DESIGN

2.1 General Specifications

Geometric, kinematic and dynamic specifications are based on system requirements established for accomplishing the mission with the manipulator arm.

General specifications are the following:

. Angular excursion	-90 deg to +90 deg
. Maximum output motor torque	10 Nm
. Holding output torque	60 Nm
. Global stiffness	7000 Nm/Rad.
. Maximum output backlash	2×10^{-3} Rad.
. Lifetime under ultravacuum	250 hours
. Vacuum level	10^{-8} torr
. Maximum output speed	5×10^{-2} Rad./sec

*Centre National d'Etudes Spatiales (CNES), Toulouse, France

2.2 Technological Definition

The electromechanical joint consists of a DC brushless motor, two optical encoders (input and output), a gearbox, and an electromechanical brake.

a) Motor

Our nominal choice consists of a space-qualified electrical DC brushless motor. The most important advantages are electronic and feedback law simplicity and good stability over a wide dynamic operating range as well as good steady state system performance. An alternative choice is based on a space-qualified stepper motor driven with a special control law. A standard mechanical interface allows comparison of the two alternatives with the same joint.

b) Sensor Modules

Two sensor configurations are possible for measuring joint parameters such as output angular position and input motor speed. System studies have shown that the angular speed of the joint must be sensed on the motor shaft and the angular position on the output housing to satisfy stability and damping criteria of the control loop.

c) Brake

A brake is placed on the motor shaft to latch the joint. A solenoid is energized to release the brake whenever the joint is to be driven.

d) Gear Transmission Description (see Figure 1)

Trade-off studies have concluded that the best solution consists of a parallel-axis back-driveable gearbox with several stages. The main advantages are:

- relatively simple manufacturing
- good potential efficiency
- low backlash
- gear geometry consistent with dry lubrication technology.

Gearbox Definition (see Figure 2). The reducer is divided into two kinematically symmetrical closed parallel branches which consist of:

- external spur gears for the first and second stage
- internal gear wheel for the output stage.

The objective of this design is to allow a fine adjustment of backlash and equal load sharing between the two branches.

- Gear material : 35NCD16 - Air hardened 875 deg C
                              - Tempered 650 deg C
- Lubrication   : PTFE coating

Geometric Characteristics

Stage	Module (mm)	Width	Teeth number	Gear ratio
1	0,5	3	20 120	6
2	0.7	7	17 86	5.06
3	1	9	17 103	6.06
Total ratio				183,90

Stage		Bending stress	Allowable bending stress	Hertzian Stress	Allowable Hertzian Stress
1	P	7.3	48.8	52.3	80.4
	W	6.4	55.5	52.3	91.9
2		12.7	54.4	74.5	91.9
		10.2	76.8	74.5	104.5
3		22	60.4	83.3	104.5
		24.8	69.3	83.8	104.5

P : pinion   Stress : 0.1 N/mm^2
W : wheel

Theoretical calculations have shown that the most important parameter is Hertzian stress which occurs on the gear teeth and could induce fast coating wear and scuffing.

2.3 Description of the Joint Test Machine (see Figure 3)

a) Joint Test Machine Objectives

The test machine, which consists of the electromechanical joint and an inertia simulation, has been designed to allow interchangeability of the electrical components while using the same joint. Several configurations are available for testing:

- Configuration 1: Brushless DC motor, angular potentiometer (output shaft) and optical encoder (input shaft)

- Configuration 2: Brushless DC motor, optical encoder for input and output angular position sensing

- Configuration 3: Stepper motor with the sensors of Configuration 2. This configuration is not presented in detail because some numerical problems must be resolved to achieve feedback law stability.

b) Test Machine Technological Definition (see Figure 4)

The choice of electrical components has been made with the main criteria, vacuum operation without significant outgassing. Also, bearing lubrication and materials have been selected to satisfy environmental conditions of low outgassing.

List of components:

- Brushless DC motor: SAGEM 23 MCM90, space-qualified
- Stepper motor SAGEM 23 PP, space-qualified
- Brake: Binder magnetic 86: 61104, vacuum rated only
- Optical encoder: Sopelem RI110: vacuum rated only

c) Inertia Simulator (see Figure 5)

The objective is to simulate an inertia of 800 Kg m^2 on the output of the joint shaft. Several solutions were examined. Step-up gearing with a tooth belt was chosen to minimize mass and reduce volume in the vacuum chamber. Moreover, it allows a good simulation of the robotic arm inertia parameters.

- Simulated inertia: 800 Kgm^2
- Step-up gearing ratio: 125
- Step-up gearing stiffness: 1950 Nm/Rad.

3. GEAR MATERIAL AND LUBRICATION PROCESSES

3.1 Material Trade-off

High contact stresses occur with the given requirements and lead to a material which provides good mechanical characteristics. These include yield strength, surface hardness, and contact pressure fatigue resistance. Good machinability is also required to obtain excellent geometric tolerances

Among different alloy and stainless steels, 35NCD16 steel was chosen, which has the following advantages:

- o Air hardened, which produces minimum deformation
- o Ease of plasma nitrided (if necessary)
- o Use after tempering without other surface treatment
- o Useful for heavy dynamic loads combined with fatigue, one of the best steels used in aircraft technology.

3.2 Lubrication Processes

In considering lubrication, the investigation was deliberately limited to dry lubricants. To ensure sufficient life and to avoid scuffing, a coating must be used which gives low friction coefficient and low wear rate.

Possibilities giving satisfaction in this case are:

- o crystalline solids with lamellar structure (MoS_2)
- o soft metals (Pb, Au, Ag)
- o anti-wear ceramics (TiN, TiC)
- o polymer materials (PTFE, polyamides, polyacetal)

To help choose the right solution, gears coated with different dry lubricants were tested on a "four square" (closed loop) gear test machine. These machines allow simulation of the speed and load environment of each reducer stage.

The selection criteria for the coatings are:

- o no scuffing in ultravacuum
- o low wear rate
- o low friction coefficient
- o no geometric modification during and after coating
- o good adhesion between coating and substrate.

Five treatments were selected for preliminary experiments at atmospheric pressure. Those which gave better performance were later tested in a vacuum (10^{-7} Torr).

223

Descriptions of coatings, test machines and results in air were given in 1985 at the 2nd European Space Mechanisms and Tribology Symposium at Meersburg, Germany.

Performance classification is as follows:

Heavy load Low speed		Medium load Medium speed
1	PTFE	PTFE
2	MoS_2	MoS_2
3	TiN	FeMoS
4	Ion nitriding	Ion nitriding
5	FeMoS	TiN

The underlined coatings give satisfactory results in air.

3.3 Vacuum Tests

In tests under vacuum, the field was narrowed to PTFE and MoS_2 for the final choice of coating (see Figure 6). The chronology of these tests was:

1) Running-in process with a small torque
2) Qualification test of 100 hours with the nominal torque
3) Qualitative examination of the teeth
4) Short test with step-by-step increasing torque from nominal to maximum value. Each step lasts five hours..
5) Endurance test of 100 hours with maximum torque to classify coatings and to evaluate their resistance under severe loading conditions.

In the case of medium speed (100 rev/min) and medium Hertzian pressure (450 N/mm^2), both coatings had satisfactory friction coefficient evolution, but PTFE provided less surface wear. In case of low speed (2 rev/min) and high Hertzian pressure (550 N/mm^2), only PTFE reached the end of the endurance test (84,000 revolutions) without damage.

It was not possible to estimate the real friction coefficient during this endurance test because of disturbances caused by debris located in the four-point angular contact ball bearings. Finally, PTFE performed best in vacuum and it had a rather good accommodation to misalignment. We finally chose it for the robotic joint application.

3.4 Robotic Joint Gears

The gears were tooled on a MIKRON milling machine. Then they were checked before and after PTFE deposition. Profile, lead, run-out, pitch deviation, and tooth thickness were controlled.

The ISO quality range after treatment was always better than or equal to 6, except for the internal ring whose quality was 7.

4. DYNAMIC ASPECTS OF THE JOINT

4.1 Feedback Loop Description

a) Configuration 1

The DC brushless motor operates as a servo actuator with an analog position feedback loop. The rotor angular position is sensed by an optical encoder, sampled and differentiated to obtain angular speed. A numerical speed loop is implemented to damp high frequency oscillations. The reducer output position is sensed by a potentiometer.

Functional Scheme

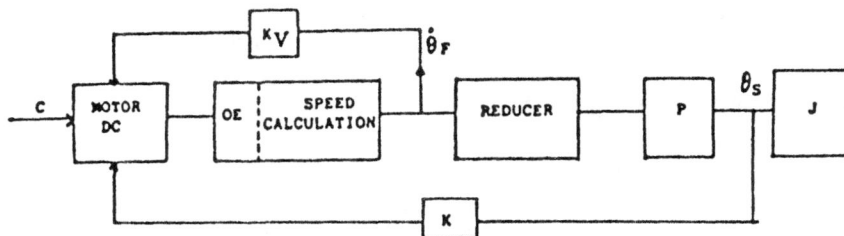

OE – Optical encoder   K – Position gain
P – Potentiometer K_V – Speed gain
J – Simulated inertia

b) Configuration 2

For this case, the potentiometer is replaced by an optical encoder which senses the output angular position.

c) Configuration 3

This configuration uses a stepper motor with a specific dynamic control law. A torque feedback control law has been simulated, but the first experiments indicated some stability problems due to the numerical speed calculation.

This configuration is not considered as a nominal solution because of the feedback complexity. Dynamic analysis and electronic improvement have been performed to prove the feasibility of the law. The results of these studies and experiments will be presented at the symposium.

4.2 Servo Loop Definition

The dynamic system can be treated as a second-order problem if rotor inertia, reducer output inertia, and the electrical constant of the motor are neglected.

The differential equation is:

$$J_s \ddot{\theta}_s + \frac{K_v \dot{\theta}_s + K\theta_s}{T_a} = T_m - T_R$$

θ_s: Output position T_a: Control torque
N: Reducer ratio T_m: Motor torque
K: Position gain T_R: Resisting torque
K_v: Speed gain J_s: Output inertia

with the usual 2nd order servo parameters:

$$\frac{J_s}{K} = \frac{1}{\omega_N^2}$$

ω_N : Undamped eigenfrequency

$$\frac{2\zeta}{\omega_N} = \frac{K_v}{K}$$

ζ : Damping rate

$$\frac{1}{K} = G$$

G : Static gain

With ω_N = 1.125 Rad./sec and ζ = 1

K and K_v are calculated to be

K   =   1000 Nm/Rad.
K_v =   1800 Nm/Rad./sec

These theoretical gains have been introduced in the experimental servo electronics. Some stability problems in the high frequency range limit speed gain efficiency.

The phenomena can be explained by the following considerations: In the backlash range, the load inertia changes between two extreme values: low rotor inertia and high simulated inertia. The digital tachometer encoder cannot track these rapid changes in motor rate without significant phase lag. Rate sampling precision is poor because the samping frequency is low (5 msec).

To improve the servo loop stability, it is possible to add a low-pass filter in series with the digital tachometer output. It appears that one of the best solutions is to use an analog tachometer for the high frequency range and a digital tachometer for the low frequency range. This mix of the two types of rate feedback provides the desired characteristics, while attenuating undesirable phenomena.

5. PROTOTYPE JOINT EXPERIMENTS

5.1 Philosophy of Testing

The purpose of the tests was to achieve confidence in the performance of the electromechanical joint under heavy loading and long operation in an ultravacuum environment.

The main objective was to assess the performance of the dry-lubricant gear reducer and electrical components. Backlash and friction coefficient evolution are the most important parameters which characterize reducer transmission.

During preliminary analysis, dynamic performance of the joint was evaluated using different types of feedback loop. The angular excursion law consists of large amplitude movements, small oscillations and locking phases. This is considered a good simulation of the approach and grappling sequence. All dynamic parameters are easily modified, utilizing a "menu" on a computer. The following parameters are automatically recorded: output angular position, motor position, winding intensity, and motor speed (deduced from sampled angular position (200 Hz)).

5.2 Parameter Measurements Before Endurance Tests

- Backlash

After assembly, it is easy to check the residual backlash by comparing the information from the two optical encoders when rotational direction is changed. However, before final assembly, backlash has been checked by measuring the axial displacement of a comparator placed on the output gear of the reducer, the motor being locked. The measured backlash in this case was:

$$j = 5 \times 10^{-4} \text{ Rad.} \simeq 1.7 \text{ Arc min}$$

- Friction Torque

Joint friction torque was measured with an angular dynamometer before final assembly with an inertia simulator. Two values were obtained by rotating the joint successively by each extremity. The results were similar (T = 0.04 Ncm), proving good joint reversibility.

The multiplier mechanism had an effective friction torque of 0.6 Ncm, but there is a dispersion in the results, depending on the belt tension. The global resisting torque was about 0.65 Ncm.

- Efficiency

At maximum speed and nominal torque, the motor efficiency is determined as follows:

$$\eta_M = \frac{T \times \omega}{U \times I}$$

T - Torque ω - Angular velocity
I - Winding current U - Voltage

$$\eta_M = \frac{\omega}{(R/K^2)T + \omega}$$

R - Winding resistance R = 10.5 ohms
K - Torque constant K = 0.06 Nm/A
  T = 0.08 Nm

so that

$$\eta_M \simeq 0.04$$

When the inertia decreases, motor efficiency rises to 0.3. Measurement was made of the global mechanical efficiency of the joint with a constant resisting torque of 0.09 Nm. A torque of 0.042 Nm on the motor shaft was obtained.

Then,

$$\eta_G = \frac{T_M \, \omega_M}{T_R \, \omega_R}$$

T_M = Motor torque ω_M = Motor speed
T_R = Resisting Torque ω_R = Output speed

228

The multiplier efficiency was also measured and found to be $\eta = 0.85$. An estimate of the reducer efficiency is then

$$\eta_\eta = \frac{\eta_G}{\eta} = \frac{0.69}{0.85} = 0.81$$

This corresponds to $\eta = 0.93$ for each meshing gear.

– Stiffness

To ensure good locking stiffness, a steel housing was placed on the joint. The joint torsional stiffness was calculated with individual shaft and gear contributions. A value of $K \simeq 1 \times 10^5$ Nm/rad was obtained. This value was checked by measuring the angular difference between the two encoders under the following conditions: the brake was locked and the output inertia mass was loaded by a static torque.

With an 11.85 Nm applied torque, an angular difference of 0.03 deg was measured. Global stiffness was deduced from the previous results to be K 23,000 Nm/rad.

The inertia simulator stiffness has been calculated to be K 2,000 Nm/rad.

– Inertia

Assuming a geometric design for the mechanical parts, we have:

- Joint output inertia : 0.84 m^2kg
- Multiplier inertia : 800 m^2kg
- Motor shaft inertia : 0.8 m^2kg
- Ratio I load / I motor : 1,000

5.3 Preliminary Joint Experiments

– Maximum Static Torque Specification

A 56 Nm static torque was put on the joint using weights with the electro-magnetic brake locked. One of the encoders was therefore fixed and the other rotated from 0 deg to 0.13 deg which confirmed the above stiffness value of about 25,000 Nm/rad. This test was repeated ten times without any angular position modification, thus demonstrating good behavior of the loaded gears and brake.

– Magnetic Brake Test

The brake was locked and the motor reversed from clockwise maximum torque to counterclockwise maximum torque. The motorshaft encoder was monitored to detect any rotation. No movement was observed.

In a second test, the brake was released for seven hours, which requires application of constant input voltage. Subsequently, after switching off the input voltage, the brake did not relock for several minutes. The reason for this may be the high temperature under vacuum, which increases the distance between the magnetic parts of the brake.

- Maximum Torque Motor Test

For this test, the output part of the joint was locked onto the housing and the motor driven from "0" to maximum current (0.17 Nm) with a linear rise during two minutes. This was repeated twenty times in succession. The motor was found reliable.

5.4 Endurance Tests

For all experiments, the minimum vacuum was 10^{-6} Torr, and was performed with a 600 ℓ/sec RIBER ion pump.

- Running in Process

With thick PTFE films (6 μm), the running-in process is very important. A low Hertzian pressure and small angular acceleration of 20 deg/sec (i.e., 1.52 Nm output torque was chosen. For run-in, the output gear was rotated 80 revolutions (40 clockwise and 40 counterclockwise).

- Cyclic Test Description

Two standard movements were recorded, one for large angular displacement with maximum dynamic performance and the other for large angular displacement combined with small oscillations to simulate a final capture.

The first standard movement, which is called A, consists of an angular excursion of 170 deg at the output of the joint. It begins with an accelerated phase (= 130 deg/sec , T = 10 Nm), followed by a constant angular rate of 500 deg/sec and finally, a decelerated phase with the same dynamic parameters. Then the joint comes back to the initial position with the same displacement law.

The second standard movement, B, has the same angular excursion and the same maximum angular rate, but the acceleration is limited to 65 deg/sec. Before coming back, four sinusoidal oscillations with a 1 deg amplitude are imposed on the joint.

- Endurance Test Procedure

The joint was tested with 1100 cycles of the A program (34 hours) and 1600 of the B program (87 hours), corresponding to a total of 121 hours lifetime.

The number of cyclic loads reached by each gear tooth was:

- Third stage internal ring   :   5400
- Third stage pinion   :   32700
- Second stage wheel   :   32700
- Second stage pinion   : 165600
- First stage wheel   : 165600
- First stage pinion   : 1987200

The average torques which were applied to the gear teeth are shown in Figure 7.

- Parameter Control

Each standard cycle was divided into several phases depending on the driving law. For each phase, a reference clock time was defined at which the position and angular rate parameters were automatically recorded by a computer program.

The program calculates, for these times, the difference between the theoretical position and the effective position of the joint and the difference between angular position at the ends of the joint. After analyzing 120 cycles, the program draws the evolution curves of the parameters and calculates the mean value and standard deviation for each one.

The following curves depict joint parameter fluctuations:

- Static position error: (Θoutput - Θtheoretical at t = 0) (see Figure 8)
- Dynamic error, recorded for the accelerated phase (see Figure 9)
- Output angular rate (see Figure 10)
- (Θoutput - Θinput x R(Ratio)): backlash variation, machined-gear-tooth deviation, and elastic deformation due to the load (see Figure 11).

   - Backlash Evolution

Several measurements of backlash were performed during the lifetime tests. The final value, after 121 hours, is j = 1.8 arc/min (Figure 12). Thus, no significant evolution could be detected. Backlash evolution is representative of output gear wear. However, it is not possible to draw any conclusions about the wear of the other gear stages.

   - Friction Coefficient Evolution

An 8 gm weight is sufficient to move the joint (T $\simeq$ 0.5 Ncm). On the other hand, the minimum motor voltage to start the joint is 0.42V, i.e., T = 0.75 cmN. These values are similar and no evolution could be detected.

   - Gear Tooth Examination

The gear housing has some circular holes that allow direct examination of gear teeth with an endoscopic cane. Unfortunately, it is possible to observe only the second stage of the reducer. The teeth looked polished, almost brilliant, near the top, and it appears that the two mat blue-gray extremeties of the pinion were never in contact after 120 hours of running. If the results of checking (wear and surface inspection) and the functional performances (output backlash, resistant torque, efficiency) are all found to be in accordance with the specifications, the endurance test duration will be increased to 150 hours of running in ultravacuum. A discussion of the final results will be presented at the symposium.

CONCLUSIONS

These robotic joint endurance experiments are a good simulation of the orbital mission. The tests have confirmed behavior of the components after three months under vacuum and 121 hours working time, except for the brake which showed some problems due to increased temperature.

The feasibility of a robotic joint of 150 mm diameter and 100 mm length, with 10 Nm dynamic output torque and 60 Nm static torque, has been demonstrated for more than 100 hours lifetime with dry lubrication. The backlash and friction torque stability indicate a potential increased lifetime without significant damage.

Some improvements in the feedback law must be implemented in the future to take into account high frequency resonance problems, inertia simulator stiffness effects and the effect of backlash.

Although the new torque requirements of the Hermes telemanipulator arm are much higher than for the joint discussed in this paper, it seems that a joint with force capability in the range 10 to 20 Nm could have several applications in orbital station manipulators.

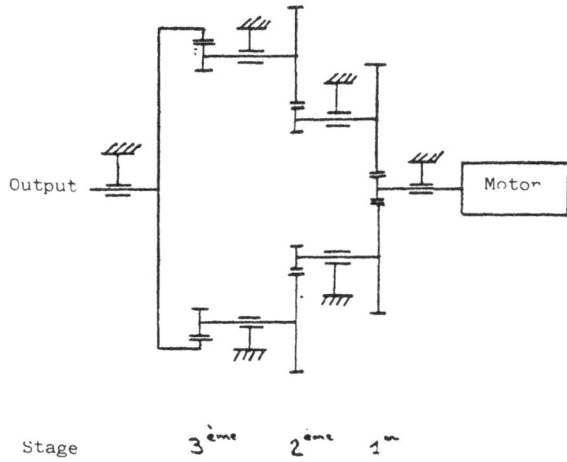

Figure 1. Gear box description

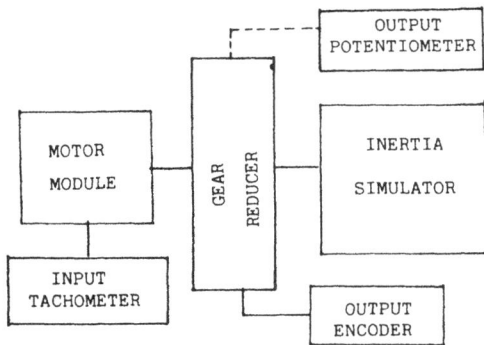

Figure 2. Gear box arrangement

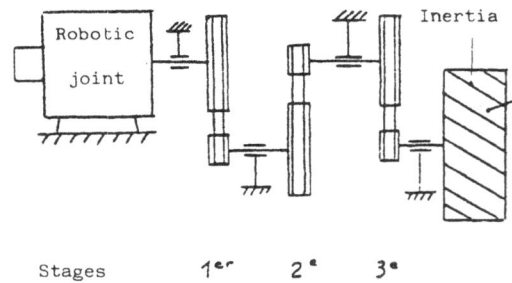

Figure 3. Joint prototype functionnal
definition

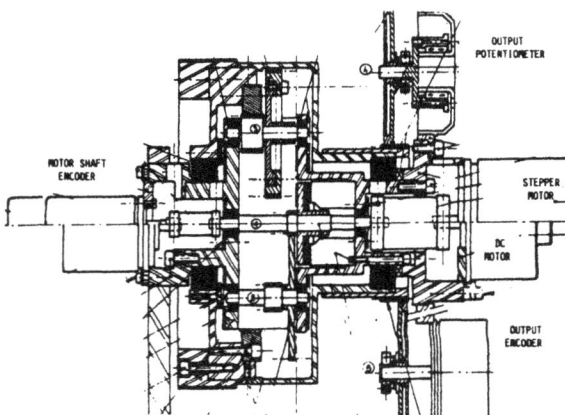

Figure 4. Prototype technological
definition joint

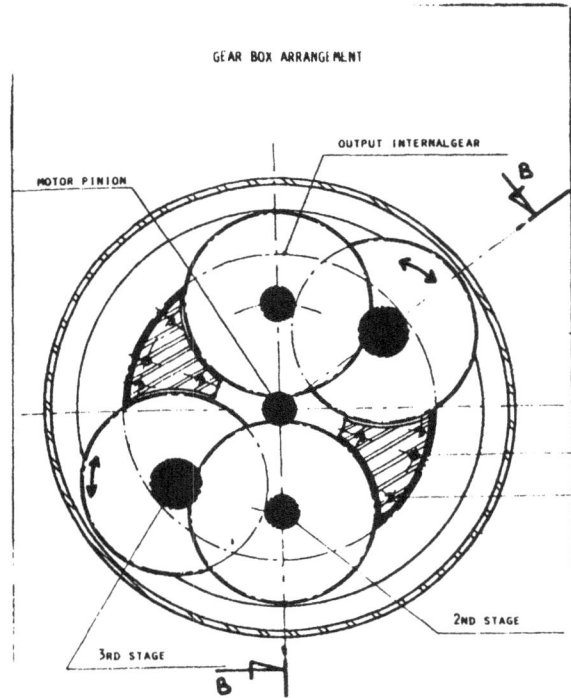

Figure 5. Inertia simulator

234

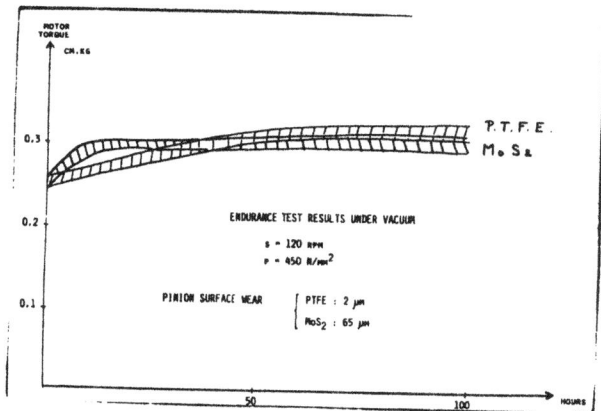

Figure 6. Endurance tests MOS2/PTFE

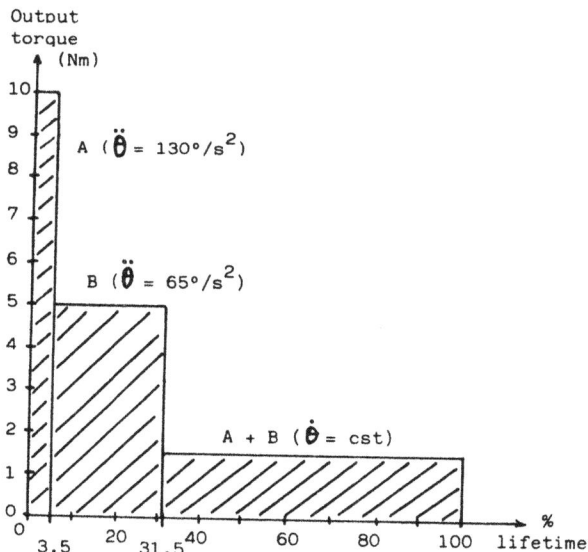

Figure 7.

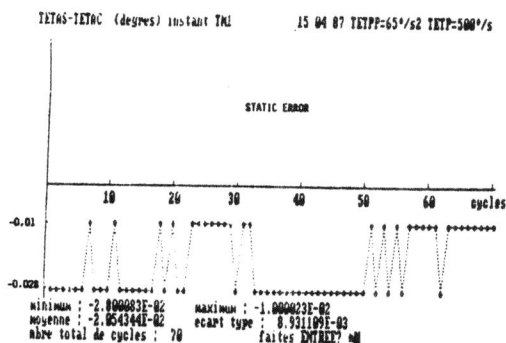

Figure 8. Static position error

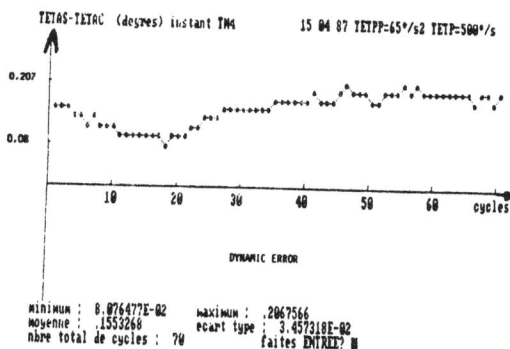

Figure 9. Dynamic error

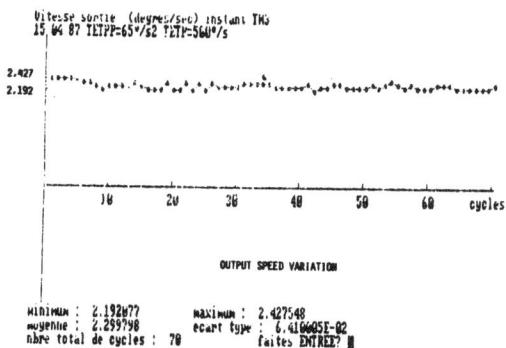

Figure 10. Output speed variation

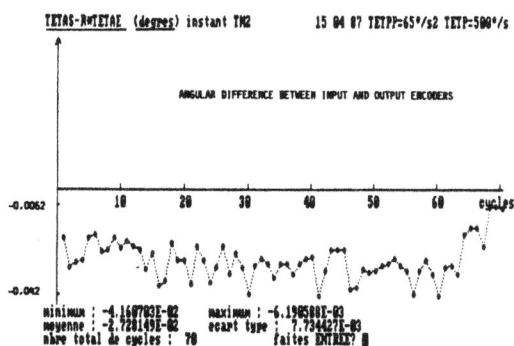

Figure 11. Backlash variation

235

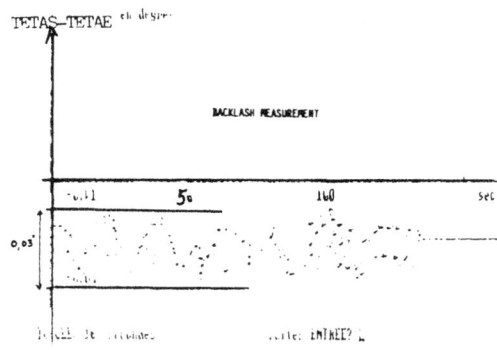

Figure 12. Backlash measurement

Chapter 5

Vacuum Mechatronics Control

Shigang Li
Steve Belinski

5.1 Introduction

Vacuum mechatronics control is concerned with how to control the active mechanical devices in an environment characterized by low pressure and limited human access. The environment can either be created by artificial means, e.g. vacuum chamber, or exist naturally, as in space. The controlling hardware (and operator, if any) is usually separated from the working space of the mechanism by a hermetic seal.

In the past few years, research in vacuum mechatronics has focused mainly on the vacuum-compatible components: actuators, sensors, lubricants, seals etc. No significant differences have been enumerated between vacuum mechatronics control and conventional mechatronics control. The problem has been seen only from the viewpoint of hostile environments. Some vacuum robots have simply been retrofitted from conventional robots however the controllers in such cases remain the same. Even some of the recently designed vacuum robots have control systems show no conceptual breakthrough. Their controllers employ new processors resulting in more computational power and more I/O capability but such improvements are still more quantitative than qualitative. These facts imply that what has been attacked is the mechanical problem and not the general vacuum mechatronics control problem.

While the lack of vacuum-compatible devices is a technical problem, and will be gradually solved by ongoing research and development within the industry, other problems essential to vacuum mechatronics are not being adequately addressed. Such problems include *restriction of control access, contamination, heat* and *reliability*. These factors affect the control concept and must be solved through new ideas specific for vacuum mechatronics control.

In a vacuum mechatronic system, the control action is restricted by limitations on actuator types and the issues of heat and contamination. The sensing ability will also be restricted by reduced sensor selection and mounting position choices. In spite of these difficulties, the performance demanded of vacuum mechatronic systems is increasing and can be summarized in the following key points:

- high overall accuracy
- good dynamic behavior
- increased dexterity
- high reliability
- multi-dimensional sensory perception
- cooperation between several robots and measuring devices
- possibility of automatic scheduling and rescheduling
- learning, contingencies handling, etc.

Thus, the difference in approach between vacuum mechatronics control and conventional control is not just the result of differences in the working environment, but the result of the limited access restriction implied by the vacuum environment and of increased performance requirements. Vacuum mechatronic systems need more sensing and maneuvering powers than conventional systems since the former are expected to perform various tasks with high reliability in a complex environment and be able to handle certain subsystem failures. However, the sensing and maneuvering power cannot be achieved by simply adding physical sensors and dexterity to the system because the technical restrictions and the reliability problems will still exist.

To solve this conflict, ways must be found to make better use of limited resources to perform given tasks. This one of the major problems in controlling vacuum mechatronic systems. "Limited resources" does not necessarily refer to a small amount but rather to less than the desired number. The actual number of sensors and degrees of freedom in a vacuum mechatronics system can even exceed those of today's conventional robotic systems. Thus, a vacuum mechatronics control method which can systematically use limited sensory information and limited control action to perform various tasks with fault-tolerance capability in the complex surroundings of the vacuum environment must be established.

Building such a controller requires integrated sensing and real-time multi-sensory data fusion. Computationally efficient world models are also important, since the controller can use them in combination with active sensing for work-space understanding and model adaptation. The world models can also be useful in expectation generation and sensory data interpretation during operation. Fast simulation (e.g., kinematic, dynamic and heat simulation) is useful for vacuum mechatronics control since optimization of a certain performance measure or verification of a certain maneuver before execution may be desirable. Verification via results of simulations is preferable to trial and error since limited accessibility makes the latter approach especially difficult for vacuum mechatronics. Conventional teaching methods are no longer adequate for vacuum mechatronics: it is very difficult for an operator to observe the working scene through windows in order to use the lead-through method i.e.,where the operator uses a teach pendant to guide the robot through desired positions. Also, in a vacuum chamber, it is impossible for the operator to perform a walk-through, where the operator physically moves the robot end-effector along the desired trajectory. Instead, program control becomes necessary, and real-time simulation capability is highly desirable in assisting program control. Automatic trajectory-planning and automatic task-planning using information from parts/tasks/knowledge databases are desirable and useful. Also, new criteria for optimal trajectory planning and task scheduling can be introduced (e.g., minimization of temperature rise).

This chapter, for obvious reasons, cannot be written like the other chapters in this book. It cannot present a history of research on the topic of Vacuum Mechatronics Control because the field has scarcely begun to emerge. It cannot include reprints of state-of-the-art research that has been accomplished, because very little has been written. The chapter is not interspersed with references;

rather a bibliography is included at the end. The chapter focuses on the *issues* that must be considered as the field of Vacuum Mechatronics Control develops.

5.2 Basic Issues

Kinematics

Kinematics is related to the geometry and geometrical motions of the mechanism's linkage. The relationship between position and orientation of the end-effector and the joint displacement, i.e. the direct and inverse kinematic problem, is one of the basic topics in this area.

The position vector specifies the end-effector location. It may be the only controlled variable in some industrial object transfer applications. In addition, the hand frame, described by orientation vectors, provides information on how the working object, measuring device or mounted camera is held by the mechanism. This information deserves more attention for applications in vacuum mechatronics. Machining and *in situ* inspection tasks require information on not only the location, but the attitude of the parts. In some applications, objects require gentle handling in order to avoid possible deformation. In such a case the information about orientation may be required so that specially designed grippers can achieve the proper grasp.

For a robot working inside a vacuum chamber, programming by means of a teach pendant is inconvenient due to limited access. Automatic or computer assisted programming is more suitable than conventional teaching methods for vacuum mechatronics. A practical method for solid modeling and collision avoidance must be developed to support such a programming method.

Dynamics

The dynamics involve the correlation between the motion of, and the forces (torques) applied to the mechanism. The direct dynamic problem of finding the trajectory of the robot under given driving forces is the key to simulation. The inverse dynamic problem, which is to find the applied forces which will drive the robot along the given trajectory, is an essential part of many advanced control algorithms. In both cases, one needs the robot dynamic model. The dynamics of the flexible arm are very complex. For vacuum mechatronics devices working in the vacuum chamber for industrial applications, the dynamic model of a rigid body with or without the joint elasticity model is most likely a sufficient approximation of the real mechanism.

Performance

The performance of the vacuum mechatronic system can be evaluated in many ways. We will divide the performance characteristics into three groups:

1. positioning (steady state)
2. motion (dynamic)
3. task (planning)

Resolution, repeatability and accuracy describe how precise the system is in the steady state. They are generally dictated by the mechanics and the encoders, as well as the controller. Stability, overshoot and smoothness represent the dynamic behavior of the system. They depend mainly on the control algorithm. Traveling time, energy consumption, temperature variation, cleanliness and reliability characterize how well the system will perform a given task. They depend on the controller's intelligence as well as the hardware (e.g., mechanisms, actuators and sensors).

Steady state performance

The resolution of vacuum mechatronics devices is often dictated by the resolution of the sensors, provided that it is not less than the smallest incremental movement the controller can allow. Since high resolution is necessary for high accuracy and maneuverability, it is especially important in vacuum mechatronics. One method of achieving high resolution is by reduction though gears. This could be an unsuitable method in a clean vacuum production environment because of the added contamination source. Microstepping, if a vacuum rated stepper motor is employed, and other driving techniques using sensor output signals can be helpful for increasing the resolution.

Repeatability describes the degree to which a robot can repeatedly return to a given position. High repeatability does not necessarily require high resolution or high accuracy. For example, a 200 step/revolution stepper motor has an angular resolution of 1.8 degrees; however, its repeatability can be 0.09 degrees or less. High repeatability enables production to have a uniform processing environment, especially for the record-playback programmed robot. The ability to repeat to a taught position is often more significant than accuracy for a conventional industrial robot. In the vacuum environment, repeatability drift due to heating can be significant. Thus, the relationship between repeatability, resolution and accuracy of vacuum mechatronics devices cannot be assumed to be the same as for conventional devices. The repeatability of a vacuum mechatronics device will most likely be dictated by its accuracy .

Accuracy describes how closely a robot can approach a given position in space, regardless of whether the robot has been there before. Accuracy may be classified as: 1) absolute global accuracy within the working volume, or 2) local accuracy relative to a reference point. For vacuum mechatronics devices, instead of manual planning methods such as walk-through and lead-through, automatic or computer-assisted planning methods are used. In such cases, accuracy is more significant than repeatability. Accuracy as good as one half of the resolution can be achieved in an ideal situation. Mechanism imperfections such as loose bearings, backlash, arm flexibility and encoder mounting errors greatly reduce both global and local accuracy. To overcome or prevent these liabilities, target-position direct sensing is required, as opposed to calculations from the robot kinematic structure and encoder readings. Calibration is pivotally important in achieving good global accuracy.

Dynamic performance

The dynamic behavior of a robot usually refers to its acceleration and deceleration ability as well as its traveling speed, overshoot, motion smoothness and general stability. Maximum acceleration/deceleration depends on many factors, such as the actuator's power, structure, dimensions, dynamic parameters, displacement and speed.

Good dynamic behavior is difficult to achieve with a simple controller. For example, it is hard to choose gains for a local PID controller. A small value of damping leads to a fast response but large overshoot; a large damping factor can eliminate the overshoot but leads to slow response. Also, the damping factor not only depends on the gains, it often varies with the position of the robot. Even if a set of satisfactory PID parameters is found for a certain trajectory, this set might not work if either the load or trajectory is changed. An advanced controller is essential in vacuum mechatronics to achieve good dynamic behavior.

Overshoot is to be avoided in most cases, especially in vacuum mechatronic devices. Since the volume of the vacuum chamber is at a premium, the working environment is usually tightly arranged, for economical reasons. Extreme overshoot can cause an unexpected collision in such a situation. Also, it usually leads to inferior positioning accuracy. Vacuum mechatronics devices should reach high speed without overshooting, using a controller with dynamic-model-based planning and suitable tracking algorithms.

"Poor smoothness" describes robot movement having high frequency oscillations around a planned trajectory and/or a jerking movement at the starting and ending points. Motion with poor smoothness can have a negative effect on the effectiveness of holding or gripping on the object. An application requiring a vacuum production environment usually demands very high precision, requiring objects to be handled in a certain way by a specially designed gripper. Oscillation and jerking can sometimes cause sliding and even scratching of the work piece. Smooth motion requires a well planned trajectory (especially to avoid jerking), as well as tracking algorithms to compensate nonlinearity and to provide proper degrees of stability.

System stability analysis is very important from the theoretical point of view. It can be fairly difficult to accomplish because of the nonlinearity of the robotic system. Many control algorithms in literature are supported only by simulation, instead of by stability analysis. Since poor smoothness and large overshoot are usually an indication of a low degree of stability, a vacuum mechatronics controller should posses a high degree of stability for good dynamic behavior.

Task related performance

Traveling time, energy consumption, temperature rise, reliability and cleanliness are related to particular tasks. Optimization of these parameters can be achieved through varying levels of controlled intelligence.

Since short traveling time is a key to productivity, it is nearly always desirable. Short traveling time requires high acceleration and deceleration and consistent speed. However these attributes (acceleration, deceleration and speed) are related to robot dynamics and restricted by the robot actuators. An intelligent controller can bring about a shorter traveling time than a simple controller by optimization of the trajectory.

Energy consumption affects the efficiency of the mechanism. Today's industrial robots generally have a low efficiency of energy transformation mainly due to mechanical braking and friction within the mechanism. Energy consumption must be considered more seriously in vacuum mechatronics, since low efficiency may lead to excessive heating of the component inside the vacuum chamber. Friction may be reduced by using direct drive methods, proper lubrication, ball bearings, and magnetically levitated bearings. Using electrical braking can decrease the energy consumption. Also, an intelligent controller can increase the system's efficiency by planning the trajectory with minimal energy consumption; it is usually compromised by the timing requirements.

Heat becomes a significant constraint for vacuum mechatronic devices. Heat problems also exist in conventional robots, affecting the motor's maximum torque and speed but manufacturers usually specify torque and speed for steady state. In a vacuum environment, since there is no convection, the motor's torque and speed at the steady state will be much lower than in air, unless special efforts are made to provide efficient cooling systems. The instantaneous torque, however, should not be affected by the lack of convection because of the integral character of the heat-temperature relation. Thus, a trajectory which will evenly heat the whole mechatronics device might make better usage of all actuators. Through careful planning, a trajectory and a work schedule can be found which minimize the cycle time while satisfying temperature variation restrictions.

Reliability represents the quality of the vacuum mechatronic system and a measure of the extent to which it can be counted on to do what is commanded. Presently, most industrial robots will fail whenever a working subsystem is not functioning properly; therefore subsystem reliability is critical. Since the robots have only two states (up and down), reliability is often indicated only in terms of the up-time. On the other hand, it is desirable to implement vacuum mechatronics systems with intelligent controllers which can still work when certain subsystem failures occur, perhaps with slower motion or poor accuracy but within the tolerance of the task specifications.

Although the reliability problem exists in both vacuum and usual mechatronics systems, the degree of reliability and method of approaching highly reliable systems can be different. In the case of vacuum mechatronics, besides putting more emphasis on each component, the controller's fault-tolerance ability must also be emphasized. Replacing a system, or even one component, is undesirable. The conventional method of attaining high reliability is through system redundancy. Since high costs are associated with redundancy, especially in vacuum mechatronics, there is a question as to how many redundancies are enough, and how to distribute them. Furthermore, since the functions of the components may overlap, a systems reconfiguration, rather than a subsystem substitution, is probably a more effective way to solve the fault-tolerance problem and achieve high system reliability. Rescheduling and trajectory redefining subsequent to certain subsystem failures can also be a means of expanding the fault-tolerance ability. In addition to the question of degree and distribution of redundancy, there is the question of how to make use of them.

5.3 Selected Papers

The literature is very rich in fields related to vacuum mechatronics control, such as robot control, perception, and artificial intelligence. However, directly applicable material is difficult to find, in spite of a recent trend towards intelligent control. The three reprint papers included in this chapter are helpful in motivating research ideas in vacuum mechatronics control.

J. Albus' paper analyzes the central nervous system as a hierarchical control system from a bionics point of view. It gives insights as to how high-level goals can be decomposed into low-level actions, and how knowledge can be acquired, stored and accessed to produce a sensory-interactive, goal-directed behavior. His "computational hierarchy" can be one of the general concepts of intelligent controllers.

The paper of T. Fukuda and S. Nakagawa concerns dynamically reconfigurable robotic systems. These consist of a set of pre-designed cells that have some mechanical functions and can combine themselves autonomously, depending on the given task. Self-configuration and self-repair are very attractive features of such advanced robotic systems.

T. Henderson's and E. Shilcrat's "logical sensor" concept is a useful framework in dealing with multisensory systems. In this concept, the logical sensors are defined as the computational processes operating on output from other logical sensors.

The reason for this particular choice of papers is that there is a lack of directly applicable material on vacuum mechatronics control, per se. On the subject of robot control or general control theory and artificial intelligence, many excellent papers can be easily found (see references), so there is no compelling reason for reprinting them here.

Bibliography

Atkinson, D. et.al., "Autonomous Task Level Control of A Robot", *Proc. ROBEXS'86,* pp.117-122, Johnson Space Center, June, 1986.

Bejczy, A.,"Task Driven Control", *IEEE Workshop on Intelligent Control '85*, pp. 38-48, Troy, New York, 1986.

Brady, M., ed., *Computer Vision*, North-Holland, 1981.

Brooks, R. A., "Planning Collision-Free Motions for Pick-an-Place Operations", *Int. J. Robotics Research*, vol.2, no.4, pp.19-44, 1983.

Davis, R., D. Lenat, *Knowledge-Based Systems in Artificial Intelligence*, New York: McGraw-Hill, 1982.

Donald, B. R., "Robot Motion Planning with Uncertainty in Geometric Models of the Robot and Environment", *Proc. IEEE Conf. Robotics and Automation*, pp.1588-1593, San Francisco, CA, Apr., 1986.

Dubowski, S., "On the Adaptive Control of Robot Manipulator", *Proc. JACC*, vol. TA-2B, 1981.

Hollerbach, J. M., "A Recursive Lagrangian Formulation of Manipulator Dynamics and a Comparative Study of Dynamics Formulation Complexity", *IEEE Trans. Systems Man and Cybernetics*, vol. SMC-10, no. 11, Nov. 1980.

Isidori, A., A. J. Krener, C. Gori-Giorgi, S. Monaco, "Nonlinear Decoupling via Feedback: A Differential Geometric Approach", *IEEE Trans. Automatic Control*, vol.AC-26, no. 4, pp. 331-345, Apr., 1981.

Kak, A. C. et al., "Knowledge-Based Robotics", *Proc. 1987 IEEE Conf. Robotics and Automation*, pp. 637-646, Raleigh, North Carolina, 1987.

Kazerooni, H., T. B. Sheridan, P. K. Houpt, "Robust Compliant Motion for Manipulators, Part I: The Fundamental Concepts of Compliant Motion", *IEEE J. Robotics and Automation*, vol.RA-2, no. 2, pp.83-92, June, 1986.

Koivo, A., T. Guo, "Adaptive Linear Controller for Robotic Manipulators", *IEEE Trans. on AC*, vol. AC-28, no. 2, pp. 162-171, Feb., 1983.

Lee, C. S., B. H. Lee, "Resolved Motion Adaptive Control for Mechanical Manipulators", *J. Dynamic Syst. Meas. Cont.* vol, 106, 1984.

Leskov, A. G. and V. S. Medvedov, "Analysis of Dynamics and Synthesis of Movement Control of Robot Manipulator Functioned Organs", *Engineering Cybernetics*, vol. 12, no. 6, pp. 56-65, 1974.

Lozano-Perez, T., "Spatial Planning: a Configuration Space Approach", *IEEE Trans. on Computers*, vol. 32, p. 108, 1983.

Luh, J. Y. S., "Conventional Controller Design for Industrial Robot: A tutorial", *IEEE Trans. Systems Man and Cybernetics*, vol.SMC-13, no.3, pp.298-316, May/June, 1983.

Luh, J. Y., M. W. Walker, R. Paul, "On Line Computational Scheme for Mechanical Manipulators", *ASME Trans. Dynamic Systems, Measurement and Control*, vol. 102, no. 2, pp.69-76, June, 1980.

Luo, R. C., M. H. Lin, R. S. Scherp, "Dynamic Multi-Sensory Data Fusion System for Intelligent Robots", *IEEE J. Robotics & Automation*, vol. 4, no. 4, pp. 386-396, Aug., 1988.

Meyestel, A.,"Intelligent Control: Issues and Perspectives", *Proc. IEEE Symp. Intelligent Control*, Troy, New York, 1985.

Paul, R. , *Robot Manipulators: Mathematics, Programing and Control*, Cambridge, Massachusetts: MIT Press, 1981.

Paul, R., H. Durrant-Whyte, M. Mintz, "A Robust Distributed Sensor and Actuation Robot Control System", in *Robotics Research*, ed. by O. Faugeras, G. Giralt, pp. 93-100, 1986.

Saridis, G. N., "Intelligent Robotic Control", *IEEE Trans. on Automatic Control*, vol AC-28, no. 5, May. 1983.

Skarr, S. B., W. H. Brockman, R. Hanson, "Camera-Space Manipulation", *Int. J. Robotics Research*, vol. 6, no.4, pp.20-32, Winter, 1987.

Unimation Inc. "User's Guide to VAL", Denbury, CT, 1980.

Vukobratovic, M., D. Stokic, *Control of Manipulation Robots, Monograph*, New York: Springer-Verlag, 1982.

Willer, W. T., "Sensor Based Control of Robotic Manipulator Using a General Learning Algorithm", *IEEE J. Robotics and Automation*, vol.RA-3, no. 2, pp.157-165, April, 1987.

Winter, H., ed., *Artificial Intelligence and Man-Machine Systems*, New York: Springer-Verlag, 1986.

Wu, C. H., R. P. Paul, "Resolved Motion Force Control of Robot Manipulator", *IEEE Trans. Systems Man and Cybernetics*, vol.SMC-12, no. 3, pp. 26-275, May/June, 1982.

Yoshikawa, T., "Analysis and Control of Robot Manipulators with Redundancy", *Robotics Research* ed. by M Brady and R. Paul, pp. 735-747, 1984.

Zavidovique, B., A. Lanusse, P. Garda, "Robot Perception Systems: Some Design Issues", *NATO, ASI Series, Vol. F42*, pp. 93-109, Springer-Verlag, 1988.

Journal of Intelligent and Robotic Systems **1** (1988) 55–72.

Approach to the Dynamically Reconfigurable Robotic System

TOSHIO FUKUDA and SEIYA NAKAGAWA
Department of Mechanical Engineering, The Science University of Tokyo, 1-3 Kagura-zaka, Shinjuku-ku, Tokyo 162, Japan

(Received: 1 September 1987)

Abstract. In this paper, a newly proposed robotic system called the dynamically reconfigurable robotic system (DRRS), is reconfigurable for given tasks, so that the level of flexibility and adaptability is much higher for a change of working environments than conventional robots which have un-metamorphic shapes and structures. This robotic system consists of many cells which have fundamental mechanical functions. Each cell is able to detach and combine autonomously, so that the system can self-reorganize depending on a task or on working environments, and can also be self-repairing. DRRS has many applications in many fields, e.g. maintenance robots, more advanced working robots, free-flying service robots in space, more evolved flexible automation, etc. This paper shows the concept of this system, the mechanism of cells, the basic experimental results of the rough approach control between cells, and the decision method of such cell-structured manipulator configurations. This method is based on the reachability of the manipulators for working points, and so is able to apply the design of ordinary manipulators.

Key words. DRRS (dynamically reconfigurable robotic system), cell/module structure, reorganization of the optimal manipulator configurations, algorithm for approach control between cells.

1. Introduction

Cells of living creatures show very complicated and quite new configurations or functions when they gather up, although one cell itself has a very simple structure and function. The idea of this study is to apply such a cell system of living creatures to a robotic system with some intelligence analogous to the biological gene. This new concept of robotic systems is based on the concept of a cell/module strucutre in which each cell of the robotic module can detach itself and combine autonomously, depending on a task such as manipulators or mobile robots or structures. So the system can reorganize the optimal total shape. unlike the robots so far developed which cannot be automatically reorganized by changing the linkage of arms or replacing some links with others or reforming shapes in order to adapt itself to a change of working environments and demands. This robot system will be able to make a reconfigurable and more flexible system which, so far, has not been proposed in any other references. Such a system, called the dynamically reconfigurable robotic system (DRRS), which can dynamically reorganize its shape and structure by employing limited available resources for a given task and strategic purpose, has many unique advantages, such as optimal shaping under circumstances. fault tolerance. self-repairing, etc. DRRS is

245

also applicable in many fields, such as maintenance robots [1], more advanced working robots, free-flying service robots in space [2], and more evolved flexible automation [3], etc. The robot based on this cell-structured concept is also considered to be a distribution system of the Vroid, as previously reported [4–6], and so has a lot of incentives for intelligent robotic technology [7].

2. Concept of the DRRS

2.1. MERITS OF THE DRRS

There have been some robots or studies based on ideas similar to those given in this study, e.g., the machining center in FMS and maintenance robots for a nuclear power plant [1, 3]. They can only replace part of their tools with other tools, depending on a given task. But the tool parts are only instruments without versatilities, so that a large variety of tools are required for each operation. Naturally, they do not have any functions that can be called intelligence and other intellectual data bases. Therefore, we propose a new robot system that satisfies the following items:

(1) The robot consists of several cells, called the 'cell structure'.
(2) Each cell has a measure of intelligence.
(3) The cells can be automatically combined and detached by each other, depending on a given task.

2.2. DEFINITION AND CLASSIFICATION OF THE CELLS

A cell is defined here as a fundamental component of a robot structure and mechanism and has a single mechanical function: joint, end-effector, adjustable part of the arm length between joints of a manipulator, mobile mechanism, and power, etc. These cells are classified according to their general purpose when they construct manipulators or mobile robots.

LEVEL 1. Joint cells (bending joint, rotation joint, and sliding joint) and mobile cells (wheel, crawler, etc.)
LEVEL 2. Branching cells in two- or three-dimensional space, adjustable cells for the arm length between joints or the transformation of the orientation and power cells for heavy duty works.
LEVEL 3. Work cell (end-effectors, etc.) and the rests which are required for a nontypical purpose.

The combination and detachment between cells is carried out by mobile cells which can be transferred by themselves, or by the joint cells which are already combined like a manipulator.

2.3. APPLICATIONS OF THE DRRS

Most of the present industrial robots have four to six degrees of freedom and a large workspace according to their general work purposes. However, in some cases, it is doubtful that those configurations are the optimum. Cells of the DRRS can construct optimal configurations, since they can freely change combinations and choose degrees of freedom for given tasks. For example, it is easy to decompose a whole robotic system and carry the cells of the robot based on a spacecraft capability when it is transported from Earth into space. This robot can flexibly adapt itself to given works by using the combination and detachment between cells with a restriction of working environments: closed space, like a tank which has only a small entrance (Figure 1), and space that the entire configuration of a mobile robot cannot pass through because of some obstacles such as pipes or handrails. Or if the same type couplers with cells are set on a floor or a wall, this robot can move along them, and can be fixed at the base of a manipulator with any position or orientation in a given situation. Figure 2 shows the serial-parallel manipulator which uses branching cells of Level 2 described above, and can generate a higher torque. This robot can replace failed cells with new ones or can cover troubled ones with normal ones in the case where some cells are suspected of failing. So this robotic system has a fault-tolerant feature which is to keep, to some extent, the original functions. There is a lot of merit in a robot which can decompose into 'parts' called cells, although such a system has not yet been proposed, because of difficulty in realization. However, such a type of intelligent robotic system as DRRS is considered to be a kind of ideal robotic system and

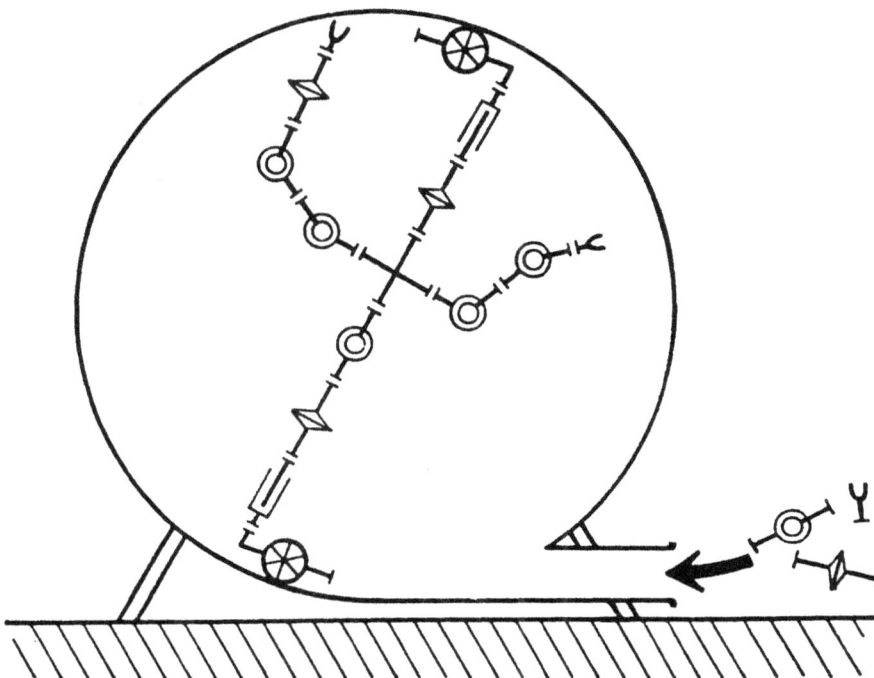

Fig. 1. An example of maintenance works in a storage tank.

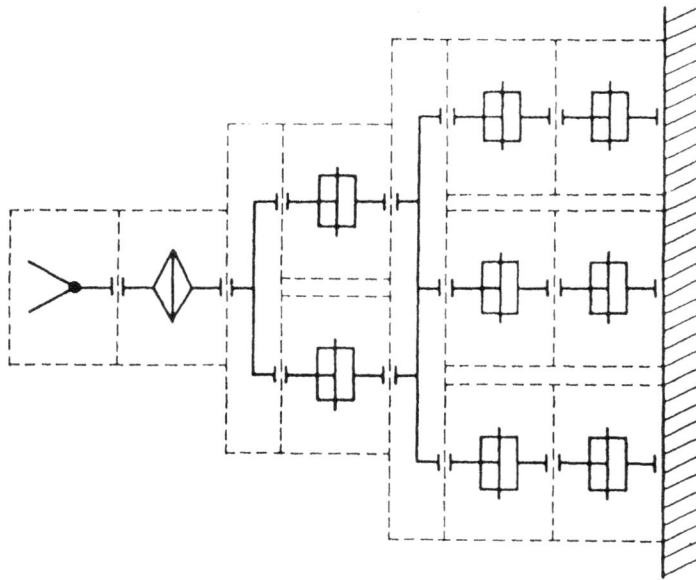

Fig. 2. An example of a parallel and serial manipulator.

will be, to some extent, realizable through advanced technology and microcomputer powers.

2.4. SYSTEM CONFIGURATION

A cell of living creatures works independently of other cells, but it cooperates with the entire cell group, although each cell is structurally and functionally autonomous. The control system configuration of this robot is also shown in Figures 3 and 4. Figure 3 shows the entire system configuration, and the main CPU may be independently set on the mobile cell or out of the cells. Figure 4 shows a cell control system. The data base in each cell has information of some of the characteristics of the cell: type of cell, connectable pair cell, admissible torques, mechanical strength, positioning accuracy.

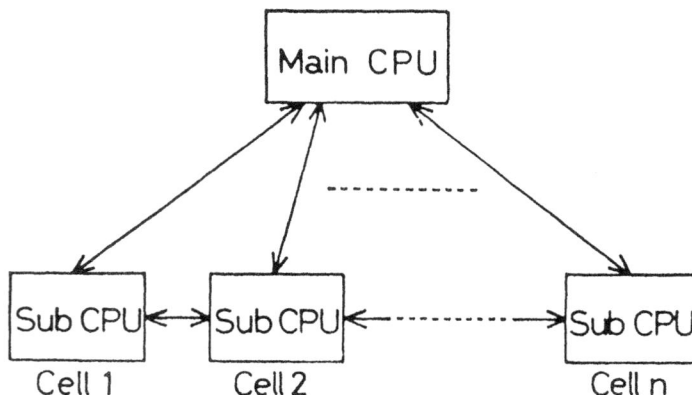

Fig. 3. System configuration.

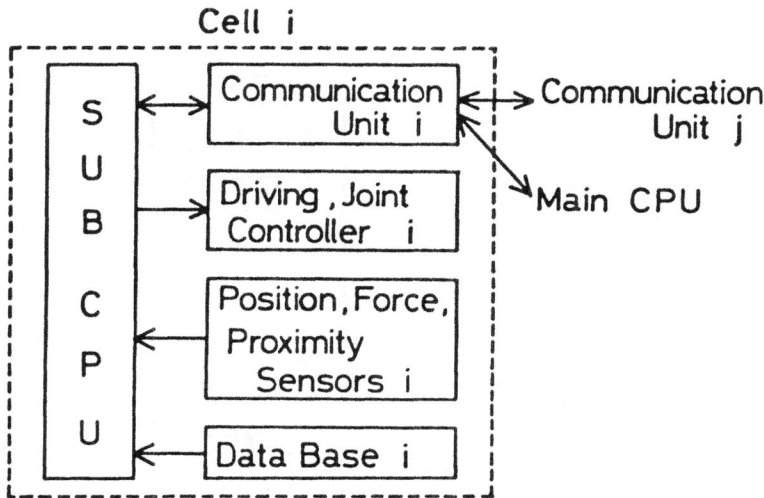

Fig. 4. System structure of a cell control system.

2.5. RESEARCH SUBJECTS

The following methods and technologies are necessary to realize such a new flexible robotic system:

(1) The discrimination and communication methods between cells.
(2) The control method of the approach between two cells.
(3) The control method of the combination and detachment between two cells.
(4) The method for making optimal configurations depending on a given task.
(5) The detection method for the abnormal cells due to troubles or degradings.
(6) The control method for the restoration and the reconstruction. In case all functional requirements cannot be met, some degrading control methods must be generated as an alternative.

In this paper, some experiments of the rough approach control between cells and a basic method of an optimal configurations for a cell-structured manipulator from DRRS are shown as a first step towards the realization of DRRS.

3. Cell Specification and Experimental Cells

Figure 5 and Plates 1 and 2 show features of three types of cells and experimental cells of Level 1, respectively: the bending joint cells, the rotation joint cell, and the wheel mobile cell. Each cell has couplers at the front and back to combine and detach at will. The actuators of the couplers are shape memory alloys (SMA) which are set on the external couplers (Figure 6). Each cell, which is tentatively considered to work in two-dimensional space for the sake of first-step experiments, has castors free from friction on the ground and joints and wheels are driven by DC motors. The frame sizes are all equal (length 90 mm × width 180 mm × height 50 mm), weights are 1.1 to 1.3 kg depending on the cells, and the movable range of the bending joint cell is between −90 and 90 degrees. The wheel mobile cell has a PSD (position sensitive

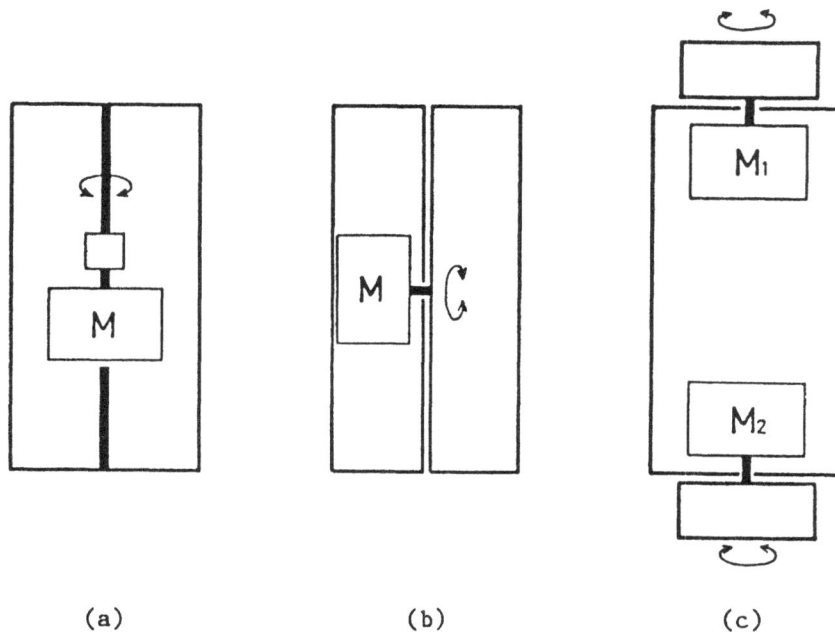

Fig. 5. Structure of cells (a: bending joint cell, b: rotation joint cell, c: wheel mobile cell).

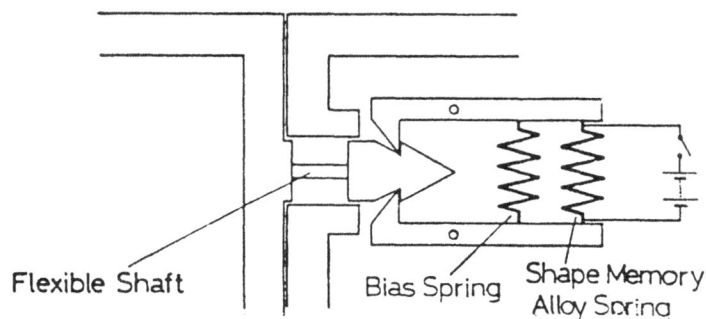

Fig. 6. Coupler structure and mechanism.

detector) sensor with some slits on the surface for the rough approach control which will be mentioned later.

4. Approach Control Between Cells

4.1. COMBINATION AND DETACHMENT METHOD BY MOBILE CELL

In this section, a two-dimensional plane is assumed to be the working environment for the combination and detachment of cells. It is only the mobile cell here that has a transferable function in a plane. By using this mobile cell, the combination and detachment method is shown. It carries any requested cells for achieving a given task from one point to some desired location, so that the system can make a desired configuration. Therefore, the following five steps are made to combine a cell with the mobile cell.

Plate 1.   Overview of the cells of DRRS.

Plate 2.   An example of a combination.

STEP 1. The mobile cell must find a desired cell and know the position of its couplers.

STEP 2. It roughly approaches the front of the couplers of the mating cell – the 'rough approach'.

STEP 3. It measures the relative position for a correction of attitude.

STEP 4. It approaches closer to the mating couplers – the 'fine approach'.

STEP 5. It combines with the couplers for unification.

4.2. EXPERIMENTS OF THE ROUGH APPROACH CONTROL

In the five steps, an experiment on the rough approach control (STEP 2) is shown. The PSD sensor with the slits, installed in the mobile cell, searches for the light emitted from the LED installed at the front of the other cell, and the mobile cell wheels are

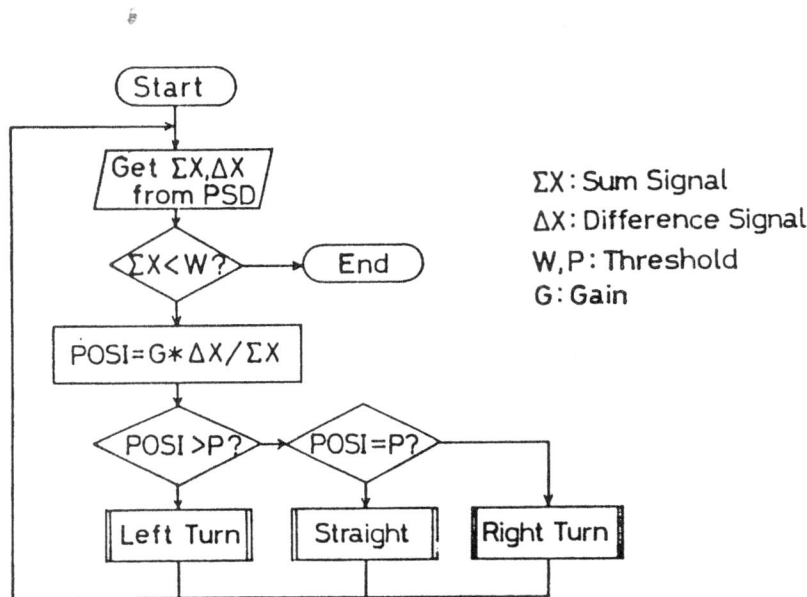

ΣX : Sum Signal
ΔX : Difference Signal
W,P : Threshold
G : Gain

Fig. 7. Algorithm of the rough approach control.

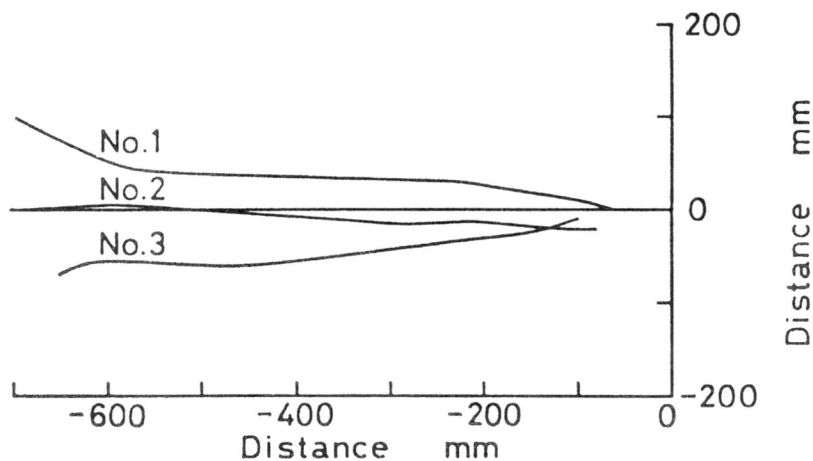

Fig. 8. Experimental results of the rough approach.

controlled by use of the PSD signal in a plane without obstacles. The control algorithm is shown in Figure 7, and some experimental results are given in Figure 8 which shows the tracks of the mobile cell in three different initial conditions. It is found that the mobile cell approaches the LED installed in the origin of the coordinate system.

5. Optimal Configurations

5.1. WORKSPACE OF MANIPULATORS

There are optimal configuration problems for given tasks when the DRRS cells configure some manipulators. To solve this problem, a criterion is required under some constraints. For a demonstrating example, manipulators with 3 degrees of freedom, which consist of only bending joint cells, rotational joint cells, and longitudinal sliding joint cells, can be designed with various joint configurations ($3^3 = 27$ types, without manipulator offsets). But the available manipulators are only 13 types after checking up on their characteristics. Figure 9 shows their workspaces, of only the top five types, under the condition that some of the adjustable cells, called 'elongated cells',

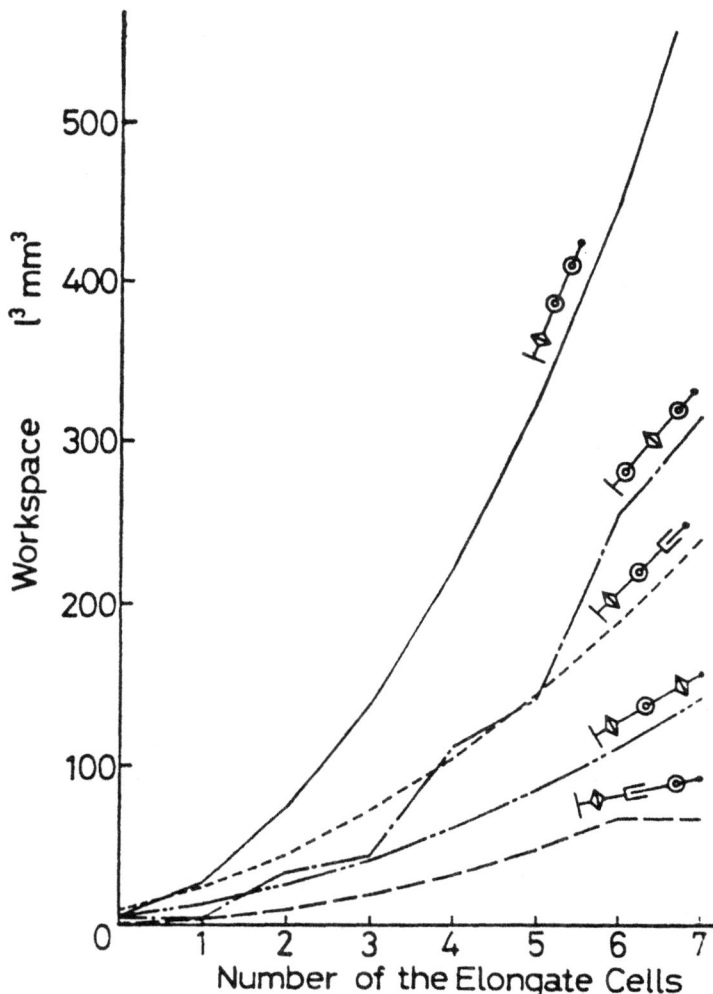

Fig. 9.   Work space evaluation for 3 degrees of freedom.

can be connected between the joint cells. If the workspace is chosen as the criterion, it is clear according to this figure that there is an optimal joint configuration, and that there is an optimal connecting position of the adjustment cells. In some cases, a number of adjustment cells change their orders.

5.2. DECISION PROCESS OF THE OPTIMAL CONFIGURATIONS

The optimal configuration problems for such cell-structured manipulators in DRRS are defined as follows:

(1) Decision of necessary and sufficient degrees of freedom.
(2) Decision of joints combinations.
(3) Decision of arm lengths between joints.

To solve this problem, it must be defined how to give tasks. A path required by the task for a manipulator-end is divided into N points in its working area, and the following six properties are given at each point:

(1) Required position and orientation for the manipulator end

$P_n(p_x, p_y, p_z, p_a, p_b, p_c)$.

(2) Required force for the manipulator end

$F_n(f_x, f_y, f_z, m_x, m_y, m_z)$.

(3) Required positioning accuracy for the manipulator end

$E_n(d_x, d_y, d_z, e_x, e_y, e_z)$.

(4) Setting position and orientation for the manipulator base

$Q_0(q_x, q_y, q_z, q_a, q_b, q_c)$.

(5) Type of an end-effector.
(6) Physical constraints in the working space

$S_m(x, y, z)$.

5.3. CANDIDATE SELECTION

5.3.1. *Assumptions*

This method is based on the structural merit of the DRRS. But now the following assumptions are set in order to simplify the problem of such cell-structured manipulator configurations:

(1) Each cell can connect and detach with complete control.
(2) Each cell has the same dimensions.
(3) The cells must connect serially.
(4) There are no offsets of manipulators.

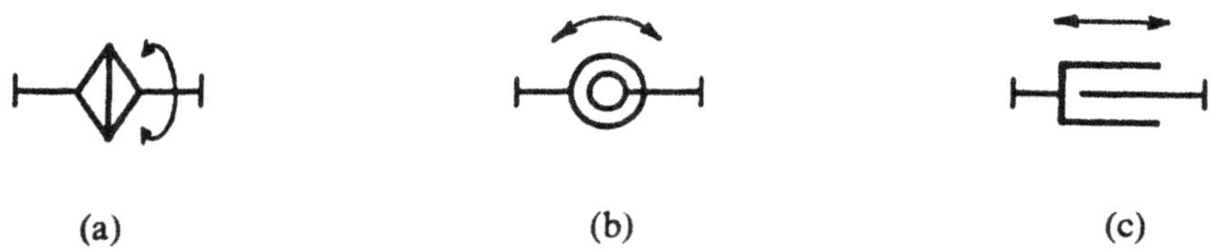

(a)   (b)   (c)

Fig. 10. Types of joints (a: rotation joint, b: bending joint, c: sliding joint).

(5) Rotation joint, bending joint, and sliding joint are only used for manipulator joint construction (Figure 10).

5.3.2. *Fundamental Idea*

Two coordinate systems are known from the previously-mentioned task. One exists at the base (Q_0) and the other exists at the end of the manipulator (P_0) at about each work point in the task. Then, two straight lines along each Z axis of the coordinate systems can be imagined. So a line P_iQ_j can be drawn because a cell, which has the same dimensions, can be added at the l (the length of a cell) intervals (Figure 11). A simple manipulator which has three basic links, Q_0Q_j, Q_jP_i, and P_iP_0 is imagined. The degree of freedom, the joints combination, and the length of links can be settled by checking the distance of the line P_iQ_j and the angle of torsion between the basic links. These criterions are shown in the next subsection. All possible combinations of P_i and Q_j must be checked, then tables of candidates for manipulator configurations about each work point can be made. After that, by checking each table, the common candidates which can reach all given work points are chosen. Finally, the necessary force and positioning accuracy at the end of the manipulator are checked, so that the desired manipulator configuration can be obtained.

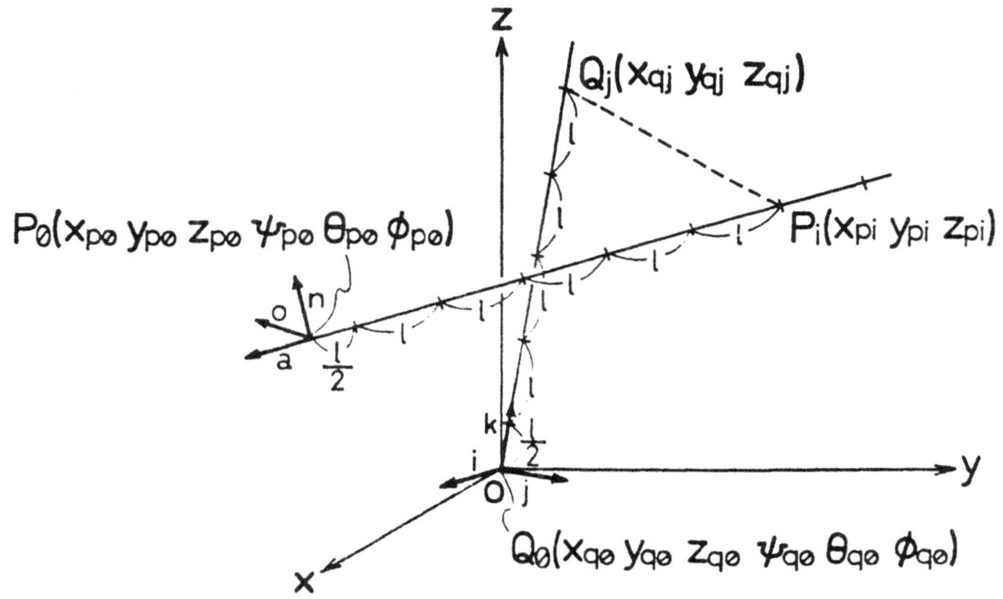

Fig. 11. Coordinate system of cell-structured manipulator.

255

5.3.3. *Criterions*

By dividing the length of a line P_iQ_j into l, the number of cells D which should be connected between P_i and Q_j can be known.

$$D^2 = \left\{ \frac{x_{p0} - x_{q0}}{l} - (n_p - \tfrac{1}{2})A_p - (n_q - \tfrac{1}{2})A_q \right\}^2$$
$$+ \left\{ \frac{y_{p0} - y_{q0}}{l} - (n_p - \tfrac{1}{2})B_p - (n_q - \tfrac{1}{2})B_q \right\}^2$$
$$+ \left\{ \frac{z_{p0} - z_{q0}}{l} - (n_p - \tfrac{1}{2})C_p - (n_q - \tfrac{1}{2})C_q \right\}^2, \tag{1}$$

where

$$A_p = C_{\phi p}S_{\theta p}C_{\psi p} + S_{\phi p}S_{\psi p}, \qquad A_q = C_{\phi q}S_{\theta q}C_{\psi q} + S_{\phi q}S_{\psi q},$$

$$B_p = S_{\phi p}S_{\theta p}C_{\psi p} - C_{\phi p}S_{\psi p}, \qquad B_q = S_{\phi q}S_{\theta q}C_{\psi q} - C_{\phi q}S_{\psi q},$$

$$C_p = C_{\theta p}C_{\psi p}, \qquad C_q = C_{\theta q}C_{\psi q},$$

$$S_{\theta i} = \sin \theta_i, \quad C_{\theta i} = \cos \theta_i, \ldots, \qquad n_p, n_q = 1, 2, 3, \ldots.$$

and l is the length of a cell.

The following three equations, which detect the angle of torsion between the basic links, are geometrically obtained by using the inner product. If a basic link satisfies them, it does not need one degree of freedom to correct the angle of torsion.

$$\mathbf{j} \cdot \overrightarrow{P_iQ_j} = 0, \tag{2}$$

$$\overrightarrow{P_0P_i} \cdot (\overrightarrow{Q_0Q_j} \times \overrightarrow{Q_0P_i}) = 0, \tag{3}$$

$$\mathbf{o} \cdot \overrightarrow{P_iQ_j} = 0, \tag{4}$$

where

$$\mathbf{j} = (0 \ 1 \ 0),$$

$$\mathbf{o} = (C_{\phi p}S_{\theta p}S_{\psi p} - S_{\phi p}C_{\psi p}$$
$$S_{\phi p}S_{\theta p}S_{\psi p} + C_{\phi p}C_{\psi p} \quad C_{\theta p}S_{\psi p}).$$

5.3.4. *Process of Candidate Selection*

D (Equation (1)) is given for all possible combinations of P_i and Q_j. If D is equal to zero or an integer, the manipulator can decrease its degrees of freedom and the joint combinations can become relatively simple. But if D is not equal to that, it has to increase the degrees of freedom by using the strategy shown in Figure 12. Equations (2)–(4) give the information whether a rotational joint cell

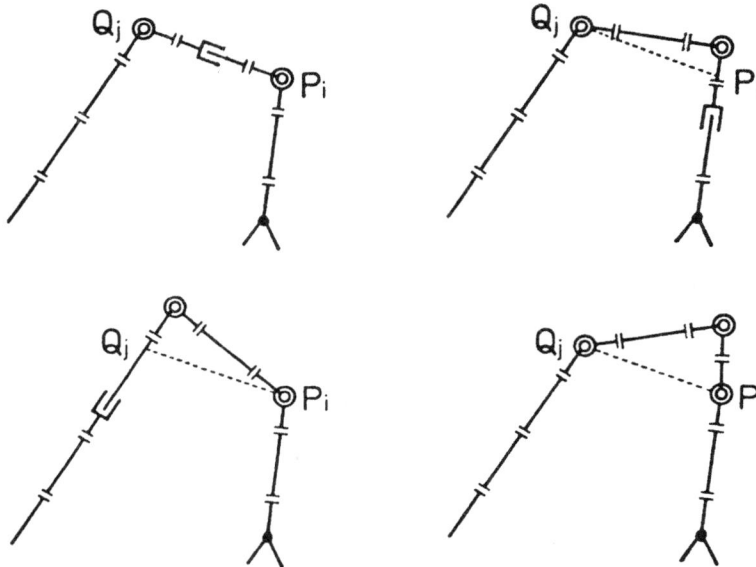

Fig. 12.   Measures when $D \neq$ integer.

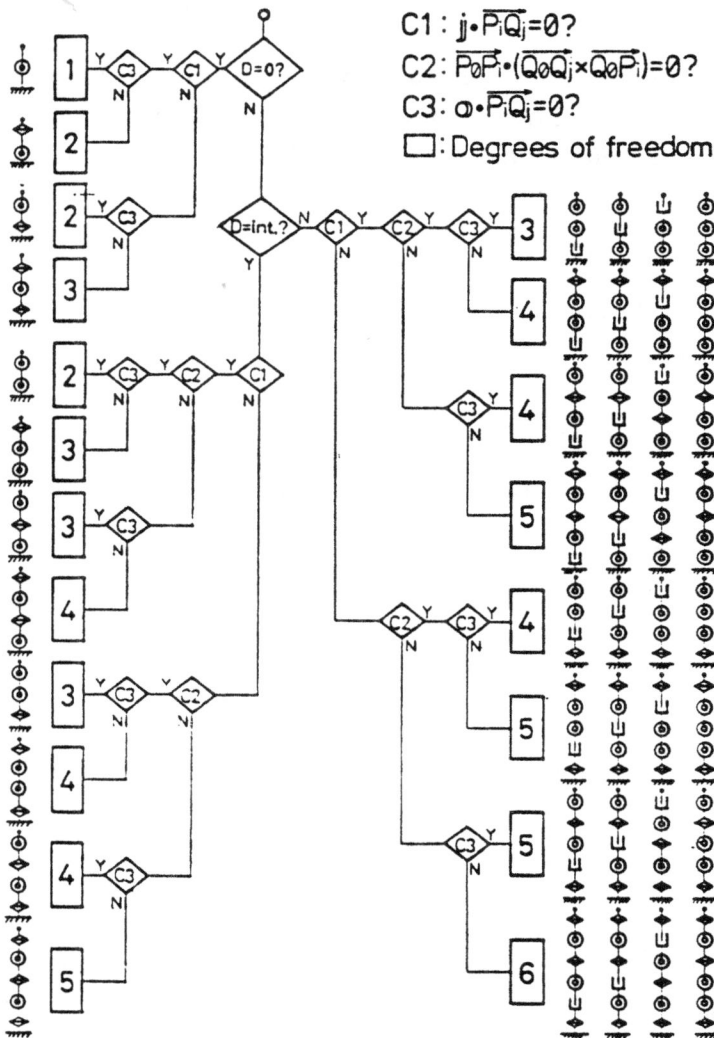

C1: $\vec{jj} \cdot \overrightarrow{P_i Q_j} = 0$?

C2: $\overrightarrow{P_0 P_i} \cdot (\overrightarrow{Q_0 Q_j} \times \overrightarrow{Q_0 P_i}) = 0$?

C3: $\vec{o} \cdot \overrightarrow{P_i Q_j} = 0$?

$\square$: Degrees of freedom

Fig. 13.   Candidate selection of manipulator configurations.

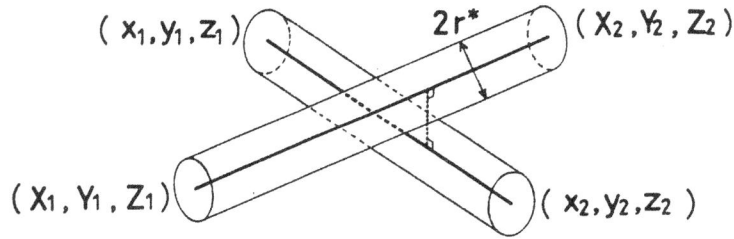

Fig. 14. Coordinates of two links.

is necessary or not. If the angle of torsion also is not zero, the degrees of freedom increase. This process is described in Figure 13. However, in this process, the candidates need to satisfy the three geometric constraints mentioned in the next section.

5.4. GEOMETRIC CONSTRAINTS

5.4.1. *Constraint for Link Interference*

If the basic links interfere with one another, the configuration must be excluded from the candidates. Each link can be regarded as a cylinder (radius: r^*) and the central axis lines of the two cylinders are shown in Figure 14. So knowing the distance between the two straight lines, the following expression is obtained. (A configuration must be excluded from the candidates when it satisfies this.)

$$(2r^*)^2 \geqslant \min_{s, t} \{(As - at + X_1 - x_1)^2 \\ + (Bs - bt + Y_1 - y_1)^2 + (Cs - ct + Z_1 - z_1)^2\}, \tag{5}$$

where

$$A = X_2 - X_1, \quad B = Y_2 - Y_1, \quad C = Z_2 - Z_1,$$
$$a = x_2 - x_1, \quad b = y_2 - y_1, \quad c = z_2 - z_1,$$
$$0 \leqslant s \leqslant 1, \quad 0 \leqslant t \leqslant 1.$$

5.4.2. *Constraint for Movable Range of Joints*

The bending and sliding joints have structurally movable limits. So if these joints are out of their limits, the configuration must be excluded from the candidates. Two neighbouring links can be described as vectors, so that the angle between them is given as

$$\theta^* = \cos^{-1}(L_1 \cdot L_2/|L_1||L_2|). \tag{6}$$

Therefore, θ^* must satisfy the following inequality in order that a configuration may be a candidate. θ_{max} is the movable angle limit of a bending joint cell.

$$|\theta^*| \leqslant \theta_{max}. \tag{7}$$

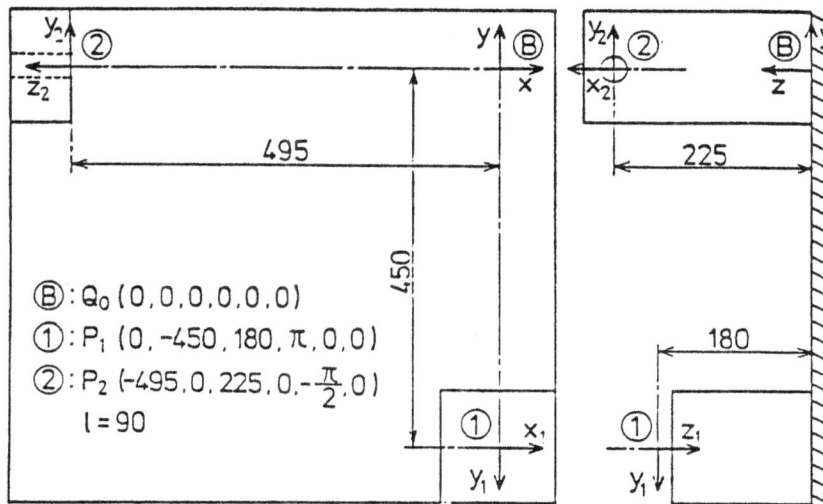

Fig. 15. Work points in a task.

B: Q_0 $(0,0,0,0,0,0)$
$①$: P_1 $(0,-450,180,\pi,0,0)$
$②$: P_2 $(-495,0,225,0,-\frac{\pi}{2},0)$
$l=90$

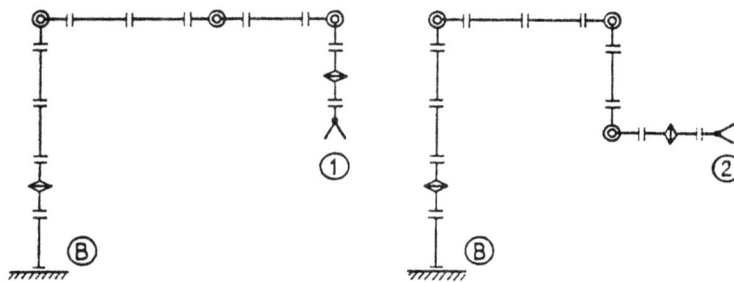

Candidate 1

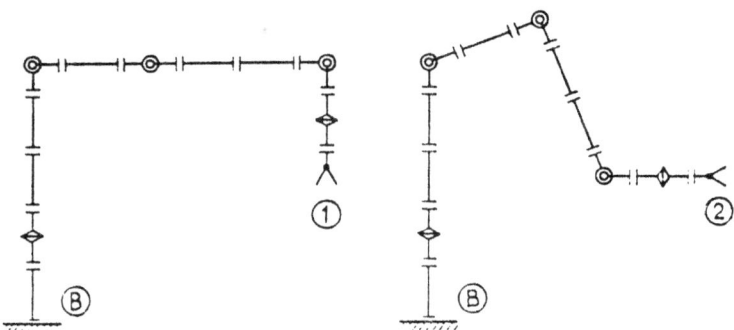

Candidate 2

Fig. 16. Common candidates about for work point.

Similarly, a displacement value of a sliding joint d^* must satisfy (8) if it can move 0 to d_{max}

$$0 \leqslant d^* \leqslant d_{max}. \tag{8}$$

259

Table I. Candidates of manipulator configurations about work point 1

No.	Joint configurations	Link length (*l*) Base (j)	Center (D)	End (i)	θ_q	θ_p	Degrees of freedom
1	–R–B–B–B–R–e	2	5.83	3	59.0	121.0	5
2	–R–B–B–B–R–e	3	5.39	3	68.2	111.8	5
3	–R–B–B–B–R–e	4	5.10	3	78.7	101.3	5
4	–R–B–B–R–e	5	5.00	3	90.0	90.0	4
5	–R–B–B–B–R–e	6	5.10	3	101.3	78.7	5
6	–R–B–B–B–R–e	7	5.39	3	111.8	68.2	5
7	–R–B–B–B–R–e	2	6.40	4	51.3	128.7	5
8	–R–B–B–B–R–e	3	5.83	4	59.0	121.0	5
9	–R–B–B–B–R–e	4	5.39	4	68.2	111.8	5
10	–R–B–B–B–R–e	5	5.10	4	78.7	101.3	5
11	–R–B–B–R–e	6	5.00	4	90.0	90.0	4
12	–R–B–B–B–R–e	7	5.10	4	101.3	78.7	5
13	–R–B–B–B–R–e	2	7.07	5	45.0	135.0	5
14	–R–B–B–B–R–e	3	6.40	5	51.3	128.7	5
15	–R–B–B–B–R–e	4	5.83	5	59.0	121.0	5
16	–R–B–B–B–R–e	5	5.39	5	68.2	111.8	5
17	–R–B–B–B–R–e	6	5.10	5	78.7	101.3	5
18	–R–B–B–R–e	7	5.00	5	90.0	90.0	4
19	–R–B–B–B–R–e	2	7.81	6	39.8	140.2	5
20	–R–B–B–B–R–e	3	7.07	6	45.0	135.0	5
21	–R–B–B–B–R–e	4	6.40	6	51.3	128.7	5
22	–R–B–B–B–R–e	5	5.83	6	59.0	121.0	5
23	–R–B–B–B–R–e	6	5.39	6	68.2	111.8	5
24	–R–B–B–B–R–e	7	5.10	6	78.7	101.3	5
25	–R–B–B–B–R–e	2	8.60	7	35.5	144.5	5
26	–R–B–B–B–R–e	3	7.81	7	39.8	140.2	5
27	–R–B–B–B–R–e	4	7.07	7	45.0	135.0	5
28	–R–B–B–B–R–e	5	6.40	7	51.3	128.7	5
29	–R–B–B–B–R–e	6	5.83	7	59.0	121.0	5
30	–R–B–B–B–R–e	7	5.39	7	68.2	111.8	5

B: bending joint, R: rotation joint, S: sliding joint, e: end-effector.

5.4.3. Constraint for Possibility of Combination in a Basic Link

As mentioned above, each basic link needs some correcting joint cells when they have an angle of torsion, or when D in Equation (1) is not an integer. So it is necessary for the basic link to have a space which can join them. This constraint is shown as

(length of the basic link)/l − 1

$$\geq \text{(number of the combining cells)}. \qquad (9)$$

A configuration must be excluded from candidates when it satisfies this.

Table II. Candidates of manipulator configurations about work point 2

No.	Joint configurations	Link length (*l*) Base (j)	Link length (*l*) Center (D)	Link length (*l*) End (i)	θ_q	θ_p	Degrees of freedom
1	–B–B–B–e	1	4.47	2	63.4	26.6	3
2	–B–S–B–e						
3	–B–B–B–e	2	4.12	2	76.0	14.0	3
4	–B–S–B–e						
5	–S–B–B–e						
6	–B–e	3	4.00	2	90.0	0.0	1
7	–B–B–B–e	4	4.12	2	104.0	14.0	3
8	–B–B–B–e	5	4.47	2	116.6	26.6	3
9	–B–B–B–e	6	5.00	2	126.9	36.9	3
10	–B–B–B–e	7	5.66	2	135.0	45.0	3
11	–B–B–B–e	1	3.61	3	56.3	33.7	3
12	–B–S–B–e						
13	–B–B–S–e						
14	–B–B–B–e	2	3.16	3	71.6	18.4	3
15	–S–B–B–e						
16	–B–S–B–e						
17	–B–B–S–e						
18	–B–e	3	3.00	3	90.0	0.0	1
19	–B–B–B–e	4	3.16	3	108.4	18.4	3
20	–B–B–B–e	5	3.61	3	123.7	33.7	3
21	–B–B–B–e	6	4.24	3	135.0	45.0	3
22	–B–B–B–e	7	5.00	3	143.1	53.1	3
23	–B–B–B–e	1	2.83	4	45.0	45.0	3
24	–B–S–B–e						
25	–B–B–S–e						
26	–B–B–B–e	2	2.24	4	63.4	26.6	3
27	–S–B–B–e						
28	–B–S–B–e						
29	–B–B–S–e						
30	–B–e	3	2.00	4	90.0	0.0	1

B: bending joint, R: rotation joint, S: sliding joint, e: end-effector.

5.5. SIMULATION

By using this algorithm, the available manipulator configurations can be chosen when, for example, two work points in a task is given, as in Figure 15. The obstacles are not considered in the work space. The results of each work point in this problem are shown in Tables I and II. However, the movable angle of the bending joint is – 90 to 90 degrees and the movable displacement of the sliding joint is 0 to 180. θ_p and θ_q is the angle between two basic links (the angle of the bending joint). Two candidates are obtained from these tables (Figure 16), which are common to the two work points and have the least number of cells (see the arrows in Tables I and II). Therefore, they are the manipulator configuration candidates that can reach each work point. It is found that five degrees of freedom are enough to carry out this task.

6. Conclusions

(1) The concept of a DRRS (dynamically reconfigurable robotic system) is proposed based on cell structurization for general types of robots.

(2) The classification of the cells, the system configurations, and the research subjects for the realization are stated concretely.

(3) Some experimental results for one of the processes of the combination-detachment between cells are shown.

(4) A method of the manipulator configuration is proposed and its availability is confirmed by simulation.

References

1. Fukuda, T., *Advanced Robotics under Hostile Environments*, McGraw-Hill (1986) (in Japanese).
2. Akin, D. L., *et al.*, Space applications of automation, robotics and machine intelligence systems (ARAMIS) – Phase II, NASA (1983).
3. Knight, J. A. G., The latest developments of FMS in Japan. *Proc. of the 1st Int. Conf. on FMS*, pp. 31–47, (1982).
4. Fukuda, T. and Kobayashi, H., A study on a flexible mobile robot with versatile shape control based on multiple link-wheel mechanism, 1st report, *Bull. JSME* **29**-256, 3545–3552 (1986).
5. Fukuda, T. and Kobayashi, H., A control configured robot (CCR) approach for a mobile robot with consideration of joint load distribution about each work point control, *Proc. Japan-USA Symp. on Flexible Automation*, pp. 309–316 (1986).
6. Fukuda, T. and Kobayashi, H., Configuration control method of a control configured robot (CCR) consisting of multiple link-wheel mobile mechanism, *Prepr. of 10th IFAC World Congress on Automatic Control*, Vol. 4, pp. 226–231, (1987).
7. Fukuda, T. and Nakagawa, S.: Cell structured robots, *3rd Robotics Society of Japan*, pp. 423–424, (1985) (in Japanese).

262

The Central Nervous System
as a Low and High Level Control System

James S. Albus
Chief, Robot Systems Division
Center for Manufacturing Engineering
National Bureau of Standards

The division of the central nervous system into high and low level control systems has long been recognized. How these two systems are interconnected and how they influence each other is one of the great mysteries of modern science. Recent attempts to produce intelligent behavior in robots and computer integrated manufacturing systems have produced insights as to how high level goals can be decomposed into low level actions, and how knowledge about the environment can be acquired, stored, and accessed by task decomposition processes to produce sensory-interactive goal directed behavior.
Research on computer brain models has also shown how networks of neuron-like elements can learn patterns and motor skills and generalize from one task to another. This paper proposes a model of how the high level understanding, evaluating, goal selection, planning and reasoning functions commonly associated with the mind are tied into the lower level sensing, filtering, recognizing, task execution, and servo control functions that are commonly associated with the mechanisms of the body.

The Brain as a Hierarchy

Specific neurological theories that assume the brain (the seat of both the mind and the motor system) to be hierarchically structured are well over a hundred years old (Jackson, 1931). In the past three decades, a large number of neurophysiological experiments have shown that the brain processes sensory information through a number of distinct hierarchical levels. It is now well established that specific functions are performed by specific neuronal structures at a variety of hierarchical levels in a number of neuronal pathways. For example, the retina of the eye is known to perform edge enhancement and spatial filtering operations on visual images produced by the optics of the eye. The lateral geniculate performs stereo matching and gating operations. The visual cortex (area 17) detects edges and corners, and measures the position and orientation of such features in the visual field. Cell clusters in the visual association areas recognize three dimensional objects, and areas in the frontal cortex perform spatial reasoning operations.

Hierarchically structured computing modules have been observed and studied not only in the ascending sensory systems but in the descending motor control pathways as well. It has been demonstrated (Evarts and Tanji 1974) that neurons in the cerebellum, thalamus, and motor cortex alter their firing rates at various intervals prior to either movement or feedback, as motor commands propagate down the task decomposition hierarchy.

NATO ASI Series. Vol. F43
Sensors and Sensory Systems
for Advanced Robots
Edited by P. Dario
Springer-Verlag Berlin Heidelberg 1988

There is much evidence that the descending pathway influences, and is influenced by, the ascending pathways. Downward flowing motor commands are known to be capable of significantly modifying the interpretation of sensory data. At many different levels, internally generated expectations are matched against experientially observed objects, relationships, and temporal events. Both correlations and differences between expectations and observations are computed.

Similarly, ascending sensory input significantly modifies downward flowing motor control signals by providing feedback at a variety of hierarchical levels. Thus, the brain does not have simply a vision system, but a vision-attention system. There is not simply a speech generating system, but a hearing-speech system. The brain is, at least at the lower levels, a strongly cross-coupled hierarchical sensory-motor system.

The Brain as a Control System

It is important to realize that the principal function of the brain is not to reason and plan, or even to sense and recognize, but to select goals and control behavior. Reasoning and planning are evolutionarily recent properties of only a tiny fraction of the brains that have ever existed. But all brains control behavior. Effective control of behavior is crucial to survival and reproduction. Behavior either succeeds or fails in achieving the ultimate goal of all living creatures: <Propagate-Genes>.

This suggests that the best approach to the study, and hopefully the artificial production, of intelligent behavior would be to build machines that must produce behavior in the real world, and then to equip these machines with sensors and control systems that enable them to carry out sensory-interactive goal-directed behavior.

Goal Selection

Any autonomous creature, be it a robot, a human, a bird, or an insect, must have some internal mechanism for selecting between alternative possible behaviors. In animals, the highest level, longest term, goals are selected by hard-wired (or PROM memory) routines, called instinct. Instinctual goals are known to be triggered into play by hormonal action, which itself is synchronized with long term external events, such as changes in the seasons, by means of sensory input.

It is also known that there are areas in the human and animal brain, primarily in the limbic regions, that evaluate the goodness or badness of situations and events. These are the neural areas that human subjects report produce feelings of pleasure-pain, joy-horror, hope-despair, love-hate, curiosity-fear, craving-revulsion, etc. The values of these "feeling variables" provide the basis for choosing between approach behavior and flight, between acceptance of other creatures as friends or rejection of them as enemies; between fighting, feeding, resting, sleeping, and caring for others; between caring for off-spring, family, peer group, or those outside the peer group. The emotional evaluation functions provided by the limbic system affect both the planning and execution of behavior at many different levels.

Goal selecting and situation evaluating functions can be provided to robots through decision-making methodologies taken from the fields of operations research, game theory, cybernetics, and artificial intelligence. Goal selection typically requires:
  1. A search through the space of possible future actions
  2. An evaluation of those futures considered
  3. A selection of the best course of action
Clearly this process requires both a means for modeling and predicting the effect of future actions, as well as a set of evaluation functions with which to compute the cost, benefit, risk, and payoff of various alternative courses of action. Both natural and artificial autonomous creatures must be able to evaluate the costs and benefits of actions, both while they are being planned and while they are being executed.

Natural creatures that possess superior high level decision making processes and low level skills tend to be more successful than their competitors in selecting and executing behavior which propagates genes. This presumably is the mechanism of natural selection which produced natural intelligence in the first place.

Workshop Goals

The goals of this workshop are to:

  1. Explore what is known in a wide diversity of fields about sensors and sensory processing systems, and
  2. Understand how sensory information can be applied to advanced robotics.

The objective of this session on the "biological model" is to correlate what is know from biology with what is known from robotics and artificial intelligence.

A Unifying Hypothesis

This paper attempts to formulate a unifying hypothesis of how the high level understanding, evaluating, goal selection, planning and reasoning functions commonly associated with the mind are tied into the lower level sensing, filtering, recognizing, task execution, and servo control functions that are commonly associated with the mechanisms of the body. Hopefully, this model will focus discussion on unifying principals that form a common thread through the wide diversity of experimental data and theoretical methodologies that apply to this most challenging and interesting problem; the study of the nature of intelligence, both natural and artificial.

The model which I am proposing is derived from neurophysiological theory and experiments as well as results from CIM (Computer

Integrated Manufacturing) experiments (Albus 1982). CIM is the attempt to integrate robots, machine tools, inspection machines, and material inventory and distribution systems into totally automated factories.

The proposed model has the general structure shown in Figure 1. It consists of a hierarchy of levels, each of which is composed of task decomposition, world modeling, and sensory processing modules. Commands flow vertically downward in a task decomposition tree; sensory information flows vertically upward in a sensory processing tree; and world model information flows horizontally between task decomposition, world modeling, and sensory processing modules at each level of the hierarchy. Computing modules at all levels of the hierarchy are "data-flow" machines, each of which itteratively executes the following control cycle:

1. Read a set of input variables,
2. Compute a mathematical function on these inputs, and (after a computational delay)
3. Produce a set of output variables.

The functions computed depend both on the input variables and on state variables internal to the modules. If such a system is modeled as a discrete-time sampled-data system, the computational modules can be represented as finite-state automata.

There are three ways that such dataflow computational modules can be interpreted:

1. If the input to any computational module is treated as a vector of variables S, the output is a vector of variables P computed by the mathematical function H, i.e.,

$$P = H(S) \tag{1}$$

2. If the input S is treated as an "IF premise", the output P is a "THEN consequent" such as is computed by an expert system rule, i.e.,

$$IF(S) \; THEN \; (P) \tag{2}$$

(Here the function H is defined by the entire set of rules in the expert system.)

3. If the input S is treated as an address of a location in memory, then the output P is the contents of the address S, i.e.,

$$P = contents(S) \tag{3}$$

(Here the function H is defined by a look-up table where for each address S there is stored a value P.)

In all three cases, the input defines an input vector S on an input space which is mapped by the function H into an output vector P on an output space.

The Functional Levels

Figure 2 shows the type of functions performed, and the type of output produced, at each level in the proposed hierarchy . The lowest level (level 0) is the servo level. At this level position, velocity, and force information is sensed, scaled, filtered, and compared with commanded values of position, velocity, and force. In a robot, the level 0 task decomposition modules compute the correct drive to the joint actuators to null the difference between commanded and observed values.

In animals, level 0 consists of the circuitry shown in Figure 3. Level 0 contains alpha and gamma motor neurons, muscles, stretch receptors, tendon tension sensors, and sensory ganglia neurons that make up the stretch reflex.

At level 0 all computations are performed in joint or muscle coordinates. Output values are typically updated every few milliseconds.

Level 1 of the proposed hierarchy computes the kinematic transformations necessary to translate from a convenient coordinate system (world, tool, or part coordinates) into joint coordinates. Commands at level 1 are defined in a coordinate system in which it is convenient to express problems of manipulation or locomotion. Sensory information is transformed into the same coordinate system so that observed positions, velocities, accelerations, and forces can be easily compared with commanded values. Level 1 commands typically are executed in a few hundreths of a second.

Level 2 in the proposed hierarchy accepts input commands defined in terms of "keyframe poses", or "key knot points", on a manipulation or locomotion elemental-move trajectory. It computes a dynamically efficient pathway, or smooth trajectory, through the keyframe poses in space/time. Level 2 also coordinates motions between closely related body parts so as to accomplish dynamically efficient movements of arms, legs, hands, and fingers. Level 2 commands typically are executed in a few tenths of a second.

In animals, level 1 and 2 functions have been hypothesized to be computed in the cerebellar cortex and in the cerebral motor cortex.

Level 3 in the proposed hierarchy transforms input commands expressed in terms of symbolic names of elemental-movements into key frame poses along trajectories in the chosen coordinate system of levels 1 and 2. Elemental-move commands are typically of the form <REACH>, <APPROACH>, <GRASP>, <LIFT>, <MOVE-TO>, etc. Level 3 modules control individual subsystems such as a single manipulator, a mobility unit, or a single machine tool. Obstacle avoidance, kinematic singularities, and limits on joint motion are handled at this level. Level 3 commands are typically accomplished in periods of a few seconds.

Level 4 of the proposed hierarchy accepts input commands expressed in terms of tasks, or operations, to be performed by an individual animal or machine on specific objects, or with respect to other individuals. Level 4 decomposes these tasks into sequences of elemental-moves that can be executed by manipulation or locomotion subsystems. Coordination between various subsystems is performed at this level: for example, manipulation-locomotion coordination, eye-hand coordination, or coordination between multiple arms takes place at this level. Precedence constraints on elemental-move sequences, as well as cost/benefit trade-offs between alternative elemental move action sequences may be computed at this level. In the manufacturing environment, this is the equipment level, i.e. the level of individual machines or stand-alone equipment on the factory floor. Level 4 commands are typically accomplished in periods of a few seconds.

Level 5 takes input commands expressed in terms of tasks to be performed on groups of objects by small tightly coupled groups, or families, of machines. Input commands often specify only the set of tasks and priorities without specific instructions as to the sequence in which the tasks must be done. Coordination between actions of various individuals, such as coordination between robots, machine tools, automatic clamping and material handling systems takes place at this level. In the manufacturing environment, this is the work station level. Level 5 commands are typically accomplished in periods of minutes to hours.

Level 6 input commands are expressed in terms of goals to be accomplished by cells, or groups of workstations. Commands at this level typically consists of production requirements and due dates for the next day to a week. Level 6 schedules production and performs batching and routing of parts and tools so that the workstations can efficiently accomplished their assigned tasks with a minimum of waiting. In the manufacturing environment, this is the cell control level. Level 6 commands are typically accomplished in a day to a week.

Level 7 takes input commands expressed in terms of goals to be accomplished by entire manufacturing shops consisting of groups of cells. Level 7 commands typically contain the entire backlog of production orders, together with priorities and due dates. Level 7 functions maintain inventory at levels necessary to meet production schedules, and assign machine tools and other production resources so as to meet promised deliveries in an efficient manner. Level 7 deals with control problems that have lead times of weeks to months.

Level 8 is where decisions are made about the product line and the engineering design of parts to be manufactured. Decisions made at this level must be based on predicted needs and requirements months to years into the future.

At each level in the model described above, there exists a task decomposition, a world modeling, and a sensory processing module. At each level there is a feedback loop whereby the world model is updated and the task decomposition process can be modified to take into account previously unknown information about the external world. The information contained in the world model is thus a synthesis of apriori knowledge and sensed information.

Time

In the generation and control of behavior, time is a crucial variable. Behavior is produced in the instant of the present, at $t = 0$. Feedback is derived from sensory data collected during the past. Plans are made for action in the future. How far into the past is sensory data relevant to current behavior? How far into the future is it reasonable to plan? The hierarchical control model in Figure 2 suggests answers to these questions. Each level plans only how to decompose its current (or next) command into a sequence of subcommands to the next lower level. Sensory data at each level is integrated over a historical time period that extends only as far into the past as the planning algorithms at that level look into the future. Hence, modules at low levels in the hierarchy plan only for the immediate future and need sensory data integrated only over the immediate past. At successively higher levels, the planning horizon extends exponentially further into the future, and the relevant historical trace extends exponentially further into the past.

The task decomposition and sensory processing systems are duals of each other. Each level in the task decomposition hierarchy produces both a spatial and temporal decomposition of tasks and goals into subtasks and subgoals. Spatial decomposition involves assigning task subelements to subordinate computing modules. Temporal decomposition requires that each subordinate computing module decompose its task subelement into a temporal sequence of activities.

Similarly, each level in the sensory processing hierarchy produces both a spatial and temporal integration of patterns and sequences into objects and events. Spatial integration involves the collection of pixels into features, features into regions, regions into objects, etc. Temporal integration involves the collection of motion trajectories (of points, features, regions, and objects) into events, sequences, and trends.

Neuronal Computational Modules

How does the brain organize nerve cells to perform the functions of the H, M, and G modules? No one really knows. Nevertheless, a number of attempts have been made to demonstrate how a neuronal

net can be organized so as to filter and process sensory data, recognize patterns, remember events, make predictions, learn, generalize, decompose tasks, and control actions (Albus 1981).

Among the earliest works in brain modeling were the simple neuron analogues of McCulloch and Pitts (1943). The concept of cell assemblies, which learn to compute behavioral functions, was introduced by D. O. Hebb (1949) as part of a neurophysiological theory of the organization of behavior. In (1961) Rosenblatt proposed a mathematical construct called the "perceptron" to model the neurodynamic functions of learning, generalization, and pattern recognition. Minskey (1967) demonstrated the ability of neural nets to compute any computable function. Later (1969) Minsky and Papert published a treatise which so effectively demonstrated the limitations of perceptrons that it had the effect of halting almost all work on brain modeling for a decade. Recently, interest in associative memory (Kohonen 1978) and brain modeling (Palm 1982) has revived.

The 1960's brought advances in the technology of single cell recording and electron microscopy. These made possible a set of experiments by Eccles and others (1967) that identified the functional interconnections between the principal components in the cerebellar cortex. This experimental data provided the basis for a theoretical model developed by Marr (1969) and Albus (1971) which suggested how a layer of cerebellar cortical tissue could learn, generalize, and compute arithmetic functions on input variables carried by a set of mossy fiber inputs. As shown in Figure 4, this model also suggested how a single precise value, such as the angle of an elbow joint, could be encoded by a multiplicity of imprecise neuronal channels.

Figures 5 and 6 show how the Marr-Albus model decodes input variables so as to select a set of synaptic weights. The model then computes by table look-up the value of functions such as described in formula (3) above. Figures 7 and 8 illustrate how the cerebellar model can produce the properties of learning and generalizing. Figure 9 illustrates how such modules can be arranged in a hierarchical architecture so as to produce sensory-interactive goal directed behavior.

Implications for Advanced Robotics

It is clear from the papers presented at this workshop that the sensors for advanced robots are being built. Sensors are being developed for proprioception (joint position, velocity, force),

touch, temperature, haptic perception, vision, hearing (vibration, speech, sonor), taste, smell, and a number of senses not available to biological creatures (capacitive, inductive, and ultrasonic).

Work is proceeding rapidly in computer graphics and image understanding on methods for modeling the geometric, topological, and structural properties of objects. Much is known about how to use this information to plan, schedule, and control machines and processes to manipulate and manufacture such objects.

270

It seems clear that we soon will see the integration of this knowledge and these capabilities into advanced robots that are capable of successful goal selection and task decomposition in complex and unpredictable environments. Such advanced robots will provide the tools for quantitative experiments into the fundamental mystery of intelligent behavior. Only through such experiments will we ever really understand the relationship between the high and low levels of control within the central nervous system.

References

Albus, J.S., (1982) "An Architecture for Real-Time Sensory-Interative Control of Robots in a Manufacturing Facility", 4th International Federation of Automatic Control Symposium, Gaithersburg, MD.

Albus, J.S., (1971) "A New Approach to Manipulator Control: The Cerebellar Model Articulation Controller (CMAC)", Journal of Dynamics Systems, Measurement and Control

Evarts, E., and Tanji, J. (1974) "Gating of Motor Cortex Reflexes by Prior Insturction", Brain Research 71: 479-494

Eccles, J.C. (1967) "The Cerebellum as Neuronal Machine", New York: Springer-Verlag

Marr, D. (1969) "A Theory of Cerebellar Cortex", Journal of Physiology 202: 437-470

Jackson, J. (1931) "Selected Writings of John Hughlings Jackson", Edited by J. Taylor, London

Minksy, M. (1967) "Computation: Finite and Infinite Machines", Prentice-Hall, Engelwood Cliffs, NJ

Hebb, D.O. (1949) "The Organization of Behavior", Wiley, NY

Kohonen T. (1978) "Associative Memory", Springer, Berlin, Heidelberg, New York

McCulloch W.S. and Pitts W.H. (1943) "A logical calculus of ideas immanent in nervous activity",

Minsky M. and Papert S. (1969) "Perceptrons MIT Press", Cambridge, Massachusetts and London

Palm G. (1982) "Neural Assemblies. An Alternative Approach to Artificial Intelligence.", Springer, Berlin, Heidelberg, New York

Rosenblatt F. (1961) "Principles of neurodynamics: Perceptrons and the theory of brain mechanisms.", Spartan, Washington D.C.

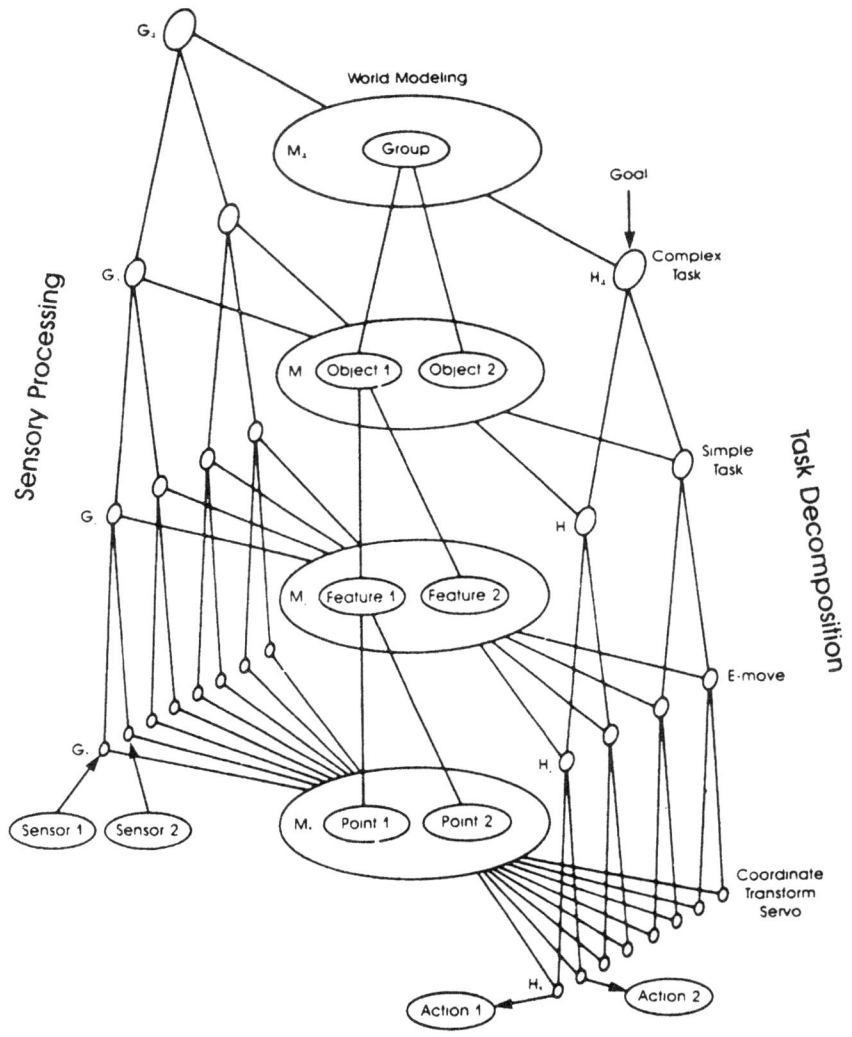

World Modeling

Sensory Processing

Task Decomposition

Goal

Complex Task

Simple Task

E-move

Coordinate Transform Servo

FIGURE 1

272

Computational Hierarchy

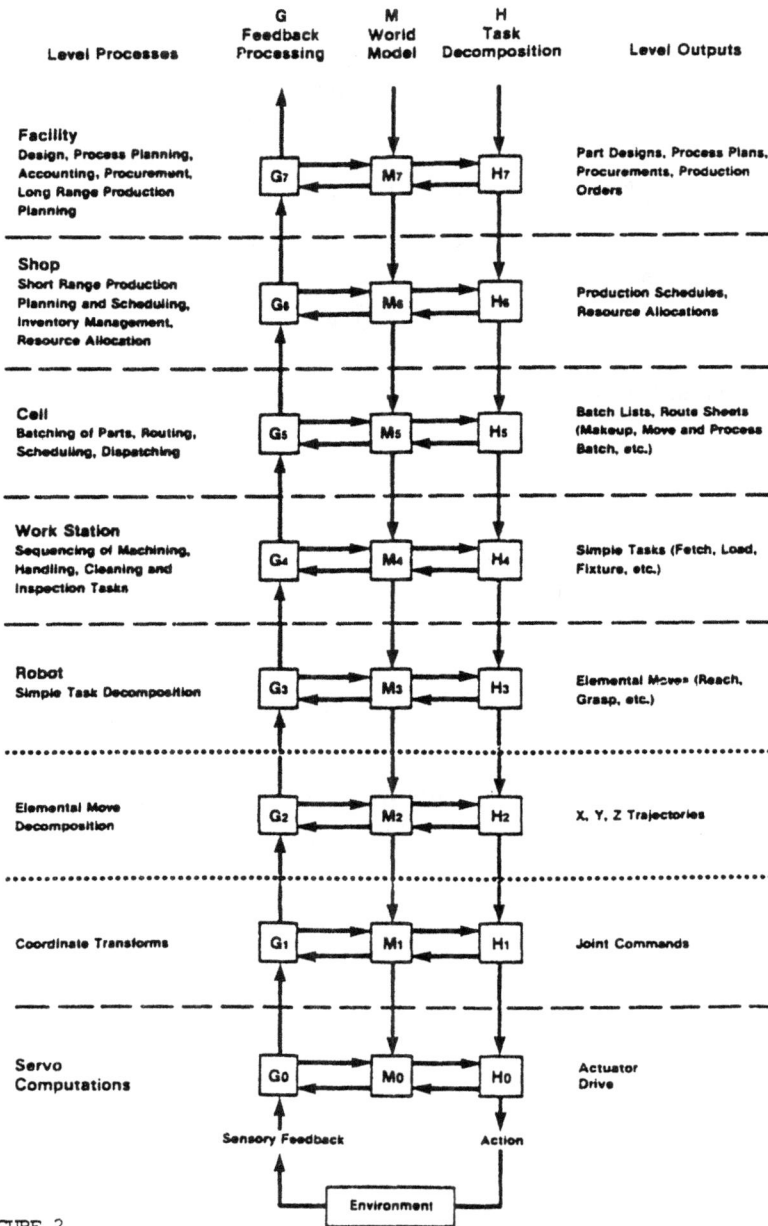

FIGURE 2

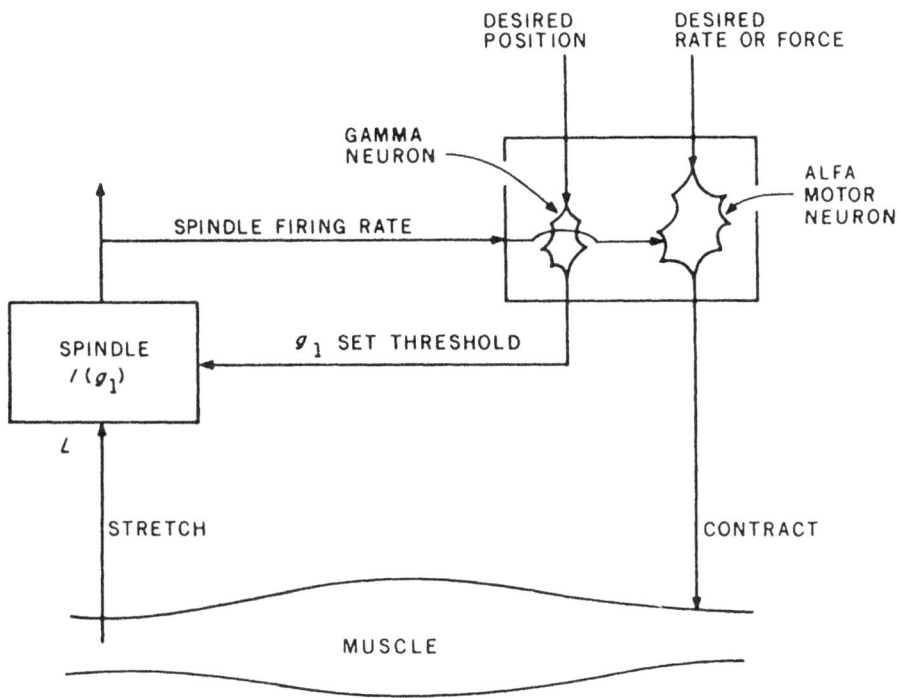

Figure 3: *A schematic diagram of the relationship between the motor neurons, muscles, stretch sensors, and input commands from higher motor centers.*

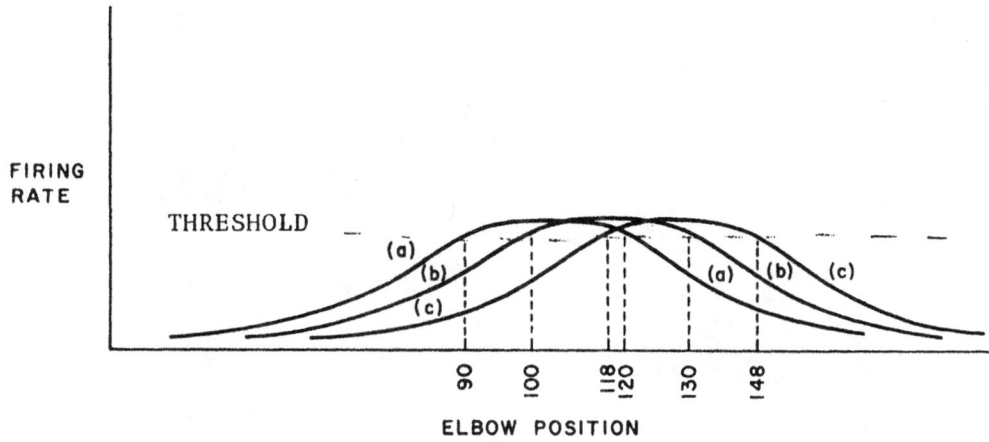

Figure 4: *Three different mossy fibers encoding a single sensory variable (elbow position). All three fibers maximally active simultaneously indicate that the elbow lies between 118° and 120°.*

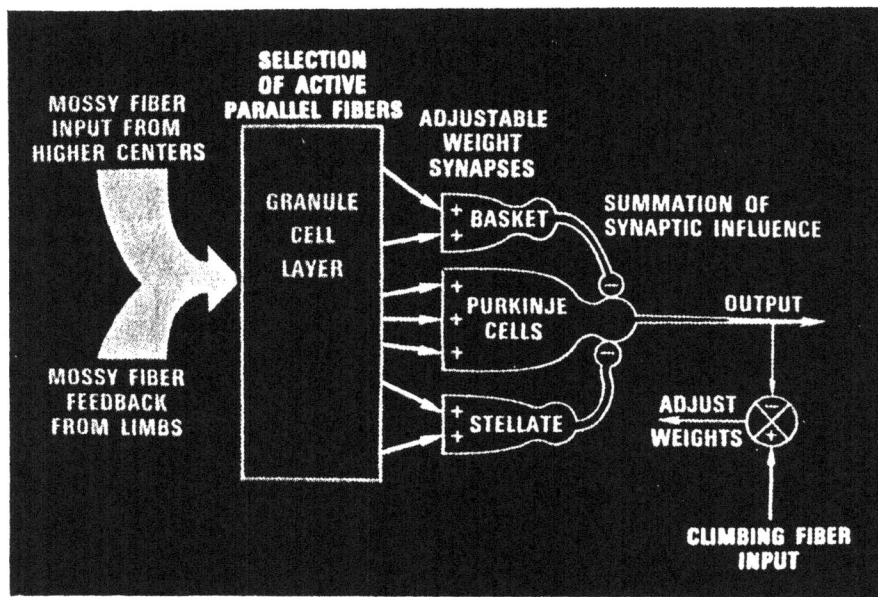

Figure 5: *A theoretical model of the cerebellum.*

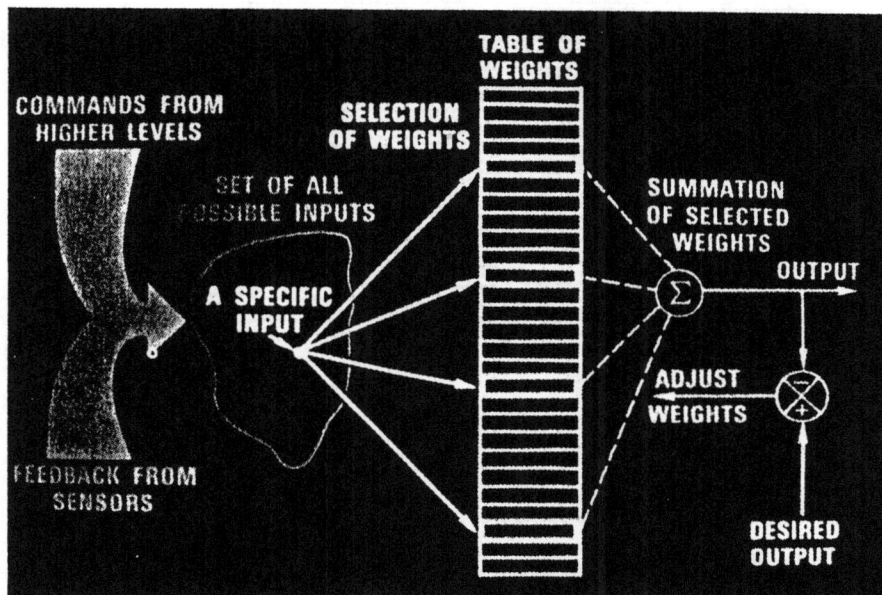

Figure 6: *A schematic representation of CMAC (Cerebellar Model Arithmetic Computer).*

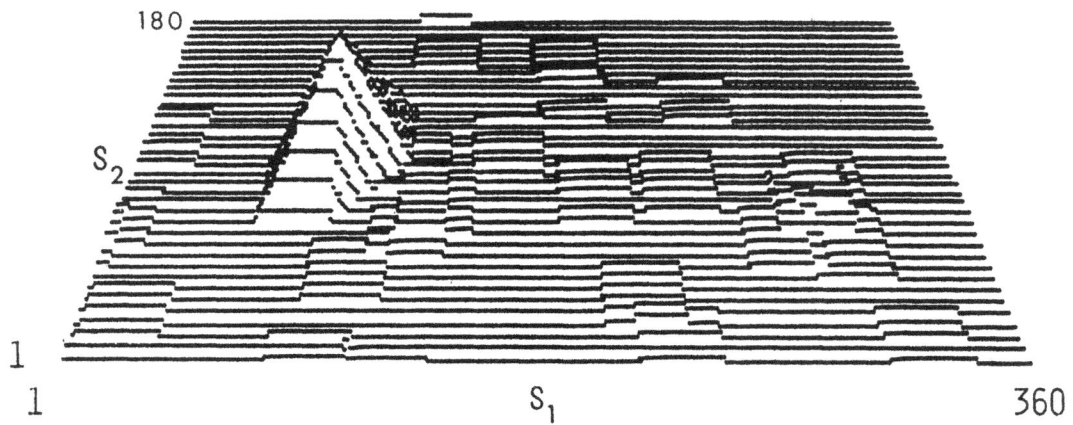

Figure 7 : *The effect of training CMAC on the function* $\hat{p} = sin\ (2\pi\ s_1/360)\ sin\ (2\pi\ s_2/360)$.
a: *One training at* $(s_1,\ s_2) = (90,\ 90)$.

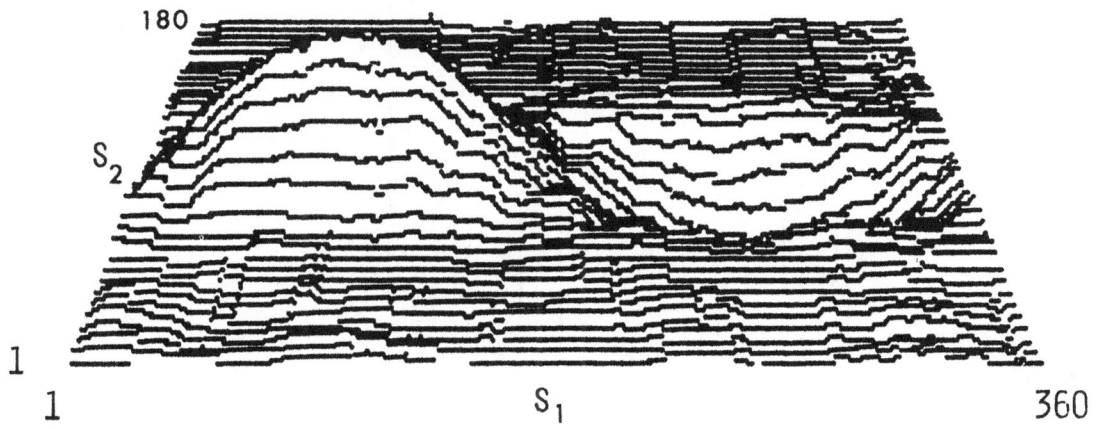

Figure 8 : *Training at 16 points along a trajectory defined by* $s_2 = 90$.

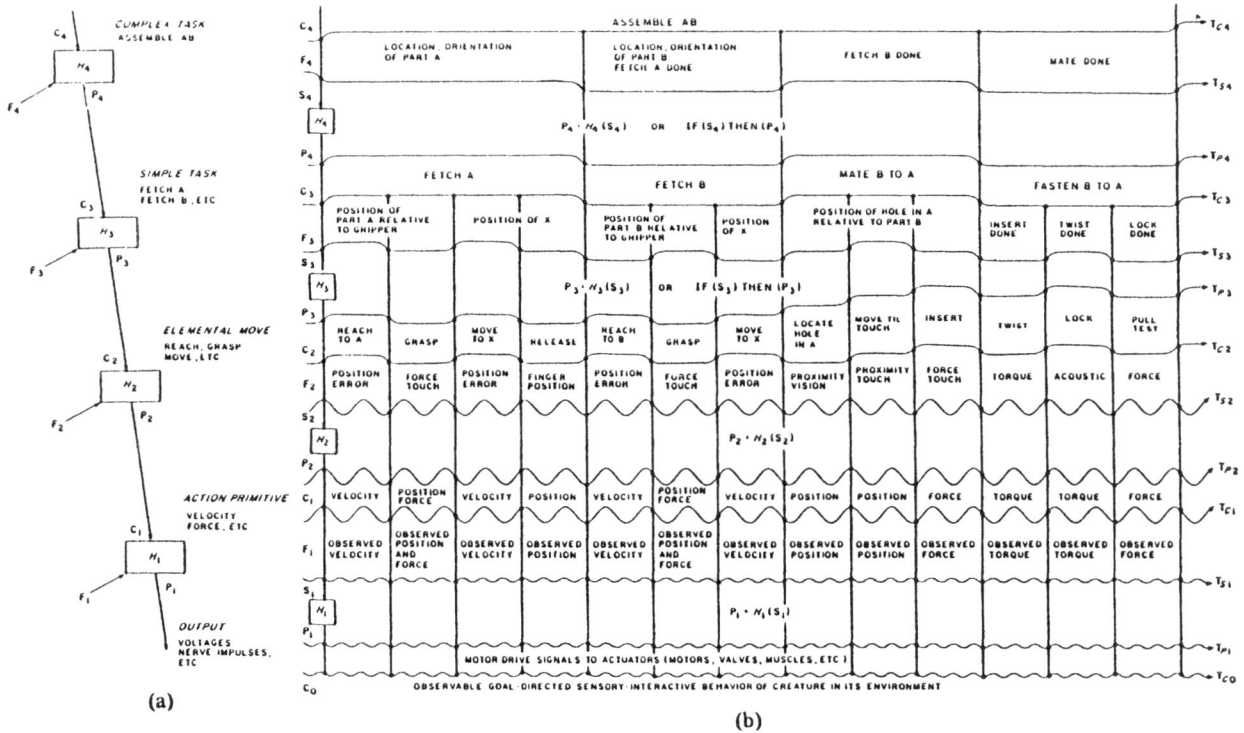

Figure 9: *A hierarchy of H operators produces sensory interactive goal-directed behavior. The highest level input command C_4 defines a goal, which in this example is <ASSEMBLE AB>. The feedback F_4 carries highly processed sensory data describing the state of environment in which the assemble command must operate, including the state of the lower level P vectors. The H_4 operator maps each input S_4 into an output P_4. As F_4, changes the goal <ASSEMBLE AB> is decomposed into a sequence of subgoals <FETCH A>, <FETCH B>, <MATE B TO A>, <FASTEN B TO A>. At each level in the hierarchy a different type of feedback data with a different rate-of-change drives the decomposition of a higher level command into a sequence of lower level subcommands. Finally, at the lowest level the P_0 vector consists of motor drive signals which actuate observable behavior C_0.*

Logical Sensor Systems*

Tom Henderson and Esther Shilcrat
*Department of Computer Science, The University of Utah, Salt Lake City,
Utah 84112*
Received March 16, 1984; accepted March 23, 1984

Multisensor systems require a coherent and efficient treatment of the information provided by the various sensors. We propose a framework, the Logical Sensor Specification
System, in which the sensors can be defined abstractly in terms of computational processes
operating on the output from other sensors. Various properties of such an organization are
investigated, and a particular implementation is described.

複数のセンサーを備えたシステムでは、それぞれのセンサーからの情報を、整然としかも
効率的に取り扱えることが必要である。ここに提案する "論理的センサー装置" では、セ
ンサーからの出力を計算処理操作の一部とみなして概念的に取り扱うことができる。セン
サー系などの各種特性について検討し、特定の応用例について述べる。

I. INTRODUCTION

We describe and motivate a particular sensor system methodology, that of
Logical Sensors, and its linguistic implementation, the Logical Sensor Specification Language. The overall goal of Logical Sensors and the Logical Sensor
Specification Language is to aid in the coherent synthesis of efficient and reliable
sensor systems.[1-3]

Both the availability and need for sensor systems is growing, as is the complexity in terms of the number and kind of sensors within a system. For example, most
robotic sensor-based systems to date have been designed around a single sensor
or a small number of sensors, and *ad hoc* techniques have been used to integrate
them into the complete system and for operating on their data. In the future,
however, such systems must operate in a reconfigurable multisensor environment; for example, there may be several cameras (perhaps of different types),
active range-finding systems, tactile pads, and so on. In addition, a wide variety

* This work was supported in part by the System Development Foundation and NSF
Grants ECS-8307483 and MCS-82-21750.

Journal of Robotic Systems, 1(2), 169–193 (1984)

of sensing devices, including mechanical, electronic, and chemical, are available for use in sensor systems, and a single-sensor system may include several kinds of sensing devices. Thus, at least two issues regarding the configuration of sensor systems arise:

(1) how to develop a coherent and efficient treatment of the information provided by many sensors, particularly when the sensors are of various kinds;

(2) how to allow for sensor system reconfiguration, both as a means toward greater tolerance for sensing device failure, and to facilitate future incorporation of additional sensing devices.

The Multisensor Kernel System (MKS) has been proposed as an efficient and uniform mechanism for dealing with data taken from several diverse sensors.[4-6] MKS has three major components: low-level data organization, high-level modeling, and logical sensor specification. The first two components of MKS concern the choice of a low-level representation of real-world phenomena and the integration of that representation into a meaningful interpretation of the real world, and have been discussed in detail elsewhere.[6] The logical sensor specification component aids the user in the configuration and integration of data such that, regardless of the number and kinds of sensing devices, the data are represented consistently with regard to the low-level organization and high-level modeling techniques that are contained in MKS. As such, the logical sensor specification component is designed in keeping with the overall goal of MKS, which is to provide an efficient and uniform mechanism for dealing with data taken from several diverse sensors, as well as facilitating sensor system reconfiguration. However, the logical sensor specification component of MKS can be used independently of the other two MKS components; for example, in conjunction with any desired low-level organization and high-level modeling technique. Thus, a use for logical sensors is evident in any sensor system which is composed of several sensors, and/or where sensor reconfiguration is desired.

The emergence of significant multisensor systems provides a major motivation for the development of logical sensors. Monitoring highly automated factories or complex chemical processes requires the integration and analysis of diverse types of sensor measurements; e.g., it may be necessary to monitor temperature, pressure, reaction rates, etc. In many cases, fault tolerance is of vital concern; e.g., in a nuclear power plant.[7] Our work has been done in the context of a robotic work station where the kinds of sensors involved include:

- cameras: an intensity array of the scene is produced,
- tactile pads: local forces are sensed,
- proximity sensors: the proximity of objects to a robot hand is sensed,
- laser range finders: the distance to surface points of objects in the scene are produced, and
- smart sensors: special algorithms implemented in hardware for detecting features such as edges.

Oftentimes, if the special hardware is not available, then some of these sensors may be implemented as a software/hardware combination which should be viewed as a distinct sensor and which ultimately may be replaced by special hardware. Other examples of sophisticated sensor systems include automatic target recognition (ATR) systems[8] and the Utah/MIT Dextrous Hand.[9] ATR systems integrate data from three (or more) sensors: microwave, FLIR, and LADAR. The Utah/MIT Hand includes a tactile sensing system which is composed of tactile element sensors gathered into tactile pads and placed on the Hand.

Other principal motivations for logical sensor specification are:

Benefits of data abstraction: the specification of a sensor is separated from its implementation. The multisensor system is then much more portable in that the specifications remain the same over a wide range of implementations. Moreover, alternative mechanisms can be specified to produce the same sensor information but perhaps with different precision or at different rates. Thus, several dimensions of sensor granularity can be defined. Further, the stress on modularity not only contributes to intellectual manageability[10] but is also an essential component of the system's reconfigurable nature. The inherent hierarchical structuring of logical sensors further aids system development.

Availability of smart sensors: the lowering cost of hardware combined with developing methodologies for the transformation from high-level algorithmic languages to silicon have made possible a system view in which hardware/software divisions are transparent. It is now possible to incorporate fairly complex algorithms directly into hardware. Thus, the substitution of hardware for software (and vice versa) should be transparent above the implementation level.

II. RELATED WORK

The work most related, in a high-level way, to logical sensor specification has been done in computer graphics. The need for some device-independent interactive system has been so widely recognized in the area of graphics that the Graphical Kernel System (GKS) is now a Draft International Standard, and is under consideration as an American National Standard. The main idea behind GKS is to provide "a means whereby interacive graphics applications could be insulated from the peculiarities of the input devices of particular terminals, and thereby become portable."[11] This was accomplished by allowing only a restricted view of an input device; the only aspect of an input device which could be viewed was the *type* of its output. Input devices so restricted are called *virtual input devices*.

Criticisms of GKS have focused on the need for virtual devices to have visible aspects other than type alone. This led to the adoption of the *logical* device concept, which is a virtual device with an enlarged view whereby other details of importance are visible.

Logical sensors are also proposed as a means by which to insulate the user from the peculiarities of input devices, which in this case are (generally) physical sensors. Thus, for example, a sensor system could be designed around camera input, without regard to the kind of camera being used. However, in addition to providing insulation from the vagaries of physical devices, logical sensor specification is also a means to create and package "virtual" physical sensors. For example, the kind of data produced by a physical laser range-finder sensor could also be produced by two cameras and a stereo program. This similarity of output result is more important to the user than the fact that one way of getting it is by using one physical device, and the other way is by using two physical devices and a program. Logical sensor specification allows the user to ignore such differences of how output is produced, and treat different means of obtaining equivalent data as logically the same.

Another related graphics interface system is SYNGRAPH.[12] This system automatically generates graphical user interfaces. The user expresses the desired interface in a modified BNF wherein a primitive input device must be declared so that a set of special features as well as output type are visible. A grammar-driven approach is favored because the syntactic description makes automated analysis of the interface possible.

The need for higher-level robotics languages has also been articulated by Donner[13] in his work on the OWL language. However, OWL is not a sensor specification language, but rather a simple programming language for describing concurrent processes to control a walking machine.

III. LOGICAL SENSORS

We have touched briefly on the role of logical sensors above. We now formally define logical sensors.

A *logical sensor* is defined in terms of four parts:

(1) A *logical sensor name.* This is used to uniquely identify the logical sensor.

(2) A *characteristic output vector.* This is basically a vector of types which serves as a description of the output vectors that will be produced by the logical sensor. Thus, the output of a logical sensor is a set (or stream) of vectors, each of which is of the type declared by that logical sensor's characteristic output vector. The type may be any standard type (e.g., real, integer), a user-generated type, or a well-defined subrange of either. When an output vector is of the type declared by a characteristic output vector (i.e., the cross product of the vector element types), we say that the output vector is an "instantiation" of that characteristic output vector.

(3) A *selector* whose inputs are alternate subnets and an acceptance test name. The role of the selector is to detect failure of an alternate and switch to a different alternate. If switching cannot be done, the selector reports failure of the logical sensor.

(4) *Alternate subnets.* This is a list of one or more alternate ways in which to obtain data with the same characteristic output vector. Hence, each alternate subnet is equivalent, with regard to type, to all other alternate subnets in the list, and can serve as backups in case of failure. *Each* alternate subnet in the list is itself composed of (a) a set of *input sources.* Each element of the set must either be itself a logical sensor, or the empty set (null). Allowing null input permits *physical* sensors, which have only an associated program (the device driver), to be described as a logical sensor,

thereby permitting uniformity of sensor treatment. (b) A *computation unit* over the input sources. Currently such computation units are software programs, but in the future, hardware units may also be used. In some cases, a special "do-nothing" computation unit may be used. We refer to this unit as PASS.

A logical sensor can be viewed as a network composed of subnetworks which are themselves logical sensors. Communication within a network is controlled via the flow of data from one subnetwork to another. Hence, such networks are *data flow* networks.

Alternatively, we present the following inductive definition of a logical sensor:

A logical sensor is an acceptance test which checks (sequentially and on demand) the output of either (base case 1):

(1) A list of computation units, with specified output type (the characteristic output vector), which require no input sources.
(2) A list of computation units, with specified output type, whose input sources are logical sensors.

Figure 1 gives a pictorial presentation of this notion. The characteristic output vector declared for this logical sensor is (x-loc:real, y-loc:real, z-loc:real, curvature:integer). We present two examples to clarify the definition of logical sensors, and in particular to show how the inputs to a logical sensor are defined in terms of other logical sensors and how the program accepts input from the source logical sensors, performs some computation on them, and returns as output a set (stream) of vectors of the type defined by the characteristic output vector. Figure 2 shows the logical sensor specification for a "camera" which happens to have no other logical sensor inputs. The specification for a stereo camera range finder called "Range_Finder" is given in Figure 3. The program "stereo" takes the output of the two cameras and computes vectors of the form (x, y, z) for every point on the surface of an object in the field of view. The idea is that a logical sensor can specify either a device driver program which needs no other logical sensor input, but rather gets its input directly from the physical device and then formats it for output in a characteristic form, or a logical sensor can specify that the output of other logical sensors be routed to a certain program and the result

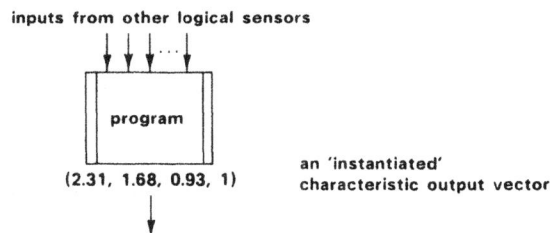

Figure 1. Graphical view of a logical sensor.

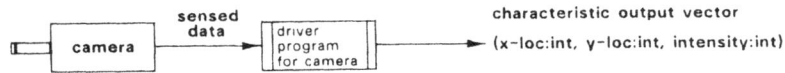

Figure 2. The Logical Sensor Specification of a "Camera."

packaged as indicated. This allows the user to create "packages" of methods which produce equivalent data, while ignoring the internal configurations of those "packages."

A. Formal Aspects

Having described how logical sensors are developed and operate, we now define a logical sensor to be a *network* composed of one or more subnetworks, where each subnetwork is a logical sensor. The computation units of the logical sensors are the nodes of the network. Currently, the network forms a rooted directed acyclic graph. The graph is rooted because, taken in its entirety, it forms a complete description of a single logical sensor (versus, for example, being a description of two logical sensors which share subnetworks). We also say that it is rooted because there exists a path between each subnetwork and a computation unit of the final logical sensor. Logical sensors may not be defined in terms of themselves, that is, no recursion is allowed, and hence the graph is acyclic.

All communication within a network is accomplished via the flow of data from one subnetwork to another. No explicit control mechanism, such as the use of shared variables, alerts, interrupts, etc., is allowed. The use of such control mechanisms would decrease the degree of modularity and independent operation of subnetworks. Hence, the networks described by the logical sensor specification language are data flow networks, and have the following properties:[14]

- A network is composed of independently, and possibly concurrently, operating subnetworks.
- A network, or some of its subnetworks, may communicate with its environment via possibly infinite input or output streams.
- Subnetworks are modular.

Since the actual output produced by a subnetwork may depend on things like hardware failures (and because the output produced by the different subnets of a logical sensor are only required to have the same type), the subnetworks (and hence the network) are also indeterminate.

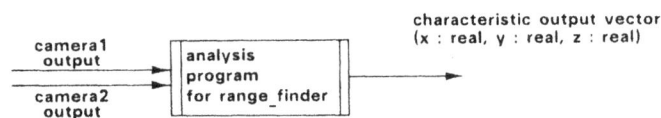

Figure 3. The Logical Sensor Specification of "Range_Finder."

B. Logical Sensor Specification Language

We have shown that a logical sensor has the following properties:

- A logical sensor is a network composed of subnetworks which are themselves logical sensors.
- A logical sensor may be defined only in terms of other, previously defined, logical sensors.
- A computation unit is an integral part of the definition of a logical sensor.
- A logical sensor produces output of the type declared by its characteristic output vector, and the declaration of the characteristic output vector is also an integral part of the definition of a logical sensor.

It should be noted that there may be alternate input paths to a particular sensor, and these correspond to the alternate subnets. But even though there may be more than one path through which a logical sensor produces data, the output will be of the type declared by the logical sensor's characteristic output vector.

With these points in mind, a language for describing the logical sensor system can be formed. We give the syntax below.

1. Syntax

(logical-sensor)	→	(logical-sensor-name) (characteristic-output-vector) (selector) (alternate-subnet-list)
(logical-sensor-name)	→	(identifier)
(characteristic-output-vector)	→	(name-type-list)
(name-type-list)	→	(identifier):(type) {; (name-type-list)}
(selector)	→	(acceptance-test-name)
(alternate-subnet-list)	→	(computation-unit-name) (input-list) {(alternate-subnet-list)}*
(acceptance-test-name)	→	(identifier)
(input-list)	→	(logical-sensor-list) | null
(logical-sensor-list)	→	(logical-sensor) {(logical-sensor-list)}*
(computation-unit-name)	→	(identifier)

2. Semantics

Below we present a high-level description of the *operational* semantics (i.e., the execution effect) for each rule of the grammar:

287

(1) A *logical sensor* declaration provides an associated name for the logical sensor used for identification purposes, a characteristic output vector to declare the type of output for that logical sensor. A selector performs the test and switch after the acceptance test and the alternate subnet list establishes the alternative ways of providing the characteristic output vector.

(2) A *logical sensor name* declaration associates a (unique) identifier for the logical sensor.

(3) A *characteristic output vector* declaration establishes the type of output for the logical sensor.

(4) A *name-type list* declaration establishes the precise nature of the output type as declared by the characteristic output vector. It consists of a cross product of types, with an associated name.

(5) A *selector* declaration specifies the order in which the alternates in the alternate subnet list will be tested by the acceptance test.

(6) An *alternate subnet list* declaration establishes a series of input sources, computation unit name tuples, thus making known which logical sensors and computation units are part of the definition of the logical sensor being declared.

(7) An *input list* declaration establishes which legal input sources (either none or a series of logical sensors) are to be used as input to the computation unit.

(8) A *logical sensor list* declaration establishes the set of logical sensors to be used as input.

(9) A *computation unit name* declaration establishes the name of the actual program which will execute on the declared input sources.

(10) A *acceptance test name* declaration establishes the name of the actual program which will be used to test the alternate subnets.

We are also currently working on providing more formal semantics for the logical sensor specification language. Many works provide *denotational* semantics (i.e., semantic schemes which associate with each construct in the language an abstract mathematical object) for general data flow networks.[14-16] When such semantics have been given for the networks represented by logical sensors, we will be able formally to prove desired network properties, such as that a network can execute forever.[15] We will also be able to prove that the output of a specified logical sensor has particular properties of interest (e.g., that its type matches that of the characteristic output vector).

C. Implementation

We currently have two implementations of the logical sensor specification language running: a C version (called C-LSS) running under UNIX, and a functional language version (called FUN-LSS). The C version has been described elsewhere[2] and produces a shell script from the specification. We give details here of the functional language version.

FUN-LSS provides a logical sensor specification interface for the user and maintains a database of s-expressions which represents the logical sensor definitions (see Fig. 4). The operations allowed on logical sensors include:

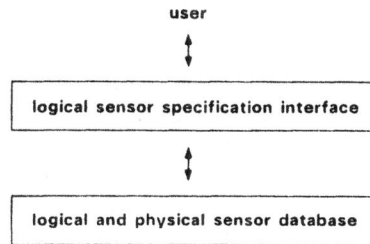

Figure 4. The Logical Sensor System Interface.

Create: a new logical sensor can be specified by giving all the necessary information and it is inserted in the database.

Update: an existing logical sensor may have certain fields changed; in particular, alternative subnets can be added or deleted, program names and the corresponding sensor lists can be changed.

Delete: a logical sensor can be deleted so long as no other logical sensor depends on it.

Display: show all parts of a logical sensor or list all logical sensor names.

Dependencies: show all logical sensor dependencies.

Appendix A gives a sample session with the logical sensor specification interface.

Once the logical sensors are specified, they are stored as s-expressions in the database. In order to actually execute the logical sensor specification, it is necessary to translate the database expressions into some executable form, e.g., to produce source for some target language, and then either interpret or compile and run that source. Our approach is displayed in Figure 5.

We have written a translator which converts the s-expressions in the database into abstract syntax trees for a Function Equation Language (FEL).[17] These are then passed to the FEL compiler which produces a function graph which can then be evaluated, using a combination of graph reduction and data flow strategies.

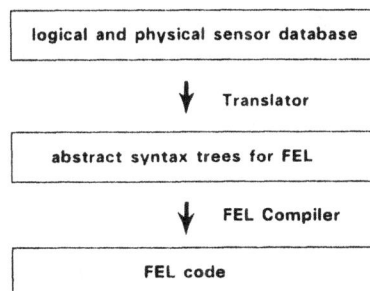

Figure 5. Steps to obtain executable code.

More on these topics can be found elsewhere.[3] In that paper we discuss a methodology for configuring systems of sensors using a functional language. The use of abstraction and of functional language features leads to a natural and simple approach to this problem. The features of a particular functional programming environment, Function Equation Language (FEL) running on the REDIFLOW simulator, are exploited to develop a scheme that avoids complicated issues of state restoration and switching protocols. Moreover, the use of reduction allows us to store that part of the alternate subnet list which is currently backup in a skeletal form. Thus, a large savings in runtime space requirements may be achieved.

IV. FAULT TOLERANCE

The Logical Sensor Specification Language has been designed in accordance with the view that languages should facilitate error determination and recovery. As we have explained, a logical sensor has a selector which takes possibly many alternate subnets as input. The selector determines errors and attempts recovery via switching to another alternate subnet. Each alternate subnet is an input source–computation unit pair. Selectors can detect failures which arise from either an input source or the computation unit. Thus, the selector together with the alternate subnets constitute a failure and substitution device, that is, a fault-tolerance mechanism, and *both* hardware and software fault tolerance can be achieved. This is particularly desirable in light of the fact that "fault tolerance does not necessarily require diagnosing the cause of the fault or *even deciding whether it arises from the hardware or software*" (emphasis added).[18] In a multi-sensor system, particularly where continuous operation is expected, trying to determine and correct the exact source of a failure may be prohibitively time-consuming.

Substitution choices may be based on either *replication* or *replacement*. *Replication* means that exact duplicates of the failed component have been specified as alternate subnets. In *replacement* a different unit is substituted. Replacement of software modules has long been recognized as necessary for software fault tolerance, with the hope, as Randall states, that using a software module of independent design will facilitate coping "with the circumstances that caused the main component to fail."[18] We feel that replacement of physical sensors should be exploited both with Randall's point in view and because extraneous considerations, such as cost, and spatial limitations as to placement ability are very likely to limit the number of purely back-up physical sensors which can be involved in a sensor system.

A. Recovery Blocks

The recovery block is a means of implementing software fault tolerance.[18] A recovery block contains a series of alternates which are to be executed in the order listed. Thus, the first in the series of alternates is the *primary* alternate. An acceptance test is used to ensure that the output produced by an alternate is correct or acceptable. First the primary alternate is executed and its output scrutinized via the acceptance test. If it passes, that block is exited, otherwise the next alternate is tried, and so on. If no alternate passes, control switches to a new recovery block if one (on the the same or higher level) is available; otherwise, an error results.

Similarly, a selector tries, in turn, each alternate subnet in the list and tests each one's output via an acceptance test. However, while Randall's scheme requires the use of complicated error recovery mechanisms (restoring the state, and so on), the use of a data flow model makes error recovery relatively easy. Furthermore, our user interface computes the dependency relation between logical sensors.[3] This permits the system to know which other sensors are possibly affected by the failure of a given sensor.

The general difficulties relating to software acceptance tests, such as how to devise them, how to make them simpler than the software module being tested, and so on, remain. It is our view that some acceptance tests will have to be designed by the user, and that our goal is simply to accommodate the use of the test. Unlike Randall, we envision the recovery block as a means for both hardware and software fault tolerance, and hence we also allow the user to specify general hardware acceptance tests. Such tests may be based, for example, on data link control information, two-way handshaking, and other protocols. It is important to note that a selector must be specified even if there is only one subnet in a logical sensor's list of alternate subnets. Without at least the minimal acceptance test of a "time-out," a logical sensor could be placed on hold forever even when alternate ways to obtain the necessary data could have been executed. Given the minimal acceptance test, the selector will at least be able to signal failure to a higher-level selector which may then institute a recovery. However, we also wish to devise special schemes for acceptance tests when the basis for substitution is replacement. While users will often know which logical sensors are functionally equivalent, it is also likely that not all possible substitutions of logical sensors will be considered. Thus, we are interested in helping the user expand what is considered functionally equivalent. Such a tool could also be used to automatically generate logical sensors.

We give an example logical sensor network in Figure 6. This example shows how to obtain surface point data from possible alternate methods. The characteristic output vector of Range_Finder is (x:real, y:real, z:real) and is produced by selecting one of the two alternate subnets and "projecting" the first three elements of their characteristic output vectors. The preferred subnet is composed of the logical sensor Image_Range. This logical sensor has two alternate subnets which both have the dummy computational unit PASS. PASS does not effect the type of the logical sensor. These alternatives will be selected in turn to produce the characteristic output vector (x:real, y:real, z:real, i:integer). If both alternates fail (whether due to hardware or software), the Image_Range sensor has failed. The Range_Finder then selects the second subnet to obtain the (x:real, y:real, z:real) information from the Tactile_Range's characteristic output vector. If the Tactile_Range subsequently fails, then the Range_Finder fails. Each subnet uses this mechanism to provide fault tolerance.

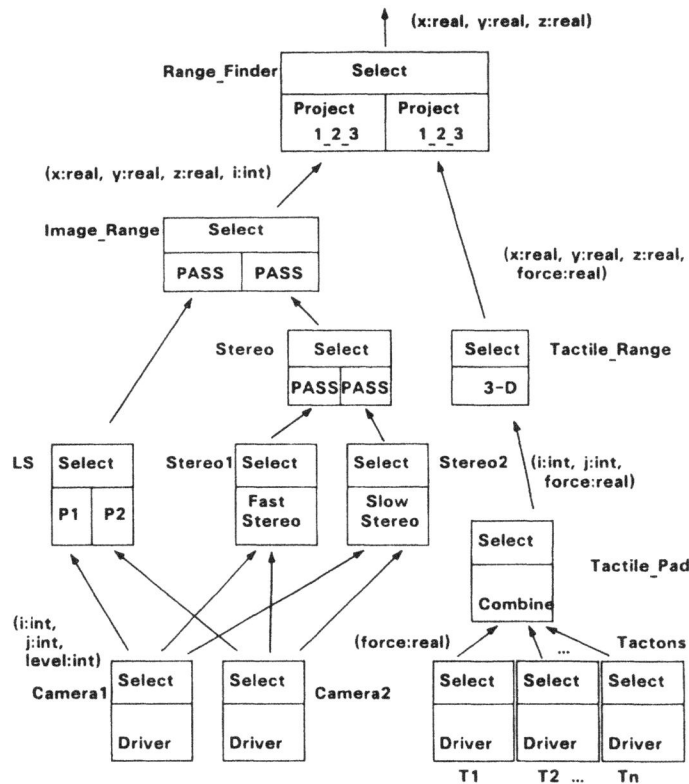

Figure 6. Logical Sensor Network for Range_Finder.

B. Ramifications of Fault Tolerance Based on a Replacement Scheme

Many difficult issues arise when fault tolerance is based on a replacement scheme. Because the replacement scheme is implemented through the use of alternate subnets, the user can be sure that the *type* of output will remain constant, regardless of the particular source subnet. Ideally, however, we consider that a replacement-based scheme is truly fault tolerant only if the effect of the replacement is within allowable limits, where the allowable limits are determined by the user. As a simple example, consider a sensor system of one camera, A, and a back-up camera of another type, B. Suppose camera A has accuracy of $\pm 0.01\%$, and camera B has accuracy of $\pm 0.04\%$. If the user has determined that the allowable limit on accuracy is $\pm 0.03\%$, then replacement of camera A by camera B will not yield what we call a truly fault-tolerant system; if the allowable limit is $\pm 0.05\%$, the replacement does yield a truly fault-tolerant system, as it will if the user has determined that the system should run regardless of the degree of accuracy.

As mentioned above, determining functional equivalence may necessitate seeing more of a logical sensor than merely its type. This example illustrates this point in that we have isolated a need to know more about leaf logical sensors (physical sensors). However, we also mentioned that the above example was simplified. Let us now assume, in addition, that the user can use a variety of algorithms to obtain the desired final output. Suppose one of those algorithms incorporates interpolation techniques which could increase the degree of accuracy over camera B's input. In this case, the user may be able to use camera B and this algorithm as an alternate subnet and have a truly fault-tolerant system, even if camera B's output is not itself within the allowable accuracy limit. Thus, when we consider a slightly more complex example, we see a general need for having features (beside type of output) of logical sensors visible, and a need to propagate such information through the system.

Feature propagation, together with allowable limit information, is needed for replacement-based fault-tolerance schemes and constitutes an acceptance test mechanism. In addition, such feature propagation has a good potential for use in automatic logical sensor system specification/optimization. For example, consider a workstation with several sensors. Once various logical sensors have been defined and stored, feature propagation can be used to configure new logical sensors with properties in specified ranges, or to determine the best (within the specified, perhaps weighted, parameters) logical sensor system. Thus, feature propagation is necessary for both fault tolerance and automatic generation of logical sensor systems, and it is our view that the basic scheme will be the same in either case.

V. FEATURES AND THEIR PROPAGATION

Our view is that propagation of features will occur from the leaf nodes to the root of the network. In sensor systems, the leaf nodes will generally be physical sensors (with associated drivers). Thus, we first discuss the important features of physical sensors.

A. Features of Physical Sensors

Our goal here is to determine whether a set of generally applicable physical sensor features exists, and then to provide a database to support the propagation mechanism. In addition, it is possible for the user to extend the set of features. Currently, the system provides a small set of generally applicable features (see below).

All physical sensors convert physical properties or measurements to some alternative form, and hence are transducers. Some standard terms for use in considering transducer performance must be defined.[19] We have selected a set of features defined by Wright which we feel are generally applicable to physical sensors.

- *Error*—the difference between the value of a variable indicated by the instrument and the true value at the input.

- *Accuracy*—the relationship of the output to the true input within certain probability limits. Accuracy is a function of nonlinearities, hysteresis, temperature variation, and drift.

- *Repeatability*—the closeness of agreement within a group of measurements at the same input conditions.

- *Drift*—the change in output that may occur despite constant input conditions.

- *Resolution*—the smallest change in input that will result in a significant change in transducer output.

- *Hysteresis*—a measure of the effect of history on the transducer.

- *Threshold*—the minimum change in input required to change the output from a zero indication. For digital systems this is the input required for 1-bit change in output.

- *Range*—the maximum range of input variable over which the transducer can operate.

Based on this set of physical sensor characteristics, the next step in arriving at a characterization of logical sensors is to "compose" physical sensor feature information with computation unit feature information.

B. Algorithm Features

There are several difficult issues involved in choosing a scheme whereby features of algorithms can be "composed" with features of physical sensors such that the overall logical sensor may be classified. As Bhanu[8] has pointed out: "The design of the system should be such that each of its components makes maximum use of the input data characteristics and its goals are in conformity with the end result."

One issue to be resolved is how to represent features and feature composition. One approach is to record feature information and composition functions separately. Thus, it would be necessary to classify an algorithm as having a certain degree of accuracy, and, in addition, provide an accuracy function which, given the accuracy of the physical sensor, produces the overall accuracy for the logical sensor which results from the composition of the physical sensor and the algorithm. A major difficulty in resolving such issues is presented by the great variety of sensor systems, both actual and potential, and the varying level of awareness of such issues within different sensor user communities. For example, experienced users of certain types of sensors may have a fairly tight knowledge of when and why certain algorithms work well, whereas other user communities may be aware in only a vague way which algorithms work well under which circumstances. Indeed, even within a sensor user community, algorithm evaluation techniques may not be standardized, hence yielding a plethora of ways in which properties of algorithms may be described. This problem is manifest in Bhanu's survey of the evaluation of automatic target recognition (ATR) algorithms.[8]

The state of the art in algorithm evaluation techniques effects the choices made regarding the use of classifying physical sensors whether we wish to simply catalog information or maximize criteria. For example, if the user cannot provide information about the degree of resolution for the algorithms being used, then an overall logical sensor resolution figure cannot be determined, even if the resolution of all physical sensors is known. Also, if such is the case, then the system cannot be used to help the user maximize the degree of resolution of the final output.

On the other hand, there are some encouraging results reported in the literature; a systematic study of robotic sensor design for dynamic sensing has recently been undertaken by Beni et al., and more of that kind of work is required if we are to achieve comprehensive sensor systems.

VI. FUTURE RESEARCH: AUTOMATIC LOGICAL SENSOR GENERATION

We are investigating ways in which to generate logical sensor systems automatically. We recognize that, considering the number of unanswered questions we listed above, we will not be able to establish a fully automatic logical sensor system, and therefore our proposal is to confine ourselves to an automatic logical sensor system of limited generality.

A. Tupling/Merging Data

We now describe some techniques to allow for dynamic specification and allocation of logical sensors. Though the kinds of logical sensors which we consider represent only simple extensions to the existing logical sensor system, this type of work is a first step toward generally extensible logical sensor systems. The goal here is to show how, given information about logical sensors which can be configured in the system, new logical sensors can be defined automatically. There are two techniques under investigation, tupling and merging data.

1. Tupling Data

Tupling data is a technique which can be used to generate automatically new logical sensors in a feature-based sensor system. In such systems, the logical sensors would be returning information about certain features found in the scene, such as number of edges, number of holes, temperature, metallic composition, and so on. The user may then request that a new logical sensor be established by specifying the name for the new logical sensor, and giving the names of the input logical sensor(s). The output of the new logical sensor will be, simply, a set of tuples (one for each object in the scene), where the tuple is composed of the Cartesian product of the features which were input from the source logical sensors. Thus, we are basically packaging together features of interest so that they will be in one output stream. For example, suppose that the features "number of edges" and "number of holes" are sufficient to determine the presence of bolts. Then a logical sensor *bolt-detector* could be created by tupling the output of the logical sensors *edge-detector* and *hole-detector*. It should be noted that we assume that the latter two logical sensors produce output of the form (object No., feature 1, feature 2, . . . , feature *N*). For the sake of simplicity, in this example we assume that logical sensor *edge-detector* produces output of the form (object No., number of edges) and logical sensor *hole-detector* produces output of the form (object No., number of holes). Logical sensor *bolt-detector* will match an object number, and produce tuples of the form (object No., number of edges, number of holes).

2. Merging Data

Another facility we are investigating dynamically incorporates, in response to a system demand, a newly defined logical sensor which outputs the merge of three-dimensional logical sensor inputs. The idea is to accommodate an interactive request to allow the output of two physical sensors to be treated as one, for example, to create a multiple-view laser range-finder logical sensor from what had been two different laser range-finder logical sensors. In this example, a logical sensor *multiview-laser* is created with input logical sensors of both laser range finders, and the inputs are merged to produce output. Thus, the user can decide, interactively, to get more views without having to reconfigure the entire system. Also, such a facility obviates the need for having multiple program units where the only difference is in the number of expected inputs.

B. Choosing Algorithms Based on Appropriateness/Reliability

Our view is that a feature propagation mechanism is useful for both fault-tolerance checking and logical sensor optimization. Some difficulties are involved in using the feature propagation mechanism in a logical sensor optimization system. From the optimization viewpoint, the task which we wish the logical sensor system to perform is not merely to produce output, but to produce output which is optimal. One difficulty is that what makes the output optimal may change from application to application, or from use to use. Hence, the logical sensor system should produce output of the specified type which is optimized according to the *user-specified optimization criteria*.

In light of the above-discussed difficulties in developing a feature propagation mechanism, we are considering optimization facilities which could also be used in the absence of a general feature propagation mechanism. Our goal is to help the user choose algorithms which maximize desired capabilities of a logical sensor system. Therefore, in addition to providing what may only constitute a catalog of physical sensor characteristics, we wish to establish a database of algorithms which can be searched to determine how to configure the optimal logical sensor system for the task at hand. Since, once again, we are forced to consider the level of information detail which the user can provide in setting up the database, we recognize that this database may or may not be part of a general feature propagation mechanism. In other words, if the user tells us only that a certain algorithm works well, for example, then this database will basically serve merely as an automatic cataloging device. On the other hand, if we can be provided with numerical estimates of certain parameters for each algorithm and with composing functions, the database can be used as part of a feature propagation mechanism. In the latter case, not only can we provide a much closer realization of the user's goal, but we may also be able to indicate which performance attributes cannot be met by any known configuration of physical sensors and algorithms; in such cases, the system may actually specify a new configuration of the parameters on an algorithm which would make the demanded performance possible.

C. Automatic Generation of Algorithm Feature Information

Several approaches to the incorporation of algorithm feature information into a logical sensor specification system have been discussed. As an extension to this idea, we intend to investigate ways in which to use a logical sensor specification system to *generate* algorithm feature information. We are looking into the use of models for algorithm evaluation, together with a database of training data, that is, sample data to be used as a standard against which algorithms are evaluated. For the ATR (Automatic Target Recognition) systems, Bhanu states that the models for algorithm evaluation should be chosen such that each part of the system should be evaluated with respect to its own figures of merit but also against its effect on the overall classification (i.e., the overall goal of the system). In this view, statistical measures of an algorithm's performance such as edge point measures and structural measures, the ability of an algorithm to make maximal use of the specific characteristics of FLIR images, and the three general parameters which are used to determine the overall performance of an ATR system (probability of target detection, probability of classification, and false alarm per frame) must all be taken into account when evaluating an algorithm. In addition, these statistical, heuristic, and parametric models are to be used in establishing the requirements of the database in terms of data collection and organization, with the end goal of generating databases of FLIR images which are increasingly representative of the real world. Thus, Bhanu envisions a training database–algorithm database interaction such that the original figures of merit for algorithms are refined, on the basis of sample data, to reflect ability to make maximal use of specific characteristics of particular physical sensor data toward the end of promoting the overall system performance. We agree with the philosophy that sensor systems should be viewed as the best source of information on how to improve themselves, and intend to investigate the use of training databases, and possible training database–algorithm database interaction schemes.

VII. CONCLUSION

We have defined a Logical Sensor Specification Language as a framework facilitating efficient and coherent treatment of information provided in multisensor systems. In addition to the issues raised when considering the language implementation itself, various extensions have been suggested. In particular, we have implemented:

(1) a Logical Sensor Specification Language compiler,
(2) general fault-tolerance features such as
    (a) a mechanism for detecting two types of sensor failure,
    (b) a technique by which switching to an alternate subnet is accomplished,
    (c) a method for determining when a sensor failure dictates top-level sensor failure,
(3) a database of physical sensors,
(4) automatic generation of tupling/merging logical sensors.

In addition, we intend to investigate formal semantics for the Logical Sensor Specification Language; features and feature propagation, in particular, how to arrive at a classification scheme for algorithm features and composing functions; the establishment of an algorithm database for at least optimization purposes; inference schemes by which to determine a need for new physical sensors; and training databases, and training database–algorithm database interaction schemes.

APPENDIX

The following session demonstrates the logical sensor specification system. Comments have been added in bold.

[PHOTO: Recording initiated Thu 8-Mar-84 1:19PM]

@psl
Extended 20-PSL 3.1, 15-Jun-83
1 lisp> (dskin "start.sl")

               Welcome to LSS!

Allowed options:

1. Create Logical Sensor
2. Update Logical Sensor
3. Delete Logical Sensor
4. Show Logical Sensor
5. Show Dependency Graph
6. Exit LSS

          First, display all the existing logical sensors. They are either physical sensors or previously defined logical sensors.

Enter option number: 4   **Show Logical Sensor**

Enter name of logical sensor to be shown, or L for list: 1

   Defined logical sensors: CAMERA1 CAMERA2

Enter name of logical sensor to be shown, or L for list: camera1
   Logical sensor CAMERA1:
         Acceptor: TIMEOUT
            COV: (I:INT J:INT LEVEL:INT)
   Alternate Subnets: ((DRIVERCAM1 NIL))

The two logical sensors CAMERA1 and CAMERA2 are already known to the system. They are both physical sensors since they are leaf nodes.

Enter option number: 4   **Show Logical Sensor**

Enter name of logical sensor to be shown, or L for list: camera2

   Logical sensor CAMERA2 :
         Acceptor: TIMEOUT
            COV: (I:INT J:INT LEVEL:INT)
   Alternate Subnets: ((DRIVERCAM2 NIL))

Next, create the logical sensor laser-range-finder as shown in Figure 6 in the text

Enter option number: 1   **Create Logical Sensor**

Enter name of logical sensor to be created: laser-range-finder

Enter name of acceptance test: timeout

Enter characteristic output vector: (x:real y:real z:real intensity:int)

Enter program name: P1

Enter the list of input logical sensors: (camera1)

Do you want to define another subnet? (yes or no): y

Enter program name: P2

Enter the list of input logical sensors: (camera2)

Do you want to define another subnet? (yes or no): n

299

Enter option number: 4

Enter name of logical sensor to be shown, or L for list: laser-range-finder
Logical sensor LASER-RANGE-FINDER:
        Acceptor: TIMEOUT
          COV: (X:REAL Y:REAL Z:REAL INTENSITY:INT)
Alternate Subnets: ((P1 (CAMERA1)) (P2 (CAMERA2)))

Next, create the logical sensor stereo of Figure 6.

Enter option number: 1   **Create Logical Sensor**

Enter name of logical sensor to be created: stereo

Enter name of acceptance test: timeout

Enter characteristic output vector: (x:real y:real z:real intensity:int)

Enter program name: ^V   **Hit wrong key. Fix later.**

Enter the list of input logical sensors: (Camera1)

Do you want to define another subnet? (yes or no): y

Enter program name: slow-stereo

Enter the list of input logical sensors: (Camera1 Camera2)

Do you want to define another subnet? (yes or no): n

Display the logical sensors defined to this point.

Enter option number: 4   **Show Logical Sensor**

Enter name of logical sensor to be shown, or L for list: l

  Defined logical sensors: CAMERA1 CAMERA2 LASER-RANGE-FINDER STEREO

Enter name of logical sensor to be shown, or L for list: stereo

  Logical sensor STEREO :
        Acceptor: TIMEOUT
          COV: (X:REAL Y:REAL Z:REAL INTENSITY:INT)
  Alternate Subnets: ((^V (CAMERA1)) (SLOW-STEREO (CAMERA1 CAMERA2)))

Note the ^V for program name.

Correct the typo made when creating the logical sensor stereo. Also, correct the input list for the first alternative.

Enter option number: 2   **Update Logical Sensor**

Enter name of logical sensor to be updated, or L for list: stereo

Add, delete or modify an alternate? (a, d or m): m

   Alternates defined for logical sensor STEREO are:
   Number   Alternate
   1   (^V (CAMERA1))
   2   (SLOW-STEREO (CAMERA1 CAMERA2))

Enter the NUMBER of the alternate you wish to modify: 1

Enter p to modify program name, i to modify input list: p

   Here is the alternate you wish to change: (^V (CAMERA1))
   Enter the new program name.

Enter program name: fast-stereo   **Correct the program name.**

More changes to this sensor? (y or n): y

Add, delete or modify an alternate? (a, d or m): m

   Alternates defined for logical sensor STEREO are:
   Number   Alternate
   1   (FAST-STEREO (CAMERA1))   **Stereo requires 2 cameras.**
   2   (SLOW-STEREO (CAMERA1 CAMERA2))

Enter the NUMBER of the alternate you wish to modify: 1

Enter p to modify program name, i to modify input list: i

   Here is the alternate you wish to change: (FAST-STEREO (CAMERA1))
   Enter the new input list.

Enter the list of input logical sensors: (camera1 camera2)

More changes to this sensor? (y or n): n

Display the updated logical sensor.

Enter option number: 4   **Show Logical Sensor**

Enter name of logical sensor to be shown, or L for list: stereo

    Logical sensor STEREO :
           Acceptor: TIMEOUT
               COV: (X:REAL Y:REAL Z:REAL INTENSITY:INT)
    Alternate Subnets: ((FAST-STEREO (CAMERA1 CAMERA2)) (SLOW-STEREO (CAMERA1 CAMERA2)))

Create the logical sensor image-range of Figure 4.

Enter option number: 1   **Create Logical Sensor**

Enter name of logical sensor to be created: image-range

Enter name of acceptance test: timeout

Enter characteristic output vector: (x:real y:real z:real intensity:int)

Enter program name: pass

Enter the list of input logical sensors: (stereo)

Do you want to define another subnet? (yes or no): n

Display the logical sensor image-range.

Enter option number: 4   **Show Logical Sensor**

Enter name of logical sensor to be shown, or L for list: image-range

    Logical sensor IMAGE-RANGE :
           Acceptor: TIMEOUT
               COV: (X:REAL Y:REAL Z:REAL INTENSITY:INT)
    Alternate Subnets: ((PASS (STEREO)))

Next, we add an alternative subnet to an existing sensor.

Enter option number: 2   **Update Logical Sensor**

Enter name of logical sensor to be updated, or L for list: image-range

Add, delete or modify an alternate? (a, d or m): a

Alternates defined for logical sensor IMAGE-RANGE are:
Number   Alternate
1   (PASS (STEREO))

Enter the NUMBER you wish for the new alternate: 1   **The switching order can be rearranged.**

Enter progam name: pass

Enter the list of input logical sensors: laser-range-finder

  Inputs must be a list. What you gave LASER-RANGE-FINDER is not

Enter the list of input logical sensors: (laser-range-finder)

More changes to this sensor? (y or n)n

Display the new version of the logical sensor.

Enter option number: 4   **Display Logical Sensor**

Enter name of logical sensor to be shown, or L for list: l

  Defined logical sensors: CAMERA1 CAMERA2 LASER-RANGE-FINDER STEREO IMAGE-RANGE

Enter name of logical sensor to be shown, or L for list: image-range

  Logical sensor IMAGE-RANGE:
          Acceptor: TIMEOUT
              COV: (X:REAL Y:REAL Z:REAL INTENSITY:INT)
  Alternate Subnets: ((PASS (LASER-RANGE-FINDER)) (PASS (STEREO))

Enter option number: 5

Dependency Table:

CAMERA1	0	0	1	1	1
CAMERA2	0	0	1	1	1
LASER-RANGE-FINDER	0	0	0	0	1
STEREO	0	0	0	0	1
IMAGE-RANGE	0	0	0	0	0

Exit the system. The sensor database is automatically updated.

303

Enter option number: 6   **Exit LSS**

Your requests have been handled!

NIL
NIL
2 lisp> (quit)
@pop

[PHOTO: Recording terminated Thu 8-Mar-84 1:29PM]

References

1. C. Hansen, T. C. Henderson, E. Shilcrat, and Wu So Fai, "Logical Sensor Specification," in *Proceedings of SPIE Conference on Intelligent Robots*, SPIE, New York, November 1983, pp. 578–583.
2. T. C. Henderson, E. Shilcrat, and C. Hansen, *A Fault Tolerant Sensor Scheme*, Computer Science UUCS 83-003, University of Utah, November 1983.
3. E. Shilcrat, P. Panangaden, and T. C. Henderson, *Implementing Multi-sensor Systems in a Functional Language*, Tech. Rep. UUCS-84-001, The University of Utah, February 1984.
4. T. C. Henderson and Wu So Fai, "A Multi-sensor Integration and Data Acquisition System," in *Proceedings of the IEEE Conference on Computer Vision and Pattern Recognition*, IEEE, New York, June 1983.
5. T. C. Henderson and Wu So Fai, *Pattern Recognition in a Multi-sensor Environment*, UUCS-83-001, University of Utah, July 1983.
6. Wu So Fai, "A Multi-sensor Integration and Data Acquisition System," Master's thesis, University of Utah, June 1983.
7. W. R. Nelson, "REACTOR: An Expert System for Diagnosis and Treatment of Nuclear Accidents," in *Proceedings AAAI-82*, August 1982, pp. 296–301.
8. Bir Bhanu, "Evaluation of Automatic Target Recognition Algorithms," in *Proceedings of the SPIE West '83*, SPIE, New York, August 1983.
9. S. Jacobsen, J. E. Wood, D. F. Knutti, and K. Biggers, "The Utah/MIT Dextrous Hand," in *MIT/SDF IRR Symp.*, August 1983.
10. N. Wirth, "On the Composition of Well-Structured Programs, in *Classics in Software Engineering*, E. N. Yourdan, Ed., Yourdon Press, London, 1979, pp. 153–172.
11. D. S. Rosenthal, J. C. Michener, G. Pfaff, R. Kessener, and M. Sabin, "The Detailed Semantics of Graphics Input Devices," *Comput. Graphics*, **16**(3), 33–38 (July 1982).
12. D. R. Olsen, and E. P. Dempsey, "SYNGRAPH: A Graphical User Interface Generator," in *SIGGRAPH '83 Conference Proceedings*, ACM, New York, July 1983, pp. 43–50.
13. M. D. Donner, "The Design of OWL: a language for walking," *ACM SIGPLAN Notice*, **18**(6), 158–165 (June 1983).
14. R. M. Keller, "Denotational Models for Parallel Programs with Indeterminate Operators," in *Formal Descriptions of Programming Concepts*, E. J. Neuhold, Ed., North Holland, Amsterdam, 1978, pp. 337–366.
15. G. Kahn, "The Semantics of a Simple Language for Parallel Programming," in *Proceedings of IFIP*, 1974, pp. 471–475.
16. G. Kahn, and D. MacQueen, "Coroutines and networks of parallel processes," in *Proc. IFIP 77*, 1977, pp. 993–998.

18. B. Randell, *System Structure for Software Fault Tolerance,* Prentice-Hall, Englewood Cliffs, NJ, 1977, pp. 195–219.

19. J. D. Wright, *Measurements, Transmission, and Signal Processing,* Van Nostrand Reinhold, New York, 1983, pp. 80–112.

20. G. Beni, S. Hackwood, L. A. Hornak, and J. L. Jackel, "Dynamic Sensing for Robots: An Analysis and Implementation," *Robot. Res.,* 2(2), 51–60 (Summer 1983).

Chapter 6

Vacuum Mechatronic Systems and Applications

Steve Belinski

Previous chapters have presented the basic issues involved in vacuum mechatronics. Topics ranging from production, measurement and maintenance of vacuum to material and lubricant properties, energy transfer, actuators, sensors, and control issues have been covered. The reprint papers and the references provide greater detail into all of these issues. This chapter on systems and applications is an introduction to the types of vacuum mechatronic systems which are currently being developed and their relation to various application areas. In addition, a description of some of the commercially available components used as building blocks in vacuum mechatronic systems is presented.

The selection of reprints for this chapter includes material from various application areas. For space applications, Aikenhead *et al.* present many of the issues involved in developing the shuttle's remote manipulator system, while H. Ludwig discusses design requirements for mechanisms used on a scanning microwave radiometer detector. Systems and devices for sample manipulation and transfer in artificial vacuum are presented in the papers by Hobson *et al.*, Clausing *et al.*, Hatsuzawa *et al.*, Saeki *et al.* and Zinck *et al.*. Borrien and Petitjean present issues on the development of robot joints for use at high vacuum.

6.1 System Components

The operating principles of the components used in vacuum mechatronic systems have already been presented. Many of these components, including bellows and various types of feedthroughs, have been commercially available for many years and operate quite reliably. Over the past few years, however, a number of more advanced system components have appeared on the market, rendering in-vacuum manufacturing a feasible prospect for a large number of manufacturing tasks.

The technology of lubrication for vacuum is very dynamic, especially for dry lubricants. These developments have been discussed in chapter 3. The following information on specific products will become less useful with the passage of time. It is therefore suggested to check with the companies listed in appendix A (and others) for information on the latest products.

Motors

A growing number of companies are marketing vacuum-compatible motors and X-Y stages, which are ideal for use as system building blocks. ARUN Microelectronics, Ltd. (AML) is marketing a four phase stepper motor described as UHV compatible to 10^{-10} Torr and bakeable to 200°C. Using microstep control, AML has achieved 3000 steps per revolution. The motors contain embedded K-type thermocouples so that temperature can be monitored continuously. Although in low duty cycle applications the typical working temperature is about 70°C above ambient, the temperature will reach 175°C after 20 minutes of continuous running at nominal phase current. At this point it is recommended that the motors be shut down. The use of embedded thermocouples is a very useful, if not necessary, component of a vacuum-compatible motor. AML can also provide a special heatsink clamp and a copper braid in an attempt to accomplish more effective heatsinking.

Empire Magnetics, Inc. also produces vacuum rated stepping motors. Three grades are available from Empire: *commercial* grade motors are rated to 10^{-6} Torr and 130°C internal temperature; *standard* grade are rated to 10^{-7} Torr and 155°C internal temperature, use a dry lubricant, and are specially cleaned; *laboratory* grade are rated to 10^{-7} Torr and 180°C internal temperature and undergo a proprietary extraction process to remove contaminants. All of the motors are available in three frame sizes. Empire offers an optional thermistor mounted inside the motor for monitoring purposes. An optional vacuum-rated encoder can also be ordered. The encoders are rated only to 80°C, but are thermally shielded from the rest of the motor. Nonetheless, great care in operation is required.

Princeton Research Instruments, Inc. provides in-vacuum stepper motors and stage assemblies. The motors are listed as UHV compatible to 4 x10^{-10} Torr and bakeable to 150°C. A unique stepper motor driver has been designed with vacuum operation in mind. Power to the motors is automatically cut off 0.4 seconds after a step, and the power levels for speed and holding torque are adjustable. This allows for a limit on power used and a corresponding reduction in the temperature of the motor. Princeton Research Instruments also markets motorized slides with standard aluminum (and optional stainless steel) tables. The slides are available in various travel lengths and step sizes ranging from 0.33 μm to 6.6 μm.

New England Affiliated Technologies, Inc. also offers a series of high vacuum positioning stages, which are driven by vacuum-compatible stepping motors and lubricated with Braycote 602. Single axis, X-Y, rotary and multi-axis stages are available with micron resolution and vacuum-compatibility to 10^{-7} Torr.

A number of companies in Japan also produce vacuum stepper motors, including Sanyo Electric, Sukegawa Electric and Yaskawa Electric Companies. Yaskawa builds three main types: axial gap, radial gap and linear pulse motors. In the axial gap motors, the rotor/stator gap is in the same direction as the axis of rotation. Yaskawa is also developing linear, magnetically levitated transport systems for operation in vacuum.

Actuators

Many simple manual actuators are currently available as feedthrough devices. Most are made using bellows for linear sealing. Although some of these have been automated, more flexible actuators which can be used completely in vacuum are required in many cases.

A piezoelectric linear actuator called the Inchworm is produced by Burleigh Instruments, Inc. The Inchworm was discussed in section 4.1. The actuators are fabricated from alumina and piezoelectric ceramic, beryllium copper, silver, and either UHV-compatible epoxy or glass frit seal. They are bakeable to 150°C. The maximum velocity is 1.0 mm/s at a full load of 1.0 Kg, and travels from 6.25 to 200 mm are standard. Burleigh also offers a translation stage system driven by Inchworms, which has 25.4mm of travel.

Micro Pulse Systems, Inc. also uses piezoelectrics and simple mechanics to provide components with rotary, linear, and multi-axis motion. According to Micro Pulse, the various devices they produce are easily adapted for use in vacuum.

Robots

A self-contained in-vacuum manufacturing system requires a high level of automation due to the characteristic limited access problem and the fact that humans are excluded by default. On the other hand, characteristics of the vacuum environment itself present numerous challenges to full automation, and robot design in particular. Self-contained systems will eventually make use of many types of sensors brought together as an intelligent sensing system. Meanwhile, innovations in vacuum technology, robotics, control, sensors, lubrication and materials have made it possible to provide advanced robotics for use in vacuum manufacturing. Examples of commercially available vacuum-compatible robots are given below. In addition, the many microelectronics companies which are developing multichamber processing have developed robots for their particular systems.

Brooks Automation markets a number of products for development of in-vacuum manufacturing, mostly for semiconductors. Their Vacu-Tran™ robot is a three axis frog-leg type used for handling wafers. The drive motors for this robot are located outside of the vacuum chamber; a differentially pumped assembly is used to seal the motions. The standard horizontal extension is 15", with a vertical lift range of only 0.2" due mainly to the mechanical feedthrough configuration. To provide for accessing wafers in a cassette, Brooks also markets a vacuum cassette elevator, a wafer indexing mechanism within a chamber. It can act as the load-lock for a system, in particular Brooks' Vacu-Tran System which combines the elevators, the robot and a "Vacuum Transport Chamber". Other vacuum mechatronic building blocks sold by Brooks are the Orbitrac™, which uses the "walking beam" concept to transport wafers, and a rectangular vacuum isolation (gate) valve for use in Brooks' and other systems.

Genmark Automation produces a vacuum-compatible robot and a vacuum elevator. The three axis robot, called the Vacubot, utilizes ferrofluidic seals to feed the motion from the atmospheric side into the vacuum chamber. As with the Brooks robot, the vertical axis stroke is limited to 0.2". The vertical axis and the main rotational axis are fully programmable, but the the second link of the "single-frog-leg" is linked to the main rotation as a "follower" axis. This second link is thus not independently programmable. Various end-effectors are available depending on the application.

The Yaskawa Electric Mfg. Co. makes use of its vacuum-compatible motors to produce a number of vacuum-compatible robot types. Placing the motors in vacuum allows more flexibility in design, but presents problems for heat dissipation, outgassing, etc. Yaskawa seems to have overcome those problems. Yaskawa's four axis SCARA-type robot has two horizontal links, with a rotational motion at each of three joints. A motor mounted at the end provides vertical end-effector motion. A cylindrical-coordinate type of robot developed by Yaskawa is ideal for most cylindrical vacuum chambers. One of these robots is the SCARF™ robot, developed in conjunction with the Center for Robotic Systems at the University of California, Santa Barbara. Use of vacuum-compatible motors with encoder feedback made it possible for the SCARF™ robot to have an 8" vertical stroke (via ballscrew), and a 26" horizontal stroke using a linear motor.

6.2 Applications for Microelectronics

Many of today's VLSI processing steps are performed at some level of vacuum. These include processes such as rapid thermal processing, molecular beam epitaxy, various types of chemical vapor deposition, sputtering, plasma deposition, ion plating, ion etching and ion implantation. Besides the fact that vacuum is *required* for many process steps, the possibility of extreme cleanliness provides motivation to implement even more steps under vacuum conditions. An increasing number of microelectronics production systems make use of vacuum mechatronics and represent significant progress towards complete self-contained manufacturing systems.

Ion Implantation

An example of the rapid progress of these types of systems may be found in ion implanters. One of the first ion implanters to be designed for complete automation was the Veeco Instruments Model VHC-120 high-current implanter. Other early efforts in automation came from Eaton and Varian. [1] The first automation steps focused on process control, including data logging, recipe control, closed-loop dose control and maintenance diagnostics. Now, even more sophisticated systems are being produced by all of the major players: e.g. Varian, Applied Materials, Eaton and Genus Ion Technology. These latest implanters, controlled by sophisticated control routines running on high-speed microprocessors, make use of a multitude of sensors for increased intelligence, and provide the operator with superior control and process feedback.

At least two implanters have moved a step further by making use of self-contained manufacturing principles. Both the Applied Materials PI-9000 Series and Varian's Extrion 1000 machines use sophisticated loadlocking and in-vacuum wafer transport to reduce contamination and improve process parameters. By maintaining the process chamber at vacuum rather than constantly cycling the pressure, the implant chamber can be maintained around 10^{-7} Torr. Cycling the pressure usually means that the implant chamber is at a higher pressure during processing, perhaps 10^{-5} Torr. These systems make use of multiple load lock chambers to improve cycle time, and an increasing number of sensors to assure safe handling of the wafers.

Multichamber Processing

In 1989, the high growth rate of multichamber processing methods is evident. Dr. T. Keith McNab has recently released a 150 page technical and market study on multiprocessors. In it he states that "the North American market for multiprocessors was $50 million in 1988 and will grow rapidly to greater than $600 million in 1993." Dr. McNab is also active in the drive to standardize such systems. In January, 1989 he brought the issue of multiprocessor interface standardization before the SEMI Equipment Automation and Interface Committee. A Multiprocessor Common Interface Task Force was formed, which presented a recommended course of action during the SEMI standards meeting at the 1989 SEMICON/West Show in San Mateo, California.

During the SEMICON/West Show, more than ten vendors displayed advanced versions of multichamber processing equipment for a number of processes. In most cases the equipment was developed from the ground up within the specific organization. In fact, the number of different vacuum-compatible robots used was surprisingly large, considering the similar geometries of the systems and the fact that there are established robot vendors attempting to market vacuum-compatible robots.

In October of 1989 the first major conference dedicated specifically to multichamber processing, entitled "Multichamber and In-Situ Processing", was held by the International Society for Optical Engineering (SPIE).

6.3 Applications for Space

Present Applications

Applications in space have contributed most of the the vacuum mechatronics technology to date. Explorer missions, satellites and the Space Shuttle Canadarm have all contributed to the technology of materials, heat transfer and lubrication for vacuum environments. The Space Station program will make use of these advances and continue the progress. Construction of Space Station itself, and various systems associated with it, will require full use of vacuum mechatronics technology.

Relationship of Space Technology to Ground-Based Self-Contained Systems

Great progress has been made over the past five years in the area of earth-based, self-contained manufacturing systems, and work over the next five years should demonstrate a high level of maturity. The technologies developed for these systems (e.g. facilities layout, integrated sensor systems, vacuum-compatible mechanisms and robotics, load-locking techniques, particulate control, intelligent system control, in-vacuum inspection, reliability and redundancy) will make a substantial contribution to the realization of space-based research and manufacturing facilities.

The microgravity condition in space is advantageous for many types of manufacturing processes. The United States Laboratory module, a part of Space Station, is being established as a national microgravity facility. It will provide facilities for basic microgravity research, and even for the development of materials production equipment. Beyond this, there is a need to establish more extensive facilities for space-based processing. These new facilities can be designed to take advantage of space vacuum, microgravity, and the availability of unlimited solar energy. Successful implementation depends on using self-contained systems.

In 1985, the NASA Advanced Technology Advisory Committee reported [6.2] that "one unique aspect of a space manufacturing facility compared to a terrestrial factory includes the inability to bring technicians and specialists in for maintenance and malfunction repair". This is also true of self-contained manufacturing systems on earth, which have zones which, under normal operating conditions, are not designed for access by maintenance personnel. The solutions for the earth-based, self-contained systems are the same ones recommended in [6.2] with regards to two space-based manufacturing design concepts proposed by General Electric [6.3]. In particular, the study by GE states:

> "...advanced automation technology requirements identified by the study are those systems required to remotely monitor and diagnose and to automatically reconfigure, maintain, and repair in the event of malfunction. These requirements embrace a broad spectrum of enabling technologies ranging from ultimate expert systems which monitor, diagnose, and reconfigure to teleoperators and robotic, manipulative systems that perform manufacturing, servicing, and repair functions under supervisory control from either Space Station or the ground."

The close relationship to self-contained systems is obvious. It is crucial to take advantage of this relationship and the currently developing technologies in order to provide new capabilities for space manufacturing.

The Space Station program has spawned many of the proposals for experimentation in space. The most outstanding characteristic of space as compared to the earth environment is microgravity, which leads to reduction of sedimentation, buoyancy, and gravity-induced deformations. Promising proposals have been made for using the microgravity of space for materials processing [6.4-6.6], semiconductor crystal growth [6.7], fluid science experiments [6.8] and containerless processing [6.9]. Fewer studies have been done to plan how to take advantage of the natural processing [6.9]. Fewer studies have been done to plan how to take advantage of the natural vacuum of space, possibly because of the relative ease of creating the vacuum environment on earth as compared to creating a microgravity environment. In many cases, it is advantageous to use both environments together during a processing task. One concept for taking advantage of these conditions is the orbiting molecular *Wake Shield Facility* [6.10]. In this concept, reduced gas molecular densities are achieved using an orbiting shield which travels faster than the mean gas molecular speed, and thus sweeping out a wake region. In a near-earth orbit, the pressure level is comparable to an equilibrium level of 10^{-12} to 10^{-13} Torr, with extremely high pumping speed. It is proposed that this shield will first be deployed by the Space Shuttle using the Shuttle's robotic arm (RMS), and be used for thin-film epitaxial growth experiments.

The United States Laboratory module on Space Station is designed to provide the necessary laboratory workspace to carry out a comprehensive microgravity research program [6.5]. A standard set of Space Station-provided laboratory support equipment is provided, along with any payload-unique support equipment. After commercial proof-of-concept under these conditions, it is envisioned that the user will be obliged to develop a separate manufacturing spacecraft. The design and successful implementation of this spacecraft should be based on the most advanced technology available. Much of the applicable technology is currently being developed and proven on self-contained manufacturing systems, which share many of the constraints of eventual space-based systems.

The learning curve involved in implementing space-based manufacturing is steep, but in the long run this type of manufacturing will likely lead to a number of scientific breakthroughs. Methods by which current technology can be applied and extended to support these programs must be continuously nurtured. David Black, Chief Scientist in the Office of Space Station has stated [6.11]: "The life, material and fluid sciences could well be the greatest beneficiaries of Space Station, but they are not as mature as their brethren space science disciplines of astronomy, Earth sciences, and solar system exploration. This lack of maturity means that much of the early activity on Space Station will be devoted to learning how best to utilize this environment."

6.4   Other Applications

Vacuum mechatronics technology is by no means limited to microelectronics processing or space applications. Many other processes will benefit from in-vacuum manufacturing when the technology to support such efforts is sufficiently mature. The Center for Robotic Systems in Microelectronics, in conjunction with the Yaskawa Electric Mfg. Co., completed a vacuum-compatible robot in 1988 for use in assembling a new type of gyroscope under ultra-clean conditions [6.12 - 6.14]. Scanning Tunneling Microscopy (STM) is another application which requires precise positioning in vacuum. Lamberton robotics has developed a robot which operates in vacuum for isothermal forging of parts for the aerospace industry [6.15].

In 1989, the Space Astrophysics Group at the Space Sciences Laboratory at the University of California, Berkeley completed a vacuum robot called the Large Tank Manipulator (LTM) [6.16]. The primary purpose of the LTM is to calibrate (on earth), each of four Extreme Ultraviolet Explorer (EUVE) telescopes. The Ultraviolet Explorer satellite, set for launch in 1991, will be used to look into space 300 light years and beyond from Earth to study white hot dwarfs, coronas, binary stars, quasars and pulsars. The LTM can handle a 1000 pound payload with accuracies as high as 0.2 arc-sec for an individual rotation and 0.0001" for linear movements. The robot is vacuum-compatible to 10^{-7} Torr, but was also designed with very stringent cleanliness requirements because it is used to test telescopes with extremely high quality mirrors.

6.5 The Future of Vacuum Mechatronics

The first experiments with vacuum were carried out in the 17th century. In an article summarizing early vacuum applications, Theodore E. Madey writes [6.17]:

> "The techniques and understanding of vacuum processes have paralleled and influenced major developments in physics, chemistry, and engineering for over 300 years. Vacuum technology of the 18th century provided a technical basis for the huge engines which made possible the Industrial Revolution; the highly sophisticated vacuum science and technology of the late 20th century have given birth to a new Revolution – the Microelectronics Revolution – with its far-reaching impact on the lives of all."

Major developments in vacuum mechatronics did not occur until space exploration required it. Since the environment in most cases could not be changed to suit the mechanisms, the mechanisms had to be engineered to operate in the vacuum of space. Now that vacuum is recognized to have properties that are desirable and/or required for many manufacturing tasks, vacuum mechatronics is again rapidly progressing. This time, however, work is not limited to a relative handful of engineers from a single field. Thus, vacuum mechatronics is now in a position to develop as its own field of study. The techniques discussed in this book for successful implementation of vacuum mechatronic systems will be required learning for a growing number of engineers.

References

[6.1] Burggraaf, P., "Today's Implanters: 'Thinking Through any Problems", *Semiconductor International*, March, 1989.

[6.2] NASA Advanced Technology Advisory Committee, "Advancing Automation and Robotics Technology for the Space Station and for the U.S. Economy", 1985.

[6.3] General Electric, "Space Station Automation Study – Automation Requirements Derived from Space Manufacturing Concepts", Vol. I,II, Contract NAS5-25182, November, 1984.

[6.4] Maekawa, S., "Toward New Materials Processing in Space", *Space Technology*, Vol. 7, No. 1/2, pp. 165-170, 1987.

[6.5] Willenberg, H.J., "Commercial Materials Processing in the Space Station", in *Advanced Topics in Manufacturing Technology: Product Design, Bioengineering and Space Commercialization*, Boston, MA, , pp. 89-102, Dec. 13-18, 1987.

[6.6] Carruthers, J.R., "Materials Science and Engineering in Space", in *Space Science and Applications: Progress and Potential*, IEEE Press, pp. 155-158, 1986.

[6.7] Viola, J.T., "Space Commercialization: An Overview by an Aerospace Corporation", in *Advanced Topics in Manufacturing Technology: Product Design, Bioengineering and Space Commercialization*, Boston, MA, pp. 53-57, Dec. 13-18, 1987.

[6.8] Napolitano, L.G., "Prospects and Problems in Microgravity Fluid Science", *Space Technology*, Vol. 7, No. 1/2, pp. 149-155, 1987.

[6.9] Monti, R., "Space Processing", *Space Technology*, Vol. 5, No. 1/2, pp. 129-138, 1985.

[6.10] Ignatiev, A., "Proposed Epitaxial Thin Film Growth in the Ultra-Vacuum of Space", in *Proceedings of the First Vacuum Mechatronics Workshop*, Center for Robotic Systems in Microelectronics, American Institute of Physics Conference Proceedings Series, University of California, Santa Barbara, Feb. 1-2, 1989.

[6.11] Black, D.C., "Science on Space Station", *Space Technology*, Vol. 7, No. 1/2, pp. 133-135, 1987.

[6.12] Belinski, S.E., W. Trento, R. Imani and S. Hackwood, "Robot Design for a Vacuum Environment", in *Proceedings of the NASA Workshop on Space Telerobotics*, Jet Propulsion Laboratory, Pasadena, CA, Jan. 20-22, 1987.

[6.13] Belinski, S.E. and S. Hackwood, "Manufacturing in a Vacuum Environment", in *Proceedings of the IEEE International Electronic Manufacturing Technology Symposium*, Anaheim, CA, pp. 239-244, Oct. 12-14, 1987.

[6.14] Shirazi, M., "Development and Testing of a Vacuum-Compatible Robot", *ASME Manufacturing Review*, pp. 259-264, Vol. 1, No. 4, December, 1988.

[6.15] Lamberton Robotics, "Robot Works in Vacuum", *Manufacturing Engineering*, May, 1988.

[6.16] Reid, N., "Space Payload Calibration Using Advanced Vacuum Robotics", in *Proceedings of the First Vacuum Mechatronics Workshop*, Center for Robotic Systems in Microelectronics, University of California, Santa Barbara, American Institute of Physics Conference Proceedings Series, Feb. 1-2, 1989.

[6.17] Madey, T.E., "Early Applications of Vacuum, from Aristotle to Langmuir", *J. Vac. Sci. Technology*, A, 2(2), pp. 110-117, 1984.

Canadarm and the space shuttle

Bruce A. Aikenhead

Teleoperator Project Office, National Aeronautical Establishment, National Research Council of Canada

Robert G. Daniell and Frederick M. Davis

RMS Division, SPAR Aerospace Limited, Toronto, Canada

(Received 18 August 1982; accepted 12 October 1982)

"Canadarm" is the remote manipulator system (RMS) given by Canada to NASA for installation on the shuttle orbiter. The RMS is to be used for deployment, retrieval, and handling of payloads up to 65 000 lb mass, 15 ft in diameter and 60 ft in length. Controllability, operating constraints, flight environments, safety, reliability, volume, weight, and power allocations were factors which influenced the design. The system uses brushless dc servomotors to drive the six joints and the end effector of the 50 ft arm in a variety of control modes, which provide both manual and automatic control of arm functions. In describing the system, this paper concentrates on the various mechanical devices which are required to operate in the orbital environment and discusses some of the problems encountered during their development. SPAR Aerospace Ltd., under contract to the National Research Council of Canada, designed, developed, manufactured, tested, and delivered the system, which at the time of this writing has been successfully demonstrated on two flights of the orbiter "Columbia."

PACS numbers: 94.80. — f

I. INTRODUCTION

"Canadarm" is the name given to the remote manipulator system of the space shuttle orbiter by the president of the National Research Council of Canada as he presented it to the National Aeronautics and Space Administration in February 1981. This system was installed in the orbiter Columbia and was successfully demonstrated in orbit in November 1981.

In 1975, Canada agreed to participate in the space shuttle program in response to an invitation from NASA and undertook to design, develop, manufacture, test, and deliver the complete remote manipulator system (RMS) as Canada's contribution. NASA, in return, agreed to purchase from Canada the manipulator systems for the remainder of the orbiter fleet; the delivery of the first of these is scheduled for November 1982.

The National Research Council of Canada (NRCC) was assigned responsibility for the project and contracted the work to a consortium of Canadian companies headed by SPAR Aerospace Ltd. of Toronto, as the prime contractor.

II. THE REMOTE MANIPULATOR SYSTEM

With the payload deployment and retrieval system office of NASA's Johnson Space Center, NRCC and SPAR established the specifications for the RMS, of which a summary is contained in Tables I and II. These requirements, to manipulate an object of such size and mass and to control it with such precision, were without precedent. The components of the system which was developed[1] are identified in Fig. 1.

The software algorithms, also developed by SPAR as one of the deliverables, enable the operator to command motion of the tip of the arm without having to give conscious thought to the motions of the six individual joints. This motion may be commanded by his use of the translation and rotation rate command hand controllers, by selection of desired endpoint position and attitude coordinates entered at the computer keyboard, or by selection of stored automatic trajectories. To save weight, the RMS uses one of the general purpose computers (GPC's) in the orbiter to perform these functions.

The display and control panel, the hand controllers, and the manipulator controller interface unit (MCIU) are line

TABLE I. Summary of RMS requirements.

Performance
- Provide arm tip velocity of 2 ft/s (0.6 m/s)
- Maneuvre and hold tip of arm within ± 2 in. (0.05m) and ± 1°
- Capture a payload moving at 0.1 ft/s relative to orbiter
- Stop a 32 000 lb (14 500 kg) payload moving at 0.2 ft/s within 2 ft.
- Deploy/retrieve 65 000 lb (29 500 kg), 15 × 60 ft (4.6 × 18.3 m) payload
- Release 65 k payload within ± 5° of specified attitude and with angular rates < 0.015°/s

Other capabilities
- Place payloads on a suitably configured & stabilized body
- Inspect orbiter and payloads via closed circuit television (CCTV)
- Assist crew in extravehicular activities (EVA)

Required features
- Six degrees of freedom and 50 ft (15.2 m) reach envelope
- Brakes, mechanical stops, software stops, CCTV, spotlight
- Fail safe
- Built-in test equipment ("BITE")
- 100 mission/10 y life

Required components
- Two identical manipulator arms (serial operation)
- Display and control panel
- Translation and rotation hand controllers
- Manipulator controller interface unit
- RMS software for orbiter general purpose computer

Weight and power budget
- 994 lb (450 kg) (one-arm system)
- 1000 W operating power, 1050 W for heaters

TABLE II. Summary of RMS environment.

Operating:
  Pressure
    Design for 14.7 psi to 1×10^{-10} Torr
    Test at ambient and 1×10^{-6} Torr
  Temperature
    As determined by:
    Solar radiation (444 BTU/ft^2/h) (1400 W/m^2)
    Earth albedo (30%)
    Earth radiation (77 BTU/ft^2/h) (243 W/m^2)
    Space sink temperature (0° Rankine)
    Orbiter structural heat transfer parameters
    Could be in range $+ 250°$ to $- 250°$F ($+ 120°$ to $- 157°$C)
  Radiation
    Solar flares
    Galactic cosmic radiation
    Geomagnetically trapped radiation
Nonoperating:
  Vibration and shock
    Design and test for:
    0.067 g^2/Hz, 80 Hz to 350 Hz, 5 min sinusoidal
    0.008 g^2/Hz, 100 Hz to 250 Hz, 48 min random
    20 g, 11 ms shock
  Acceleration
    3 g peak
  Humidity
    100%, including 1% salt fog

replaceable units (LRU's) which are installed in the cabin of the orbiter. Although these units have many interesting features, they will not be discussed further in this paper, which concerns mechanical motion in vacuum, and, as we all have seen in the televised pictures from Columbia, a shirt-sleeve environment is provided in the cabin.

A. The manipulator arm assembly and its installation in the orbiter

The manipulator arm, which weighs approximately 850 lb (390 kg) and is 50 ft (15.2 m) long, contains a number of mechanisms which are required to operate in the vacuum conditions of earth orbit. The arm assembly is shown in Fig. 2. Designed for installation on either longeron, the first arm was installed on the port side. It is stowed in the small space

which exists between the inside of the closed cargo bay door and the outside of the dynamic envelope of a stowed 15-ft (4.6 m)-diam payload.

The interface at the shoulder is with the pedestal of the manipulator positioning mechanism (MPM), which is one of the systems provided as part of the orbiter by Rockwell International. The MPM also contains the provisions for jettison of the arm, should this be dictated by complete failure of the prime and backup control systems of the RMS. Pyrotechnically driven gas generators would release mechanical clamps and operate a guillotine cutter to sever the cable bundle at the shoulder.

The MPM also provides restraint for launch and landing at three additional points, one at the elbow and two at the wrist. Two hooks at each location engage a striker bar pulling the arm into a cradle, with a force of about 2000 lb (9000 N).

After the cargo bay doors have been opened, the shoulder MPM and the three retention latch housing assemblies are driven in synchronism through a swing-out angle of about 30° to an outboard position by dual three-phase ac motors. These drive through a differential gear mechanism to rotate a long torque rod and gear box system. With both motors operating, the unstowing of the arm takes 34 s. Should either motor fail, the operation would require approximately 68 s. Similar redundancy is provided for release of the three latches.

To alleviate the launch loads on the shoulder pitch-joint gear-train assembly, a shoulder brace (Fig. 3) is located between the shoulder pitch-joint electronics compartment and the shoulder yaw-joint housing. This is the first mechanism in the arm which is operated on orbit. Before flight, a pin is driven into the expanding section of a collet, thereby locking it into a fixed socket. Before using the arm, the astronaut operates a switch to apply power to a 115 V 400 Hz motor-driven linear actuator. The actuator, which has a stroke of about 1 in. (0.025 m), withdraws the pin, allowing the collet's segments to contract and release the brace. At the end of travel, a microswitch stops the motor and provides feedback to the operator. The operation requires approximately 30 s.

The astronaut next commands the retention latch motors to operate. When the retention latch hooks have released the arm, he uses a single-joint control mode to lift the

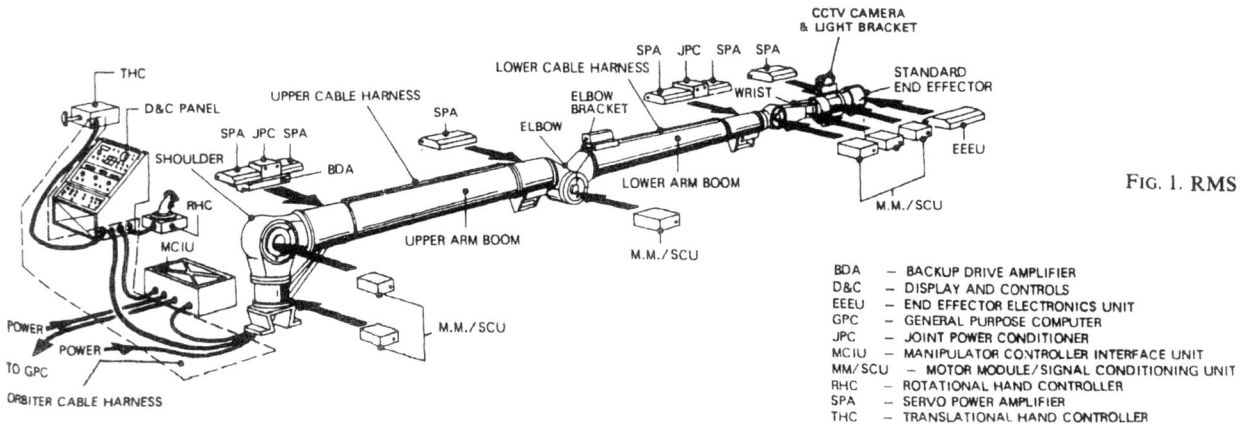

FIG. 1. RMS components.

BDA — BACKUP DRIVE AMPLIFIER
D&C — DISPLAY AND CONTROLS
EEEU — END EFFECTOR ELECTRONICS UNIT
GPC — GENERAL PURPOSE COMPUTER
JPC — JOINT POWER CONDITIONER
MCIU — MANIPULATOR CONTROLLER INTERFACE UNIT
MM/SCU — MOTOR MODULE/SIGNAL CONDITIONING UNIT
RHC — ROTATIONAL HAND CONTROLLER
SPA — SERVO POWER AMPLIFIER
THC — TRANSLATIONAL HAND CONTROLLER

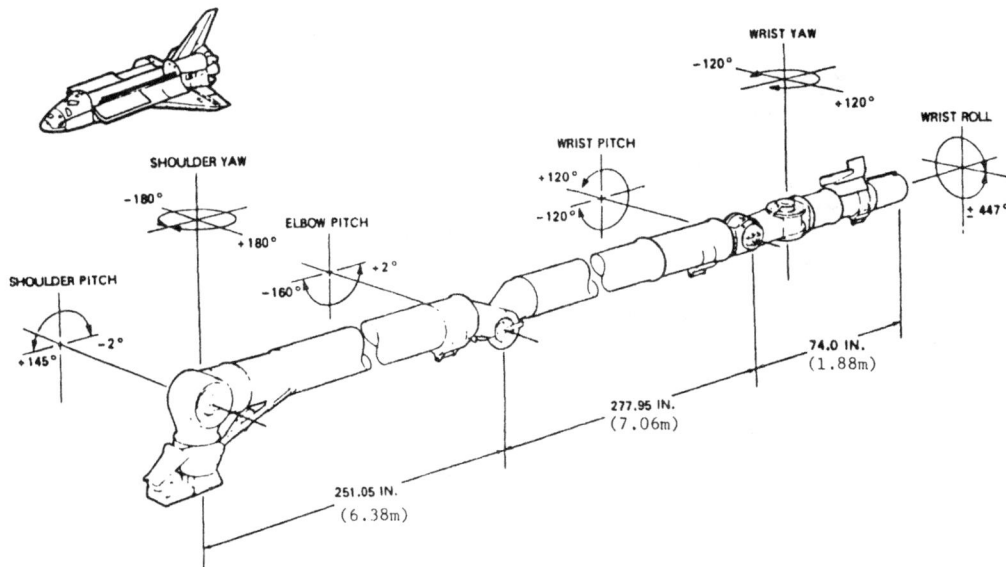

FIG. 2. Mechanical arm.

wrist pitch joint and then the shoulder pitch joint, thus clearing the arm of the orbiter sill fittings.

B. The joint assemblies

The motion available at each of the six joints is shown in Fig. 2. Some of the details typical of the joint assemblies can be seen in Figs. 3 and 4. The joints are driven through gear trains by motor modules. These are fractional horsepower brushless dc servo motors which were developed by SPAR for this application, and some of their details are shown in Fig. 5.

The motor uses a 16-pole samarium–cobalt permanent magnet rotor and a delta-connected 16-pole stator whose windings receive excitation from a servo power amplifier (SPA) dedicated to that joint or from a backup drive amplifier, which may be selected if the prime system fails. Motor

current commutation is provided by an optical commutator system. This uses photodiodes to sense the light reflected from a black and white pattern on the edge of the commutator disc. LED's provide the light source and fiber optics conduct the light from the LED to the disc and back to the photodiodes which generate the switching signals for the motor windings. Current limits can be programmed to set motor torques to desired levels.

Motor speed is sensed by an inductosyn tachometer whose rotor induces a Doppler shift in the signal coupled into one set of windings of the stator. The other set of stator windings carries a 25-kHz excitation signal. Electronic circuits in the SPA sense the frequency shift which is proportional to angular speed and direction, digitize it, and pass it back to the GPC along with other data concerning the joint such as the joint angle. The joint angle is obtained as a digital signal from a 16-bit optical position encoder giving a

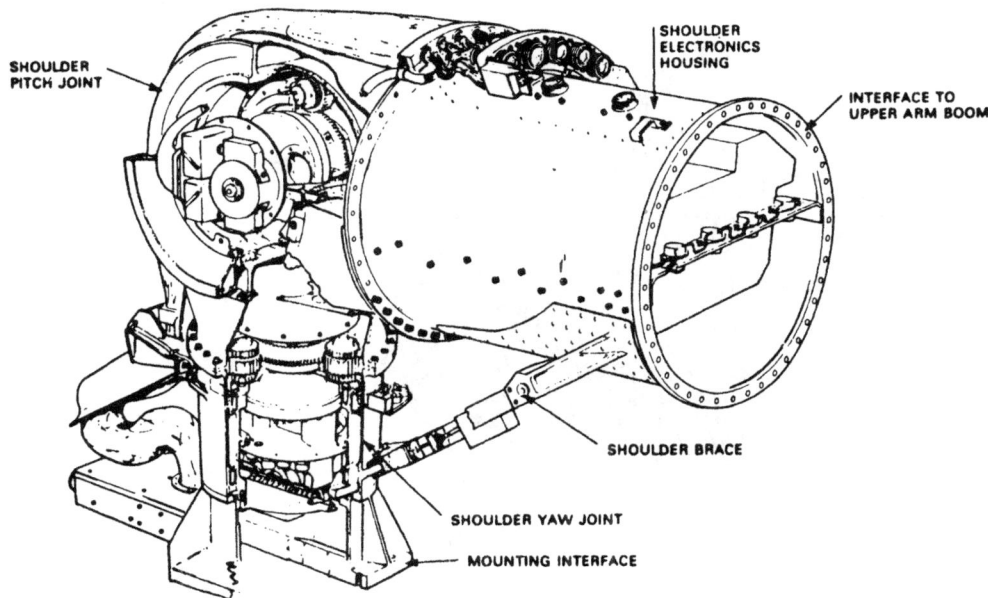

FIG. 3. Shoulder joint.

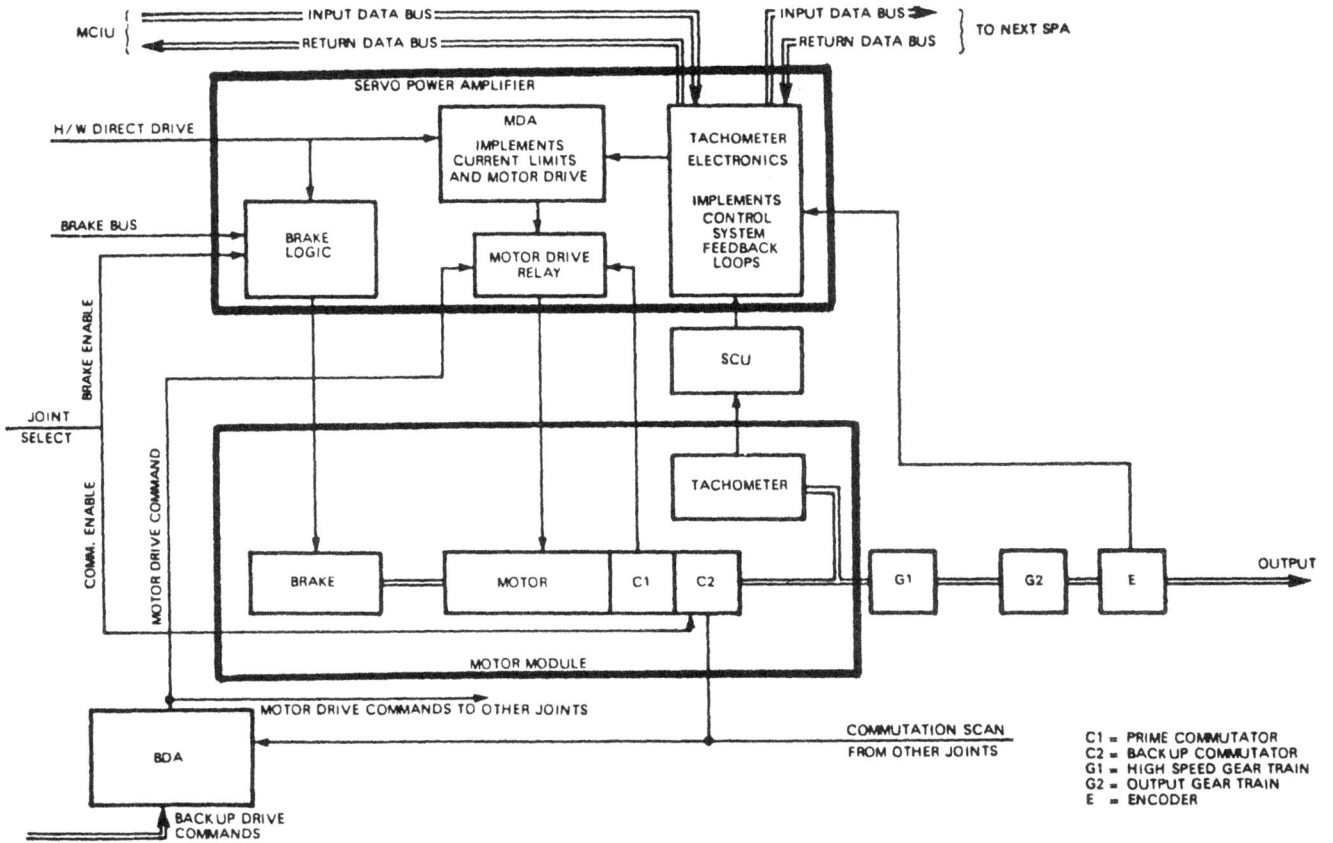

FIG. 4. Typical joint block diagram.

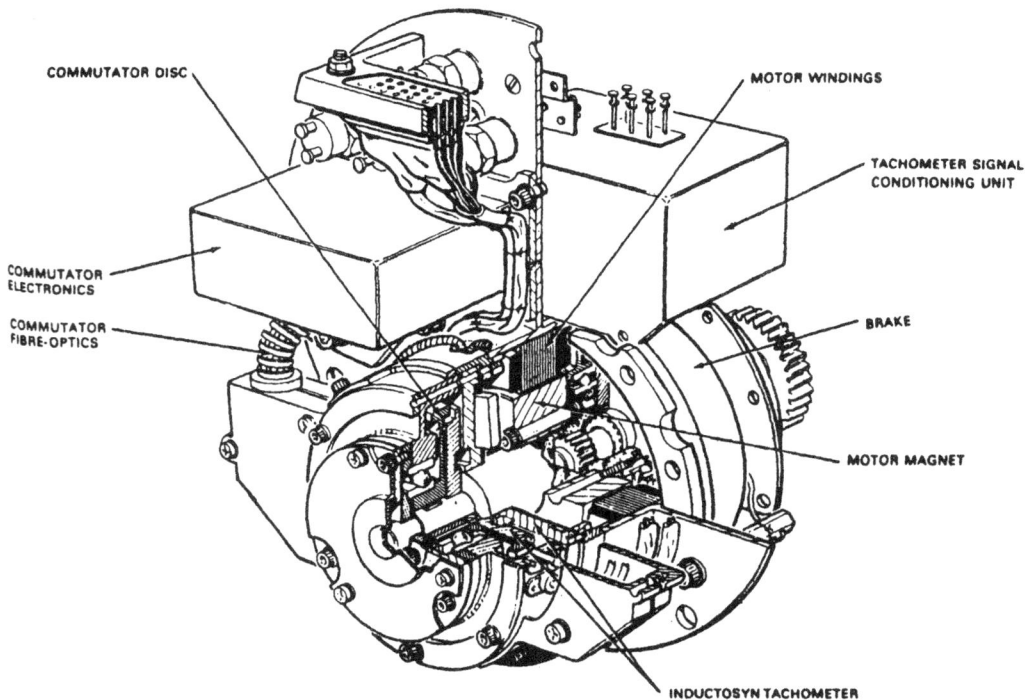

FIG. 5. Motor module.

resolution of 0.01°.

The motor module also includes a spring-loaded disk brake. A solenoid is energized to release the brake whenever the joint is to be driven. Removal of power from the system thus applies the brakes automatically on all joints in compliance with the fail-safe requirement.

The systems design that evolved[2] required that the gear trains have very low backlash, that they be backdriveable, and that they have high gear ratios. For example, the shoulder pitch joint has a gear ratio of 1843:1, an efficiency of approximately 80%, a backlash of 0.1° and can develop a torque of 1158 ft lb (1575 Nm).

The illustrations also show the electronics units which drive the motor modules. All of the arm power and signal wiring originates at the connector box at the shoulder yaw joint and runs along the outside of the arm, branching off to the connectors at the joints, until it reaches the end effector.

The upper and lower arm booms are thin-walled multiply graphite–epoxy tubes approximately 13 in. (0.33 m) in diameter and fitted internally with stiffening rings and with aluminum end rings for attachment to the joints. Despite the fact that together they weigh less than 120 lb (54 kg), their stiffness, combined with the stiffness of the joints, provides an overall arm stiffness of about 10 lb/in. (1760 N/m) deflection at the tip.

C. The end effector

The standard end effector,[3] which was delivered with the arm, is the most complex of the mechanisms of the RMS. It appears in Fig. 6 and the payload's grapple fixture, with which it interfaces, is also shown. The snare wires of the end effector capture the grapple fixture shaft, center it, and then draw it into the open end of the end effector until its base is pulled firmly against the end ring with a tension load on the shaft of 800 lb (3600 N). The reaction of the three hemispherical cams of the grapple fixture against the three cutouts in the end ring overcomes roll misalignment. With this arrangement, successful capture and rigidization of the payload can be achieved, despite initial misalignments of 10° in roll, 15° in pitch and yaw, and radial misalignment of 4 in. (0.1 m).

The three snare wires are actually 5/32-in. (4.0 mm)-diam

steel rope aircraft cables anchored at one end to a fixed ring and at the other end to a rotating ring such that they lie concealed in grooves until capture is commanded. Then they close rapidly, like the iris of a camera. The set of rings is then drawn by the action of three ball screws which pull the snared grapple fixture shaft deeper into the end effector until the whole combination is rigid. The torque to drive the snare rings and the ball screws is provided by another brushless dc motor module and gear train. An electronics unit produces the motor drive voltage pulses in response to the operator's commands, and also processes the signals returned by tension-sensing microswitches to provide feedback to the GPC and to the operator. When commanded, the payload release sequence begins with derigidization followed by opening of the snare, thus imparting no tipoff motion. The astronaut then backs the end effector away from the stationary payload.

The standard end effector mechanism also includes a simple, weight-effective, backup release system which will allow separation from the payload, should the prime system fail after a payload has been captured. The release motor is, in fact, a negator spring which is wound by the gear train during the capture process. The backup release command opens a clutch between the primary drive system and the snare ring gear train, allowing the backup release motor to drive the snares open, regardless of the state of rigidization.

It was recognized early in the program that at some time an as-yet-unidentified mission task would require capabilities which the standard end effector could not provide and that during such missions it would be necessary to use some form of special purpose end effector. This could be installed, before flight, instead of the standard end effector or, alternatively, could be picked up in flight by the standard end effector with power and control transfered by an automatically engaged electrical connector. Provision for either approach has been made in the design.

D. The wrist roll joint

It was adoption of this concept in end effector design which led to simplification of the wrist roll joint, which initially was to have been capable of continuous rotation. As can be seen, this would have meant provision of many slip rings to carry power, control data, and television signals. In a

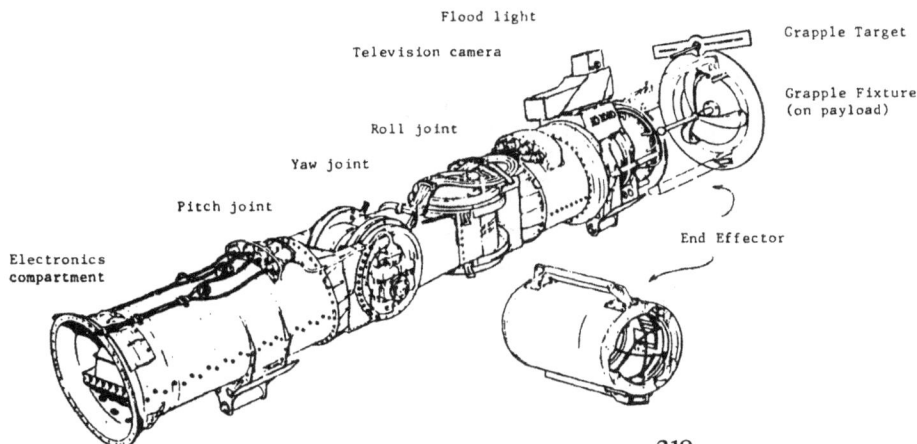

FIG. 6. Wrist joint, end effector, and grapple fixture.

319

compromise solution a wrist roll rotation of $\pm 450°$ was chosen. The conductor cable bundle through the roll joint lies on the roll axis and fans out into a short funnel-like housing, which permits the required cable motion, without inducing strain on wires or connectors.

The wrist roll joint, like the others, is fitted with mechanical stops which could absorb the energy of a 32 000-lb. (14 500 kg) payload rotating at 0.5°/s, should the software-controlled automatic stop fail.

The mechanical stops on the other joints are designed to absorb the energy of this design-case payload moving at 0.2 ft/s (0.06 m/s).

III. ALLOWANCE FOR VACUUM ENVIRONMENT

To ensure reliable and repeated operation after reaching the vacuum environment of earth orbit (Table II) it was necessary to allow for several factors, chiefly, the rapid change of pressure during ascent and re-entry, the behavior of lubricants in vacuum, and the loss of convective cooling.

Suitably sized venting apertures were provided in every closed structural assembly and electronics unit to limit pressure gradients which otherwise could lead to bulging of the enclosure during ascent or crushing during the repressurization of the reentry phase. Particular attention was paid to the venting of the arm booms and of the thermal blankets to avoid risk of damage through ballooning of the blankets or risk of their detachment, resulting in possible mechanical interference with joint motion as well as loss of thermal protection.

Fluid lubricants for the gears and rolling element bearings were considered and rejected because of potential contamination problems arising from evaporation and migration to adjacent surfaces. A solid lubricant (Lubeco 905), previously used in space-qualified mechanisms, was then chosen. This has a molybdenum–disulphide base and after completion of a specially developed running-in procedure it has the necessary load-bearing capability while providing high-performance lubrication of the surfaces for many bearing applications.

It was found, however, during the Canadarm hardware test program that certain bearings subjected to rapid acceleration and abrupt stops suffered degradation of this lubricant and it was finally concluded to be unsuitable for some applications.

To overcome this problem of rapid wear of the solid lubricant, where skidding of the balls in a bearing could occur, the use of fluid lubricant was reexamined. A light grease (Braycote 3L-38RP), based on a perfluorinated polyether, was found to have both excellent lubricant qualities and high stability over a wide range of environmental conditions. The quantity of grease used is 15% of the bearing cavity, calculations and tests indicate that the loss by evaporation will be insignificant over the 10 y/100 mission period (Table I) while at the same time posing no threat to neighboring surfaces.

Dissipation of electrical power can rapidly raise the temperature of an assembly in the absence of convective cooling unless adequate conductive and radiative paths are provided for the heat energy. In most parts of the arm this was readily achieved. The floodlight mounted on top of the wrist TV camera, however, presented a problem since the incandescent lamp dissipates approximately 130 W. Only a small amount could be conducted into the mounting bracket, since this also supports the camera whose heat input had to be limited. Nor could much heat be allowed to be conducted into the wrist joint itself. The conduction path was therefore deliberately restricted by the use of small mounting feet. Calculations by the lamp assembly contractor (ILC Sunnyvale, California) showed that the internal lamp temperatures would soon become excessive and result in burnout. The problem was solved by the use of a large sapphire window on the front of the unit to allow most of the heat to be transmitted out of the unit as part of the floodlight beam.

IV. VERIFICATION

Proof of compliance with specified requirements, and certification of the readiness of the hardware for flight, involved many steps and was, without doubt, the largest cost element in the entire program, as compared with the design and fabrication. Table III identifies most of the techniques employed.

Every electronic unit, and each of the joint assemblies, was tested separately, and again after integration. These tests included thermal-vacuum tests. Because of its size, however, it was judged impractical to attempt to conduct thermal-vacuum testing of the completely assembled arm. These tests were therefore limited to the joint level. Most of the environmental testing was performed in the David Florida Laboratory of the Communications Research Center in Ottawa. Typically, in these tests, the chamber was held near 10^{-6} Torr while the temperature of the unit under test was cycled, by the use of heat lamps, betwen upper and lower limits which were at least 10 °C above and below the predicted operational temperature limits. Six test cycles were customary with each cycle requiring 24 h for the larger units.

After the arm had been completely assembled by integration of all of its component subassemblies, it was in turn integrated with the cabin equipment units to form the total RMS system. This was then subjected to a lengthy series of end-to-end performance tests to ensure compatibility of all units and software. The arm was then supported on a near-friction-free air bearing system which permitted a limited, but adequate simulation of the microgravity condition of

TABLE III. Summary of RMS verification.

Verification:
- The process to prove compliance with requirements

    Techniques:
- Review of design
- Analysis
- Development tests
- Inspection
- Nonreal-time simulations
- Real-time (man-in-the-loop) simulations
- Certification tests (qualification tests)
- Acceptance tests
- Major ground tests
- Orbital flight tests

    Constraints:
- System can only be partially tested in a 1–g environment
- Oribital flight tests will not verify all requirements,
    e.g., biggest test payload is smaller and lighter than
      specified max payload

earth orbit. These tests provided final assurance that the system was ready for delivery.

V. CONCLUSION

Since then, this first Canadarm system has been flown, tested, and used on orbit during Columbia flights STS-2, STS-3, and STS-4. Crew reports and postflight analysis of instrumentation data have provided confirmation that the requirements have been met and that the system is capable of performing the intended tasks.

[1]P. Kumar, P. Truss, and C. G. Wagner-Bartak, Proceedings, Fifth World Congress on Machines and Mechanisms, Montreal, Canada, 1979.

[2]J. A. Hunter, T. U. Ussher, and D. M. Gossain, presented at the AIAA/ASME/ASCE 23rd Conference of Structures, Structural Dynamics and Materials, May, 1982.

[3]R. G. Daniell and S. S. Sachdev, presented at the 16th Aerospace Mechanisms Symposium at NASA-KSC, May, 1982.

Titles of referenced papers were omitted:

1. System Design Features of the Space Shuttle Remote Manipulator.

2. Structural Dynamic Design Considerations of the Shuttle Remote Manipulator System.

3. Design and Development of an End Effector for the Shuttle Remote Manipulator System.

J. Vac. Sci. Technol. A, Vol. 1, No. 2, Apr.–June 1983

321

SIX MECHANISMS USED ON THE SSM/I RADIOMETER

Howard R. Ludwig*

ABSTRACT

Future USAF Block 5D Defense Meteorological Satellites will carry a scanning microwave radiometer sensor known as SSM/I. SSM.I senses the emission of microwave energy and returns to earth data used to determine weather conditions, such as rainfall rates, soil moisture, and oceanic wind speed.

The overall design of the SSM/I radiometer was largely influenced by the mechanisms. The radiometer was designed to be stowed in a cavity on the existing spacecraft. The deployment of the sensor is complex due to the constraint of this cavity and the need for precision in the deployment. The radiometer will continuously rotate, instead of oscillate, creating the need for a bearing and power transfer assembly (BAPTA) and a momentum compensation device. The six mechanisms developed for this program are described in this paper.

1. INTRODUCTION

Future spacecraft of the USAF Defense Meteorological Satellite Program (DMSP) will carry a passive microwave radiometric system known as the special sensor microwave/imager (SSM/I). It is a seven channel, four frequency, linearly polarized, passive system. The SSM/I measures ocean, atmospheric, and land surface brightness temperatures. The Air Force Air Weather Service and the Naval Oceanography Command will process these data to obtain precipitation maps, oceanic wind speed, sea ice morphology, and soil moisture percentage. Figure 1 illustrates the SSM/I integrated into the DMSP satellite.

2. SSM/I OVERALL MECHANICAL DESIGN

The existing DMSP Block 5 spacecraft had only two possible locations to stow the sensor. Potential designs were made to use these locations. For each of these designs, mechanisms were selected. The final selection was based on an efficient, low-risk system. Figure 2 illustrates the selected sensor in stowed and deployed configurations. The mechanisms had a major influence in the overall design of the sensor. These account for approximately 30 percent of the weight of the SSM/I. The stowed volume is approximately 20 percent of the deployed on-orbit swept volume.

One of the major decisions was in the manner of scanning. It was decided to continuously rotate the sensor at 31.6 rpm to provide 103° of uninterrupted earth scan. As the sensor rotates beyond this range, the sensor passes a hot load and a cold space reflector that calibrate the sensor. This method is more efficient and accurate than a back and forth or oscillating reflector. Continuous rotation of the sensor adds uncompensated momentum to the

* Hughes Aircraft Company, El Segundo, California

spacecraft. Previous studies (Ref. 1) suggested use of gear or belt driven counterrotating masses to compensate for sensor momentum. A trade-off study for providing compensating momentum determined that an existing momentum wheel would be more reliable and efficient than the previously proposed devices. Developing a geared or belted system for a 3-4 year life was felt to be too risky. The momentum wheel could be located in a separate compartment of the spacecraft to help its balance and not take up volume in the location chosen for the sensor.

Deployment of the SSM/I is unique. First, the whole sensor is deployed from the spacecraft. The reflector is then released and rotated 170° by the reflector deployment mechanism. At that point a four-bar linkage is engaged which translates the reflector into its final location (Figure 3). This sequence does not interfere with the view of the DMSP's other sensors. Six mechanisms are used on the SSM/I: bearing and power transfer assembly, momentum wheel, three deployment mechanisms, and pyrotechnic devices.

3. MECHANISMS

3.1 Tribological Design

In the design of mechanisms there are two general philosophies (usually unstated) on the tribological aspects. One is to perform the electrical and mechanical designs and to then add the lubrication method. The other approach, and the one used on the SSM/I mechanisms, is to consider the tribological aspects of the design from inception. The general type of lubrication is chosen early in the design, and compatible materials are selected. The final lubricant is based on requirements such as temperature, load, and speed.

3.2 Bearing and Power Transfer Assembly

Continuous rotation of the SSM/I sensor is provided by the BAPTA (Figure 4). The BAPTA developed for this program has some unique features, but it was based on existing technologies. The BAPTA provides rotating mechanical and electrical interfaces between the spinning and stationary sections of the sensor. The physical characteristics are shown in Table 1.

The tribological design of the SSM/I BAPTA uses state of the art technology. A pair of CEVM 440C angular contact bearings with patented nitrileacrylic copolymer retainers (Refs. 2-3), in an open cell configuration, are used. Stainless steel sideplates provide the required mechanical stiffness and strength. The copolymer material has a large capacity for oil storage (Figure 5).

Each bearing is lubricated with HMS 20-1727 oil, which is a highly refined, low vapor pressure, mineral oil with a 5 percent lead naphthanate extreme pressure additive. The nitrile-acrylic copolymer with this oil is now at the 9 year mark in a continuing life test and has been successfully flown on the Pioneer Venus, GMS II, and GOES programs.

The electrical power and signal transfers are through slip rings made of silver and brushes of a self-lubricating material made of 85 percent silver, 3 percent graphite, and 12 percent molybdenum disulfide. Based on the wear rate from a 10 year life test, these brushes will last more than 100 years.

3.3 Deployment Mechanisms

3.3.1 General Design Philosophy. There are two general approaches to the design of spring-driven deployment mechanisms for delicate appendages or devices that have to point precisely. One method is to use a deployment spring that is only slightly stronger than the resistive torque. This is done to avoid shock damage at the end of stroke. A great deal of testing and adjustments are usually associated with this approach. The second approach is to employ very large torque margins and to rate-limit the deployment. This is accomplished by viscous or eddy current damping. The bottom line of the trade-off study indicates that the high torque margin rate limited method is more reliable and easier to test at the expense of weight. Although the simple spring mechanism is less costly as a unit, the overall cost is usually higher due to the complexity of testing and need for exotic offloaders for system level testing.

The use of the high torque margin was chose for SSM/I so that there would be no risk of a failure to deploy. Additionally, all tests were performed in 1G without offloaders. This was accomplished by mounting the sensor with the deployment hinge in line with the gravity vector for each deployment.

3.3.2 Viscous Damping. The deployment mechanisms for SSM/I are multispeed. There was a system requirement that all deployments be made in a 10 minute period and also a self-imposed rate requirement to keep the precision and repeatability. The two were not compatible with a single speed device. In order to make the mechanism multispeed, the damping gap is varied. The radiometer deployment mechanism (RDM) has two speeds and the antenna deployment mechanism (ADM) has four speeds (two fast and two slow) to meet the time requirement at cold condition.

The two main deployment mechanisms for SSM/I are rotary viscous damped. These devices combined the features of the pivot hinge and bearings, deployment springs, viscous damper, and hard stops into one unit. The mechanisms are powered by redundant laminated spring sets. They feature redundant spring-loaded tapered latch pins to provide a positive lock at latch-up and adjustable hard stops to ensure positioning accuracy. These actuator designs have evolved from the basic JPL concept originally developed for Mariner '71 and Viking Orbiter to their current compact third generation design (Ref. 4).

The actuator mechanism consists of a housing, two end plates containing O-ring seals, and a stator shaft with integral damping paddles. These are all made of aluminum. A take-up reel is provided for the deployment springs. The dimethyl silicone fluid of 0.1 m^2/s kinematic viscosity provides damping. A reservoir compensates for the change of fluid volume with temperature. Figure 6 is a cutaway view of an actuator.

325

Double paddle wheels are used for deployment angles less than 130° and a single paddle is used for angles greater than 130°. The housing rotates about the slator shaft. Torque is supplied by the constant torque deployment springs. Damping is provided by flow of the damper fluid from the high pressure side of each paddle, through the gaps between each paddle and the housing, into the low pressure side. The majority of the damping comes from the flow of fluid through these orifices (the gap between each paddle and the housing), while the remainder is due to viscous shear.

The main pivot ball bearings are selected to provide proper stiffness and strength. Lubrication of the deployment devices is a combination of wet and dry. The main bearings are in the silicone fluid. The laminated springs, take-up reel bearings, and latch pins are lubricated with sputtered MoS_2 (Ref. 5). The tips of the latch pins are made of a self-lubricating plastic material.

3.3.3 <u>Radiometer Deployment Mechanism</u>. Deployment of the SSM/I radiometer from the stowed position is accomplished by the RDM. The pyrotechnic devices at the base plate are fired, releasing the radiometer. The RDM rotates the sensor 90° about the deployment axis, bringing the BAPTA spin axis parallel to the local vertical. Tapered latch pins drop into place to brace the radiometer in the deployed position. Completion of the deployment is indicated by a normally closed microswitch which opens as the latch pin drops into place. Adjustable hard stops control the deployment angle and repeatability of the deployment angle. Figure 7 shows the RDM. As can be seen, the mounting base for the radiometer was made integral with the RDM. Table 2 provides test results for the engineering, qualification, and the first flight models. Figure 8 shows a typical deployment angle versus time plot for the RDM. As can be seen, the first (70°) is at a rate of 0.4 rad/s and the last (20°) at a rate of 0.04 rad/s.

3.3.4 <u>Antenna Deployment Mechanism</u>. After the radiometer has been deployed from the spacecraft, the antenna is released by firing another set of pyrotechnic devices. The antenna is then driven 170° about the first antenna deployment axis by the deployment springs. As the antenna reaches this fully extended position, a four-bar linkage is engaged. The antenna is then translated into the operating position. Again, tapered latch pins are engaged and microswitches are closed to indicate latch-up (Figure 3). The deployment scheme is a result of a customer imposed requirement that other sensors on the spacecraft not have their fields of view obscured by SSM/I during its deployment. Table 3 summarizes the test results.

3.3.5 <u>Bias Momentum Wheel Assembly</u>. The specification for the SSM/I radiometer required no unbiased momentum from the radiometer system. The choice of using a continuously rotating antenna, together with the unbiased momentum specification, led to incorporating a bias momentum wheel into the design.

The momentum compensation system, an independent open loom momentum wheel and a continuously rotating radiometer, was compared with mechanically coupled momentum wheels through gear or belt drives. These alternatives were eliminated as a result of radiometer jitter caused by the coupling system, dry lubricant life limitations, and volume constraints. Gear or belt drive linking the radiometer to the compensation wheel would feed back any jitter to the radiometer spin control loop inducing rate errors. It would also result in an excessive weight penalty (for gear ratios near unity) or excessive number of revolutions of the compensation wheel (for gear ratios of 50 or more). Dry film gear lubricant is not compatible with a 3-4 year life requirement. Wet lubrication could be employed, but would require an additional sealed housing, adding complexity and weight. Other self-lubricating gear materials would require extensive development and life testing. Belt drives similar to those used on JPL designed radiometers were considered, but again additional development/life testing would be necessary to meet reliability requirements. Discussions with other aerospace company personnel have indicated the critical nature of belt drive details to reliability. It was unlikely that a belt drive system development program could be satisfactorily accomplished during the 14 month SSM/I development period.

The momentum required to compensate for the SSM/I in nominally 3.0 N · m. There were several commercial momentum wheels in this size range. The final momentum specified can be adjusted over the range of 2.75 to 4.25 N · m in 205 increments by final adjustment of the drive electronics. The desired speed of the momentum wheel is programmed by using 10 jumpers on a programming connector plug. This system allows final selection and momentum adjustments to match the momentum of the radiometer as the two are integrated. Table 4 delineates the characteristics of the momentum wheel, and Figure 9 shows the momentum wheel with its drive electronics.

3.4 Simple Deployments

When precision pointing or end of stroke shock are not important, simple undamped deployments are ample. The fact that a deployment is without rate control does not mean that the details of design and lubrication are not important. The materials should be selected with lubrication in mind and the drive springs should be redundant. In the case of SSM/I, there are two such deployments.

3.4.1 Lubrication Aspects. The general lubrication technique used on simple deployment mechanisms is to choose shaft and bushing or pivot materials that will not gall. For example, a hard stainless steel pin and a hard anodized aluminum bearing have to be used on most of the SSM/I pivot joints. All rubbing surfaces and springs are then sputtered with MOS_2 to give a low coefficient of friction. This combination of materials and lubricants has been reliable and trouble-free throughout the development and hardware programs.

327

3.4.2 Launch Support Bracket Deployment. The SSM/I antenna is held in the stowed position with a pyrotechnic device. The launch support bracket is a swing-away structural member. The loads from the antenna are passed through this bracket into the radiometer structure. Upon firing the pin pullers, the launch support bracket pivots 135° and is held in place by the redundant deployment springs. The springs and part of the bracket can be seen in Figure 10.

3.4.3 Cold Sky Reflector Deployment. The SSM/I radiometer is calibrated each revolution. An ambient hot load and reflected cold space are viewed by the sensor each revolution. The cold sky reflector can be seen in Figure 10. It is the small reflector on the top surface of the sensor body. During launch it is stowed under the larger antenna. It is kept from deploying by a strip of self-lubricating plastic material attached to the back of the main reflector. As the main reflector deploys approximately 45°, the cold sky reflector is no longer restrained and rapidly rotates to its deployed position.

3.5 Pyrotechnic Release Devices

The selection of pyrotechnic release device types is dictated by locations and mass properties of the items to be released and the amount of preloading required. For SSM/I an additional requirement of pyrotechnic redundance was specified.

The SSM/I is secured to its base plate by two independent redundant pyro lock systems, and is preloaded against four posts attached to the spacecraft interface mounting plate (Figure 11). Launch locking in this fashion provides two distinct advantages. First, pulling down the spinning portion of the sensor against the base provides rigid nesting without carrying undue loads through the radiometer deployment mechanism bearings of the BAPTA bearings. Second, the launch loads, spread over a large area of the spacecraft panel, place no local moment loading on this panel.

The two pyro systems with redundancy provisions have been fully qualified and flown on the NASA Pioneer Venus program. The same hardware was used. Redundant pint pullers tie down a "whiffle tree" fitting so that retraction of either pin will release the load. The pins are in double shear. With this dual system, retraction of either pin in each of the two circuits will deploy the radiometer.

The antenna lock system uses the same hardware as the base lock system, except that two whiffle trees are in the double shear clevis to effect a dual element release. The antenna support arm is locked to the support arm as shown in Figure 10. Both of these elements are independently preloaded to achieve an overall rigid system.

328

There were two structural failures during development test in the launch restraint devices. The whiffle tree pivots failed structurally under the 454 kg load. The whiffle tree material was then changed from aluminum to stainless steel. The second failure occurred when the bracket holding the pin pullers on the antenna support arm broke off as the pin pullers were fired. It was theorized that the shock load caused by the pistons in the pin pullers pulled the bonded-on bracket loose. This bracket was redesigned to be a bolted-on bracket. After these two redesigns were made, no further failures occurred.

4. CONCLUSION

Six mechanisms were successfully designed, developed, and qualified for the SSM/I radiometer. The major accomplishment was in the successful development of high precision, repeatable, multispeed deployment mechanisms.

5. REFERENCES

1. NASA-CR-144853, "Scanning Mechanism Study for Multi-Frequency Microwave Radiometers," General Electric Company.

2. Christy, R. I., "Evaluation of a Nitrile Acrylic Copolymer As A Ball Bearing Retainer," ASLE preprint 74 AM5A-1, 1974.

3. U.S. Patent 4,226,484 (1980), Bearing Retainer.

4. JPL Drawing Package, Deploy/Damper/Latch Assembly Solar Panel 10040060, Rev D, 1972.

5. Christy, R.I. and Ludwig, H.R., "RF Sputtered MoS_2 Parameter Effect on Wear Life" Thin Solid Films, 64 (1979) 223-229.

Table 1. BAPTA Characteristics

Weight	9.1 kg
Dimensions	19.4 cm max OD x 27.5 cm length
Bearings	
Type	Angular contact
Size	60 mm bore
Class	ABEC Class 9
Material	CRES type 440C CEVM
Lubricant	HMS 20-1727
Retainer	Nitrile-acrylic copolymer
Slip Rings	
Power rings	4 ea
Power circuits	2 ea
Circuit capacity	1.5 A
Brushes per power ring	4 ea
Current density	700 A/m^2 (2 brush contact)
Signal rings	36 ea
Brushes per signal ring	2 ea
Ring material	Silver
Brush material	85% Ag, 3% C, 12% MoS_2
Lubricant	MoS_2 in brushes
Pulse Generators	
Master	1 pulse per revolution
Encoder	45 pulses per revolution
Motor	
Type	Brushless dc resolver commutated
Torque constant	0.1100 Mkg/A
Winding resistance	15Ω
Back EMF	0.983 V/rad/sec

330

Table 2. RDM Test Results

Parameter	Requirement	Model		
		Engineering	Qualification	Flight
Repeatability, $^\circ$	±0.05	+0.0003 −0.00015	+0.001 −0.000	+0.003 −0.000
Spring torque, N · m	11 to 12	11.08	11.0	11.8
Torque margin, %	400	433	636	585
Deployment time, s	4 to 80			
22°C		15.7	9.2	8.0
−34°C		50.0	42	17.0 at −11°C
71°C		7.1	4.1	5.6 at 41°C

Table 3. ADM Test Results

Parameter	Requirement	Model	
		Qualification	Flight
Repeatability, $^\circ$	±0.04	+0.00 −0.02	+0.00 −0.01
Spring torque, N · m	3	3.6	4.3
Torque margin, %	200	329	215
Deployment time, s	<161		
22°C		17.8	29.0
−32°C		89.0	118.5
69°C		10.0	16.8
Deployment rate into stop, $^\circ$/s	<10	8.3	4.9

Table 4. SSM/I Momentum Wheel Design Data

Unit angular momentum	3.0 N · m at 1300 rpm
Unit weight	5.5 kg without electronics
Rotor inertia	0.022 kg · m^2
Rotating weight	3.64 kg
Rotor diameter	18 cm
Envelope dimensions	20 cm dia by 28.5 cm high excluding mounting feet and connector
Spin motor type	AC squirrel cage induction
Synchronous speed	2400 rpm (theoretical) at 400 Hz
Unit internal atmosphere	98% HE, 2% O_2, 50% ATM
Friction and windage torque	58 gm · cm at 1400 rpm
Bearing type	Deep groove R8, double shielded
Lubrication type and quantity	MIL-L-6085A, +5% TCP, 20 to 25 mg
Tachometer type	Permanent magnet pulse Generator - 12 pulses/rev

Figure 1. DMSP satellite with SSM/I

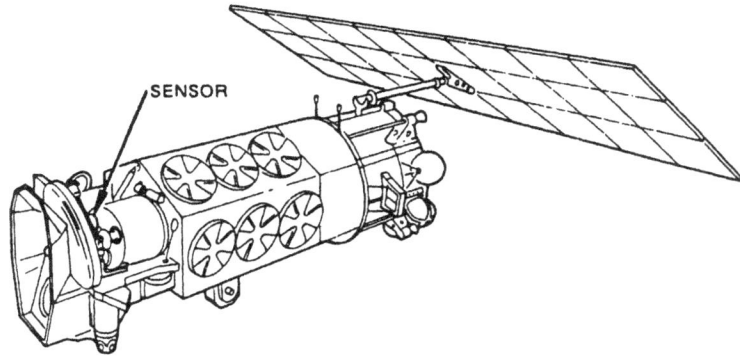

SENSOR

a) Stowed

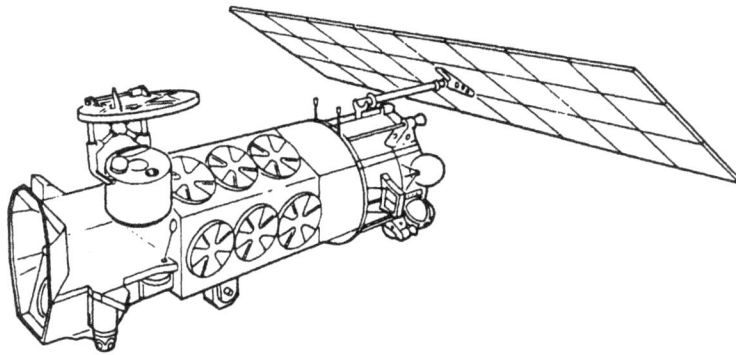

b) Deployed

Figure 2. Sensor location

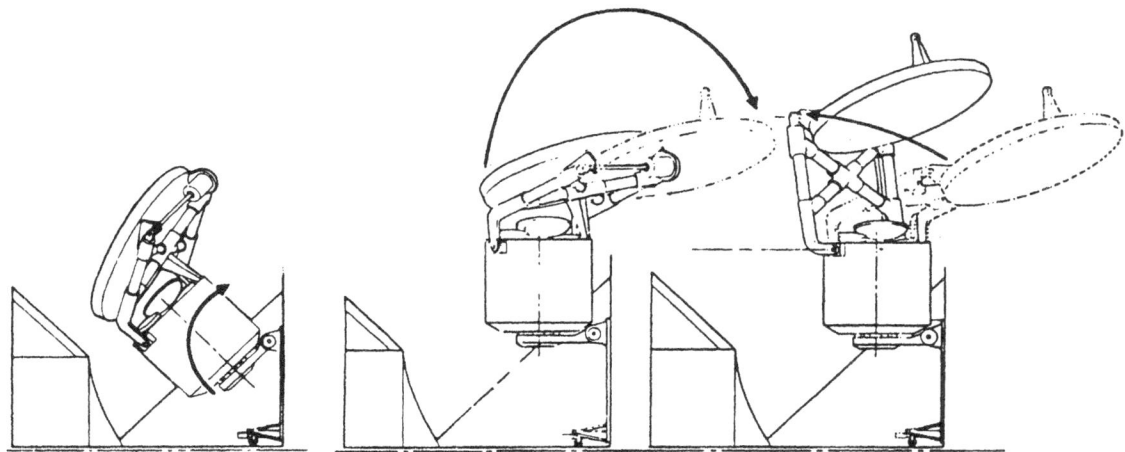

a) Instrument pivoting   b) Antenna reflector rotation   c) Antenna reflector translation

Figure 3. Deployment sequence

Figure 4. SSM/I BAPTA

Figure 5. Retainer lubricant capacity

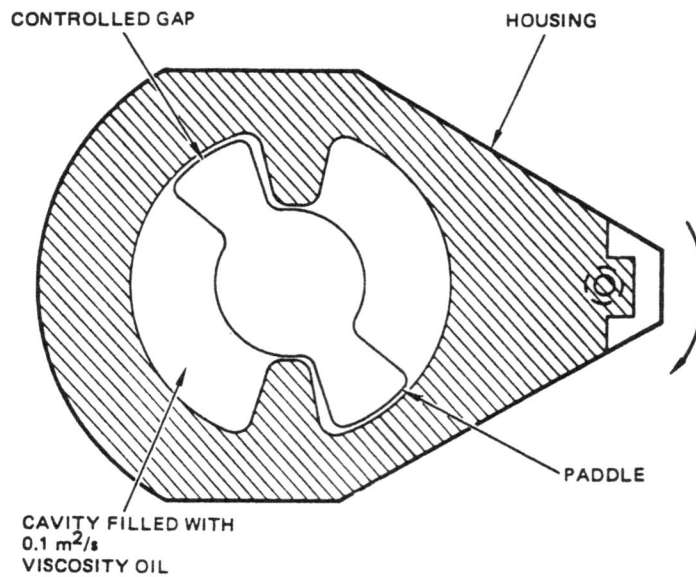

CONTROLLED GAP

HOUSING

PADDLE

CAVITY FILLED WITH
0.1 m^2/s
VISCOSITY OIL

— PADDLE IS PINNED TO SUPPORT STRUCTURE
— HOUSING ROTATES WITH RESPECT TO PADDLE
— FLUID IS FORCED THROUGH CONTROLLED GAPS

Figure 6. Deployment mechanism cross section

Figure 7. Radiometer deployment mechanism

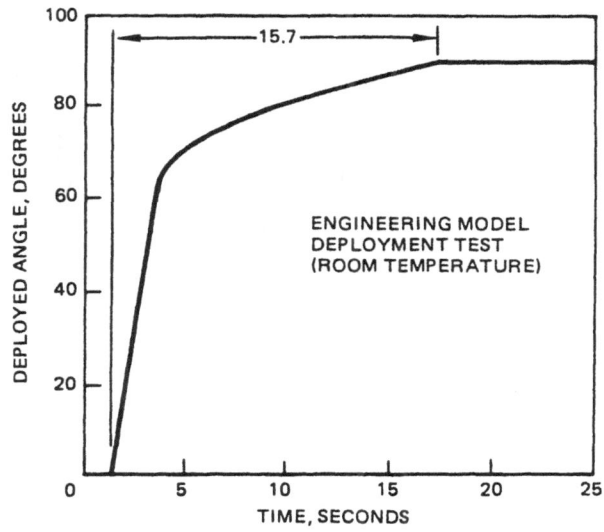

Figure 8. RDM deployment angle versus time

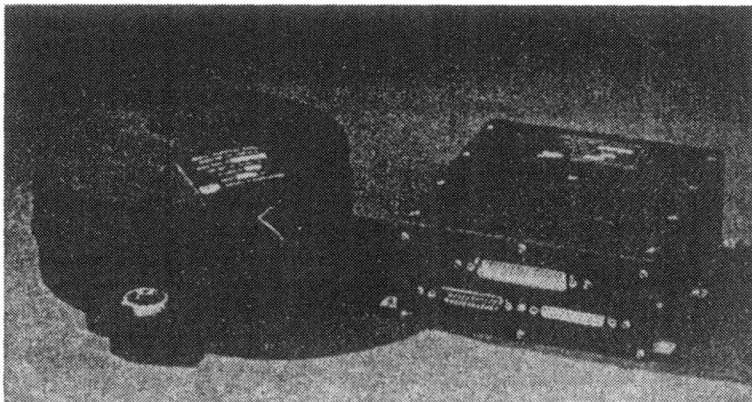

Figure 9. Momemtum wheel assembly and drive electronics

337

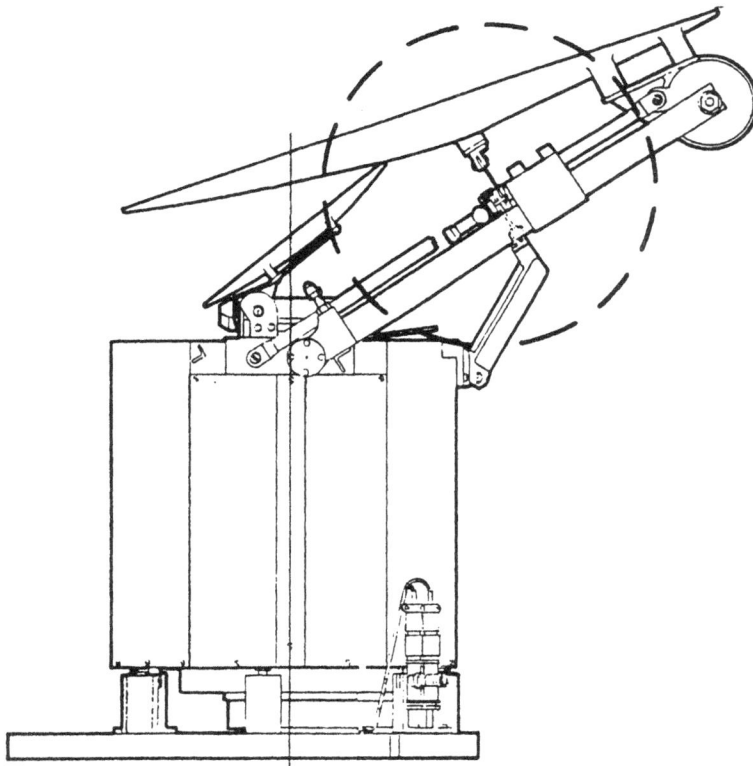

Figure 10. Launch support bracket deployment

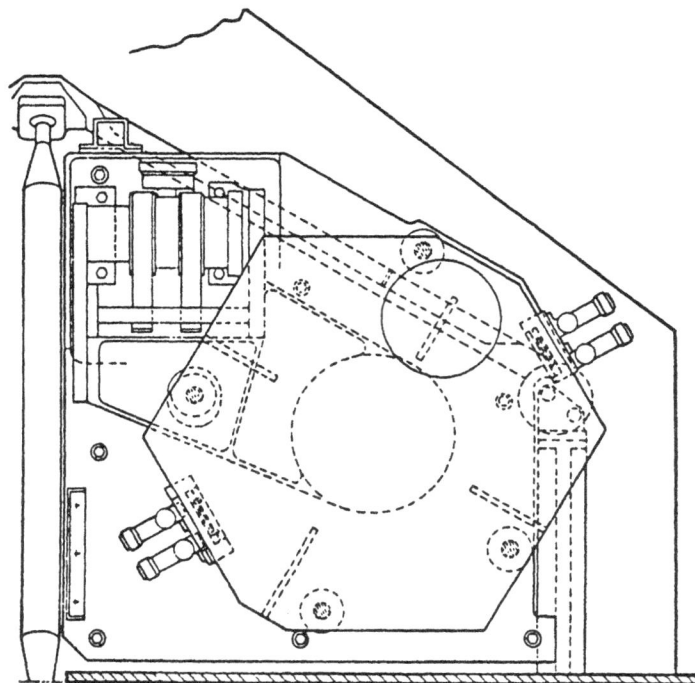

Figure 11. Base launch lock system

UHV technique for intervacuum sample transfer

J. P. Hobson and E. V. Kornelsen

Division of Electrical Engineering, National Research Council of Canada, Ottawa, Canada K1A 0R8

(Received 25 September 1978; accepted 15 November 1978)

An alternative to the modern UHV multiport surface analytic system is a transfer technique for intervacuum sample transfer which maintains the sample at UHV ($< 10^{-9}$ Torr) *at all times*. Major advantages are (1) economical use of available facilities by a number of experimentalists (2) instrument and surface standardization (3) complementary experiments by remote specialized laboratories (4) preparation of samples in specialized facilities (e.g., space shuttle or thermonuclear reactor) followed by a transfer to surface analytic facility. Such a transfer system has been built and tested. Transfers currently take three to four hours between nearby stations. Intercontinental transfers are feasible using a battery operated ion pump during transport. For a transfer a UHV system must have a suitable gate valve (35-mm bore, 10^{-13} ls^{-1} helium closed conductance) and a platform capable of accepting a standard 22.8–mm-diam ring, which carries the sample itself.

PACS numbers: 07.30. − t, 47.45. − n

I. INTRODUCTION

Recently various aspects of an ultrahigh vacuum technique for intervacuum sample transfer have been published.[1-3] The object of the present paper is to summarize the project to date, to describe in more depth its background, its potential applications, and its operational results.

II. BACKGROUND

One of the major prerequisites for modern surface science using clean well-characterized surfaces is the routine availability of UHV systems in which measurements may be made, usually with some combination of molecular beams, electron beams, ion beams, or photon beams. The modern trend in the realization of uhv surface analytic systems has been toward multiport systems, each port admitting a different type of beam or measuring service, generally directed on to a sample located in a central geometrical position. Such systems are available from all the major suppliers in the field, the main measuring chamber being a single vacuum system. The resulting apparatus is large, expensive, difficult to service without breaking vacuum, and requires at least 24 hours and often longer to be pumped from atmosphere to an acceptable pressure in the range of 10^{-10} Torr (1.3×10^{-8} Pa) or below. Coupled with a carousel for the switching of samples in situ or for moving a sample to face another port this type of apparatus is well suited to an in-depth study of true surface properties, but is limited in general to a single experimenter with adequate time at his disposal. An advance in flexibility at some cost to the quality of the vacuum is achieved by adding a sample entry system[4,5] permitting the entry and exit of a sample from the atmosphere to the vacuum. These sample entry accessories are becoming more widely available as options on multiport systems. Speed of experimentation is greatly increased, but frequently the sample will require some surface conditioning after entry before the desired measurements are

made. Variants on the combinations described to date are clearly possible, e.g., the sealing off of one part of a system from another with the sample being transferred from one to the other for various purposes.[6] The systems described so far might be termed multiport systems with accessories.

A transfer system with rather different properties results if a sample can be transferred between two remote ultrahigh vacuum systems maintaining the sample at UHV ($<10^{-9}$ Torr) *at all times*. In this case the sample does not "know" that it has been transferred. It is such an UHV technique for intervacuum sample transfer that will be described in this paper. With reference to Fig. 1 consider sample I in the conventional UHV surface analytic system called C which has one port dedicated to an UHV gate valve (2) whose closed conductance is so low that atmospheric leakage through it is negligible ($<10^{-13}$ l s^{-1} for He). Consider that another uhv system called a UHV transfer device (UHVTD) equipped with its own internal pumps and gate valve (1) can be coupled to C at the mating flange. Assume that means exist to exhaust the intermediate volume between the gate valves to UHV. Assume that under this condition the gate valves can be opened and sample I extracted from system C and placed in the UHVTD while another sample II replaces it in system C. The gate valves are then closed, the two systems decoupled at the mating flange and the UHVTD moved to a second system D, also equipped with an UHV gate valve, in which sample I is deposited. If sample II originated in system D then systems C and D have exchanged samples with minimal disturbance to either the two vacua or the two samples. If the UHVTD is designed to be portable then the distance between C and D becomes in principle very large; earth-scale distances of thousands of miles have been shown to be practical,[2] and there seems no reason why transfers between earth and space should not also be practical.[7]

Before describing the apparatus and its performance a section will be devoted to possible applications of the UHVTD,

339

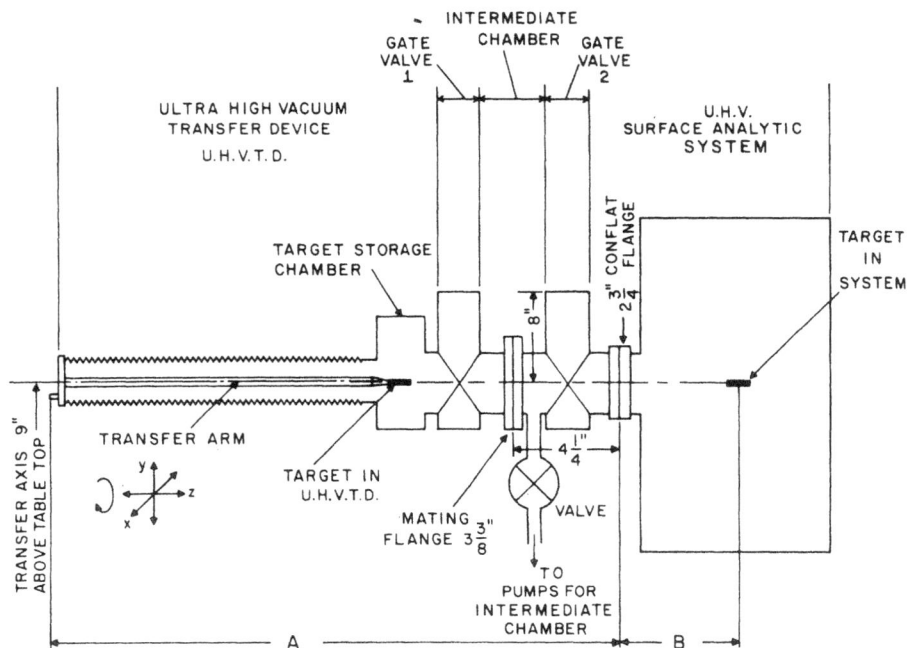

FIG. 1. Schematic showing relationship of UHVTD to an UHV surface analytic system. At present A = 39.5 in. (100 cm), B_{max} = 4.25 in. (10.8 cm). In the next model A = 49.75 in. (126 cm), B_{max} = 9.25 in. (23.5 cm). Other dimensions will remain as shown.

since its availability tends to alter many of the current logistics and capabilities of multiport surface analytic systems. Ideally, of course, it would be desired to transfer targets between any two UHV systems anywhere, at zero pressure, in zero time. This ideal has been kept in mind in our design and Sec. IV below describes the degree to which these objectives have been achieved. Also ideally, it is visualized that a sample and/or an UHV system which is a component in the transfer network can stay under UHV forever. In reality only the necessity to expose such a component to higher pressures for tactical reasons negates this objective.

III. APPLICATIONS OF THE UHVTD

A. Accessory to a multiport system

The use of the UHVTD would require: (a) one dedicated port ($2^3/_4$ in. in our design) with (b) an UHV gate valve and (c) suitable mechanical arrangements for landing and takeoff of the transferred sample, (d) one port with a window for visual observation of the transfer process, and (e) some means present locally for pumping the intermediate chamber between the UHVTD gate valve and the gate valve of the UHV system after coupling of the two systems (Fig. 1). This pumping arrangement must be capable of reaching UHV, but its design is flexible and the local pumps of the multiport system could be adapted to this task. Alternatively a mobile UHV pumping system dedicated to this task throughout a laboratory could be used, as in our laboratory.

As an accessory to a multiport system the UHVTD would extend its capabilities in ways described in more detail below. The UHVTD can be thought of as increasing the number of available ports in a multiport system. It is quite practical to think of retrofitting the UHVTD to existing multiport systems,[3] both commercial and home built.

B. Central element in a network of compatible UHV systems

If the availability of the UHVTD is assumed before a network of surface analytic systems is designed then the design considerations of these systems are influenced. The same requirements as noted in Sec. IIIA (a)–(e) above are present but the retrofitting requirement is lifted. Since in many respects the UHVTD merely increases the effective number of ports in a multiport system, individual systems can be designed to be smaller, simpler, and less expensive. They become easier to service individually and new systems can be added to the network in building block fashion without disturbing those already in service. This is the route we have followed in our laboratory where we first built an ion bombardment system[8] and the UHVTD itself,[1] and we modified an existing UHV system to accept transfers as a test bed. Three gate valves were built initially for test and development purposes. Essentially the results given below will be for this initial phase of two systems accepting transfers, the UHVTD itself and the first three gate valves which have been operational for about two years. We are currently well along in a second phase of building an UHV system with Auger–RHEED capability, an UHV system with UPS capability, a system for permitting entry of a sample from atmosphere to the transfer network, and three more gate valves. Two Canadian universities, Laval in Quebec City and McMaster in Hamilton, Ontario have taken steps which envisage target exchanges between themselves and our laboratory, although their operations should be classed as retrofits and are at an early stage.

The simplification of individual multiport systems is of value in a laboratory with several experiments. All simplified systems can be operated simultaneously. Redundancy in equipment can be reduced, the whole transfer network implying considerable sharing of available resources. The cost of each apparatus (i.e., the cost per experimenter) is reduced. The space requirement per experimenter is reduced. Systems can be designed with standardized overall dimensions permitting the use of standard baking ovens, etc.

Transfers do introduce some technical complexity and risk, although in two years we have had only one serious failure of vacuum leading to a rebake of the UHVTD. This was caused by careless operating procedures. It is clear however that if the UHVTD is to be used as a routine instrument then it must present no greater a risk than other instruments in daily use in the laboratory.

Problems of scheduling transfers have not to date presented any difficulties although we have not yet reached the stage of full tactical use of the UHVTD. At present we estimate one UHVTD would service about six UHV systems in daily tactical use.

The UHVTD represents of course, a physical invasion of an experimenter's privacy and in suggesting its use we have encountered from time to time a reaction which might be termed an example of the scientific territorial imperative. This is a nontrivial barrier, whose magnitude we cannot yet assess, to the widespread use of the UHVTD.

C. Transfer between distant systems

Apart from the physical and administrative problems of transporting the UHVTD in an automobile or in an aircraft,[2] transfers between distant systems present no qualitatively new problems in addition to those found in local transfers. In general, it will not be desirable to transport the UHV pumping station for the intermediate chamber over great distances and the local equivalent of this pumping unit will be required.

One of the reasons for long-distance transfers will simply be a desire to examine the same sample in two widely separated laboratories for any of the number of reasons. Some of the more probable of the reasons are specified in more detail below.

D. Exchange of surface standards and intercomparison of instruments

This procedure is already taking place in the Auger–ESCA round robin exchange of samples coordinated by Powell et al.[9] of the U.S. National Bureau of Standards. The exchanges are being made for the intercomparison of instruments using nominally reproducible surfaces. The application of the UHVTD to such an exchange[3] is almost self-evident. The UHVTD would replace the U.S. Postal Service as the transfer mechanism. Much tighter control of surface conditions would be an immediate result, with complete elimination of surface processing at the receiving site being possible. Under UHV storage standard samples could be maintained indefinitely with virtually no surface contamination. Standards of all sorts may be visualized: near perfect crystals, crystals with measured amounts of impurities, surfaces with standard degrees of roughness, etc.

E. The transfer of a sample from a unique facility to a surface analytic facility

Several topical examples of this application come to mind.

(a) The preparation of a sample in an accelerator or a neutron generator and its subsequent transfer to a surface analytic facility. This type of application is already underway in connection with the first-wall problem of a fusion reactor.[10] The sample is exposed in the facility and the permanent damage is examined in the analytic facility.

(b) The transfer of a sample from a highly dedicated specialized measuring instrument to a more routine surface analytic facility for purposes of surface preparation and/or characterization. A single professor with several graduate students would be able to perform first-rank fundamental studies in surface science by building up a dedicated specialized instrument without the need of investing in the more routine surface analytic tools, which would be supplied by a central facility. Transfer could be made at the beginning and end of the main experiment to establish that surface conditions had not changed.

(c) The preparation or exposure of a sample in the space shuttle followed by a transfer to earth for examination.[11] This implies that the surface characteristics acquired in space are stable under UHV, a situation which will generally be true for crystallization of metallurgical samples under zero gravity.

(d) The preparation or manufacture of a sample in an apparatus such as an ion doping or molecular-beam-epitaxy facility followed by transfer to a surface analytic facility.[12]

F. As a tool under ultrahigh vacuum

The present transfer mechanism is simply the equivalent of a pair of pliers which can be inserted into UHV systems without destroying the vacuum. It could be (and has been) used as a welding electrode, it could be (and has been) used as a device for changing the azimuth of the sample in incremented steps, it could be (but has not been) modified to act as a screw driver or as a wrench, it could be (but has not been) used as a diagnostic aid in establishing the location of an open connection, etc.

IV. PERFORMANCE OF UHVTD

A. General description

Figure 2 shows the UHVTD sitting atop the mobile laboratory pumping station developed specially for its use. Figure 3 shows a line drawing of this pumping station. The lower part is a conventional mechanical pump plus ion pump system which can pump the intermediate chamber of the UHVTD on top. The assumption of the design was that a conventional unit equivalent to this would exist in every laboratory to which a transfer was contemplated. Alternatively the intermediate chamber can be pumped by the portable UHV section. This section was made portable in the event that transfers were being made between moderately distant points such as between two adjacent buildings, where one was not equipped with a suitable UHV pumping station. The portable UHV section is readily carried by two people including all electronics, but does presently require ac power for operation. The UHVTD itself sits atop the pumping station and rolls freely on three ball bearings on the smooth aluminum table top. When in the desired position it is clamped. All systems de-

341

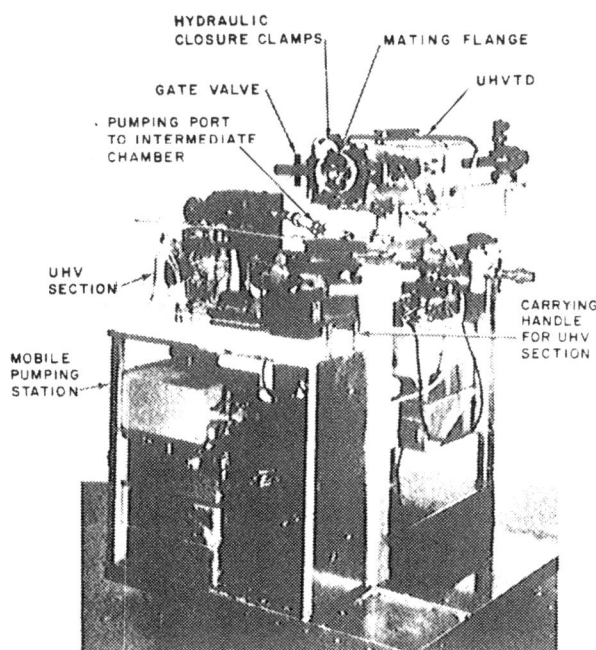

FIG. 2. Photograph of UHVTD sitting atop the mobile laboratory pumping station. The pumping port to the intermediate chamber couples to the vertical pumping stem of Fig. 1. The mating flange of the UHVTD is coupled to an identical mating flange on the surface analytic system as shown in Fig. 1. The UHV pumping section together with all its electronics can be easily decoupled and carried manually by two people with the aid of carrying handles, one of which is shown. The UHVTD itself rolls freely on the table top and is clamped when in position. It is fully demountable and is shown being carried in Fig. 4.

signed specially for the UHVTD have a transfer axis 44 in. from the floor, but this is not a hard constraint. Our test system, which is a retrofit system, has a transfer axis 53 in. from the floor and has given no difficulty because of the portability of the UHVTD. The portability of the UHVTD for internal use is shown in Fig. 4 (it can be powered entirely by six flashlight batteries for a period of 10 days); and its portability for long-distance travel[2] is shown in Fig. 5. The weight of the package in Fig. 5 is 76 lb, a formidable but not unmanageable load. In internal use the pumping station is wheeled up to a system and locked in position. The miniflange pumping connection (see Fig. 2) is first made and then the UHVTD is rolled into position to capture a gold wire O-ring, and the mating flange bolts are tightened up. The positioning of this O-ring, has turned out to be one of the weaker points of the design, but a leak-tight joint is quickly tested by the conventional pumping station with no damage to the vacua of any of the UHV systems. At this point the pumpdown of the intermediate chamber can commence once a band heater has been installed on the mating flange.

B. Overall vacuum performance

Background pressures achieved in the UHVTD on its internal pumps have been as low as 1.5×10^{-11} Torr. It is pumped by a small portable magnetron pump containing an evaporable getter of titanium[13] as well as by a separate titanium getter. Internal motions and external blows produce pulses of gas which are measured to be 95% methane (CH_4),

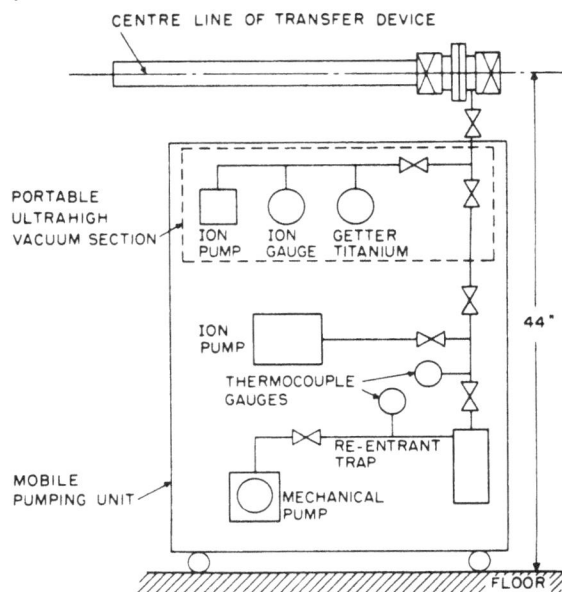

FIG. 3. Schematic drawing of mobile laboratory pumping station whose photograph is shown in Fig. 2.

a gas not readily adsorbed on many materials at room temperature. These pulses are pumped down with a time constant of a few seconds. It is practical to maintain with care the peaks of these pulses below 10^{-9} Torr at all times. Pressures in the sealed-off UHVTD are estimated from the current in the magnetron pump which has a sensitivity of about 1 A Torr^{-1}. Following evaporation of titanium in the magnetron pump, necessary perhaps once a month, it is often required to use a Tesla coil to raise the leakage resistance of the pump to a value of $\sim 10^{15}$ Ω needed for current measurements at the lowest pressures.

Background pressures achieved in the ion bombardment system have been particularly low at 5×10^{-12} Torr,[8] while those in the test system have been typically 2×10^{-10} Torr dominated by hydrogen H_2. Detailed pressures during a transfer are given in Ref. 1. Figure 6 gives additional details about the steps used in pumping the intermediate chamber to UHV. The results shown are for an early simulation of the intermediate chamber, but current practice with the intermediate chamber of Fig. 1 is very similar. Volume of the intermediate chamber is about 0.15 l. After coupling of the mating flanges the pumping valve ($3/4$ in.) is opened ($t = 2$ min) and the chamber pumped with the mechanical pump for three minutes. During this period a heater wrapped circumferentially around the two mating flanges has been turned on to provide a "minibake" lasting about 15 minutes and reaching 200°C. Heater tapes on the pumping line to the uhv pumping station are activated to achieve similar conditions. Between $t = 5$ min and $t = 21$ min the intermediate chamber is pumped by vacion and molecular sieve pumps (the latter were not used in the final design). At $t = 21$ min these pumps are cut out, the UHV pumps are introduced and a titanium getter in the intermediate chamber itself is flashed repeatedly (20 s ON, 40 s OFF). At $t = 43$ min the pressure drops below 10^{-9} Torr. Cooling the intermediate chamber with liquid N_2 vapor causes a further pressure drop below 10^{-10} Torr at $t =$

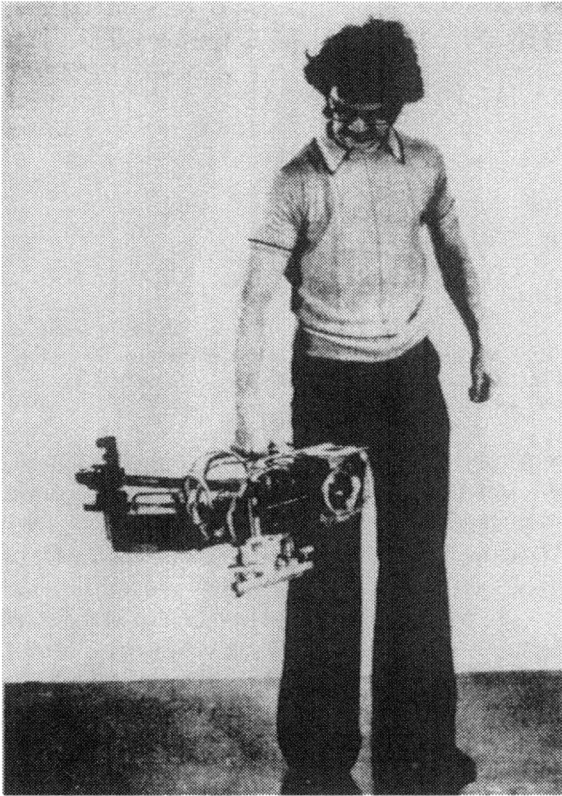

FIG. 4. UHVTD being carried manually internally in the laboratory.

FIG. 5. UHVTD being carried manually in its case for long-distance travel.

51 min. In the UHVTD setup of Fig. 1, gate valves 1 and 2 can now be opened for transfer of targets. In Fig. 6, closing the pumpout valve at t = 47 min caused the pressure to rise in the intermediate chamber pumped only by its internal titanium getter. The gas in this leakup was essentially all methane (CH_4). The actual configuration of the intermediate chamber is more complex than the simulated configuration but the

same pumping procedures generate essentially the same pressures as Fig. 6 in an elapsed time of 90 min rather than 60 min. The gate valves themselves have an axial length of $3\frac{1}{4}$ in. and some thermal breakthrough is experienced during the minibake. To suppress this, liquid N_2 vapor is directed at the downstream sides of the gate valves during the minibake.

Complete target transfers currently take 3–4 h, and a clean

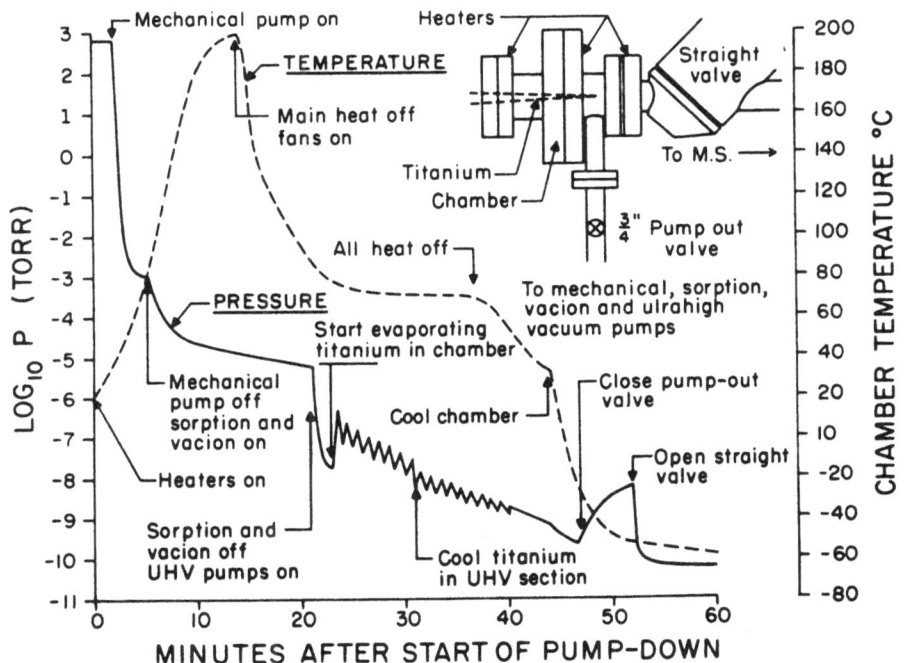

FIG. 6. Pump-down results of an early simulation of the intermediate chamber. These results are representative of current performance, except that lapsed time for a pump-down is typically 90 min rather than 60 min.

343

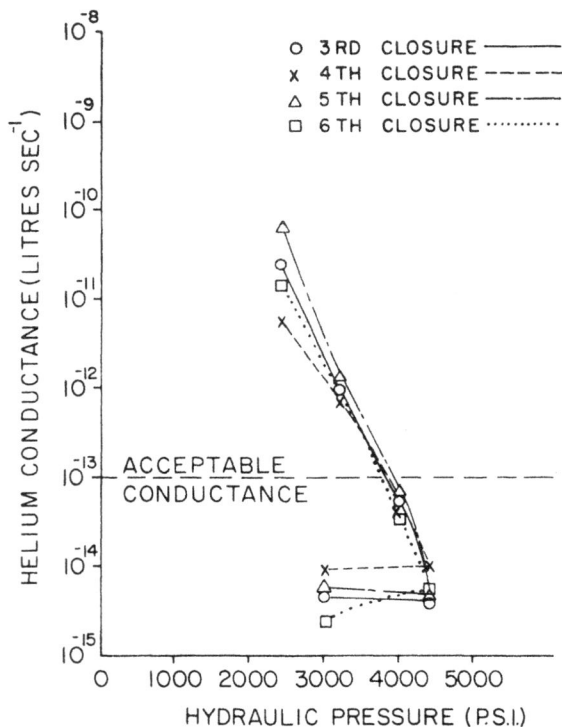

FIG. 7. Conductance vs hydraulic pressure for the gate valve operating correctly. Note that 3000 psi is insufficient to close the valve to an acceptable conductance, but is more than adequate to keep it closed after a short period (~1 min) at a higher hydraulic pressure.

target of W acquires a monolayer or less of hydrogen during transfer.[1]

C. Gate valves

The gate valves are constructed entirely of stainless steel 304 and are bakable in the open position to at least 400°C. Tests at temperatures higher than this have not been carried out. The mating flanges of the gate valves are $3\frac{3}{8}$ in. in diameter. The basic axial length of the gate valves is $2\frac{1}{2}$ in., but this requires a simple but special $2\frac{3}{4}$ in. flange at the downstream end. An adaptor of length $1\frac{3}{4}$ in. converts to a standard $2\frac{3}{4}$ in. conflat as shown in Fig. 1. When open, the internal free hole has a diameter of $1\frac{3}{8}$ in. or 35 mm. When closed a gold plated nose cone (Granville–Phillips $1\frac{1}{2}$ in. Auroseal nose piece) is squeezed against a stainless-steel seat by three mechanical clamps having two pressure points each. Force is applied hydraulically through oil up to a pressure of 5000 psi. The mechanical linkage is relatively complex but it is estimated that the total force in pounds squeezing the valve in the closed position is 1.2 times the hydraulic pressure giving a closing force in the range of 1000 pounds per linear inch of seal. To open the valve the force is released and the nose cone displaced manually in a radial direction leaving the full aperture free. Hydraulic pressure is applied conveniently through a $\frac{3}{4}$ in. nut on a master cylinder, which is carried with the UHVTD or attached to an UHV surface analytic system. If a gate valve is left closed for a long period the hydraulic pressure needs "topping up" about once a month. Figure 7 shows the conductance of a properly working gate valve as a function of applied hydraulic pressure. Detailed repro-

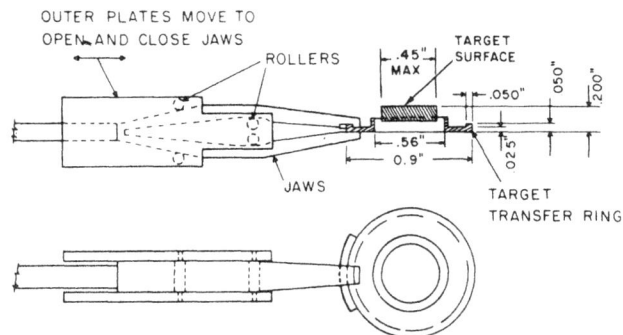

FIG. 8. Detail of target transfer ring. The items physically transferred are shown cross-hatched. Only the outer portion of the ring where it is gripped by the transfer jaws is the same for all targets. The mounting of the target on its ring is arbitrary within limits.

ducibility is not found. Note that there is a gross hysteresis effect which is useful. As the valve closes a conductance of 10^{-12} l s^{-1} is obtained at 3000 psi but a conductance of $<10^{-14}$ l s^{-1} is obtained after the valve has been taken to 4500 psi for a short period (one minute is sufficient) and returned to 3000 psi. We use 3000 psi arrived at in this way as the standard hydraulic pressure for closure. Conductances have been stable under these conditions for several months. Random measurements indicate that hydraulic pressures as low as 1500 psi are sufficient for closure provided the valve had been squeezed first. On one of our valves we have about 100 closures with no sign of deterioration but our statistics on this point are not extensive as yet. Valves are tested with helium and are accepted when an atmosphere of helium can be applied upstream with a closed conductance of $\leq 10^{-13}$ l s^{-1}. It is found that such a valve will sustain a normal air atmosphere with no visible sign of a leak downstream at a background pressure of 10^{-11} Torr.

D. Overall mechanical considerations

For internal use dimensions A and B (Fig. 1) of 39.5 and 4.25 in., respectively, are adequate and these are the current dimensions the UHVTD. However for more general applicability in retrofit situations B has been raised to 9.25 in. causing A to be increased to 49.75 in. These are the dimensions to which the second generation UHVTD's are being made.[14] The transfer arm carrying the target ring (see below) can move a relatively long distance in the axial z direction by means of a long bellows. There is 360° rotation about the z direction permitting the transfer ring to be "landed" on a platform in any orientation provided the plane of the platform contains the transfer axis. There is limited motion in the x, y directions perpendicular to the axial direction permitting the target center to be moved over a square of side about $\frac{7}{8}$ in. (or 2.2 cm), when in the fully inserted position. While all transfers to date have been made along a horizontal axis the UHVTD has been designed to permit transfers along an axis of any orientation. It has been established[2] that the UHVTD will operate in any orientation.

E. The transfer ring

In our design the sample to be transferred is mounted on a standard transfer ring. The transfer arm grips this transfer

344

ring which is the physical object actually transferred. Detailed dimensions are given in Fig. 8. These dimensions were chosen as having universal utility and it is hoped that those contemplating UHV transfers will be able to adopt them so that universal transferability of samples will be established. A bank of characterized samples has begun to be assembled mounted on these standard rings, which have azimuthal markings capable of being observed from outside the system permitting the target azimuth to be measured at any time. Eventually it is planned to store this bank of samples under UHV indefinitely. In our design the rings are made of molybdenum so that the targets can be heated to elevated temperatures (up to 2700 K), but the material of the ring could be something else if these high temperatures were not demanded. Any platforms in the UHV system designed to accept the target ring are satisfactory. All motions and operations after the target has been "landed" are supplied by the surface analytic system, the jaws being withdrawn and used elsewhere. At present our target rings are held in place by three tungsten springs, which roll over the lip of the ring. The hole in the middle of the target ring serves as a centering device on the platform and allows the target to be heated by electron bombardment on the back. All "landing" and "take-off" operations are presently done visually requiring a window in the UHVTD and in the analytic system. In the UHVTD there are at present two storage platforms and a third target can be carried in the jaws to empty systems. A design is being developed with 15 target storage positions in the UHVTD. Transfers require concentration but have become routinely successful.

V. CONCLUSIONS

An UHV technique has been described for intervacuum sample transfer which permits samples to remain under UHV at all times. The results to date indicate the practical possibility of a network of UHV systems (analytical or manufacturing) anywhere on earth and perhaps even in space, among which samples could be transferred. Such a network would broaden the capabilities of existing multiport systems and would permit the construction of simpler multiport systems. The exchange of surface standards and intercomparison of instruments, along with the storage of surface standards all under UHV indefinitely, would become possible. The use of highly specialized facilities in conjunction with more routine surface analytic facilities could readily be envisaged.

ACKNOWLEDGMENT

The authors gratefully acknowledge the aid of many of their colleagues, both professional and technical, who contributed their skills to this project.

[1]J. P. Hobson and E. V. Kornelsen. Proceedings of the 7th International Vacuum Congress and 3rd International Conference on Solid Surfaces, Vienna 1977, Vol. 3, p. 2663.
[2]J. P. Hobson, J. Vac. Sci. Technol. 15, 1609 (1978).
[3]J. P. Hobson and E. V. Kornelsen, Proceedings of Pittsburgh Conference on Analytic Chemistry and Applied Spectroscopy, Cleveland, March 1978. (in press).
[4]H. W. Breedlove, Rev. Sci. Instrum. 41, 1537 (1971).
[5]V. W. Steward and C. H. A. Syms, J. Sci. Instrum. E 6, 14 (1973).
[6]C. A. Crider, G. Cisneros, P. Mark and J. D. Levine, J. Vac. Sci. Technol. 13, 1202 (1976).
[7]J. P. Hobson, J. Vac. Sci. Technol. 14, 1279 (1977).
[8]E. V. Kornelsen and D. L. Blair, J. Vac. Sci. Technol. 14, 1299 (1977).
[9]C. J. Powell, N. E. Erickson, and T. E. Madey, Proceedings of Pittsburg Conference on Analytic Chemistry and Applied Spectroscopy, Cleveland, March 1978. (in press).
[10]J. E. Robinson, D. A. Thompson, K. K. Wittann, S. W. Poehlman, Plasma Surface Interactions Conference, Culham, England. 1978 (unpublished).
[11]J. W. Patten, J. Vac. Sci. Technol. 14, 1289, 1977.
[12]J. W. Robinson and M. Ilegems, Rev. Sci. Instrum. 49, 205 (1978).
[13]P. A. Redhead, J. P. Hobson and E. V. Kornelsen, The Physical Basis of Ultrahigh Vacuum (Chapman and Hall, London, 1968), p. 409.
[14]The manufacture of the UHVTD has been licensed to Canadian Vacuum Equipment Ltd., 2755 De Miniac, St. Laurent, Montreal, Canada, H4S 1E5.

Versatile UHV sample transfer system

R. E. Clausing, L. Heatherly, and L. C. Emerson

Metals and Ceramics Division, Oak Ridge National Laboratory, Oak Ridge, Tennessee 37830

(Received 6 October 1978; accepted 29 January 1979)

A vacuum transfer system has been developed that allows samples to be inserted from air into an Auger analyzer. Following an initial analysis they may then be moved to and from a separate vacuum chamber for other studies. The system is constructed of standard UHV components and can be fabricated to cover any desired transfer distance. All-metal sealed valves isolate the various chambers so that the vacuum integrity of each is maintained. The sample size is limited only by the smallest constriction within the components, which is 2 cm in the system described. The transfer is rapid and is capable of being controlled remotely for use in a hostile environment.

PACS numbers: 07.30.Hd, 82.80.Pv, 81.70. + r

I. INTRODUCTION

Surface analysis techniques such as Auger Electron Spectroscopy (AES) are very powerful techniques but are often frustrated in application because it is difficult or impossible to transfer a sample with a potentially interesting surface condition to the analysis chamber, without contaminating it to the point where no useful information can be obtained. In order to preserve the surfaces for analysis it is usually necessary to handle them quickly under UHV conditions. We have developed a system for this purpose which is versatile and relatively inexpensive.

This paper describes a UHV transfer system developed for use in tokamaks which are devices used in fusion energy research. It is necessary to transport samples used to study plasma–wall interactions from the plasma experiment to a special chamber where AES and other surface analytical techniques are available. Sample surfaces are studied to determine what changes are caused by exposure to the high temperature plasmas in the fusion device. Although the equipment was developed for this specific use, the technique may find other applications where it is desirable to transfer samples from one environment to another under UHV conditions through valves and/or over long distances. The apparatus can also be used as a simple vacuum lock to introduce samples from air or controlled atmospheric environment into a UHV system. Typical applications might include the transfer of samples from corrosive or harsh environments to a surface analytical facility where AES, SIMS, ESCA, ISS, or other surface analysis techniques are available. Samples might also be transferred from a manufacutring facility or surface treatment facility to an analysis chamber without exposure to air.

The reader may be interested in other types of UHV vacuum sample locks described in the literature or used commercially.[1-5] Each of these was judged unsuitable for our purpose because it failed to meet one or more of the design criteria listed below.

II. DESIGN CRITERIA

The equipment to be described below was designed to meet the following criteria:

(1) Samples must be transferred over distances of 1–5 m from the fusion experiment to the surface analysis station without altering their surface compositions significantly. This dictated the use of UHV techniques. Base pressures in the 10^{-8} Pa range are desirable.

(2) The transfer device should also permit transfer through valves and permit closing the valves with the sample on either side of them so as to allow isolation of the sample in any one of several areas of the system.

(3) The transfer system must be compatible with operation in a tokamak environment which means it must be able to resist vibration and baking to 250°–400°C; it should not include magnetic materials.

(4) It is desirable to be able to transfer the sample into separate chambers for sample preparation, analysis or storage and to position the sample accurately inside the tokamak for exposure to the plasma environment.

(5) The system should provide for the transfer of the sample to a remote location under UHV in a detachable module.

(6) It must be possible to insert or remove samples from the system without exposing the analysis chamber, sample preparation region or tokamak to air.

(7) Commercial components were to be used in so far as possible for economy and convenience.

III. DESIGN CONCEPT

The most difficult of the above criteria are 1 and 2, i.e., transfer over long distances and through valves with the option of closing any valve with the sample on either side. These criteria were met through the design of a carriage (sample carrier) which rides on a simple track from one region of the system to another. The carrier or carriage is able to span gaps in the track to permit the insertion of suitable gate valves. The carriage receives its mobility through the use of drive stations mounted on either side of the valves. The carriage may be moved by various driving mechanisms including belts, wheels or gears. With the last arrangement, the sample may be moved any required distance if enough drive stations are provided. Figure 1 shows a schematic arrangement for such

ORNL/DWG/FED 78-956R

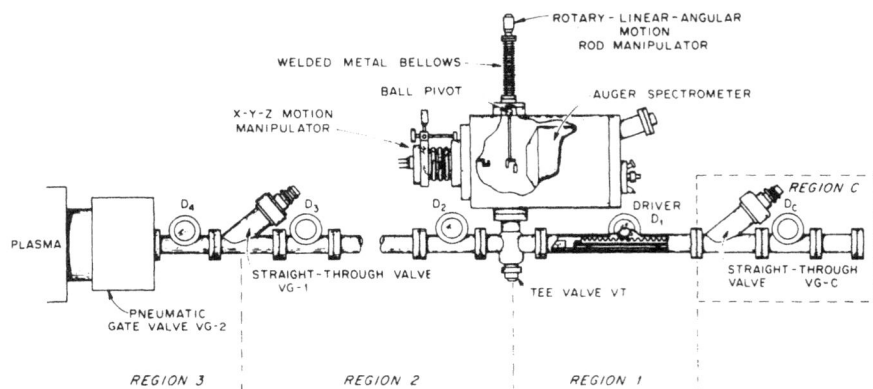

FIG. 1. Schematic drawing of the sample transfer system designed for research related to plasma–wall interactions in fusion energy. With the addition of a pump not shown here, region C, which is detachable, becomes a module suitable for transfering samples to remote locations under UHV conditions. Parts D_1, D_2, D_3, D_4 and D_c are carriage drive stations; VG-1 and VG-C are manually operated straight through valves, and VG-2 is a pneumatic gate valve.

a transfer system. Any required number of work areas may be included.

If suitable means are provided, the sample (or samples) may be removed from the carriage in any location for individualized preparation or examination. In our system, this is accomplished through the use of a rod manipulator. The rod passes through a ball pivot and may be pivoted in any direction, moved back and forth through the ball, or rotated about its axis. The end of the rod is fitted with a screw thread which may be screwed into a suitable threaded hole in the sample holder and the sample and holder removed from the sample carriage. The sample may remain on the rod manipulator for examination or other operations, or it may be placed in a work station on a carousel or a suitable precision manipulator, the screw thread disengaged, and the sample examined by any suitable surface analysis technique.

IV. APPLICATION

Figures 1, 2, and 3 illustrate this concept applied to our specific problem. In these illustrations the sample is shown on a sample holder, which may be placed on end of a carriage, shown in region 1 of the system. The carriage may be moved by the drive station D_1 through an open valve VT into region 2 and, if valve VG-1 and VG-2 are open, into region 3 where driver D_4 may be used to maneuver the carriage back and forth and to position the sample carefully at the edge of the plasma. Valve VG-1 may be closed during exposure of the sample to the plasma. After the sample is exposed to the plasma, valve VG-1 may be opened, the sample moved back through region 2 into region 1 and picked up by the rod manipulator and placed on the precision XYZ manipulator in the sample analysis area. The rod manipulator reaches into the tee-valve to engage the sample holder. The tee-valve VT may be closed as soon as the carriage is moved slightly from the transfer position. While the sample is isolated in the analysis chamber, new samples may be introduced into region 1 by opening that region to air, removing the carriage, and placing a new sample on the carriage.

Samples may also be removed to a remote location for examination without exposure to air if the assembly enclosed in dotted lines is available. In this event, the carriage with the sample in position is moved into region C and the valve closed. The carrier then, with its own separate vacuum pump, may

be removed at the connecting flange. The detached system C may be then carried to any desired location, reattached to a suitable port on another vacuum system and the sample moved into the new system for further operations.

Figure 2 shows a cross section of the transfer tube. The carriage rides on ball bearings which are guided by the channel–track assembly. The carriage is driven by a pinion gear mounted on the shaft of a rotary feedthrough.

Figure 3 shows an enlarged view of the carriage, the sample holder, and the carriage drive station. The sample is mounted on a removable holder which fits into a pocket on the front of the carriage. The sample and holder are easily removable from the pocket in front of the carriage and transferred to a pocket on the precision XYZ manipulator shown in Fig. 1 by means of the rod manipulator.

Most of the components shown in Figs. 1 and 2 are easily recognizable as standard commercial ultrahigh vacuum equipment designed for use with one and one-half inch OD tubing. The gears, racks, and bearings are also standard commercial items. Sizes may be chosen to suit the particular needs of the apparatus.

V. COMMENTS ON PERFORMANCE

The system shown schematically in Fig. 1 has been developed during the past several years. It is possible to obtain vacuums in the 10^{-8} Pa range with suitable pumping and bakeout practice. We have discovered, however, that for many purposes sample transfer at 10^{-6} Pa is satisfactory if accomplished in a reasonably short time (the major residual

ORNL/DWG/FED 78-957

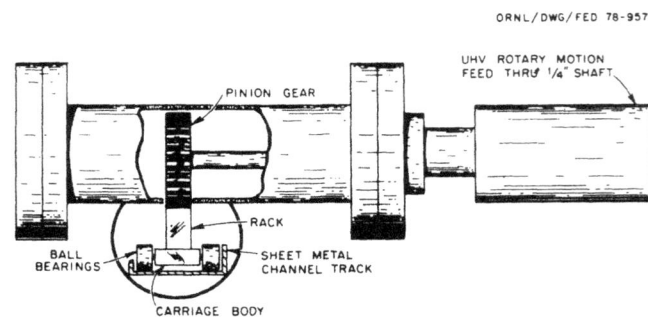

FIG. 2. Cross section of the sample transfer system showing the pinion gear drive.

ORNL/DWG/FED 78-958

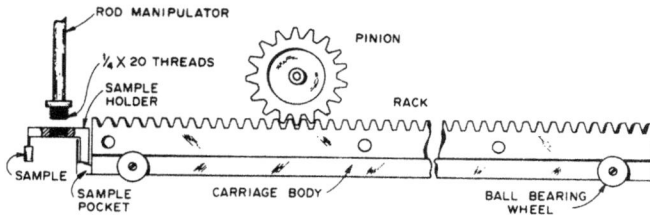

FIG. 3. Diagram of the sample carriage, sample holder, pinion gear drive, and rod manipulator showing the functional relationships.

gas in our system is usually hydrogen). Some degree of mechanical aptitude is required to learn to use the rod manipulator since in our system a mirror must be used to see the sample holder when it is in the tee valve. This has not been enough of a problem to warrant adding a view port in the side of the valve.

The sample holder is usually held on the carriage with the help of a spring clip and detent. For our purposes, it is sometimes necessary to design complex samples and sample holders involving mechanical motion, electrical connections, and special materials. Such variations are possible and in our case have been both practical and very useful.

The rod manipulator provides a wide variety of ways to manipulate the samples. They can be rotated, bent, fractured, rubbed, pushed, scratched, moved close to a window for a closer look at color or some strange artifact, attached to electrical plugs, or placed on a stage or in a storage area.

We have been very pleased with the reliability, flexibility and overall performance of this system which has operated successfully on several tokamaks, ISX-A in Oak Ridge, T-12 at the Kurchatov Institute in Moscow and Doublet III at General Atomic in San Diego.

ACKNOWLEDGMENTS

Research sponsored by the Office of Fusion Energy, U.S. Department of Energy, under contract W-7405-eng-26 with the Union Carbide Corporation.

[1]Von H. D. Polaschegg and E. Schirk, Vak. Tech. **24**, 136 (1975).
[2]B. S. Prahallada Rao and S. R. Halbe, J. Phys. E **9**, 205 (1976).
[3]G. Tauber and H. Viefhaus, Rev. Sci. Instrum. **47**, 772 (1976).
[4]C. A. Crider, G. Cisneros, P. Mark, and J. D. Levine, J. Vac. Sci. Technol. **13**, 1202 (1976).
[5]J. W. Robinson and M. Ilegems, Rev. Sci. Instrum. **49**, 205 (1978).

Piezodriven spindle for a specimen holder in the vacuum chamber of a scanning electron microscope

Takeshi Hatsuzawa, Yoshihisa Tanimura, Hirofumi Yamada, and Kouji Toyoda

National Research Laboratory of Metrology, 1-1-4 Umezono, Sakura-mura, Tsukuba 305, Japan

(Received 23 May 1986; accepted for publication 8 August 1986)

A rotating specimen holder device for use in the vacuum chamber of a SEM (scanning electron microscope) is developed by using a piezodriven mechanism. The mechanism consists of piezoelectric elements, clamps, and actuators that work as feeders to rotate the spindle. It is specially designed to eliminate the magnetic field influence of standard motors, and the holder can be placed just under the electron probe beam. In this article, the basic ideas, design, and instrumentation techniques are described. Then a servosystem is constructed for positioning the spindle. The angular resolution is 0.9° and the maximum rotation speed is 9.4 s/rotation.

INTRODUCTION

The holder device described in this article was developed for use as a positioning mechanism in the vacuum chamber of a SEM to observe profiles of LSI patterns. The simplest procedure would be to use a standard motor for vacuum use. However, because a motor's magnetic field affects the probe beam, it was necessary to develop a special mechanism that works in a SEM chamber without creating a magnetic field.

Some applications of the inchworm mechanism for STM (scanning tunneling microscope) actuators[1,2] and semiconductor inspection systems[3,4] are reported to move specimen tables precisely. These mechanisms, however, move basically in a horizontal direction, while our holder device rotates in the vertical plane. Our device is also an application of an inchworm system, but it rotates a spindle with a small flat stage for the specimen. The rotation is caused by piezoelements giving alternate clamping and feeding motions to the spindle. A positioning system is constructed by using a digital servocontrol with an angular resolution of 0.9°.

I. BASIC IDEAS

A sketch of the spindle is shown in Fig. 1. The spindle is held by bearings at both ends and freely turns around the axis. Two mechanical components, the clamp and the actuator, make contact with the spindle. The clamp moves up and down to stop or permit rotation of the spindle. The actuator has a feeding motion of swinging back and forth as well as a clamping motion.

When the components are moved according to the timing chart, as shown in Fig. 2, it starts rotating. The driving cycle consists of six phases. In phases 1 and 2, the actuator moves up and forward while the clamp remains disengaged, and the spindle is forced to rotate by the frictional force between the spindle and the actuator. In phase 3, the clamp moves up and remains engaged. Next, the actuator moves down and returns to the home position during phases 4 and

5. Last, in phase 6, the actuator moves up and the clamp disengages to prepare for the next feeding.

The direction of the rotation is easily changed by reversing the excitation timing of the actuator. A smooth and continuous rotation can be obtained by repeating this cycle rapidly, while a minute angle of rotation can also be produced by a fine feeding. This mechanism is useful for giving both coarse and fine rotational movements without using any reduction gears. These movements are similar to the mechanism developed by Binnig et al.[5]

II. DESIGN AND INSTRUMENTATION

Figure 3 is a diagram of the holder device. A spindle (6 mm diam) is supported between two bearings on a base plate having a diameter of 10 cm. The spindle is the most important component of the holder. A very fine runout is required for the spindle, or it will be touched by the actuator shoes even when the clamps are disengaged.

The actuator units are symmetrically placed on both sides of the spindle. The clamping and feeding actuators are laminate piezoelectric elements with dimensions of $5 \times 5 \times 9$ mm and $5 \times 5 \times 18$ mm, respectively. When 100 Vdc is supplied to them, the shorter of the two expands about 7 μm and

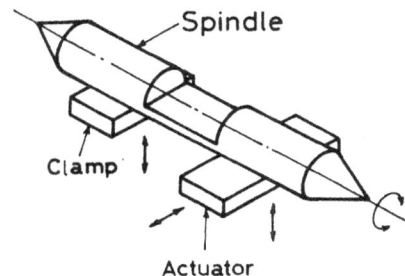

FIG. 1. Basic construction of the holder device.

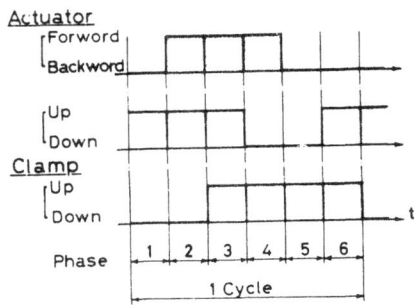

FIG.2. Timing chart to drive the spindle of Fig. 1.

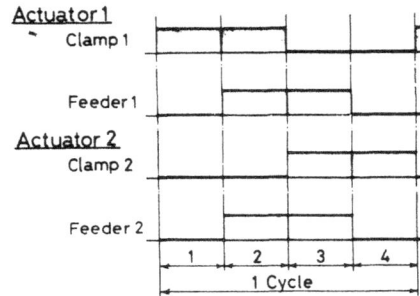

FIG. 4. Timing chart for the holder device equipped with two actuator units. Piezoactuators are excited during the hatched intervals.

the other about 15 μm. These values are too small to be used directly as a clamp or a feeder because the clearance between the clamp and the spindle is very narrow. Moreover, the small feeding length of the feeder results in very slow rotation speed. To reduce these difficulties, a complex lever mechanism is utilized. It also satisfies the height limitation due to the small size of the vacuum chamber.

The movement of the clamping piezo is amplified by the clamping lever with an adjusting screw and a leaf spring (0.4 mm thickness, stainless steel) as a fulcrum. This mechanism is also applied to the feeding piezo and the feeding lever. The magnification ratio of the clamp is approximately 2 and that of the feeder is about 4. By using these levers, the working distance of the clamp is increased to 20 μm and the feeder to 90 μm when used at 150 Vdc. These levers are made of brass plated with nickel, while the spindle is stainless steel. A small Teflon ball (2.5 mm diam) is placed between the adjusting screw and the clamping piezo to make the lever slide more easily.

The actuator units are driven according to the timing chart illustrated in Fig. 4. In this configuration, there are two of the clamp-actuator mechanisms shown in Fig. 1, and the sequence is composed of two pairs of the mechanisms. When actuator 1 is fed with clamp 1 on, actuator 2 is also fed but with clamp 2 off. Next, actuator 1 contracts with clamp 1 off, while actuator 2 contracts with clamp 2 on. Therefore, the feeding speed is twice as fast as the configuration shown in Fig. 2. The direction can be changed by reversing the timing of clamps 1 and 2.

A high-speed switching circuit was designed, as shown in Fig. 5. It has a powerful charging and discharging ability because the piezoactuators are regarded as capacitor elements. The circuit is driven by a TTL level input, while dc power is supplied by an external stabilized power source. The totem pole transistors, usually used for the horizontal deflection in a TV set, charge and discharge the piezoelement through a limiting register. This circuit gives good performance up to 150 Hz with a 2-μF capacitor that corresponds to the longer piezoelement.

The driving characteristic of the mechanism is shown in Fig. 6. The angular velocity increases as the frequency becomes higher. When the frequency reaches about 45 Hz, the velocity levels off due to the slow response of the complex lever mechanism and slippage of the actuator shoes. The average feeding length of the actuator per driving cycle is estimated to be about 40 μm. This is equal to an angular resolution of 7.6″/cycle.

The maximum angular velocity is 38.3°/s corresponding

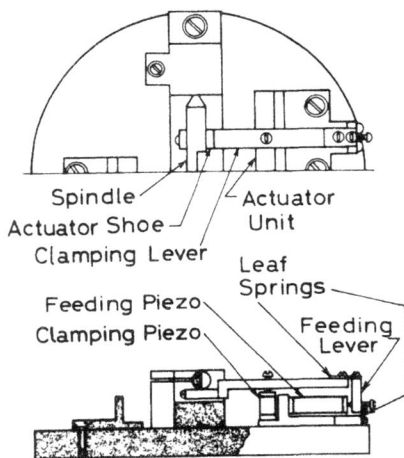

FIG. 3. Half-top view and side view of the holder device. The height is 27 mm and the diameter of the base plate is 10 cm.

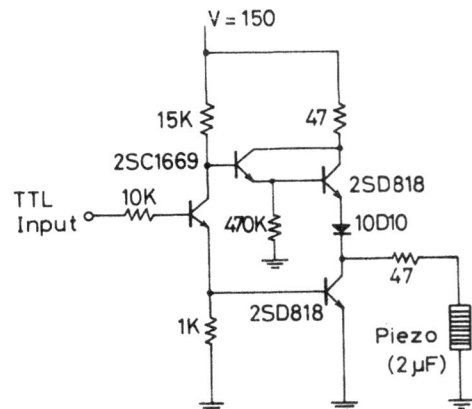

FIG. 5. High-speed switching circuit for a piezoelement.

Piezodriven spindle

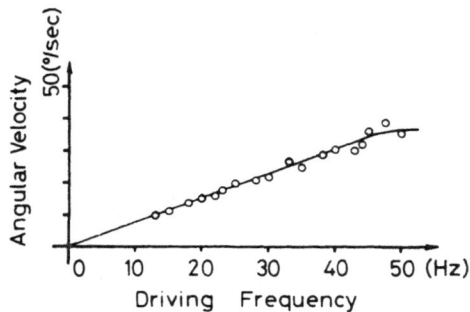

FIG. 6. Driving characteristic of the spindle.

to 9.4 s/rotation at a driving frequency of 48 Hz. This speed was sufficient for our purposes.

III. POSITIONING SYSTEM

Figure 7 shows the holder device with a rotary encoder for positioning. The encoder has an angular resolution of 1.8° (200 pulses/rotation) which is increased twofold by an electronic circuit. The digital servosystem is constructed with a microcomputer that is based on an 8086 CPU and an 8087 coprocessor. The construction of the system is illustrated in Fig. 8. The computer commands the oscillator to generate driving signals to the driver, and reads the spindle position from the rotary encoder through the counter loop. The system has an angular resolution of 0.9°. The oscillator has a basic frequency $f_0 = 750$ Hz and supplies the three signals shown in Fig. 4 to the driver. The frequency f is controlled by an 8-bit command d from the computer according to

$$f = f_0/d, \quad (d \neq 0), \tag{1}$$

through a frequency divider in the oscillator.

The performance of the system for proportional control is illustrated in Fig. 9. This result is obtained with an input command for 90° clockwise rotation with a 10-ms sampling

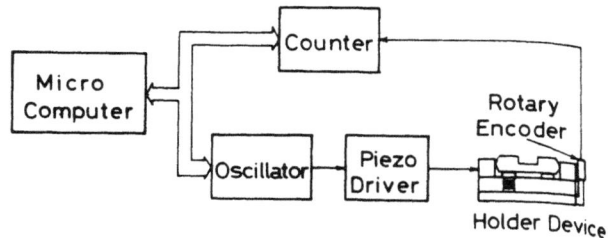

FIG. 8. Schematic construction of the digital servosystem.

time and 120 μs of overhead time for the control program. When the gain reaches about 100, the system becomes vibratory due to the limit cycle movement of the actuators. It also increases the settling time.

Figure 10 is a typical step response curve for a 90° command with a controller gain of 80. It settled within 2.69 s without overshoot. This performance is satisfactory for use as a holder for our present purposes.

IV. DISCUSSION

A piezodriven spindle for a specimen holder device in the vacuum chamber of a SEM is developed. It rotates in the vertical plane. The direction of rotation is changed by alternating the clamping timing in the driving sequence of the piezoactuators, and the speed of rotation is controlled by the driving frequency. The maximum speed is 9.4 s/rotation at 48 Hz. Average feeding length is about 40 μm, which corresponds to 7.6″/cycle in angular resolution. Using a digital servocontrol, a positioning system is constructed with an angular resolution of 0.9° under the condition of a sampling time of 10 ms and controller gain of 80.

Motion in a vacuum is being researched at our laboratory. At the same time, we are also studying how to cope with the outgassing of the device components that pollute the chamber environment. As soon as these problems are solved, we will combine this holder device with a rotary table rotat-

FIG. 7. Photograph of the holder device to be manufactured.

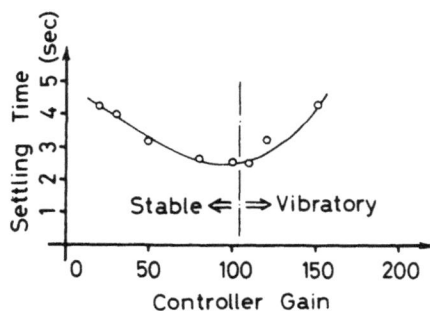

FIG. 9. Settling characteristic for proportional control with a sampling time of 10 ms.

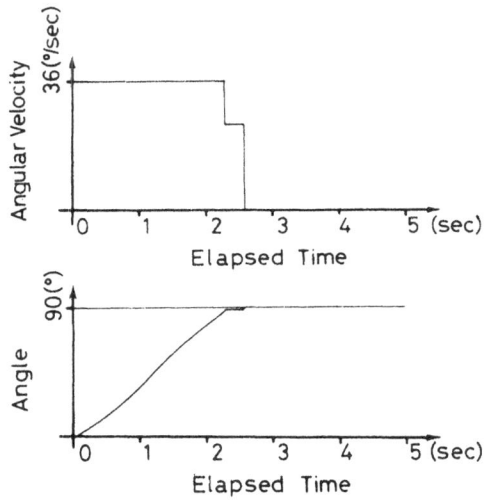

FIG. 10. Typical step response in the case of 90° clockwise rotation command and controller gain of 80.

ing in the horizontal plane, driven by an ultrasonic motor, in order to construct a two-axis positioning system.

ACKNOWLEDGMENTS

The authors wish to express their appreciation to S. J. Schvaneveldt, a visiting researcher from the Tokyo Institute of Technology, for assistance with the English translation, and Y. Kurita for his encouragement. We also wish to thank Tokyo Seimitsu Co., Ltd. for cooperating in the spindle manufacture.

G. Binnig and C. Gerber, IBM Tech. Disc. Bull. **23**, 3369 (1980).

[2]G. Binnig et al., U.S. Patent, Appl. no. 406653 (1986).

[3]K. Sugihara and T. Tojo, Proc. Jpn. Soc. Precision Eng. **85**, 305 (1985) (in Japanese).

[4]T. Tojo and K. Sugihara, Proc. Jpn. Soc. Precision Eng. **85**, 423 (1985) (in Japanese).

[5]G. Binnig and C. Gerber, IBM Tech. Disc. Bull. **22**, 2897 (1979).

A precision table for ultrahigh vacuum made of aluminum alloys[a]

H. Saeki and J. Ikeda

Matsushita Electric Industrial Co. Ltd., Matsuba-cho 2-7, Kadoma-shi, Osaka-fu, 571, Japan

H. Ishimaru

National Laboratory for High Energy Physics, Oho 1-1, Tsukuba-shi, Ibaraki-ken, 305, Japan

(Received 18 September 1987; accepted 12 April 1988)

Recently an ultrahigh vacuum environment has been necessary to reduce impurities in the production of high-density integrated circuits. In particular, tables permitting precise positioning with low outgassing rates have been required for thin-film processing equipment and surface analysis systems using spot beams. An X–Y–θ three-axis table made of aluminum alloys with a special surface treatment for ultrahigh vacuum environments has been developed. The stroke of the X and Y axes is 75 mm, and the rotational range of the θ axis is 360°. The X and Y axes are driven by built-in stepping motors, connected directly to aluminum alloy ballscrews. The stator of the motor is made using stainless-steel sheath cable. The screws and nuts of the ballscrews are coated with TiC, and the balls are coated with Ag. The characteristics of the table have been investigated in atmosphere before and after baking. The measured error motions for the X axis were pitch ± 2.3 μm/75 mm and yaw ± 1.4 μm/75 mm; and for the Y axis were pitch ± 0.78 μm/75 mm and yaw ± 0.22 μm/75 mm. The resolution and repeatability for the X and Y axes was 15 μm ± 2 μm/pulse, and for the θ axis was 1.8° ± 0.09°/pulse. The mechanical characteristics were maintained after baking. The table moved in an ultrahigh vacuum environment of the order of 10^{-11} Torr as it had in atmosphere.

I. INTRODUCTION

With recent technological advances in semiconductors, higher precision and reliability in thin-film processing equipment and surface analysis systems are required for the production of high-density integrated circuits. Therefore, ultrahigh vacuum fabrication environments are necessary to improve the quality of the films and to eliminate impurities. In particular, tables permitting precise positioning with low outgassing rates are required for thin-film processing equipment and surface analysis systems using electron beams or focused ion beams.

Regarding low-outgassing-rate materials, aluminum alloys with a special surface treatment have been reported.[1,2] These aluminum alloys are satisfactory except in environments where corrosive gases are present. Additional studies which seek improvements in corrosion resistance are presently being made.

This paper reports on a precision table made of aluminum alloys with built-in driving sources and control devices, which was designed to achieve greater accuracy with better maintenance capabilities, and a low outgassing rate.

II. APPARATUSES

This section describes the newly developed X–Y–θ three-axis precision table made of aluminum alloys, and an ultrahigh vacuum chamber made of aluminum alloys to test the table.

A. X–Y–θ three-axis precision table made of aluminum alloys

Figure 1 shows the newly developed X–Y–θ three-axis precision table made of aluminum alloys. Table I lists the specifications of the table. This table is equipped with a θ axis with a 360° rotational range, and is mounted on an orthogonally crossed X–Y table, with a stroke of 75 mm for each axis.

The dimensions of the table are 460 mm(L) × 310 mm(W) × 338 mm(H). The linear sliding guides for the Y-axis table consist of cross-ball-type guides made of quartz and 34 balls (17 balls for each guide) made of SUS440C, 9.525 mm in diameter. The guides are pressed against reference planes by springs, in order to prevent thermal distortion and cracking. The linear sliding guides for the X axis consist of cylindrical guides which are coated with CrN (thickness of ∼1.5 μm) and rollers which are coated with TiC (thickness of ∼1.5 μm). The guides and the rollers are made of aluminum alloys. All the rotational bearings were made of SUS440C, and the balls were coated with Ag (thickness of ∼0.3 μm). The main material for the table is A2219-T87 or T852 aluminum alloy with a special surface treatment,[1] machined to a surface roughness of < 1 μm.

The driving sources for the table are built-in stepping mo-

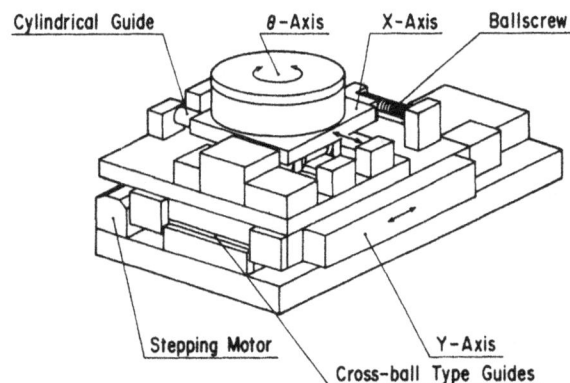

FIG. 1. X–Y–θ three-axis precision table.

TABLE I. The specifications of the table.

X – axis	Main material	A2219–T87 or T852 aluminum alloy
	Guide materials	A2219–T87 aluminum alloy coated by CrN (cylindrical guide) A2219–T87 aluminum alloy coated by TiC (roller)
	Stroke	75 mm (Max. 100 mm)
	Repeatability	± 2 μm (step size is 15 μm / pulse)
	Speed	Max. 6 mm/sec
Y – axis	Main material	A2219–T87 or T852 aluminum alloy
	Guide materials	Quartz (cross–ball type guide) SUS440C (ball)
	Stroke	75 mm (Max. 100 mm)
	Repeatability	± 2 μm (step size is 15 μm / pulse)
	Speed	Max. 6 mm/sec
θ – axis	Main material	A2219–T87 or T852 aluminum alloy
	Guide material	SUS440C
	Rotation range	360 degree
	Repeatability	± 0.09 degree (step size is 1.8 degree / pulse)
Rotational bearing	Material	SUS440C (Balls are coated with Ag)
Ballscrew	Main material	A2219–T87 aluminum alloy coated with TiC (Balls are coated with Ag)
	Type	constant pre-load type
Stepping motor	Main materials	SUS316 stainless steel (sheath cable) Silicon steel
	Step angle	1.8 degree/pulse
	Maximum starting torque	1.7 kg·cm (@ 10 pulses / sec)

tors. The motors consist of rotors made using samarium–cobalt magnet and silicon steel, and stators made of silicon steel and stainless-steel sheath cable (Mivereil Insulated Cable) of 0.5 mm in diameter.

The ballscrews for the X and Y axes are directly connected to the rotors of the stepping motors. The ballscrews and their nuts are made of aluminum alloy and coated with TiC. The SUS440C balls were coated with Ag. The material for the ballscrews is A2219-T87 aluminum alloy and electrochemical buffing was performed on the surfaces. The lead of the ballscrew is 3 mm. The ballscrew is a constant pressure type using springs. The nuts of the ballscrews are connected to the X and Y tables with the joints moved at two planes which orthogonally crossed to each axis of the screw.

The limit switch for the stroke consists of a glass-sealed switch operated by a samarium–cobalt magnet. The origin-detection switch is a Kovar contact switch which was insulated with Al_2O_3. Both switches were used together with the steel cable coated with Al_2O_3.[2]

All the fasteners (bolts, nuts, and washers) were stainless steel and were treated with acids and degassed before assembly at 300 °C for ~24 h.

B. Ultrahigh vacuum chamber made of aluminum alloys

Figure 2 schematically shows an experimental ultrahigh vacuum system made of aluminum alloys for testing various devices. The chamber has an inside diameter of 570 mm, a length of 900 mm, a total surface area of 2.1×10^4 cm^2, and a volume of ~240 l. The walls of the chamber were made of A1050-H24 aluminum alloy. The flanges were made of A2219-T87 aluminum alloy with a special surface treatment.[1] Two types of vacuum gauges were used: a modulation Bayard–Alpert gauge (MBA) and an extractor gauge (EG). The MBA type is IMR103 (Balzers Co., Ltd.). The EG type is IE511 (Leybold–Heraeus Co., Ltd.).

The vacuum pumping system consisted of a 250 l/s turbo-molecular pump (TMP) with magnetic floating bearings, a 250 l/s ion pump (IP), and a titanium sublimation pump (TSP) with a liquid-nitrogen shroud. The TMP and the vacuum chamber were connected through an L valve (LV) made of aluminum alloys. The valve was closed when the

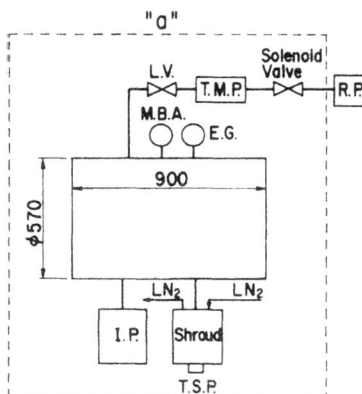

FIG. 2. Vacuum system used in this work.

356

pressure reached 6×10^{-10} Torr, and thereafter the IP and the TSP were operated.

Only aluminum gaskets (A1050-H18) and Helicoflexes (for the large flanges) were used for the seals between flanges.

The ultimate pressure on the chamber alone, during a preliminary experiment, was 6.5×10^{-12} Torr as shown in Fig. 3. This system easily reaches a pressure of the order of 10^{-11} Torr. Furthermore, the components enclosed within the dashed line "a" in Fig. 2 were mounted on an automatic stage to isolate them from vibrations. The TMP and the rotary pump (RP) were stopped occasionally to determine the accuracy of the measurements.

III. EXPERIMENTS AND RESULTS

A. Accuracy

The mechanical accuracy of the table was measured using an electronic micrometer (minimum resolution of 1 μm) and a laser measuring instrument (minimum resolution of 0.01 μm). The straightness of the motions was measured using an autocollimator (minimum resolution of 0.3 s).

As a result of the measurements, the pitching of the Y axis was ± 0.78 μm/75 mm, the yawning was ± 0.22 μm/75 mm, and the incremental repeatability was ± 2 μm for a feed step of 15 μm/pulse. The pitching of the X axis was ± 2.3 μm/75 mm, the yawning was ± 1.4 μm/75 mm, and the incremental repeatability was ± 2 μm for a feed step of 15 μm/pulse. The incremental repeatability of the θ axis was $\pm 0.09°$ for a step of 1.8°/pulse.

The entire table was then placed in the vacuum chamber and was baked at 140 °C for ~ 24 h. After that, the accuracies of the table were measured in the atmosphere again, and no changes were found.

Figure 4 shows the dynamic characteristics measured by a laser measuring instrument when the Y axis was driven for 3 mm in an ultrahigh vacuum environment. A laser oscillator and a measuring device were placed in the atmosphere. A mirror was placed at the center of the table in the vacuum

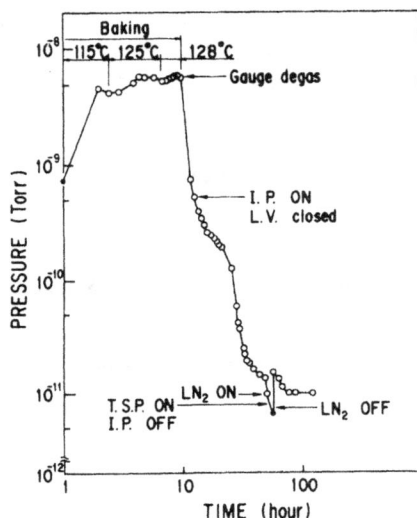

FIG. 4. Dynamic characteristics of the Y axis in a pressure of the order of 10^{-11} Torr.

environment. The measurements were carried out every 25 ms through a view port made of sapphire (flatness better than 3 μm). The dynamic characteristics of the X axis were nearly equal to those of the Y axis.

B. Pressure

1. Table stationary

Figure 5 shows the pumpdown curve of the chamber, with the table stationary throughout.

At first, the system was pumped down from 1 atm by the turbomolecular pump, and then baked at 130 °C for ~ 60 h. After a cooling period of ~ 50 h, the system was again baked at 144 °C for ~ 50 h. The vacuum gauges were degassed at the end of the second baking process. When a pressure of 6×10^{-10} Torr was attained, the IP operation was started

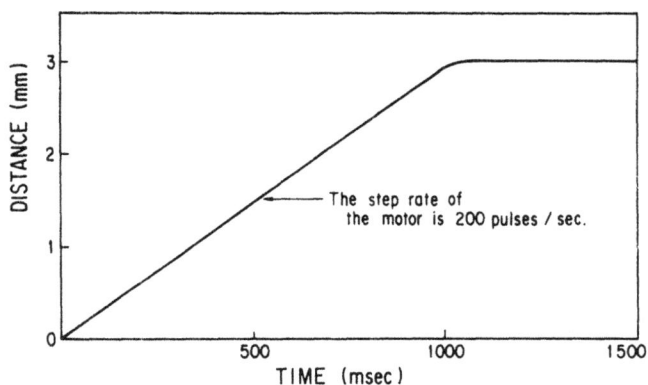

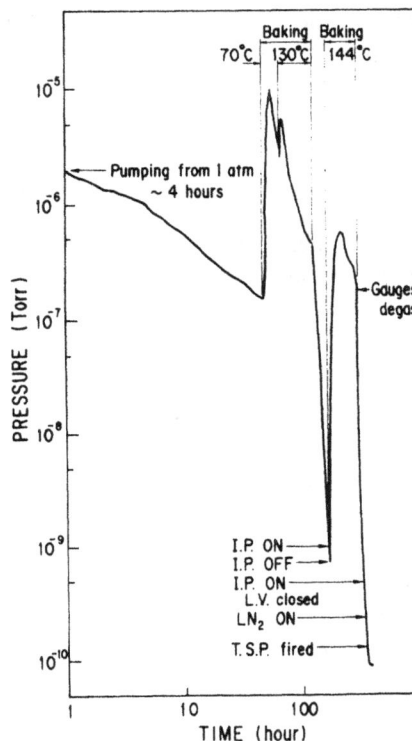

FIG. 3. Pumpdown curve. The ultimate pressure of the chamber alone is 6.5×10^{-12} Torr.

FIG. 5. Pumpdown curve. The final pressure of the chamber, with the table stationary was 9×10^{-11} Torr.

357

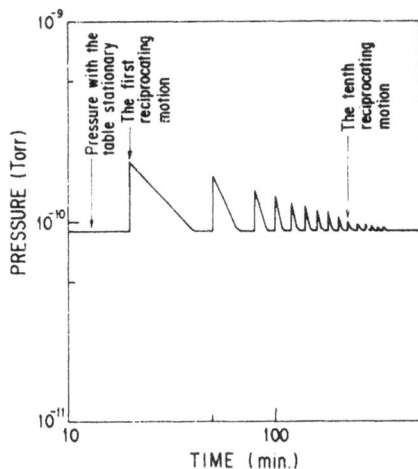

FIG. 6. Variation of the pressure with the table translating. The pressure gradually approached 9×10^{-11} Torr.

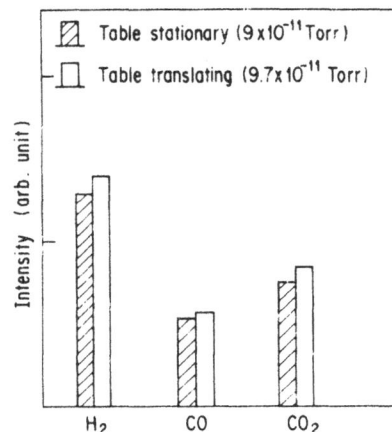

FIG. 7. Residual gas components with the table stationary and with it translating.

after the L valve was closed. The liquid nitrogen was then introduced into the shroud, and the TSP was also fired. After that, the two vacuum gauges indicated a pressure of the order of 10^{-11} Torr.

2. Table translating

Figure 6 shows the variation of the pressure when the table made a reciprocating motion of 30 mm after the static condition (1) was established. After the tenth cycle, the pressure dropped into the 10^{-11} Torr range, and approached that achieved with the table stationary.

Figure 7 shows the variation of the residual gas spectrum with the table stationary with it translating. The data were measured using a quadrupole mass filter. Note that the pressure of H_2, CO, and CO_2 increased.

IV. DISCUSSION

When the table was in the chamber, a pressure of the order of 10^{-11} Torr was attained. However, since the table was set right above the exhaust port of the chamber, this resulted in a conductance decrease. Therefore, an even lower ultimate pressure could be expected if the conductance was increased.

Better mechanical accuracies for the table can be expected when further adjustment of the springs for the preload is carried out.

While the table is being driven, it can be expected that adsorbed gases which covered the sliding surfaces and gases contained in materials (Ag, Al_2O_3) would be desorbed by the pressure of the mechanical contact.

From Figs. 6 and 7, it can be seen that H_2, CO, and CO_2 were desorbed from the materials by the pressure of the mechanical contact.

Consequently, if the gas content of the materials used for the sliding surfaces had been reduced, better results could be expected.

V. SUMMARY

(i) An experimental $X-Y-\theta$ three-axis table made of aluminum alloys with a special surface treatment, with built-in motors and control devices, was produced to be used in an ultrahigh vacuum environment. The mechanical accuracies of the table were adequately maintained after the baking processes.

(ii) The pressure in the chamber containing the table reached to the order of 10^{-11} Torr (ultimate pressure: 9×10^{-11} Torr). The motion of the table in this environment was equivalent to its motion in the atmosphere.

(iii) The durability of the table has not been confirmed, but tests are presently being carried out.

ACKNOWLEDGMENTS

The authors would like to thank to Prof. Horikoshi and Dr. Momose for their many helpful suggestions. This work was supported by Hakudo Co., Ltd., Fuji Bellows Co., Ltd., Musashino Seiki Co., Ltd., Sukegawa Electric Industrial Co., Ltd., Koyo Co., Ltd., Nichizo Ultra Finish Technology Co., Ltd., and Dipsol Chemicals Co., Ltd.

[a] This paper was presented at the 34th National Symposium of the AVS, Anaheim, CA, 1987.
[1] H. Ishimaru, J. Vac. Sci. Technol. A 2, 1170 (1984).
[2] M. Miyamoto, Y. Sumi, S. Komaki, K. Narushima, and H. Ishimaru, J. Vac. Sci. Technol. A 4, 2515 (1986).

Extended travel ultrahigh-vacuum sample manipulator with two orthogonal, independent rotations

J. J. Zinck and W. H. Weinberg

Division of Chemistry and Chemical Engineering, California Institute of Technology, Pasadena, California 91125

(Received 16 October 1984; accepted for publication 14 February 1985)

A UHV sample manipulator with greater than 22 in. of vertical translation and 360° of independent polar and azimuthal sample rotation is described.

INTRODUCTION

Ideally, an ultrahigh-vacuum (UHV) compatible sample manipulator should allow all degrees of rotational and translational freedom of the sample, independently, and without inducing strain on any part of the manipulator, provide heating and cooling for the sample, and be easily demountable from the UHV chamber. In practice, it is often difficult to satisfy all these conditions, and some aspects of design are favored preferentially, according to the needs of a particular experiment.[1-5]

We have designed a manipulator compatible with our ultrahigh-vacuum molecular beam apparatus[6] which has the following features: resistive heating and liquid-nitrogen cooling of the sample, $\frac{1}{2}$ in. of *XY* (horizontal) and 2 in. of *Z* (vertical) translation provided by a commercial (Huntington #PM-275-XYZ) precision translator, an additional 22 in. of *Z* translation via a separately supported welded bellows (Standard Bellows # 249-153-25-EE), a gimbals assembly which provides a tilt for the manipulator shaft, and mechanisms for 360° of independent polar (θ) and azimuthal (φ) sample rotation. In addition, the sample holder is sufficiently compact to be withdrawn into a load lock of $1\frac{1}{2}$ in. i.d., which may be isolated from the UHV chamber by a gate valve of the same dimension.

I. DESCRIPTION OF APPARATUS

The mechanism of polar rotation of the sample has been adapted from the design of Unwin *et al.*[7] and consists of two sliding seals of a type first described by Wilson.[8] In this design, a rotary shaft handle makes a seal against two Viton gaskets mounted in an external housing. The interstitial space between the Viton gaskets is pumped differentially by a mechanical pump to a pressure on the order of 50 μ. This differential pumping reduces the gas leakage into the UHV chamber to a negligible load when the rotary shaft handle is turned. A stainless-steel tube, which acts as a cryostat for the sample and to which the sample holder is attached, makes a UHV seal to the rotary shaft handle.

Our manipulator incorporates the same differential pumping arrangement as in the design of Unwin *et al.*,[7] but has the following modifications to the cryostat and sample holder. The main shaft of the cryostat is constructed from a $\frac{3}{4}$-in. o.d. in.×$\frac{5}{8}$-in. i.d. stainless-steel tube that is 3 ft. in length. A 1-ft. length of $\frac{9}{16}$-in. o.d. ×$\frac{3}{8}$-in. stainless-steel tube

is welded to one end of the main shaft. The smaller tube is then sealed to the rotary shaft handle via a series of knife edge seals as shown in Fig. 1. The outer diameter of the double knife-edge ring, the small copper gasket, and the spacer ring have been machined 0.010 in. smaller than the inner diameter of the rotary shaft handle to facilitate pumping of interstitial air. This arrangement provides us with a means of separating the manipulator shaft from the differential pumping housing by allowing us to drop the shaft beneath the housing once the UHV seals have been broken. This offers the advantage that a conflat flange may be welded to the bottom of the cryostat, and any commercial feedthrough may be selected to provide the desired connections to the sample holder. We are currently using a single thermocouple (W/5% Re,W/26% Re) power feedthrough mounted on a miniconflat flange (Ceramaseal #808BB8871-2). We have encountered no problems with respect to vacuum integ-

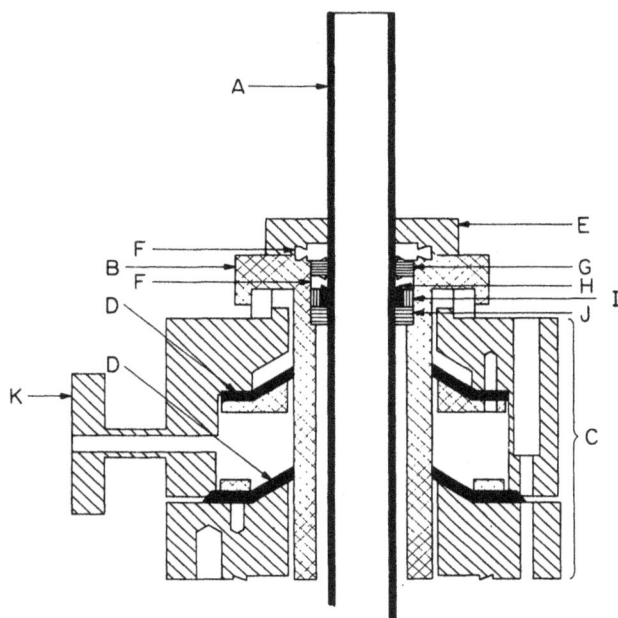

FIG. 1. Upper manipulator housing and shaft seals: (A) Cryostat tube. (B) Rotary shaft handle. (C) Differential pumping housing. (D) Viton gasket. (E) Miniconflat flange. (F) Copper gasket. (G) Double knife-edge ring. (H) Knife edge machined on cryostat tube. (I) Spacer ring. (J) Split ring. (K) Miniconflat flange connection to mechanical pump.

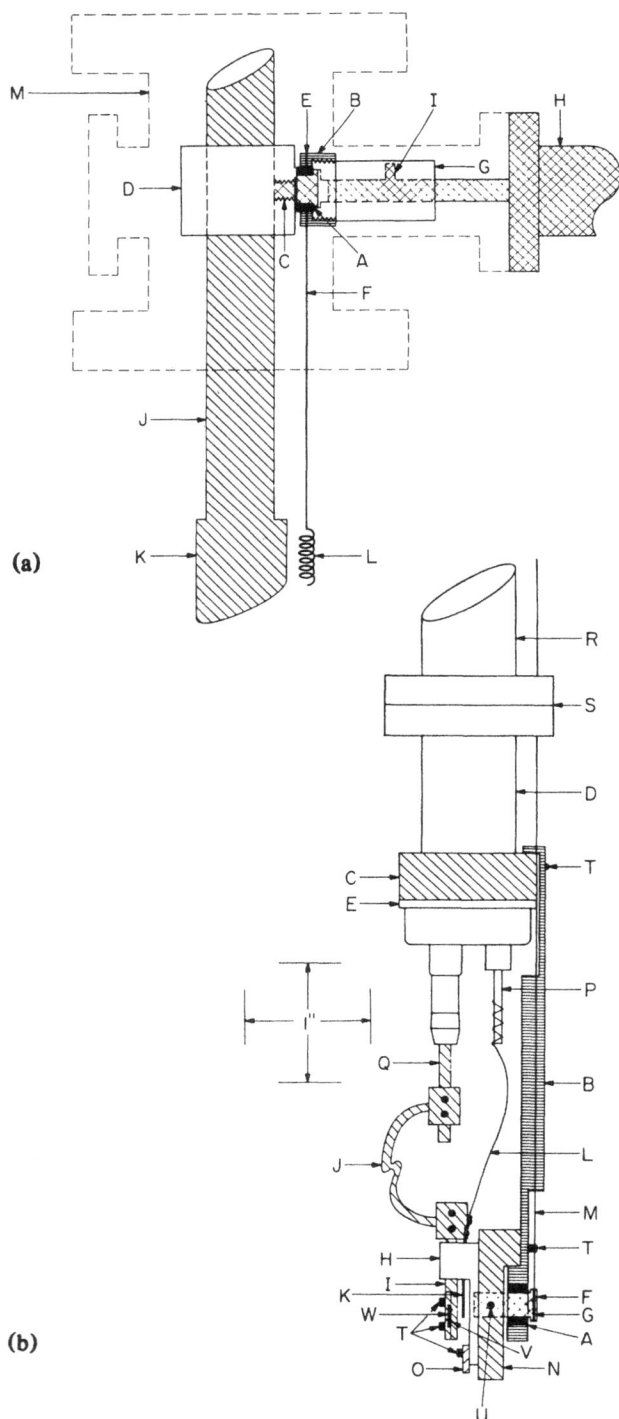

FIG. 2. (a) Upper assembly of azimuthal rotation mechanism: (A) Miniature bearing. (B) Upper bearing housing disk. (C) Fillister screw. (D) Bushing. (E) Raceway. (F) Be–Cu Wire. (G) Coupling. (H) Rotary motion feedthrough. (I) Pin. (J) Cryostat shaft ($\frac{9}{16}$ in. o.d.). (K) Cryostat shaft ($\frac{3}{4}$ in. o.d.). (L) Be–Cu spring. (M) Vacuum fitting. (b) Lower assembly of azimuthal rotation mechanism and sample holder: (A) Miniature bearing. (B) Plate for lower bearing housing and sample holder support. (C) Split ring clamp. (D) Ceramaseal feedthrough. (E) Weld lip. (F) Spindle. (G) Raceway. (H) Macor Disk. (I) Cu rod (0.094 in. diameter). (J) Flexible Cu braid. (K) Thermocouple wire (0.010 in. diameter). (L) Thermocouple wire (0.003 in. diameter). (M) Be–Cu wire. (N) Graded dial. (O) Vernier. (P) Feedthrough thermocouple post. (Q) Feedthrough heating/cooling rod. (R) Cryostat shaft. (S) Miniconflat flange copper gasket seal. (T) Cap Screw. (U) Set screw. (V) Ta heating wire (0.010 in. diameter). (W) Crystal sample.

rity of the copper conflat seal after extended and repeated exposure of the seal to liquid nitrogen.

The azimuthal rotation mechanism consists of an upper and lower assembly connected via a spring-loaded Be–Cu wire. The lower assembly is subject to the restriction that the sample holder must pass through a $1\frac{1}{2}$-in. i.d. tube.[6] As a consequence the assembly is very compact, as can be judged from the scale in Fig. 2(b).

In the upper assembly [Fig. 2(a)], a miniature bearing (MPB #S6316RHHE5P28LD) is press fit to a disk [the upper bearing housing of Fig. 2(a). A filister screw captures the inner race of the bearing and secures the disk to a bushing which is mounted on the manipulator shaft. A raceway is machined in the circumference of the disk for a Be–Cu $\frac{3}{4}$ hard wire. The inner diameter of the disk is threaded to receive a coupling which connects the upper assembly to a rotary motion feedthrough (Huntington #VF-106). The coupling is bored and slotted so that the shaft of the rotary motion feedthrough may be slip fit to the coupling inner diameter. A pin fixed on the feedthrough shaft lies in the slot of the coupling to effect the transfer of force, which will move the assembly when the rotary motion feedthrough is turned. A lock nut on the feedthrough prevents the rotation of the azimuthal mechanism when the manipulator is moved in any of its other degrees of freedom

The upper assembly is positioned on the cryostat shaft at the level of a vacuum fitting such that a port with a mini-conflat flange is available for mounting the rotary motion feedthrough, which subsequently holds the upper assembly in place.

The lower assembly contains the same type of miniature bearing as the upper assembly. In this case, the miniature bearing is press fit to a stainless-steel plate which supports the sample holder. The plate is fastened with three screws to a stainless-steel split ring clamp mounted on the Ceramaseal feedthrough above the weld lip. A stainless-steel spindle captures the inner race of the bearing. On one end of the spindle a raceway approximately $\frac{1}{32}$-in. wide is machined, in which the Be–Cu wire travels. The ratio of the circumference of the lower to upper raceways is 1:2. Although the bearings of the upper and lower assembly are not lubricated, we have experienced no difficulty with seizing of the mechanism.

The sample holder consists of a Macor[9] disk and sample supports. The macor disk is slip fit to the outer diameter of the spindle on the end opposite the raceway and held in place by set screws. Tantalum heating wires of 0.010 in. diameter spot welded to the back of the crystal are clamped to copper rods of 0.094 in. diameter mounted in the Macor disk and held in place by set screws. Flexible copper braid connectors 0.08 in. wide by 3 in. long provide the conduction path from the feedthrough to the crystal. Thermocouple wires 0.010 in. in diameter are anchored in the Macor disk via set screws. The crystal thermocouple of 0.003 in. diameter wire is spot welded to the back of the crystal and spot welded to the 0.010-in. diameter thermocouple wire mounted in the Macor. Thermocouple wire of 0.003 in. diameter is also connected from the feedthrough thermocouple posts to the 0.010-in. wire anchored in the Macor with sufficient slack to allow $\pm$ 180° of azimuthal rotation.

Crystal manipulator

360

The upper and lower assemblies are connected by a #30 Be–Cu $\frac{3}{4}$ hard wire which is loaded by two springs mounted approximately 3 in. below the upper assembly. The springs are constructed from #22 Be–Cu $\frac{3}{4}$ hard wire. They have an outer diameter of 0.23 in., are $2\frac{1}{2}$ in. long, and contain 70 coils. After winding and before mounting, the springs are stress relieved by annealing them at 300 °C for 3 h. When mounted the springs are extended in length to $3\frac{1}{2}$ in. (approximately 70% of their maximum deflection), corresponding to a load of about $1\frac{1}{2}$ lbs.

Turning the rotary motion feedthrough turns the coupling and the upper bearing housing. Tension on the wire provided by the springs succeeds in transferring the force to the lower assembly, and in turning the spindle and the Macor disk. Due to relaxation of the spring and wire with thermal cycling, such as bakeout of the vacuum chamber, and slippage of the wire in the raceways, angular displacements will not necessarily correlate with displacement of the rotary motion feedthrough. In order to measure the relative change in azimuthal angle, a dial graded in degrees is mounted on the lower bearing housing which encircles the Macor disk while leaving it free to rotate. A vernier is mounted on the Macor disk providing an angular resolution of approximate-ly 0.2°. The dial may be read using a telescope or cathetometer. An absolute determination of the azimuthal angle may be obtained from low-energy electron diffraction, when optics are available.

We have been sucessfully employing this manipulator for 1 yr. in our UHV system, where the base pressure is typically less than 2×10^{-10} Torr after a 24-h bakeout at 120 °C.

ACKNOWLEDGMENT

This work was supported by the Army Research Office under Contract No. DAAG29-83-K-0094.

[1]C. J. Russo and R. Kaplow, J. Vac. Sci. Technol. **13**, 487 (1976).
[2]P. H. Citrin, P. Eisenberger, and R. C. Hewitt, Phys. Rev. Lett. **45**, 1948 (1980).
[3]K. D. Jamison and F. B. Dunning, Rev. Sci. Instrum. **55**, 1509 (1984).
[4]P. H. Thiel and J. W. Anderegg, Rev. Sci. Instrum. **55**, 1669 (1984).
[5]J. A. Strocio and W. Ho, Rev. Sci. Instrum. **55**, 1672 (1984).
[6]J. J. Zinck and W. H. Weinberg (in preparation).
[7]Von R. Unwin, K. Horn, and P. Geng, Vak. Tech. **29**, 149 (1980).
[8]R. R. Wilson, Rev. Sci. Instrum. **12**, 91 (1941).
[9]Macor is a machinable glass ceramic manufactured by Corning Glass Works, Corning, NY 14831.

Vacuum Mechatronic Components and Companies

ACTUATORS

Burleigh Instruments, Inc.
Burleigh Park
Fishers, NY 14453 716-924-9355

Micro Pulse Systems, Inc.
3950 Carol Avenue
Santa Barbara, CA 93110 805-962-6023

FEEDTHROUGHS

A & N Corp.
Box 878
Inglis, FL 32649 904-447-2411

Ability Engineering Tech
16140 S. Vincennes
South Holland, IL 60473 312-331-0025

Ceramaseal
Box 260
New Lebanon, NY 12125 518-794-7800

Cooke Vacuum Products
13 Merritt St.
Norwalk, CT 06854 203-853-9500

Douglas Engineering Co.
14 Beach St.
Rockaway, NJ 07866 201-627-8230

High Vacuum Apparatus Mfg., Inc.
1763 Sabre St.
Hayward, CA 94545 415-785-2744

Huntington Laboratories
1040 L'Avenida
Mountain View, CA 94043 415-964-3323

Insulator Seal, Inc.
23874 Cabot Blvd., Ste B
Hayward, CA 94545 415-887-8664

Lesker, Kurt J., Co.
1515 Worthington Ave.
Clairton, PA 15025 412-233-4200

MDC Vacuum Products Corp.
23842 Cabot Blvd.
Hayward, CA 94545 415-887-6100

MTM Cryo Tech Lab
9221 S. Kilpatrick
Oak Lawn, IL 60453 312-425-9080

National Electrostatics Corp.
Graber Rd., Box 310
Middleton, WI 53562 608-831-7600

Nor-Cal Products, Inc.
Box 518
Yreka, CA 96097 916-842-4457

Omley Industries, Inc.
63345 Nels Anderson Road
Bend, OR 97701 800-541-3355

Perkin-Elmer Corp.
761 Main Ave.
Norwalk, CT 06859 203-762-1000

Physicon Corp.
221 Mount Auburn Street
Boston, MA 02138 617-491-7997

Sparrell Engineering
Bristol Road
Damariscotta, ME 04543 207-563-3224

Surface/Interface Inc.
476 Ellis Street
Mountain View, CA 94043 415-965-8205

Thermionics Laboratory, Inc.
22815 Sutro Street
Hayward, CA 94540 415-538-3304

Thermionics Northwest, Inc.
231-B Otto St.
Port Townsend, WA 98368 206-385-7707

Veeco Instruments, Inc.
Terminal Drive
Plainview, NY 11803 516-349-8300

LUBRICANTS

Ball Corporation Aerospace Systems
P.O. Box 1062
Boulder, CO 80306-1062 303-939-4000

Castrol Bray Products Division
16815 Von Karman Avenue, Suite 202
Irvine, CA 92714 714-660-9414

Dicronite Dry Lube
816 East Edna Place
Covina, CA 91723 818-967-3729

Dixon Industries Corp.
386 Metacom Ave.
Bristol, RI 02809 401-253-2000

Dupont
Wilmington, DE 19898 800-441-9442

Helvart Associates
P.O. Box 2536
La Habra, CA 90631 .... (213) 697-1196

Rogers Corp.
Main Street
Rogers, CT 06263 203-774-9605

Sputtered Films, Inc.
P.O. Box 4700
Santa Barbara, CA 93103 805-963-9651

MOTORS, X-Y STAGES

ARUN Microelectronics, Ltd.
10-12 Fitzalan Road
Arundel, West Sussex
BN18 9JP, ENGLAND 0903-884119

Anorad Corporation
110 Oser Ave.
Hauppauge, NY 11788 516-435-1612

Empire Magnetics
1318-C Ross Street
Petaluma, CA 94952 707-765-9343

Linear Industries, Ltd.
1850 Enterprise Way
Monrovia, CA 91016 800-821-2875

New England Affiliated Technologies
620 Essex Street
Lawrence, MA 01841 800-227-1066

Princeton Research Instruments, Inc.
P.O. Box 1174
Princeton, NJ 08542 609-924-0570

Sanyo Electric Co.
1-15-1 Kita Otsuka
Toshima - Ku
Tokyo, Japan 170 03-917-5151

Sukegawa Electric Co.
3119-5 Namekawa Honcho
Ibaragi - Ken
Hitachi City, Japan 317 029-421-5181

UHV Instruments, Inc.
P.O. Box 12
1951 Hamburg Turnpike
Buffalo, NY 14218 716-833-7534

Yaskawa Electric Mfg. Co. Ltd.
2346 Fujita
Yahatanishi - Ku
Kitakyushu-shi, Japan 806 093-641-3111

ROBOTS

Brooks Automation
One Executive Park Drive
North Billerica, MA 01862 617-667-0211

Genmark Automation
150 S. Wolfe Road
Sunnyvale, CA 94086 408-720-0667

Yaskawa Electric Mfg. Co. Ltd.
2346 Fujita
Yahatanishi - Ku
Kitakyushu-shi, Japan 806 093-641-3111

Unit Conversion Factors

The following tables provide conversion factors between units in the U.S. Customary system, the metric system and the International System (SI). To convert a quantity expressed in a unit in the left-hand column to the equivalent in a unit in the top row of a table, multiply the quantity by the factor listed as common to both units. Numbers followed by an asterisk are definitions of the relation between the two units.

Units of Pressure

Units	Pa (N·m^{-2})	dyn·cm^{-2}	bar	atm	mmHg (torr)	in. Hg	lbf·in.$^{-2}$
1 Pa (N·m^{-2}) = 1	10	10^{-5}	9.869×10^{-6}	7.501×10^{-3}	2.953×10^{-4}	1.450×10^{-4}
1 dyn·cm^{-2} = 0.1	1	10^{-6}	9.869×10^{-7}	7.501×10^{-4}	2.953×10^{-5}	1.450×10^{-5}
1 bar = 10^5*	10^6	1	0.9869	750.0617	29.530	14.504
1 atm = 101325.0*	1013250	1.013250	1	760	29.9213	14.6959
1 mmHg (torr) = 133.3224	1333.224	1.333×10^{-3}	1.316×10^{-3}	1	0.0394	0.0193
1 in. Hg = 3386.388	33863.88	0.03386388	0.03342105	25.4	1	0.4911541
1 lbf·in.$^{-2}$ = 6894.757	68947.57	0.06894757	0.06804596	51.71493	2.036021	1

Units of Length

Units	μm (micron)	cm	m	mil	in.	mile
1 μm (micron) = 1	10^{-4}	10^{-6}	0.03937	3.937×10^{-5}	6.2137×10^{-10}
1 cm = 10^4	1	0.01*	3.937×10^2	0.3937	6.2137×10^{-6}
1 m = 10^6	100	1	3.937×10^4	39.3701	6.2137×10^{-4}
1 mil = 25.4	2.54×10^{-3}	2.54×10^{-5}	1	0.001	1.5783×10^{-8}
1 in. = 2.54×10^4	2.54*	0.0254	1000	1	1.5783×10^{-5}
1 mile = 1.6093×10^9	1.6093×10^5	1.6093×10^3	6.336×10^7	6.336×10^4	1

Units of Area

Units	μm^2	cm^2	m^2	mil^2	$in.^2$	$mile^2$
1 μm^2	= 1	10^{-8}	10^{-12}	1.550×10^{-3}	1.550×10^{-9}	3.861×10^{-19}
1 cm^2	= 10^8	1	10^{-4}*	1.550×10^5	0.1550	3.861×10^{-11}
1 m^2	= 10^{12}	10^4	1	1.550×10^9	1550	3.861×10^{-7}
1 mil^2	= 645.16	6.452×10^{-6}	6.452×10^{-10}	1	10^{-6}	2.491×10^{-16}
1 $in.^2$	= 6.452×10^8	6.452*	6.452×10^{-4}	10^6	1	2.491×10^{-10}
1 $mile^2$	= 2.590×10^{18}	2.590×10^{10}	2.590×10^6	4.014×10^{15}	4.014×10^9	1

Units of Volume

Units	m^3	cm^3	liter	$in.^3$	ft^3	qt
1 m^3	= 1	10^6	10^3	6.103×10^4	35.3147	1.0567×10^3
1 cm^3	= 10^{-6}	1	10^{-3}	0.06103	3.532×10^{-5}	1.0567×10^{-3}
1 liter	= 10^{-3}	1000*	1	61.0237	0.0353	1.0567
1 $in.^3$	= 1.639×10^{-5}	16.3871*	0.0164	1	5.787×10^{-4}	0.0173
1 ft^3	= 2.832×10^{-2}	28316.85	28.31685	1728*	1	2.9922
1 qt	= 9.464×10^{-4}	946.353	0.9464	57.75	0.0342	1

Units of Mass

Units	g	kg	oz	lb	metric ton	ton
1 g	= 1	10^{-3}	0.0353	2.2046×10^{-3}	10^{-6}	1.1023×10^{-6}
1 kg	= 1000	1	35.2740	2.2046	10^{-3}	1.1023×10^{-3}
1 oz	= 28.3495	0.0283	1	0.0625	2.8350×10^{-5}	3.125×10^{-5}
1 lb	= 453.5924	0.4536	16*	1	4.5359×10^{-4}	0.0005
1 metric ton	= 10^6	1000*	35273.96	2204.623	1	1.1023
1 ton	= 907184.7	907.1847	32000	2000*	0.9072	1

Units of Density

Units	$g \cdot cm^{-3}$	$g \cdot L^{-1}$ ($kg \cdot m^{-3}$)	$oz \cdot in.^{-3}$	$lb \cdot in.^{-3}$	$lb \cdot ft.^{-3}$	$lb \cdot gal^{-1}$
$g \cdot cm^{-3}$	= 1	1000	0.5780	0.0361	62.4280	8.3454
$g \cdot L^{-1}$ ($kg \cdot m^{-3}$)	= 10^{-3}	1	5.7804×10^{-4}	3.6127×10^{-5}	0.0624	8.3454×10^{-3}
$oz \cdot in.^{-3}$	= 1.7300	1729.994	1	0.0625	108	14.4375
$lb \cdot in.^{-3}$	= 27.6799	27679.91	16	1	1728	231
$lb \cdot ft.^{-3}$	= 0.0160	16.0185	9.2592×10^{-3}	5.7870×10^{-4}	1	0.1337
$lb \cdot gal^{-1}$	= 0.1198	119.8264	4.7495×10^{-3}	4.3290×10^{-3}	7.4805	1

The CRSM and Artech House wish to thank the following publishers for their permission to reprint the following articles in *Vacuum Mechatronics* :

American Vacuum Society
335 E. 45th St.
New York, New York 10017

© American Institute of Physics

R.J. Miller et al, "Mechanisms of contaminant particle production, migration and adhesion", *J. Vac. Sci. Technol.*. A6 (3), May/June, pp. 2097-2102, 1988.

H. Saeki, J. Ikeda, "Precision table for ultrahigh vacuum made of aluminum alloys", *J. Vac. Sci. Technol.* A6 (5), Sept/Oct, pp. 2883-2886, 1988.

K.G. Roller, "Lubricating of mechanisms for vacuum service", *J. Vac. Sci. Technol.* A6 (3), May/June, pp. 1161-1165, 1988.

A.P. Jardine, M. Ahmad et al., "A simple ultrahigh vacuum shape memory effect shutter mechanism", *J. Vac. Sci. Technol.*. A6 (5), Sept/Oct , pp.3017-3018, 1988.

W.F. Egelhoff, "Ultrahigh vacuum leak sealing with a silicon resin product", *J. Vac. Sci. Technol.* A (4), pp. 2584-2585, Jul/Aug , 1988.

M. Miyamoto, et al, "Aluminum alloy ultrahigh vacuum system for molecular beam epitaxy", *J. Vac. Sci. Technol.* A4 (6), pp. 2515-2519, Nov/Dec, 1986.

J. F. O'Hanlon, "Advances in vacuum contamination control for electronics materials processing", *J. Vac. Sci. Technol.* A6 (4), pp. 2067-2072, Jul/Aug , 1987.

J. F. O'Hanlon, "Vacuum systems for microelectronics", *J. Vac. Sci. Technol.*, A 1 (2), pp. 228-232, Apr.- June, 1983.

C.R. Tilford, S. Dittmann, "The National Bureau of Standards primary high-vacuum standard", *J. Vac. Sci. Technol.* A6 (5), pp. 2853-2859, Sept/Oct., 1988.

R.E. Clausing et al, "Versatile UHV sample transfer system", *J. Vac. Sci. Technol,* 16(2), pp. 708-710, Mar/Apr , 1979.

J.P. Hobson and E.V. Komelsen, "UHV technique for intervacuum sample transfer", *J. Vac. Sci. Technol.*, pp. 701-707, Mar/Apr , 1979.

Bruce A. Aikenhead, "Canadarm and the space shuttle", *J. Vac. Sci. Technol*, A 1(2), pp. 126-132, Apr/June, 1983.

Pergamon Press
Maxwell House
Fairview Park
Elmsford, NY 10523

© Pergamon Press Ltd.

G.F. Weston, "Ultra-high vacuum line components", *Vacuum*, Vol. 34 (6), pp. 619-629, 1984.

M. Moraw, "Analysis of outgassing characteristics of metals", *Vacuum,* Vol. 36 (7-9), pp. 523-525, 1986.

T.J. Patrick, "Space environment and vacuum properties of spacecraft materials", *Vacuum*, Vol. 31, (8/9) pp 351-357, 1981.

G.F. Weston, "Materials for ultrahigh vacuum", *Vacuum*, Vol. 25 (11/12), pp. 469-484, 1975.

NASA
Science & Technical Information Branch
600 Independence Ave. SW
Washington, DC 20546

H. M. Briscoe, M.J. Todd, "Considerations of the Lubrication of Spacecraft Mechanisms", *Proc. 17th Aerospace Mechanisms Symposium*, NASA CP-2273, Jet Propulsion Laboratory, Pasadena, CA, p. 19, May 5-6, 1983.

H.R. Ludwig, "Six Mechanisms Used on the SSM/I Radiometer", *Proc. 19th Aerospace Mechanisms Symposium*, NASA CP-2371, NASA Ames Research Center, Moffett Field, CA, p. 347, May 1-3, 1985.

A. Borrien and L. Petitjean, "Robotic Joint Experiments Under Ultravacuum", *Proc. 22nd Aerospace Mechanisms Symposium*, NASA CP-2506, Hampton, VA, p. 307, May 4-6, 1988.

Canon Communication, Inc.
2416 Wilshire Boulevard
Santa Monica, CA 90403

© Canon Communications, Inc.

R.A. Bowling, G.B. Larrabee, "Particle Control for Semiconductor Processing in Vacuum Systems", *Microcontamination Conference Proceedings,* as presented at the Micro-contamination Conference and Exposition, pp. 161-168, November, 1986 .

R.B. Lachenbruch, O. Gomez, "Control of Particulate Emissions from Plasma-Etching Systems", *Microcontamination*, January, 1987.

American Institute of Physics
335 E. 45th St.
New York, NY 10017

© American Institute of Physics

J.J. Zinck, W.H. Weinberg, "Extended travel ultrahigh-vacuum sample manipulator with two orthogonal, independent rotations", *Rev.Sci. Instrum.*,56(6), pp. 1285-1287, June, 1985.

T. Hatsuzawa, Y. Tanimura, et al, "Piezodriven spindle for a specimen holder in the vacuum chamber of a scanning electron microscope", *Rev. Sci. Instrum.*, 57 (12), pp. 3110-3113, December, 1986.

Spring Verlag Publishers
Heidelberger Pl 3, D-1000
Berlin 33, POSTF 311340
Germany

© Springer-Verlag Heidelberg

J. Albus, "The Central Nervous System as a Low and High Level Control System" *NATO ASI Series*, Vol F43, pp 3-29, Heidlelberg: Springer-Verlag, 1988.

IEEE
Director of Publishing Services
345 E. 45th St.
New York, NY 10017

© 1987 IEEE

S. Belinski, S. Hackwood, "Manufacturing in a Vacuum Environment", *Proceedings of the Third IEEE/CHMT International Electronic Manufacturing Technology Symposium*, Anaheim, CA, pp. 239-244, October 12-14, 1987.

Kluwer Academic Press
PO Box 163
33000 AD Dordrecht
The Netherlands

© 1988 by Kluwer Academic Publishers

T. Fukuda, S. Nakagawa, "Approach to the Dynamically Reconfigurable Robotic Systems", *J. Intelligent and Robotic Systems*, Vol. 1, no. 1, pp 55-72, 1988.

John Wiley and Sons
605 Third Ave.
New York, NY 10158

© John Wiley and Sons Publishers

T. Henderson, E. Shilcrat, "Logical Sensor Systems", *J. Robotic Systems*, Vol. 1, no. 2, pp 169-183, 1984.

AIAA
370 L'Enfant Promenade
Washington, DC 20024

© American Institute of Aeronautics and Astronautics

C.V. Madhusudana, "Contact Heat Transfer-The Last Decade", *AIAA Journal,* Vol. 24, pp. 510-523, March, 1986.

Lake Publishing Corporation
PO Box 159
17730 Peterson Rd.
Libertyville, IL 60048

© Lake Publishing Corporation

J.A. Freeman, "How to Select High Vacuum Pumps", *Microelectronics, Mfg. & Testing (MMT)*, pp. 1-10, October, 1985.

www.ingramcontent.com/pod-product-compliance
Lightning Source LLC
Chambersburg PA
CBHW080706220326
41598CB00033B/5323